International Yearbook of Soil Law and Policy

Series editor
Harald Ginzky
German Environment Agency
Dessau, Germany

The *International Yearbook of Soil Law and Policy* is a book series that discusses the central questions of law andpolicy with regard to the protection and sustainable management of soil andland. The Yearbook series analyzes developments in international law and newapproaches at the regional level as well as in a wide range of nationaljurisdictions. In addition, it addresses cross-disciplinary issues concerningthe protection and sustainable management of soil, including tenure rights,compliance, food security, human rights, poverty eradication and migration.Each volume contains articles and studies based on specific overarching topicsand combines perspectives from both lawyers and natural scientists to ensure aninterdisciplinary discourse.

The *International Yearbook of Soil Law and Policy* offers a valuable resource for lawyers, legislators, scholars andpolicymakers dealing with soil and land issues from a regulatory perspective.Further, it provides an essential platform for the discussion of new conceptualapproaches at the international, national and regional level.

More information about this series at http://www.springer.com/series/15378

Harald Ginzky • Elizabeth Dooley •
Irene L. Heuser • Emmanuel Kasimbazi •
Till Markus • Tianbao Qin
Editors

International Yearbook of Soil Law and Policy 2017

Editors
Harald Ginzky
German Environment Agency
Dessau, Germany

Elizabeth Dooley
Farmer Center
Iowa State University
Urbandale, Iowa, USA

Irene L. Heuser
Sustainable Soils and Desertification
IUCN World Commission on
Environmental Law
Kleinmachnow, Germany

Emmanuel Kasimbazi
Kampala, Uganda

Till Markus
Research Centre for European
Environmental Law
University of Bremen
Bremen, Germany

Tianbao Qin
Research Institute of Environmental Law
Wuhan University
Wuhan, China

ISSN 2520-1271 ISSN 2520-128X (electronic)
International Yearbook of Soil Law and Policy
ISBN 978-3-319-68884-8 ISBN 978-3-319-68885-5 (eBook)
https://doi.org/10.1007/978-3-319-68885-5

Library of Congress Control Number: 2017960848

Printed on acid-free paper

This Springer imprint is published by Springer Nature
The registered company is Springer International Publishing AG
The registered company address is: Gewerbestrasse 11, 6330 Cham, Switzerland

Preface

Soil has often been referred to as the neglected medium. This book series, the *International Yearbook of Soil Law and Policy*, is intended to draw more attention to it and provide a platform for discourse on soil governance topics as a useful tool by and for academics, legislators, and policy makers. Thus, it is supposed to underline the notion that society acknowledges the indispensable services provided by soils.

One major driver of detrimental effects on soils is unsustainable agriculture. Numerous soil threats, such as erosion, loss of organic carbon, and biodiversity, as well as compaction and salinization, are often the result of unsustainable agricultural practices. A substantial percentage, however, of land is dedicated for agriculture, which is in particular true for developing countries whose economies depend on cultivation.

Accordingly, the second volume of the *International Yearbook of Soil Law and Policy* is dedicated to "Soil and Sustainable Agriculture." Part I begins with introductory greetings, while Part II, entitled "The Theme," comprises nine chapters on the topic of agriculture and soils from very different perspectives.

Volume 2 of the *International Yearbook of Soil Law and Policy* continues further with the following three parts:

- Part III: Recent Developments of Soil Regulation at International Level
- Part IV: National and Regional Soil Legislation
- Part V: Cross-Cutting Topics

This general structure, beginning with Part I with Introductory Greetings and Part II on "The Theme,“ a compilation of various chapters from different perspectives and disciplines, followed by the fore-mentioned parts will also be used for the upcoming volumes. This structure of the Volumes permits the presentation of one specific "Theme" in more depth and to offer factual information with regard to developments both at international level and in national or regional contexts. In the part "Cross-Cutting Topics," papers with novel, controversial, or—one could say—courageous themes will feature. Therefore, this latter part should, in particular, provide food for thoughts for ongoing discussions.

In Part II of the second volume, the first chapter by Robert Rees *et al.* explores the suitability of the concept of "sustainable intensification," which aspires to cope with the need of growing food production without increasing the land and soil input. In the second chapter, Rainer Baritz *et al.* explain the Voluntary Guidelines for Sustainable Soil Management, which was recently adopted by Food and Agricultural Organization. This Guideline informs about conceptual approaches for sustainable agriculture that are generally applicable. The following three chapters provide a different angle and outline the current legislation in three Asian jurisdictions, namely China, Kyrgyzstan, and Tajikistan on soil-related provisions for sustainable agriculture. The authors of these three chapters are Bob Zhao (China), Maksatbek Anarbaev (Kyrgyzstan), and Murod Ergashev and Islomkhudzha Olimov (Tajikistan). Next, Ian Hannam looks at a specific topic of sustainable agriculture: how to effectively regulate pastoralism, taking into account environmental needs, as well as ensuring effective tenure of land. The three remaining chapters consider more general but nonetheless important aspects. Jesse Richardson demonstrates, using the example of US law, how tenure rights could impede sustainable agriculture. In their chapter, Luca Montanarella and Panos Panagos stress the need for data on sustainable agriculture, which are especially pertinent to determine suitable indicators and to establish an effective monitoring program. Finally, Andrea Schmeichel analyzes, from a legal perspective, whether and how import restrictions could be employed to incentivize sustainable agricultural practices abroad.

Part III contains a contribution by Stephanie Wunder *et al.* on how to implement the objective of "land degradation neutrality" in Germany, covering information on general approach and the selection of indicators.

Part IV on national and regional soil legislation put together a variety of interesting contributions, starting with reports on national soil legislation in Iceland and Greece, followed by several contributions on specific legal approaches in the various legislation, and finally an analysis of the soil conservation protocol of the Alpine Convention, which highlights why such an innovative regulation was possible to be achieved.

In Part V on "Cross-Cutting Topics," Robert Kibugi questions whether an African instrument could be useful and how it should be put in place. Next, Anja Eikermann puts forward a comparison of international forest law and international soil protection law, where she concludes that a more coherent and coordinated approach of the existing regimes is of need. Harald Ginzky, then, provides arguments that the sustainable management of soils should be regarded as common concern of humankind. The chapter of Irene Heuser demonstrates the development of soil awareness in Europe and other regions, arguing for an ethical approach that takes into account the crucial importance of soil function. Detlef Grimski *et al.* conclude this part with a chapter on the European research project "INSPIRATION," aimed at defining a strategic research agenda for Europe.

As the editors of the second volume of the *International Yearbook of Soil Law and Policy*, we hope that both the structure of this volume as well as the selections of topics and perspective meet the interest and expectations of our readers and that this volume contributes to raising soil awareness and to identifying appropriate solutions at international, regional, and national levels.

We would like to thank all the authors for their contributions, the members of the advisory board for helping us with the review process, and finally the publishing house SPRINGER for technical assistance. Finally, it remains to be announced that the theme for volume 3 of the *International Yearbook of Soil Law and Policy* will be "urbanization and sustainable management of soil," which covers another very significant driver of soil degradation.

Dessau, Germany Harald Ginzky
Urbandale, IA, USA Elizabeth Dooley
Kleinmachnow, Germany Irene L. Heuser
Kampala, Uganda Emmanuel Kasimbazi
Bremen, Germany Till Markus
Wuhan, China Tianbao Qin

Contents

Part I
Greetings

Greetings

Monique Barbut

Fertile soils and productive land are limited—finite—resources. But there is an ever-growing demand for fertile land to feed a growing global population. Meeting this demand is becoming a massive global challenge due to two major factors. There is competition for productive land for other uses, such as wood, fiber, bioenergy, and urban growth. Then there is land degradation and the added risk of climate change impacts, such as droughts, floods, and the rising global temperature. The need to manage the trade-offs between rising demand and falling supply is pressing.

The Sustainable Development Goals (SDGs) agreed in 2015 set a target for governments to achieve Land Degradation Neutrality (LDN) by 2030. The purpose is to ease these trade-offs by controlling the annual loss of, on average, 12 million hectares of productive land and by recovering some of the over two billion hectares of land that is degraded. This SDG target itself marks significant progress. Until recently, the loss of fertile land was hardly recognized as a serious global issue.

To achieve the target, however, two things must happen concurrently. On the one hand, we must use land sustainably to avoid degrading new land, and on the other, we must fix and restore degraded land back to health. Taken together, these two actions—avoiding land degradation and recovering degraded land—are a pathway to achieving land degradation neutrality. Of course, some land degradation will occur. The key is to ensure that we maintain a healthy balance of fertile land that will serve us now, and serve future generations as well.

The potential for change is huge.

Africa intends to restore more than 100 million hectares of degraded land by 2030. More than 20 million hectares of degraded land is targeted for restoration in Latin America and the Caribbean by 2020. Under the Bonn Challenge, the

M. Barbut (✉)
UNCCD, Bonn, Germany
e-mail: secretariat@unccd.int

H. Ginzky et al. (eds.), *International Yearbook of Soil Law and Policy 2017*,
International Yearbook of Soil Law and Policy,
https://doi.org/10.1007/978-3-319-68885-5_1

international community is aiming to restore 150 million hectares of degraded forests by 2020 and 350 million hectares by 2030. The United Nations Convention to Combat Desertification is helping to build the foundation for this change to happen, with more than 100 countries around the world now committing to concrete targets to achieve land degradation neutrality by 2030. And, through a public–private partnership it initiated, a private sector Land Degradation Neutrality Fund will be launched in Fall 2017 to secure the investment needed to support land restoration and rehabilitation around the world.

Equally critical are public engagement, effective legal instruments and good governance that can foster a conducive policy environment. In a very real sense, this second volume of the International Yearbook of Soil Law and Policy is an important forum for discussion. It should enable legislators, lawyers, and policy makers to share knowledge, build solidarity, and disseminate the most effective regulatory concepts and approaches at the international, regional, and national levels.

Policy makers and consumers are ready for change, but progress has been awfully slow. The time for change is now. It is my hope that the International Yearbook of Soil Law and Policy will develop into an indispensable communication tool and forum through which consumers, governments, activists, and socially ethical enterprises can continue to engage in the pursuit of land degradation neutrality.

Part II
The Theme: Soil and Sustainable Agriculture

Sustainable Intensification of Agriculture: Impacts on Sustainable Soil Management

Robert Martin Rees, Bryan S. Griffiths, and Alistair McVittie

1 Introduction

Soil plays a critical role in underpinning agricultural production. The importance of soils in supporting agriculture was apparent to the earliest human civilisations, and it was therefore no coincidence that as our distant ancestors spread from the African continent, they moved first into the Fertile Crescent of the Middle East and to the productive soils and plains along China's great river systems. Today, soils continue to play a critical role in supporting sustainable food production systems, and careful stewardship of soils has never been more important to securing successful economic, environmental and production outcomes in agriculture.

The explosive growth of human populations during the twentieth century was associated with a large increase in food production. Two factors were particularly important in contributing to our ability to produce more food: first, the manufacture of nitrogen fertilisers, which began in the 1920s and then accelerated rapidly in the second half of the century. Nitrogen is the mineral nutrient required in largest quantities by plants, yet despite its prevalence in the atmosphere, it can only be used once it has been converted into a fixed or reactive form. The industrial process that makes this possible was discovered by Harber and Bosch in 1911 and is currently responsible for the global manufacture of 100 Tg N annually (Fowler et al. 2015). It has been estimated that 40% of the world's population is now dependent upon this manufacturing process (Erisman et al. 2008) since without it, inputs of nitrogen to agricultural systems would be largely dependent on biological nitrogen fixation and would not likely be able to supply the quantity of nitrogen currently demanded by agricultural production systems.

R.M. Rees (✉) • B.S. Griffiths • A. McVittie
Scotland's Rural College (SRUC), Edinburgh, UK
e-mail: Bob.Rees@sruc.ac.uk

H. Ginzky et al. (eds.), *International Yearbook of Soil Law and Policy 2017*,
International Yearbook of Soil Law and Policy,
https://doi.org/10.1007/978-3-319-68885-5_2

The second transformation that took place during the twentieth century was the cultivation of large areas of previously uncultivated land, including forests, natural grasslands and wetlands. Today, agricultural production covers 40% of the earth's terrestrial surface, with much of the remainder covered by deserts, ice or areas with high conservation value such as tropical rainforests. It is anticipated that during the twenty-first century, human populations will rise to between 9 and 11 billion, and as a consequence, it is estimated that food production will need to increase by 40% relative to 2000 in order to accommodate this population increase (Godfray et al. 2010). The old model of bringing more land into agricultural production will no longer be applicable since there are only relatively small areas that could be potentially used. Indeed, other constraints such as climate change, soil degradation and water shortages may well lead to reductions in agriculturally productive land in some areas. It is against this background that the third concept of increasing productivity called sustainable intensification was developed. All three approaches can be considered conceptually in terms of the resource input and land area required to support demand for increasing productivity (Fig. 1).

The term sustainable intensification represented by (c) in Fig. 1 first became popular following the publication of a highly influential UK report published by the Royal Society called Reaping the Benefits (Baulcombe 2010; Baulcombe et al. 2009).

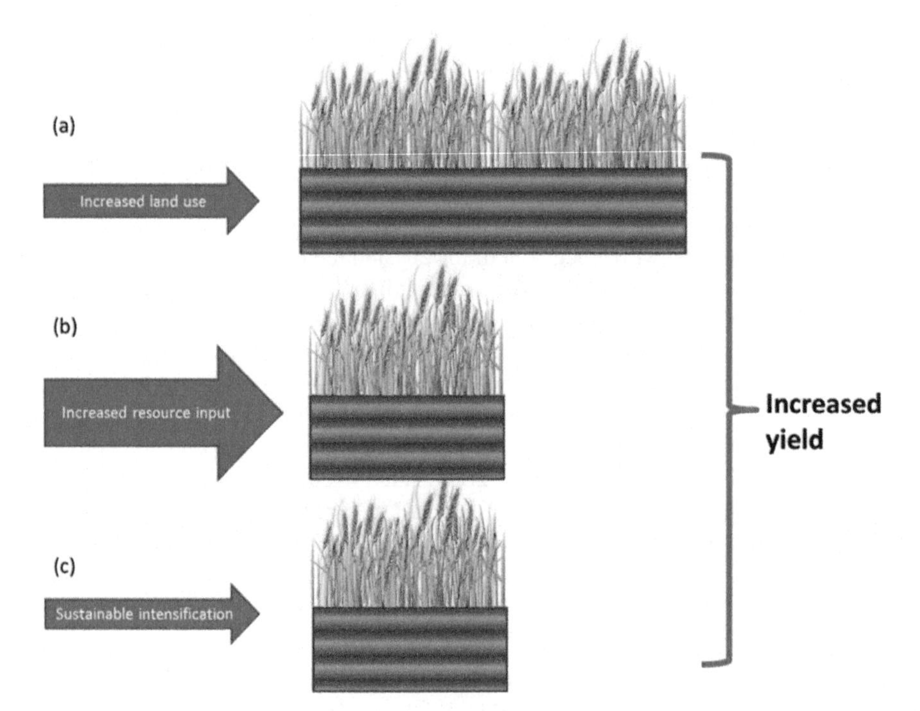

Fig. 1 A conceptual representation of alternative approaches to increasing productivity in agricultural systems; (**a**) increasing land area under cultivation, (**b**) increased resource input (e.g., nitrogen), (**c**) increasing resource use effiency (sustainable intensification)

This was followed by a series of more academic research papers defining the concept of sustainable intensification and its implication for food production systems (Godfray et al. 2010, 2011). In essence, sustainable intensification can be considered to represent a change in production systems that deliver stable increasing levels of output whilst reducing environmental impacts. Notable features of this concept are, firstly, that it can be considered to be a process of change rather than an endpoint itself and, secondly, that the term is not prescriptive about the type of production system. There are many different ways of improving productivity within systems that can both increase productivity and reduce environmental impact, and these are likely to vary between production systems and geographical locations. The term sustainable intensification has become widely supported in policy development, and, for instance, in the UK it underpins the rationale for supporting agricultural research programmes funded by national research councils.

In many ways, the requirement for sustainable intensification in agriculture reflects an inevitable analysis of the need to feed a growing global population without access to increased areas of land and against a background of adverse environmental conditions. The role of soil in supporting sustainable intensification is critical, given the need to increase productivity per unit area of land. This requires the maintenance of high quality soils that can help to produce increasing levels of output with constant or decreasing levels of input.

However, the term is not universally popular. The components of sustainable intensification include terms that are not easily compatible. Sustainability is a widely used and yet poorly defined concept. It implies the ability for continuity of activity or system. Many aspects of agricultural production threaten sustainability by damage to air, soil and water, which represent the natural capital available to support long-term productivity (Foley et al. 2011). However, some definitions of sustainability include a wider assessment of environmental, social and economic factors that contribute to long-term stability. This lack of clarity in the definition of sustainability can hinder the concept of sustainable intensification.

The intensification component of sustainable intensification has also been criticised. The twentieth century changes whereby increasing inputs were used to produce more crops with relatively little concern about environmental impact could be considered to be a process of intensification (Rees et al. 2016). Critics of the term sustainable intensification argue that this is simply a way of continuing intensification with no more than a green wash. The argument therefore becomes focused on what the relative contributions of sustainability and intensification in the sustainable intensification concept are. A number of other concepts have been defined that have parallels with sustainable intensification, such as climate-smart agriculture, organic agriculture, precision farming and ecological intensification.

A further criticism of sustainable intensification is provided by analysis of future demand for food production. It is argued that the projections of increased demand are based very much on extrapolation of current trends (Audsley et al. 2009). However, this often assumes large increases in resource-intensive production systems leading to unrealistic expectations of changes in supply-side approaches. Changing dietary demands in future projections would lead to lower requirements for production increases and reduce pressure on land-use change (Westhoek et al. 2014). It is therefore likely that both supply-side and demand-side changes will be required to meet future food requirements in order to achieve sustainable production systems. Despite the lack of a clear definition, the sustainable intensification concept is likely to be useful in guiding broad patterns of change in agricultural production. It has been recognised that the changes required will be greater than anything that has been seen in recent human history; therefore, considering the wider impacts of sustainable intensification on our environment and particularly on soils is of critical importance.

2 Soil Quality

Human societies are highly dependent upon healthy soils for the delivery of ecosystem goods and services, including provisioning (food, fibre, timber and fuel), regulation (climate, disease and natural hazards), waste treatment, nutrient cycling and cultural services (Reid et al. 2005). Many of the key functions supporting these ecosystem services depend to a large extent upon the diversity, abundance and activity of organisms that inhabit the soil. This diversity varies in terms of its taxonomic richness, relative abundance and distribution according to soil type, climatic conditions, vegetation and land use. Against this background, soil biodiversity is also subject to various threats associated with human activity, including soil erosion, organic matter decline, contamination, salinisation, sealing, compaction of soil and climate change; all these threats impair soil biodiversity and functioning with negative consequences on ecosystem service delivery (Gardi et al. 2013; Hooper et al. 2005; Wall et al. 2015). Increasing agricultural intensity, for example, has been shown to generally reduce soil biodiversity (e.g., Tsiafouli et al. 2015), although this response is likely to be non-linear given the variation in management practices and soil conditions across sites and regions and differences in the sensitivity of soil organism groups to management intensity. Intensive farming practice generally relies on the extensive use of chemical fertilisers, which can lead to a long list of serious environmental and health problems (Tilman et al. 2002; Wall et al. 2015), especially for developing countries, where most farmers face the challenges of increasing cost of fertiliser, environmental degradation and food quality deterioration. Thus, sustainable intensification (the use of environmentally and sustainable crop production practices with renewable resources) has attracted worldwide attention (Ju et al. 2009). Sustainable intensification of agriculture aims to minimise pressures on the environment in terms of water, energy and fertiliser use, and on maintaining soil as a sustainable resource, each balanced against the necessity of maintaining, or increasing, crop productivity. Soil health, which is often considered synonymous with soil quality, is integral to maintaining successful, sustainable agricultural systems and has been defined as:

"the capacity of soil to function as a vital living system, within ecosystem and land-use boundaries, to sustain plant and animal productivity, maintain or enhance water and air quality, and promote plant and animal health" (Doran and Zeiss 2000).

Success in meeting these objectives will require better understanding and management of the biological processes and interactions that underpin the functioning of plant-soil systems. Improving soil quality can reduce on-farm inputs with significant concomitant environmental benefits and reduce production costs. The maintenance of global food security mediated by sustainable intensification of agriculture is recognised as inherently complex and multidisciplinary in nature. Soil is a key asset of natural capital, providing goods and services that sustain life through regulating, supporting and provisioning roles, having impacts beyond agricultural systems. However, as a consequence of past intensification, degradation threats to soils are numerous (Powlson et al. 2011). The UN Environment Programme has supported the development of a global assessment of human-induced soil degredation (GLASOD) using a range of physical, chemical and biological indicators to assess the severity of degredation. This identifies five classes of degradation:

0. No degradation.
1. Light: the terrain has somewhat reduced agricultural suitability but is suitable for use in local farming systems. Restoration to full productivity is possible by modifications of the management system. Original biotic functions are still largely intact.
2. Moderate: the terrain has greatly reduced agricultural productivity but is still suitable for use in local farming systems. Major improvements are required to restore productivity. Original biotic functions are partially destroyed.
3. Strong: the terrain is non-reclaimable at farm level. Major engineering works are required for terrain restoration. Original biotic functions are largely destroyed.
4. Extreme: the terrain is irreclaimable and beyond restoration. Original biotic functions are fully destroyed.

This assessment demonstrates the widescale and significant degradation of soils that has ocurred at a global scale (Bridges and Oldeman 1999). Compounded with an ever-burgeoning global population, the area of soil usable for cultivation has declined from 0.32 to 0.25 ha per capita between 1975 and 2000 (FAO 2011). A reduction of soil fertility has contributed to crop yields stagnating since 1996. Soil changes slowly, and our understanding of the biological, chemical and physical components of the soil system is incomplete. To prevent further degradation of agricultural soils, programmes to monitor soil quality to help address the current and future challenges of food security are under way.

3 Regional Models of Sustainable Intensification and Impact on Soil Quality

As already discussed, there is no clear single definition of the concept of sustainable intensification, and indeed it cannot be considered as a template for production systems. However, it is likely that sustainable intensification would vary significantly between production systems and regions, and impact on soil quality would therefore also be variable. In this section, we consider some case studies according to different variables. Production levels and sustainability metrics have been used to describe the modifications to farming systems that fall under the heading of sustainable intensification in different climatic zones, and we identify the impacts that those system level changes have on soil quality.

Across Europe and North America, there has been a long history of intensification in agriculture and the application of technologies, including genetic improvement of crop and livestock products, increased fertiliser and pesticide use and, more recently, the development of precision farming approaches. Resource use input in this region is now stabilising, and more efficient use of resource inputs (particularly nitrogen) is emerging as a priority in response to concerns about pollution and development of sustainable farming systems (Sutton et al. 2011). Approaches to help achieve this include more extensive use of nutrient management plans, decision support tools to optimise nutrient use efficiency and regulatory controls that limit nutrient applications.

Across the continent of Africa, there has been widespread soil degradation, and this, alongside climate change and population growth, has placed severe strains on production systems. Crop yields have shown little or no increase in recent decades, and in a survey of 37 countries covering an area of 200 M ha, it has been shown that soil reserves of N, P and K have declined by 660, 75 and 450 kg/ha, respectively, as a consequence of nutrient mining (Sanchez et al. 1997). New approaches to develop more sustainable patterns of land management are being developed though the concept of climate-smart agriculture (FAO 2013). This provides both mitigation and adaptation to climate change and supports improvements in soil quality and fertility. Building soil fertility through increasing soil organic carbon stocks forms a particularly important component of the climate-smart agriculture approach and therefore provides direct benefits in terms of improving soil quality.

South America has also seen widespread degradation of soils in recent decades due to large-scale land use change linked to the creation of pastures and croplands from previously forested land or native prairies. This has resulted in soil organic matter loss and loss of soil fertility, with low rates of productivity. The need to address soil degradation in South America is now widely recognised, and agroecological techniques, including the development of agroforestry systems and a more extensive use of legume-based pastures that provide a biological input of nitrogen, are seen as promising approaches (Tittonell 2014). Such nitrogen addition is additionally known to be associated with increased storage of soil organic matter.

Across China and other regions in SE Asia, a very different pattern of soil management has emerged. Here, recent increases in agricultural productivity have been delivered by very large increases in resource inputs (particularly nitrogen fertilisers). This has led to severe problems of environmental degradation and loss of soil quality. New approaches to soil management are being developed based upon an integrated crop-soil management system that uses a detailed knowledge of agroecology and biogeochemistry to match more closely the demand for nutrients with nutrient supply (Chen et al. 2014).

It is generally the case that soils are poorly protected by legislation, when compared with laws protecting air and water. Legal frameworks that do protect soils generally tend to focus on specific areas, such as contamination by heavy metals or toxins and nutrient loadings. The absence of an overarching policy to protect and monitor soil quality makes it difficult to assess the impacts of soil management on long-term changes. It is also apparent from the foregoing analysis that there is no single blueprint for sustainable intensification that could apply across all regions of the globe. It is important that solutions to sustainable development are tailored to regional and local needs, taking into account social, economic and environmental considerations.

4 Economic and Behavioural Aspects of Managing Soil Quality

There are a number of management approaches that can be adopted to enhance soil quality; these can enhance both the private benefits to farmers and wider social impacts. Approaches that enhance soil organic matter (with a potential to mitigate climate change) can contribute to improved soil fertility, improved soil structure and better water availability. For farmers, these can potentially improve crop yields or reduce costs through reducing input requirements such as fertiliser and the fuel and time requirements for cultivation. More complex trade-offs might arise where yield reductions are accompanied by lower costs. This may also create an important behavioural barrier to implementation where high yield is seen as an indicator of 'successful farming' (Ingram et al. 2016) and highlights further barriers such as lack of awareness and 'tradition' with respect to changing established practices.

The improved soil quality that results from increasing soil organic matter contents can be associated with economic benefits. In an analysis of different management strategies that compared crops with (CCleg) and without (CCNoleg) legumes, zero tillage (ZeroTill), minimum tillage (MinTill) and residue management (ResMan), an economic analysis was used to compare gross margins (Glenk et al. 2015). Cover crops and residue management were shown to reduce margins across all farm types; these impacts are higher for smaller farms, although the impact of residue management diminishes after 10 years (Fig. 2). Minimum tillage offers small but increasing benefits over time, whilst zero tillage has a negative

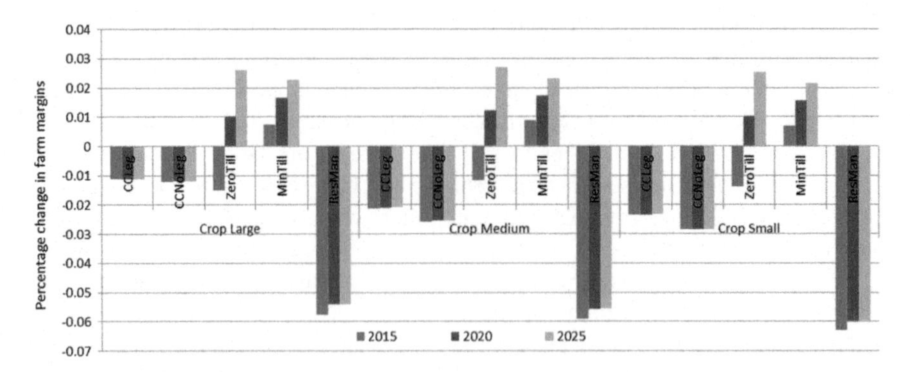

Fig. 2 Change in farm gross margin under different soil organic matter options compared to the baseline for Scottish farm groups (source: Glenk et al. 2015)

impact in the first five years before becoming positive and outperforming minimum tillage after 15 years. These results reflect a small initial reduction in yield associated with the tillage measures followed by increases after 5 years. Additionally, cultivation costs are reduced for these measures.

Economic analyses of soil quality improvement highlight that the benefits to farmers vary across combinations of management measure, crop type and location. For instance, use of cover crops at the higher latitudes of northern Europe with a short growing season may displace more valuable crops. Additionally, crop residues (e.g., straw) can have commercial value, so their use for soil improvement involves a loss to farmers where they have a market or alternate utility (e.g., bedding). Some measures such as reduced or zero tillage may take several years before benefits are realised.

These economic and behavioural dynamics of agricultural production and soil quality highlight the continuing role for policy intervention not just to ensure public benefits but also to spread the understanding of private benefits to farmers. Sustainable intensification aimed at maintaining or enhancing soil quality will need to negotiate economic and behavioural hurdles to implementation through understanding how farm systems operate and emphasising productivity benefits.

5 Conclusions

Maintaining good soil quality is critically important to sustainable food production. Historical land use changes have led to large increases in the amount of land used for agriculture; however, soils in many areas of world are now associated with significant levels of degradation. The need to increase food production in the twenty-first century will have to be accompanied by an increasing focus on maintaining and enhancing soil quality. This can be achieved by making more efficient use of resource inputs (sustainable intensification) and enhancing soil organic matter stocks, within a framework that recognises good soil quality as an essential prerequisite for sustainable food production systems. Regional approaches to managing soil quality will vary according to prevailing conditions. In areas of the world where significant intensification has already occurred (America, Europe, Asia), more efficient use of existing inputs will take priority, whilst in regions that have suffered from extensive soil degradation, building soil fertility and organic matter stocks is essential. Uptake of management approaches to improving soil quality needs to be supported by clear economic assessments that quantify the wider benefits of improvements in soil quality, and monitoring both is needed to improve understanding and deliver more sustainable production systems.

Acknowledgements The authors gratefully acknowledge funding from the Scottish Government's Rural Affairs Food and Environment Strategic (RESAS) Research programme for funding.

References

Audsley E, Brander M, Chatterton J, Murphy-Bokern D, Webster C, Williams A (2009) An assessment of greenhouse gas emissions from the UK food system and the scope for reduction by 2050. How low can we go? WWF-UK and the FCRN

Baulcombe D (2010) Reaping benefits of crop research. Science 327:761

Baulcombe D, Crute I, Davies B, Dunwell J, Gale M, Jones J, Pretty J, Sutherland W, Toulmin C (2009) Reaping the benefits: science and the sustainable intensification of global agriculture

Bridges EM, Oldeman LR (1999) Global assessment of human-induced soil degradation. Arid Soil Res Rehabil 13:319–325

Chen X, Cui Z, Fan M, Vitousek P, Zhao M, Ma W, Wang Z, Zhang W, Yan X, Yang J, Deng X, Gao Q, Zhang Q, Guo S, Ren J, Li S, Ye Y, Wang Z, Huang J, Tang Q, Sun Y, Peng X, Zhang J, He M, Zhu Y, Xue J, Wang G, Wu L, An N, Wu L, Ma L, Zhang W, Zhang F (2014) Producing more grain with lower environmental costs. Nature 514:486–489

Doran JW, Zeiss MR (2000) Soil health and sustainability: managing the biotic component of soil quality. Appl Soil Ecol 15:3–11

Erisman JW, Sutton M, Galloway JN, Klimont Z, Winiwarter W (2008) How a century of ammonia synthesis changed the world. Nat Geosci 1:636–639

FAO U (2011) The state of the world's land and water resources for food and agriculture. Managing systems at risk. UN FAO, Rome, p 50

FAO (2013) Climate-Smart agriculture sourcebook. FAO

Foley JA, Ramankutty N, Brauman KA, Cassidy ES, Gerber JS, Johnston M, Mueller ND, O'Connell C, Ray DK, West PC, Balzer C, Bennett EM, Carpenter SR, Hill J, Monfreda C, Polasky S, Rockstrom J, Sheehan J, Siebert S, Tilman D, Zaks DP (2011) Solutions for a cultivated planet. Nature 478:337–342

Fowler D, Steadman CE, Stevenson D, Coyle M, Rees RM, Skiba UM, Sutton MA, Cape JN, Dore AJ, Vieno M, Simpson D, Zaehle S, Stocker BD, Rinaldi M, Facchini MC, Flechard CR, Nemitz E, Twigg M, Erisman JW, Butterbach-Bahl K, Galloway JN (2015) Effects of global change during the 21st century on the nitrogen cycle. Atmos Chem Phys Discuss 15:13849–13893

Gardi C, Jeffery S, Saltelli A (2013) An estimate of potential threats levels to soil biodiversity in EU. Glob Change Biol 19:1538–1548

Glenk K, Shrestha S, Topp CFE, Sanchez B, Ingesias A, Dibari C, Merante P (2015) Modelling constraints and trade-offs in optimizing SOC. Deliverable D3.4. Smart Soil Report

Godfray HCJ, Beddington JR, Crute IR, Haddad L, Lawrence D, Muir JF, Pretty J, Robinson S, Thomas SM, Toulmin C (2010) Food security: the challenge of feeding 9 billion people. Science 327:812–818

Godfray HCJ, Crute IR, Haddad L, Lawrence D, Muir JF, Pretty J, Robinson S, Toulmin C (2011) The future of food and farming; challenges and choices for global sustainability. The Government Office for Science. Foresight, London, UK

Hooper DU, Chapin FS, Ewel JJ, Hector A, Inchausti P, Lavorel S, Lawton JH, Lodge DM, Loreau M, Naeem S (2005) Effects of biodiversity on ecosystem functioning: a consensus of current knowledge. Ecol Monogr 75:3–35

Ingram J, Mills J, Dibari C, Ferrise R, Ghaley BB, Hansen JGn, Iglesias A, Karaczun Z, McVittie A, Merante P (2016) Communicating soil carbon science to farmers: Incorporating credibility, salience and legitimacy. J Rural Stud 48:115–128

Ju XT, Xing GX, Chen XP, Zhang SL, Zhang LJ, Liu XJ, Cui ZL, Yin B, Christie P, Zhu ZL (2009) Reducing environmental risk by improving N management in intensive Chinese agricultural systems. Proc Natl Acad Sci 106:3041–3046

Powlson DS, Gregory PJ, Whalley WR, Quinton JN, Hopkins DW, Whitmore AP, Hirsch PR, Goulding KW (2011) Soil management in relation to sustainable agriculture and ecosystem services. Food Policy 36:S72–S87

Rees RM, Barnes AP, Moran D (2016) Sustainable intensification: the pathway to low carbon farming? Reg Environ Change 16:2253–2255

Reid WV, Mooney HA, Cropper A, Capistrano D, Carpenter SR, Chopra K (2005) Millenium ecosystem assessment synthesis report. Island Press, Washington, DC

Sanchez PA, Shepherd KD, Soule MJ, Place FM, Buresh RJ, Izac AM, Mokwunye AU, Kwesiga FR, Ndiritu CG, Woomer PL (1997) Soil fertility replenishment in Africa: an investment in natural resource capital. In: Replenishing soil fertility in Africa, pp 1–46

Sutton MA, Howard CM, Erisman JW, Billen G, Bleeker A, Grennfelt P, Grinsven HV, Grizzetti B (2011) The European Nitrogen Assessment; sources, effects and policy perspectives. Cambridge

Tilman D, Cassman KG, Matson PA, Naylor R, Polasky S (2002) Agricultural sustainability and intensive production practices. Nature 418:671–677

Tittonell P (2014) Ecological intensification of agriculture – sustainable by nature. Curr Opin Environ Sustain 8:53–61

Tsiafouli MA, Thébault E, Sgardelis SP, Ruiter PC, Putten WH, Birkhofer K, Hemerik L, Vries FT, Bardgett RD, Brady MV (2015) Intensive agriculture reduces soil biodiversity across Europe. Glob Change Biol 21:973–985

Wall DH, Nielsen UN, Six J (2015) Soil biodiversity and health. Nature 528:69–76

Westhoek H, Lesschen JP, Rood T, Wagner S, De Marco A, Murphy-Bokern D, Leip A, van Grinsven H, Sutton MA, Oenema O (2014) Food choices, health and environment: effects of cutting Europe's meat and dairy intake. Glob Environ Change 26:196–205

Voluntary Guidelines for Sustainable Soil Management: Global Action for Healthy Soils

Rainer Baritz, Liesl Wiese, Isabelle Verbeke, and Ronald Vargas

1 Soil and the Sustainability Agenda

1.1 Introduction

Soils are deteriorating worldwide and are thus becoming less fertile, providing less nutrients to plants, animals, and people who rely upon them. Recent inventories such as the Status of the World's Soil Resources report (FAO and ITPS 2015) have indicated the extent to which soils have suffered from unsustainable management: 25% of land is highly degraded, and a further 44% is slightly or moderately degraded due to the erosion, salinization, compaction, and chemical pollution of soils. Seventy-five percent of the African territory has serious soil fertility problems (Toenniessen et al. 2008); of those, 40% are characterized by inherent poor soil fertility, exacerbated by degradation from nutrient mining. This negatively affects soil productivity and leads to serious nutrient deficiencies in harvested crops. Nutrient depletion and erosion are key forms of soil degradation and can only be mitigated through specific management actions.

Population growth, food insecurity, nutrient deficiencies, soil degradation, and climate extremes are often dramatically concentrated in hot spots such as parts of the African continent (Thiombiano and Tourino-Soto 2007). The water storage capacity and nutrient reservoir of soils in these areas are of particular importance since the agriculture area can only be expanded at high cost in Latin American

R. Baritz (✉)
European Environment Agency, Kopenhagen, Denmark
e-mail: rainer.baritz@eea.europe.eu

L. Wiese · I. Verbeke · R. Vargas
Food and Agriculture Organization of the United Nations, Rome, Italy
e-mail: Liesl.Wiese@fao.org; Isabelle.Verbeke@fao.org; Ronald.Vargas@fao.org

H. Ginzky et al. (eds.), *International Yearbook of Soil Law and Policy 2017*,
International Yearbook of Soil Law and Policy,
https://doi.org/10.1007/978-3-319-68885-5_3

grasslands and African Savannahs (Bruinsma 2009; Lambina and Meyfroidt 2011) and depends on future market prices. In many developed countries, cropland area is diminishing (Kuemmerle et al. 2016). This demonstrates that agricultural intensification (including degraded land improvement and soil restoration) is currently the only solution to meet increasing food demands, requiring careful management and optimization of the available natural resources soil and water. While much scientific and technical knowledge about proper soil management practices exists, it does not sufficiently reach extensionists and practitioners.

Political initiatives are needed to stimulate and facilitate action on the ground. For this reason, the Global Soil Partnership (GSP), through the Food and Agriculture Organization of the United Nations (FAO) and its member countries, advocates for the importance of soils and sustainable soil management. After first revising the World Soil Charter, the GSP has then developed the Voluntary Guidelines for Sustainable Soil Management (VGSSM), recently endorsed by all FAO members (FAO 2017). Next, technical guidelines for implementation need to be developed, which will compile and offer technical knowledge about sustainable soil management considering regional and local conditions.

1.2 Soils and the Environmental Sustainability Agenda

Food security and balanced nutrition of the growing human population are the major driving forces to better manage and protect land and soils. Climate change and land degradation are well-known factors that undercut these objectives, and various initiatives and strategies have been created to improve the soil's resilience and potential to fulfill its manifold functions.

Following the World Summit on Sustainable Development (WSSD) in September 2002, FAO has launched the Sustainable Agriculture and Rural Development (SARD) program (2002–2007). It included the formulation of Good Agricultural Practices (GAP), actions that promote sustainable agriculture and natural resource management contributing to food security. Environmentally, locally appropriate GAP codes of practice can help improve soil fertility, increase the efficiency of water use and plant protection, and conserve biodiversity. Sustainable soil management as one of the concerned sectors was specifically addressed (Poisot et al. 2007). The SARD program has then been followed by various subsector programs, such as "Save and Grow," the framework program for sustainable crop production intensification, Sustainable Forest Management, Global Soil Partnership, Climate-Smart Agriculture, and others. FAO's current integrative approach to build sustainable agriculture across sectors (forestry, agriculture, fisheries) sets out five key principles that balance the social, economic, and environmental dimensions of sustainability; one of them is the efficient use of resources, which explicitly includes soils (FAO 2014a). Sustainable agriculture must minimize negative impacts on the environment while optimizing production by protecting, conserving, and enhancing natural resources and using them efficiently.

The Sustainable Development Goals (SDGs) impressively demonstrate the complexity of cross-sector knowledge needs and approaches, which have to be considered in the goal to sustainably manage the world's natural resources in relation to food security and nutrition. Among others, the SDGs identify the need to restore degraded soils and improve soil health. Of the 17 SDGs—addressing issues such as hunger and the fight against poverty, halting desertification, and addressing the decline of species diversity—ten goals directly or indirectly affect soils, of which four (4) contain targets related to soils.

Sustainable soil management is especially relevant to the achievement of SDGs 2, target 2.4 and 15, target 15.3, which stipulates the following:

SDG 2. End hunger, achieve food security and improved nutrition, and promote sustainable agriculture.

> 2.4 By 2030, ensure sustainable food production systems and implement resilient agricultural practices that increase productivity and production, that help maintain ecosystems, that strengthen capacity for adaptation to climate change, extreme weather, drought, flooding and other disasters, and that progressively improve land and soil quality.

SDG 15. Protect, restore and promote sustainable use of terrestrial ecosystems, sustainably manage forests, combat desertification, and halt and reverse land degradation and halt biodiversity loss.

> 15.3 By 2030, combat desertification, and restore degraded land and soil, including land affected by desertification, drought and floods, and strive to achieve a land-degradation neutral world.

SDG 2 recognizes that food security and nutrition requires the establishment of effective sustainable agricultural production, which, in turn, is impossible without maintenance of soil quality. The latter can be provided only through sustainable soil management practices that would ensure stable or increasing production from arable lands, pastures, and agroforestry systems.

Both extensive and intensive agricultural production can lead to soil degradation due to various processes. Combating soil degradation therefore requires the introduction of sustainable soil management systems that address the challenges of SDGs 2 and 15.

1.3 Global Soil Partnership: Action for Sustainable Soil Management

The Global Soil Partnership (GSP) (Fig. 1) is a global, multilateral initiative launched in 2012 by FAO, its member countries (currently 135 GSP focal points in governments), as well as other governmental and nongovernmental partners (>200 institutions), with the aim to promote sustainable soil management, avoid fragmentation of efforts, and halt the increasing rate of misuse, degradation, and loss of soils. The GSP consists of a global network of contributors through nine regional soil partnerships, focusing on five pillars of action that address the

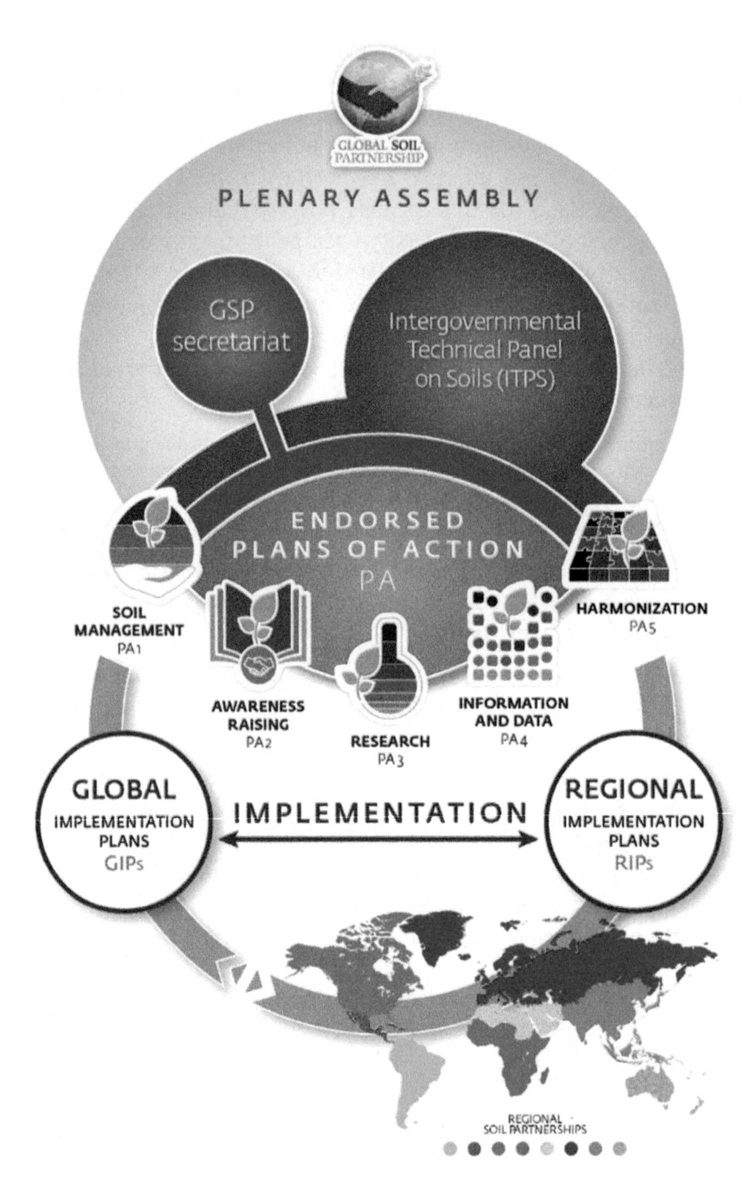

Fig. 1 The structure and governance of the Global Soil Partnership (GSP)

development and implementation of sustainable soil management systems (Pillar 1), raise awareness and funding (Pillar 2), promote soil research (Pillar 3), and build the Global Soil Information System (GLOSIS) (Pillars 4 and 5). As part of GLOSIS and the FAOSTAT family, a global monitoring system of soil-related indicators (SoilSTAT) will be developed.

Since its creation, the GSP has become an important partnership providing a forum where regional and global soil issues are discussed and addressed by multiple stakeholders. Key outputs already demonstrate that the partnership is successfully taking on the urgent challenge for advocacy about soils, including the promotion of sustainable soil management.

2 Action for Soil Protection and Rehabilitation

2.1 Soil Functions

Soils perform manifold functions that are critical for plant growth, as well as animal and human life. These functions result from interacting chemical, physical, and biological soil properties and processes that shape the environment to produce goods and services through agriculture, forestry, and other land uses (Fig. 2 provides an overview of these functions).

The understanding of the dynamics between management, soil properties, and other biophysical landscape properties (climate, slope, and parent material) is a key requirement in order to effectively target management interventions (Sauerbeck

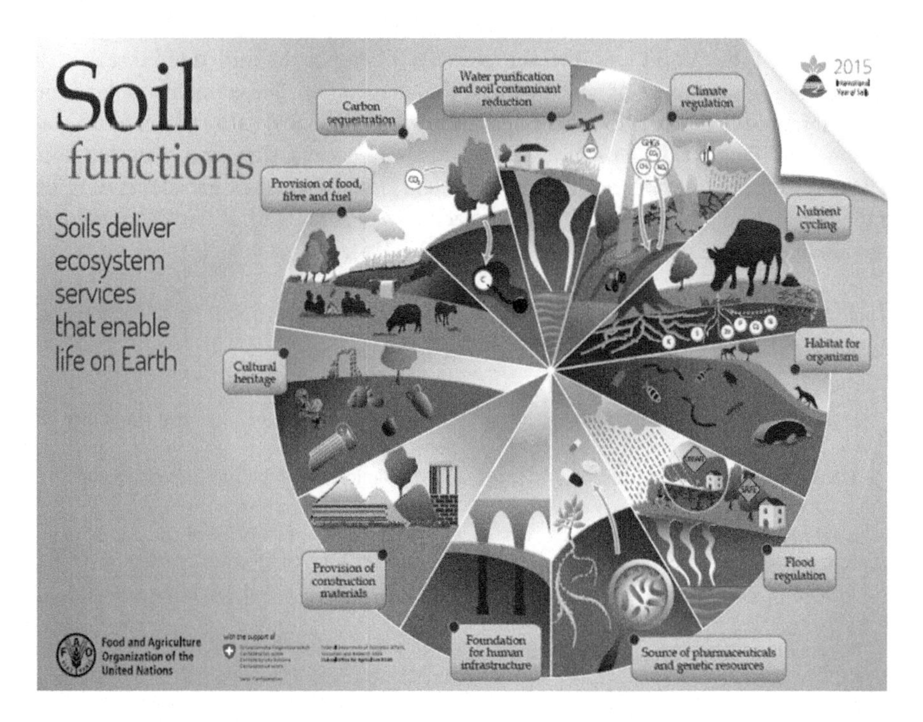

Fig. 2 Soil functions—overview (source: FAO)

1994). The planning of land use and management starts with an evaluation process of the types and distribution of soils present, and soil experts are often consulted about the kind of crops that can be cultivated on a particular soil, yields to be expected, fertilizers to use, irrigation practice, and management practice (Sys 1985). Table 1 presents characteristics of soil quality and soil functioning required for land evaluation and management planning. It includes knowledge of natural soil fertility and the current soil health and degradation status. This is the reason why the VGSSM eventually need to be technically specified in a local and regional context.

Natural factors such as climate, as well as management, affect soils in its capability to perform its functions. Considering the diversity and heterogeneity of the environment, a well-functioning, healthy soil can only be defined on the basis of its local conditions that determine different stages of equilibrium, stability, and resilience. The understanding of these conditions and the effect of abrupt environmental change on soils (e.g., through cropland management or extreme weather events such as drought) allow the definition of disturbance and degradation and how this can be counterbalanced through proper management actions.

2.2 *Target Areas for Action: The Status of the World's Soil Resources (SWSR) Report*

The Status of the World's Soil Resources (SWSR) report identified ten threats that hamper the achievement of sustainable soil management. These threats are erosion by water and wind, organic carbon loss, nutrient imbalance, salinization, contamination, acidification, loss of biodiversity, sealing, compaction, and waterlogging. These different threats vary in terms of extent, intensity, and trend depending on geographical context, though they all need to be addressed in order to achieve sustainable soil management (FAO and ITPS 2015).

Table 1 Soil characteristics and land evaluation (based on Sys 1985; UNEP 2016)

Aspects of land use planning	Soil characteristics
– Choice of crops	– Natural soil fertility
– Land use pattern on marginal soils	– Agricultural value of soils, soil suitability of marginal soils
– Management of specific soils	– Mechanical workability of clay-rich soils (Vertisols)
– Land improvement works (drainage, irrigation, leveling and grading, erosion control)	– Susceptibility of soils for degradation processes (e.g., erosion by wind, erosion by water)
– Type of irrigation	– Soil permeability and soil water holding capacity
– Type and quantity of fertilizer to be used	– Soil fixation rate for P, exceedance of critical thresholds for N
– Optimization of ecosystem services	– Soil carbon status, nutrients, CEC, etc.

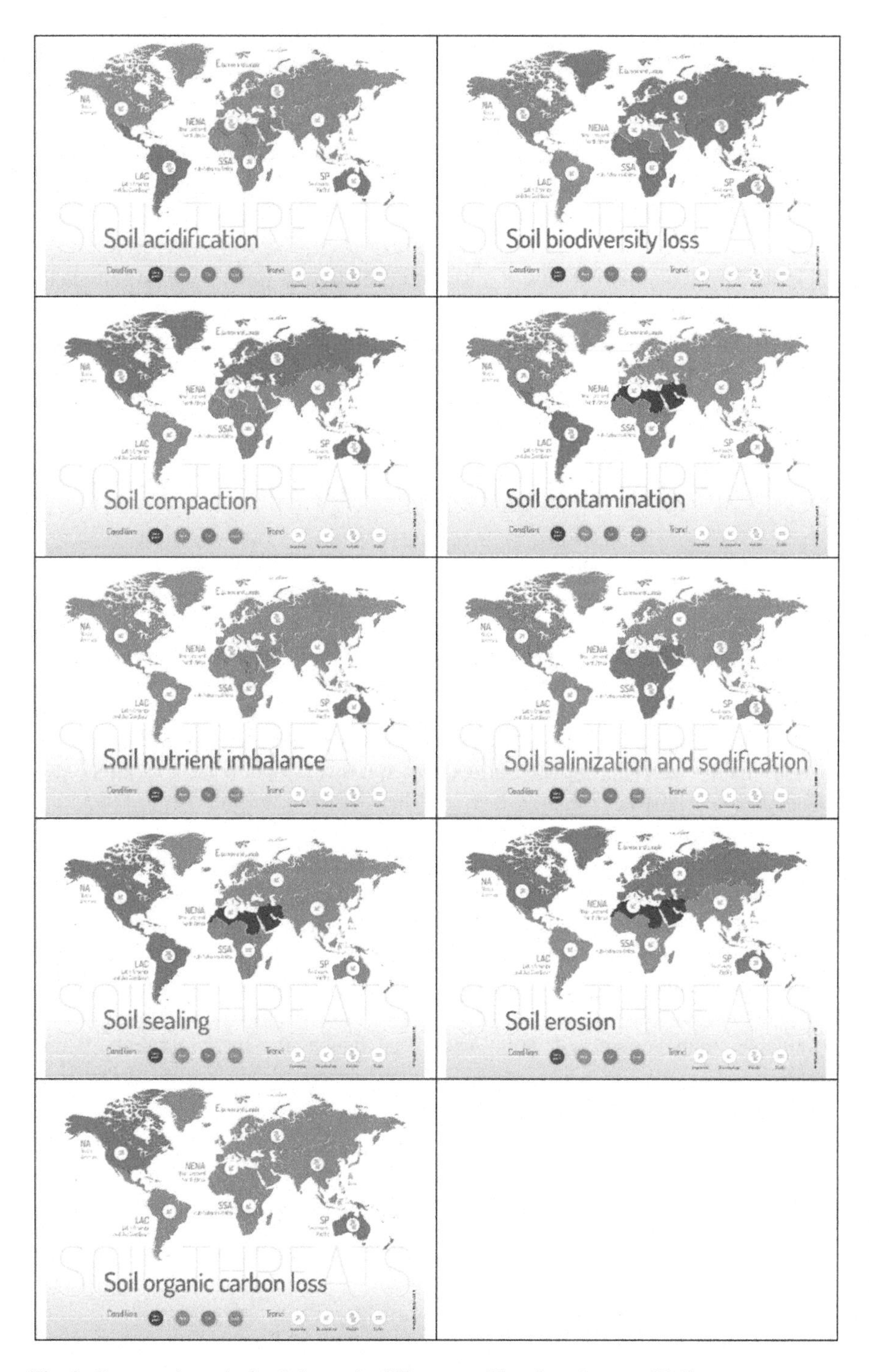

Fig. 3 Status and trend of soil threats in different world regions (source: FAO)

Based on the SWSR report and other studies, about 33% of global soils are moderately or highly degraded, and one of the main causes for this is unsustainable soil management. According to the VGSSM, soil degradation should be minimized using sustainable soil management practices, especially through soil conservation approaches that have been proven to be successful. Soil rehabilitation and/or soil restoration should also be a priority in restoring degraded soils, especially in historically sound agricultural or other production systems currently under threat.

Figure 3 presents an overview of the current soil threats (based on FAO and ITPS 2015). The information about the status and trend of soil threats is are often based on coarse estimates and expert judgments. The overview demonstrates that the extent of the soil threats and associated trends are highly region specific.

2.3 Revised World Soil Charter: Foundation for the VGSSM

Soils are a fundamental part of terrestrial ecosystems: the services that ecosystems can provide are directly dependent on the condition of the soil and how sustainably it is managed.

The World Soil Charter (WSC) provides the key principles and guidance for action against soil degradation. It has been recently revised by the Intergovernmental Technical Panel on Soils (ITPS) under the aegis of the GSP. Supported by the Committee on Agriculture of FAO at its 24th Session (September 29–October 3, 2014), the revised Charter was endorsed by the FAO Council at its 150th Session (December 1–5, 2014).

According to the revised WSC (FAO 2015a), "soil management is sustainable if the supporting, provisioning, regulating, and cultural services provided by soil are maintained or enhanced without significantly impairing the soil functions that enable those services or biodiversity." Thus, the Charter prepares the grounds for policies and action for sustainable soil management. Since the WSC's principles are quite broad and generic in nature, it was recognized that the Charter may be usefully complemented by the preparation of more detailed guidelines for the sustainable management of soil resources, hence the development of the VGSSM.

3 Development and Adoption of the VGSSM

3.1 Consultation Process

3.1.1 Initial Process: Concept Note for VGSSM

The origin of the VGSSM can be traced back to the 24th session of the FAO Committee on Agriculture (COAG[1]) (September 29–October 3, 2014), which "noted the proposal for developing a concept note on sustainable management of soil resources with support from the Intergovernmental Technical Panel on Soils, for submission to the next GSP Plenary Assembly and the FAO Council."[2]

A concept note was subsequently prepared by the GSP's Intergovernmental Technical Panel on Soils. Building on the just-approved, updated WSC, the intent was to prepare more detailed, essentially technical guidelines, which then developed into the Voluntary Guidelines for the Sustainable Management of Soil Resources.

The third Plenary Assembly of the GSP in June 2015 welcomed the concept note and supported the development of voluntary guidelines for the sustainable management of soil resources. Many GSP members commented and then engaged in an e-mail consultation process before a revised concept note was presented to the FAO Council. In December 2015—during the celebration of the International Year of Soils—the 153rd FAO Council supported the concept note to develop Voluntary Guidelines for Sustainable Soil Management (VGSSM). The ITPS was then tasked to develop a zero-order draft.

3.1.2 Zero-Order VGSSM and Public E-Consultation

The zero-order draft was subjected to a comprehensive e-consultation process with all interested partners and stakeholders. This public consultation was conducted through the Global Forum on Food Security and Nutrition (FSN Forum), 08.02.2016–07.03.2016. As a result, 96 extensive contributions were received, summarized as a report, and published.

3.1.3 First-Order VGSSM and Open-Ended Working Group

The results from the e-consultation directly fed into the first draft of the VGSSM as prepared by the ITPS and endorsed by its members on March 18, 2016. The first draft was then submitted to an Open-Ended Working Group for finalization.

---

[1]COAG, Committee on Agriculture, is FAO's main technical advisory committee, reviewing specific matters related to agriculture, food and nutrition.

[2]FAO Council is the supreme governing body of FAO in which each member is represented.

Fig. 4 Context of information and action toward VGSSM

The Open-Ended Working Group was compiled based on nominations from the permanent representatives to FAO. It was composed of 40 members, of whom 27 were representatives from FAO Member States and 13 were nonstate GSP partners representing civil society, the private sector, and academia/research institutions.

3.2 Adoption of the VGSSM

The VGSSM were adopted by the fourth GSP Plenary Assembly (Rome, May 23, 2016), subsequently approved by the 25th session of the FAO Committee on Agriculture (Rome, September 28, 2016), and finally endorsed by the 155th session of the FAO Council (Rome, December 5, 2016). These guidelines provide technical and policy recommendations on how sustainable soil management can be achieved. The successful implementation of these guidelines should pave the way to halter soil degradation and to boost soil health (Fig. 4).

4 Nature of the VGSSM

4.1 Objectives of the VGSSM

The VGSSM are of a voluntary nature and therefore are legally nonbinding. They focus on technical and biological aspects of the soil while facilitating possible strategic choices; policy-specific guidance is excluded.

The guidelines address sustainable management of soils in all types of agriculture systems and the maintenance or enhancement of the ecosystem services they provide, such as food production, climate regulation, and the regulation of water quality and quantity. The VGSSM are globally relevant without entering into details of specifying actions at the local scale.

The VGSSM may be used for a variety of purposes, notably:

- to inform and encourage the design and implementation of programs and projects aimed at the sustainable management and conservation of soil resources, land evaluation, the rehabilitation/restoration of degraded soils, as well as sustained ecosystem services provided by soils;
- to provide technical support for the development and/or activation of strategies, policies, laws, and regulations pertaining to soil governance;
- to assist in the framing of investment decisions for the responsible use of soils;
- to support awareness raising, advocacy, and capacity development;
- to contribute to the design of training programs, targeted at a range of actors engaged in agriculture, forestry, and other land uses, and the development of educational curricula, from elementary schools to universities, to provide basic knowledge of the importance of soils and of SSM principles and practices for different soils and contexts;
- to promote soil-relevant programs and research;
- to inform and promote the development of indicators to monitor and map the status of soil resources and the implementation of SSM;
- to contribute to the implementation of the Land Degradation Neutrality (LDN) voluntary objective mentioned in the Rio Declaration: "Our Common Future."

4.2 Content of the VGSSM

The VGSSM provides context and guidance for actions related to the major threats to soil functioning; hence, actions are closely related to soil threats/degradation types:

- minimize soil erosion;
- enhance soil organic matter;
- foster soil nutrient balances and cycles;
- prevent, minimize, and mitigate soil salinization and alkalinization;
- prevent and minimize soil contamination;
- prevent and minimize soil acidification;
- preserve and enhance soil biodiversity;
- minimize soil sealing;
- prevent and mitigate soil compaction;
- improve soil water management.

For each of the action areas, key practices for sustainable soil management are mentioned (Table 2 contains examples):

Table 2 Effects of different management practices (examples)

Management practice		Effects on soil properties
No till		Soil erosion reduction, water evaporation reduction, improved soil biodiversity and biological activity, use of fossil fuel reduction
No till + good agricultural practices	Crop rotation Balanced nutrition Integrated pest management Responsible agro-chemical management	Better physical, chemical, and biological soil conditions – Increased soil biological activity and biodiversity – Improved soil fertility and nutrients cycling – Higher water use efficiency (less consumption and plus storage) – Carbon sequestration – Higher yields and increased yield stability

- rotation systems: crop rotation, mixed cropping, agroforestry, intercropping, relay cropping;
- minimum soil disturbance by no tillage, reduced tillage, direct planting;
- compost, farmyard manure;
- mulching, crop residues left on the field;
- green manure/cover crops;
- improved fallow;
- soil and water conservation practices, including irrigation, and watershed protection
- cropping against the slope;
- nutrients: intensify agriculture based on using external and internally recycled nutrient resources; efficient plant nutrition.

Table 2 provides an example of how individual management practices affect soil properties and soil functioning. It can be seen that individual practices have combined effects on different soil threats, such as no tillage on the protection of soils from erosion and loss of soil organic matter (SOM).

Table 3 lists recommended management practices for VGSSM-action areas, addressing the specific soil threats of erosion and soil organic matter loss. It can be seen that the recommendations are still fairly generic and that additional local information is needed before recommendations can be implemented in the field. For example, the local implementation of balanced fertilizer application, as it is recommended to increase biomass production (which then enhances soil organic matter), requires detailed planning based on soil type and soil nutrient capacity, climate, target crop, and predicted production.

Besides the consideration of local conditions, there are also region-specific expectations and priorities to the VGSSM; for example, for Africa, priorities for sustainable soil management are the following (FAO 2015a, 2016):

Table 3 VGSSM management practices (examples: minimize erosion, enhance soil organic matter)

Degradation type	Principle	Measures
Erosion	Maintain cover of the soil surface with plants and or residues	Mulching, minimum tillage, no till by direct seeding, cover crops, agroecological approaches, controlled traffic, continuous plant cover and crop rotation, strip cropping, agroforestry, shelter belts, appropriate livestock densities
	Reduce water surface run-off	Strip cropping, contour planting, crop rotation, intercropping, agroforestry, cross slope barriers, terrace construction, grass-land waterways, vegetated buffer strips
	Riparian zone protection	Buffer strips, wetlands, water harvesting, cover crops
	Wind erosion	Vegetative and artificial wind breaks
Soil organic matter (SOM)	Increase biomass production	Maximize water-use efficiency by proper irrigation systems, minimize soil erosion and nutrient leaching by cover crops, balanced fertilizer applications, organic amendments, agroforestry and alley cropping, reforestation, afforestation
	Protect SOM-rich soils	For example, peatlands, forests, pasture lands
	Increase SOM	Crop residues, organic farming, integrated soil fertility management, integrated pest management, animal manure, compost, mulching
	Avoid fire	
	Optimize use of organic inputs	Animal manure, properly processed human waste
	Ensure organic cover of soils	Cover crops, improved fallow plant species, reduced or no tillage, live fences
	Decrease organic matter decomposition	Minimum or no tillage
	Improve crop diversity	Crop rotation, legumes

- soil organic matter/soil carbon;
- maximization of organic/mineral input efficiencies;
- education in soil science;
- platform for information exchange;
- increased voice of soil scientists at policy level.

5 Other Initiatives That Include Sustainable Soil Management

The practices listed in the VGSSM are identified to improve soil quality and thus soil functioning. Considering the need for food security through agricultural intensification and increased production where ecologically and economically feasible, the VGSSM offer the framework so that these objectives are implemented in a sustainable manner, preventing soils from continued degradation and loss.

Various individual practices mentioned in the guidelines are already considered and technically described in other activities, for example, related to sustainable land and forest management. Due to their importance for the VGSSM, some of these activities are briefly introduced here. As mentioned before, the VGSSM will be followed by technical guidelines. These guidelines will utilize experience and knowledge gained from the below-mentioned activities.

5.1 Conservation Agriculture

"Sustainable intensification" refers to productive agriculture, which avoids ecological degradation as a seemingly unavoidable side effect from agriculture until just a few decades ago. The combination of highly productive and efficient agriculture with the maintenance of ecosystem services without negative environmental side effects falls under Conservation Agriculture (CA). Based on these definitions, both approaches seem interchangeable, and both involve practices for managing agroecosystems for improved and sustained productivity, increased profits and food security while preserving and enhancing the resource base and the environment.

Regarding soils, CA is characterized by three linked principles, namely:

- continuous minimum mechanical soil disturbance;
- permanent organic soil cover;
- diversification of crop species grown in sequences and/or associations.

For these principles, much technical guidance and extension methodologies have been already developed for various applications, e.g. Bunch (2012) for green manure/cover crop management. However, challenges persist since hundreds of different implementation conditions affect the decision making and options for farmers, such as local food preferences, local and regional market conditions, established dominant cropping systems, access to inputs, environmental conditions such as soils and climate, and land ownership patterns (see also FAO 2013a).

5.2 *Organic Farming/Organic Agriculture*

Organic farming, a prototype approach for sustainable agriculture, intends to sustainably optimize agroecosystem productivity, for crops and livestock, excluding synthetic inputs such as pesticides, fertilizers, genetically modified organisms, antibiotics, and growth hormones. Organic agriculture, according to FAO, describes "a unique production management system which promotes and enhances agro-ecosystem health, including biodiversity, biological cycles and soil biological activity (...)."

One of the key principles is to minimize soil degradation and erosion, and to decrease pollution (permitted substances); soil fertility must be maintained on a long-term basis by optimizing conditions for biological activity within the soil. Organic residues and nutrients produced on the farm are recycled back to the soil. Cover crops and composted manure are used to maintain soil organic matter and fertility. Focus is on renewable resources in locally organized agricultural systems (see also Martin 2009).

Organic farming practices include the following:

- Crop rotations and cover crops are used, focusing on ecological antagonists in plant protection.
- Organic residues and nutrients produced on the farm are recycled back to the soil.
- Cover crops and composted manure are used to maintain soil organic matter and fertility.
- Soil conservation systems are valuable tools on an organic farm.

One example for guidance and practices is the National Organic Program of the United States Department of Agriculture (USDA). "However, there are many information gaps and knowledge on technical details is often scarce, especially in developing countries. Technical information needs to be very location- and product-specific" (FAO Organic Agriculture Programme).

5.3 *Climate-Smart Agriculture*

Climate-Smart Agriculture (CSA) seeks to improve food security and nutrition in the face of climate change, focusing on agricultural, forestry, and fishery practices; food systems; and policies. Reduction and/or prevention of greenhouse gas emissions associated with agriculture are important effects. Outreach to the small holder farmers is especially important.

CSA combines the need for a more productive agriculture with resilience to climate change risks, shocks, and long-term variability. The mitigation of climate change and the preservation of the natural resource base and ecosystem services are core CSA elements (FAO 2013b). CSA has been presented by FAO in 2010 and has already sought to support the Sustainable Development Goals. CSA mainstreams climate change action into sustainable agriculture.

In practice, CSA requires site-specific assessments to identify suitable agricultural production technologies and practices. Regarding soils, CSA focuses on management practices that increase soil organic carbon (SOC). As with various other agricultural programs mentioned above, soil-related CSA action fully complies with the VGSSM. Important are all technical solutions that prevent land degradation, such as erosion; improving water storage and soil structure are important technical aspects of soil-related actions as well (FAO 2013b).

An impressive example regarding CSA is Brazil's ABC plan for establishing a low-carbon agriculture, which focuses on mitigation technologies and which builds strong interfaces with soil carbon stocks, soil pollution, impact on plant protection products on soils, soil erosion, and nutrients.

5.4 Integrated Soil-Fertility Management

Agriculture has an inherent impact on human nutrition through crop production and income generation (Sanginga and Woomer 2009). The nutrient content of harvested crops is significantly influenced by nutrients provided by soils. For example, due to its natural conditions, African soils have low natural fertility, and due to unsustainable management practices, a large proportion is currently in the process of degradation. One of the main soil degradation issues is nutrient loss, which negatively affects the soil productivity in terms of quantity (crop yields), but it also leads to nutrient deficiencies of crops.

Technical options discussed mainly include synthetic fertilization, organic fertilizer applications, selection of deeply rooting crops, intercropping, cover crops, agroforestry, crop diversification (Buresh et al. 1997). A recently discussed management option is the biofortification of soils, which supplements conventional farming practices with the additional application of micronutrients to the soil (de Valença and Bake 2016).

5.5 Sustainable Land Management

Sustainable land management (SLM) refers to the use of land resources, including soils, water, animals, and plants, for the production of goods to meet changing human needs while ensuring the long-term productive potential of these resources and the maintenance of their environmental functions (SSSA 2008). SLM practices involve the sustainable management of crops (including trees and forage trees) and livestock. While not forgetting economic and social benefits from the land, SLM seeks to maintain or enhance the ecological support functions of the land resources; SLM measures and practices are adapted accordingly, protecting, conserving, and sustainably using the natural resources (soil, water, and biodiversity) and restoring them where degraded (Liniger et al. 2011). The climate-smart agriculture sourcebook as well as tools for sustainable soil management are used as references for recommending SLM action (FAO 2013b).

In agreement with all of the above-mentioned initiatives and programs, healthy and fertile soils are the foundation for land productivity. Soil organic matter, nutrients, and soil structure are the main factors influencing soil properties. SLM practices attempt to establish a balance between soil organic matter and the nutrient cycle, by eliminating carbon losses and replenishing soil nutrients to stabilize or improve the soil's fertility.

The following soil-related practices are commonly mentioned in SLM guidelines (Liniger et al. 2011):

- green manuring: planting of fast-growing species into a fallow;
- postharvest cover of soils with residues;
- application of improved compost and manure;
- use of deeply rooting plants (trees, grasses) as nutrient pumps from the subsoil;
- application of inorganic fertilizer;
- minimum soil disturbance: minimum/reduced/zero tillage systems;

5.6 *Agroecology*

Similar to conservation agriculture, agroecology defines itself as a sustainable alternative to conventional and industrial agriculture. It is a science that bridges ecological and socioeconomic aspects of agricultural management. At the same time, it is also a movement promoting biological principles in agricultural systems (Third World Network and SOCLA 2015). Agroecology advocates for a transition toward a greener agriculture and food systems; it emphasizes the role of family farm agroecology. Limited supplies of fossil fuels, and increasing prices, cause inputs like fertilizers to become too expensive for many family farmers. The term agroecology closely relates to sustainable agriculture, ecological agriculture, low-external input agriculture or people-centered agriculture.

The agroecological approach specifically addresses soils, by restoring soil life in order to reestablish and/or enhance the multiple soil-based biological processes (Lal and Stewart 2013). This specifically requires increasing and monitoring soil organic matter. It is important to build on local farmers' knowledge, establishing participatory scientific approaches to soil ecosystem management (e.g., farmer field schools), thus informing farmers to better understand and locally adapt agroecological measures.

Agroecology applies specific strategies, including the following:

- polycultures and agroforestry systems;
- cover crops;
- crop-livestock integration;
- mixed farming systems (including drought-tolerant multipurpose legume).

5.7 *Sustainable Forest Management (SFM)*

Soils of forests and woodland ecosystems are an essential contribution to agricultural production and global food security; they help to regulate important ecosystem processes, such as nutrient uptake, decomposition, and water availability. Sustainable soil management is closely connected to sustainable forest management, including restoration. Sustainably managed forests help to control soil erosion and to conserve soil: it stabilizes ridges, hills, and mountain slopes and provides the soil with the necessary mechanical structural support to prevent shallow movements of land mass. Main measures are to introduce or maintain forest cover on erosion-prone soils and run-off pathways. Forest restoration, especially in dryland areas, is vital for soil protection (FAO 2015b). FAO has developed a toolbox for SFM in order to make collective knowledge and experiences more accessible to forest managers and other stakeholders, especially supporting SFM implementation on the ground.

6 Challenges to Implement the VGSSM

The VGSSM foresee that specific technical guidelines will be needed and developed in order to allow for their implementation. The Global Soil Partnership is an important forum allowing stakeholders to share experiences and to learn from each other's experiences. The GSP secretariat, as well as the ITPS, is expected to promote and support VGSSM follow-up by stakeholders.

Pillar 1 of the GSP promotes sustainable management of soil resources for soil protection, conservation, and sustainable productivity (FAO 2014b). Despite the advanced level of development of sustainable soil management technologies over the last few decades and their documented success in reducing soil degradation and improving soil functions, the rate of adoption by land users and impact on soil improvement globally is low. The GSP Plan of Action, developed for Pillar 1, provides more background about the causes for the lack of acceptance and best-practice implementation, for example lack of local infrastructure; inadequate technology or knowledge for field application; lack of adequate equipment; overgrazing, which leads to soil exposure and the use of crop residues for livestock feed; or inadequate access to support services. A clear understanding of these challenges is needed to effectively promote and implement sustainable soil use.

Technical guidance for SSM, specific to certain soil threats, needs to address different challenges at all levels of implementation: economic, technical, social, policy, investment, and partnership challenges (see also FAO 2014b). While there is much scientific knowledge available for different aspects of sustainable soil management, there is still a huge challenge to make this knowledge available to policy makers and practitioners (for example, see UCSUSA 2017). Extension-, education-, and awareness-based services, which facilitate political incentives and which transport technical knowledge, need to be improved and provided more effectively.

The adoption of the VGSSM will be followed by an implementation process, which begins with the development of technical guidelines. Such guidelines will have a strong focus on local and regional conditions and will be targeted to specific soil hazards, as suggested by the Intergovernmental Technical Panel on Soils (ITPS). The development and implementation of technical guidance and best practices will be accompanied by various instruments of the Global Soil Partnership, for example Pillar 2 action on education, awareness, and extension.

References

Bruinsma J (2009) By how much do land, water and crop yields need to increase by 20150? The Resource Outlook to 2015. FAO Expert Meeting, 24–26 June 2009, Rome on "How to Feed the World in 2050. Food and Agriculture Organization of the United Nations, FAO

Bunch R (2012) Restoring the soil: a guide for using green manure/cover crops to improve the food security for smallholder farmers. Canadian Foodgrains Bank, 2012. ISBN 978-0-9688546-4-8

Buresh RJ, Sanchez PA, Calhoun F (eds) (1997) Replenishing soil fertility in Africa. Soil Science Society of America Special Publication No 51. Madison, Wisconsin, USA

de Valença AW, Bake A (2016) Micronutrient management for improving harvests, human nutrition, and the environment. Wageningen University and Research. Food & Business Knowledge Platform, The Hague, The Netherlands

FAO (2013a) Policy support guidelines for the promotion of sustainable production intensification and ecosystem services. Integrated Crop Management Vol. 19-2013. Food and Agriculture Organization of the United Nations, Rome, 2013. ISBN 978-92-5-108019-1

FAO (2013b) Climate smart agriculture sourcebook. Food and Agriculture Organization of the United Nations, Rome, 2013. ISBN 978-92-5-107720-7

FAO (2014a) Building a common vision for sustainable food and agriculture – principles and approaches. Food and Agriculture Organization of the United Nations, Rome, 2014. E-ISBN 978-92-5-108472-4 (PDF)

FAO (2014b) Plan of action for pillar one of the global soil partnership: promote sustainable management of soil resources for soil protection, conservation and sustainable productivity. Food and Agriculture Organization of the United Nations, Rome, 2014. http://www.fao.org/3/az898e (last accessed 22 June 2017)

FAO (2015a) Boosting Africa's soils: from the Abuja Declaration on Fertilizers to a sustainable soil management framework for food and nutrition security in Africa by 2030. FAO Regional Office for Africa, Accra, Ghana. I5532E/1/03.16

FAO (2015b) Global guidelines for the restoration of degraded forests and landscapes in drylands: building resilience and benefiting livelihoods. FAO Forestry Paper 175. Rome, Food and Agriculture Organization of the United Nations

FAO (2016) Regional implementation plan for the African Soil Partnership. Global Soil Partnership. Food and Agriculture Organization of the United Nations, Rome, 2016. http://www.fao.org/3/a-bl209e.pdf (last accessed 22 June 2017)

FAO (2017) Voluntary guidelines for sustainable soil management. Food and Agriculture Organization of the United Nations, Rome, 2017. http://www.fao.org/3/a-bl813e.pdf (last accessed 22 June 2017)

FAO and ITPS (2015) Status of the world's soil resources. Food and Agriculture Organization of the United Nations, Rome, 2015. ISBN 978-92-5-109004-6

Kuemmerle T, Levers C, Erb K, Estel S, Jepsen MR, Mueller D, Plutzar C, Stuerck J, Verkerk PJ, Verburg PH, Reenberg A (2016) Hotspots of land use change in Europe. Environ Res Lett 11: 064020

Lal R, Stewart BA (2013) Principles of sustainable soil management in agroecosystems. June 10, 2013 by CRC Press. Series: Advances in Soil Science, 568 pp. ISBN 9781466513464

Lambina EF, Meyfroidt P (2011) Global land use change, economic globalization, and the looming land scarcity. PNAS 108(9):3465–3472

Liniger HP, Mekdaschi Studer R, Hauert C, Gurtner M (2011) Sustainable land management in practice – guidelines and best practices for Sub-Saharan Africa. TerrAfrica, World Overview of Conservation Approaches and Technologies (WOCAT) and Food and Agriculture Organization of the United Nations (FAO)

Martin H (2009) Introduction to organic farming. Factsheet – Queen's Printer for Ontario. ISSN 1198-712X

Poisot AS, Speedy A, Kueneman E (2007) Good agricultural practices – a working concept Background paper for the FAO Internal Workshop on Good Agricultural Practices. Rome, Italy 27–29 October 2004. FAO GAP Working Paper Series 5

Sanginga N, Woomer PL (eds) (2009) Integrated soil fertility management in Africa: principles, practices and development process. Tropical Soil Biology and Fertility Institute of the International Centre for Tropical Agriculture (TSBF-CIAT), Nairobi, Kenya. ISBN: 978-92-9059-261-7

Sauerbeck DR (1994) Soil management, soil functions and soil fertility. Results and recommendations of an interdisciplinary workshop sponsored by the Robert Bosch Foundation, Stuttgart. Zeitschrift für Pflanzenernährung und Bodenkunde (now: J Plant Nutr Soil Sci) 157(3): 243–248

SSSA (2008) Glossary of soil science terms. Madison, WI, USA

Sys IC (1985) Land evaluation. State University of Ghent

Thiombiano L, Tourino-Soto I (2007) Status and trends in land degradation in Africa. In: Sivakumar MVK, Ndiang'ui N (eds) Climate and land degradation. Springer, Berlin, Heidelberg, New York, pp 39–51. ISBN 10 3-540-72437-0

Third World Network and SOCLA (2015) Agroecology – key concepts, principles and practices. ISBN 978-967-0747-11-8

Toenniessen G, Adesina A, Devries J (2008) Building an alliance for a Green Revolution in Africa. Ann NY Acad Sci 1136(1):233–242

UCSUSA [Union of Concerned Scientists] (2017) Rotating crops, turning profits: how diversified farming systems can help farmers while protecting soil and preventing pollution: www.ucsusa.org/RotatingCrops

UNEP (2016) Unlocking the sustainable potential of land resources: evaluation systems, strategies and tools. A Report of the Working Group on Land and Soils of the International Resource Panel. Herrick JE et al. UNESCO. ISBN: 978-92-807-3578-9

Soil Degradation Through Agriculture in China: Its Extent, Impacts and Implications for Environmental Law Reform

Xiaobo Zhao

1 Introduction

China has some of the most intense and widespread soil degradation problems in the world. China is a large country with significant spatial variation of natural/climatic conditions and diverse socioeconomic characteristics, and it is highly reliant on the agriculture sector to feed its huge population. With substantial population growth and rapid economic development in recent decades, soil degradation in China has been getting worse. This does not only affect the vast western dryland areas of China but also put significant impacts on densely inhabited agricultural regions in the middle and southern China. Mitigating soil degradation problems to achieve sustainable soil environment management has accordingly attracted high national priority in China's environmental protection agenda.[1]

Soil degradation is the decline in soil quality caused by its improper use, usually for agricultural, pastoral, industrial or urban purposes. The gravity of soil degradation and the possibility to remedy it depend on the type of degradation process, and in cases where soil erosion and salinification are very serious, they can drive farmers to abandon the land or face very high management costs to keep cropping it.[2] Symptoms of soil degradation include soil erosion, nutrient depletion, salinity, water scarcity, pollution, disruption of biological cycles and loss of biodiversity. It has been a global development and environment issue recognised by the *UN*

[1]Len et al. (2006).

[2]Tiziano (2016).

X. Zhao (✉)
School of Law and Justice, University of Southern Queensland, Toowoomba, QLD, Australia

Research Institute of Environmental Law (RIEL), Wuhan University, Wuhan, China
e-mail: bob.zhao@usq.edu.au

H. Ginzky et al. (eds.), *International Yearbook of Soil Law and Policy 2017*,
International Yearbook of Soil Law and Policy,
https://doi.org/10.1007/978-3-319-68885-5_4

Convention to Combat Desertification, the *Conventions on Biodiversity and Climatic Change*, the *Millennium Goals* in the early 2000s. The question of soil degradation has also been addressed by the more recent UN Sustainable Development Goals (SDGs) set by a major proposal prior to the Rio+20 Conference 2012.[3] In the Rio+20 Outcome Document the *Future We Want*, paragraphs focusing on desertification, degradation and drought were included.[4]

Soil degradation covers physical, chemical and biological deterioration. As a serious global environmental pandemic,[5] soil degradation is the result of complex interactions among many factors, including natural causes and human activities. Some natural causes like climate change and soil erosion by water and wind play important roles in the process of soil degradation, and complex interactions between human and natural factors make this process much more complicated.[6] However, although causes, consequences and solutions for land degradation problems are not limited to agriculture alone, cropland protection is at the heart of China's land degradation problems and should be the starting point for a comprehensive solution for reducing poverty, enhancing food security, promoting rural development by addressing land degradation.[7]

In comparison to other regions, land degradation afflicts China more seriously in terms of the extent, intensity, socioeconomic impacts and number of affected population.[8] With just 7.2% of the world's cultivated land area, China needs to feed 22% of the world's population. Agricultural production is therefore a critical issue relating to national economy and the livelihood of its citizens.[9] In addition, more than 50% of the total cultivated land has experienced degradation, which further exerts more pressure on the economic benefits of agricultural production and food security.

Over the past decades, environmental laws and regulations were developed in China to address the soil environmental protection issue, and a number of fragmental mechanisms have therefore been introduced. However, that has proven to be far from sufficient to address the challenges raised by soil degradation and to tackle the food quality and safety problems facing China. In the context that integrated ecosystem management (IEM) is recommended as best practices to restore, sustain and enhance the productive capacity of the degraded soil/land, prompt steps should be taken by China to reform its environmental regulatory frameworks to accommodate IEM for soil degradation control and sustainable soil environment management.

---

[3]UNCED (1992) and UNEP (2006). The SDG's are intended to replace the *Millennium Goals* and converge with the post-2015 development agenda. Goal 15 of the SDGs states: 'Protect, restore and promote sustainable use of terrestrial ecosystems, sustainably manage forests, combat desertification, and halt and reverse land degradation and halt biodiversity loss.' See Boer and Hannam (2015).

[4]UN (2012).

[5]DeLong et al. (2015).

[6]Hua et al. (2013).

[7]Ephraim et al. (2016).

[8]Bai and Dent (2009).

[9]Deng et al. (2008).

The concept of IEM originated in the ecological sciences and has evolved into a series of guidelines and regulations for the management of international environmental resource activities. On various occasions, IEM is identical to 'ecosystem approach' or 'ecosystem management'. The PRC-GEF Capacity Building Project defines IEM as

> a holistic approach to address the linkages between ecosystem functions and services (such as carbon uptake and storage, climatic stabilization and watershed protection, and medicinal products) and human social, economic and production systems.[10]

In international environmental law, IEM has been confirmed by various international environmental conventions, such as CBD and UNCCD. Article 2 of the CBD defines 'ecosystem' as 'a dynamic complex of plant, animal and micro-organism communities and their non-living environment interacting as a functional unit'.[11] UNCCD also calls for actions based on integrated and sustainable management of natural resources in its main text and annex.[12]

The evolution of IEM in China built onto international experience in environmental policy and the application of environmental assessment and management tools. The concept of IEM was formally introduced into the PRC in early 2000 at the time China started to develop a framework of integrated natural resources management through the country.[13] Today, IEM in China has evolved to be an underlying philosophy and set of principles, supported by a range of environmental assessment and management tools. The State Environmental Protection Administration of China has advocated ecosystem approaches for development, the essence of which is to adopt integrated management and follow the goal of ecological protection to maintain ecosystem structure, ecosystem functioning and the integrity of ecological processes.[14]

By definition, IEM emphasises the interrelationship between different aspects of the soil environment, the interdependency between the natural and the human environment and economic production. It takes into account the relationship between soil bodies as living ecological communities and the broader environmental and landscape context. With these matters in mind, IEM has become a significant guide for the improvement of law and policy for the prevention and control of soil degradation in China.[15]

Against the above background, this chapter seeks to facilitate this process by exploring the following issues: Sect. 1 examines the definition of soil degradation

---

[10]Jiang (2006).

[11]Hannam and Qun (2011).

[12]Article 2 states that to achieve the objective of the Convention will involve 'long-term integrated strategies that focus simultaneously, in affected areas, on improved productivity of land, and rehabilitation, conservation and sustainable management of land and water resources, leading to improved living conditions, in particular at the community level'.

[13]Hannam and Qun (2011).

[14]Ibid.

[15]Ibid.

and some frequently used alternatives such as soil erosion and land degradation in various contexts. Section 2 analyses some important features of soil degradation in China, under which the typologies, contributing factors and impacts are discussed. The status quo and potential causes for the five most distinct forms of soil degradation, namely soil erosion, desertification, salting, sterility and pollution, are summarised respectively. Section 3 reviews the legal frameworks governing soil degradation in China. Laws and regulations concerning farmland protection, desertification prevention and control, soil erosion prevention and control, grassland and forestry protection and soil pollution are compared under this section. Section 4 examines international experiences and provides a couple of law reform recommendations for China to bridge the gaps between the current environmental regulatory frameworks and the IEM approach in terms of soil degradation prevention and control. Section 5 briefly concludes.

2 Defining Soil Degradation

In this chapter, terms like 'soil' and 'soil degradation' are used. How we define these terms has changed over time, and the definitions differ slightly from one discipline to another, or according to the expertise of different scholars. In its traditional meaning, 'soil' is the natural medium for the growth of plants.[16] Soil has also been defined as a natural body consisting of layers (soil horizons) that are composed of weathered mineral materials, organic material, air and water. More specifically, soil is 'the natural dynamic system of unconsolidated mineral and organic material at the earth's surface', which is made up of organic matter, clay, silt, sand and gravel 'mixed in such a way as to provide the natural medium for the growth of land plants'.[17]

'Soil degradation', according to the Food and Agriculture Organization of the United Nations (FAO), was described as 'a result of one or more processes which lessen the current and/or potential capability of soil to produce (quantitatively and/or qualitatively) goods or services'.[18] A prominent feature of degraded soils is that they 'do not provide the normal goods and services of the particular soil in its ecosystem'.[19] In this definition, two significant factors are employed to indicate the healthy status of the soil: the (ecosystem) 'goods' and 'service'. Ecosystem goods refer to, according to FAO, the absolute quantities of land products having an economic or social value for present and future generations. They include animal and vegetal production, land availability and soil health, and water quality and quantity. 'Ecosystem services' concern more qualitative characteristics, and their

[16]FAO (2015).

[17]Houghton and Charman (1986), p. 115; Elizabeth (2013), p. 4.

[18]FAO (1977).

[19]FAO (2017).

impact on beneficiaries and the broader environment factors include biodiversity and maintaining hydrological and nutrient cycles. This definition also includes an explicit reference to a time over which degradation is assessed.[20]

Many other terms or words are used, often interchangeably, to describe the issue of 'soil degradation'. Among others, land degradation is the most frequently used alternative, which in general also describes the loss in productivity of the land and its ability to provide quantitative or qualitative goods or services. The United Nations Convention to Combat Desertification (UNCCD) defines 'land degradation' as

> ... the reduction or loss, in arid, semi-arid and dry sub-humid areas, of the biological or economic productivity and complexity of rain-fed cropland, irrigated cropland, or range, pasture, forest and woodlands resulting from land uses or from a process or combination of processes, including processes arising from human activities and habitation patterns, such as: (i) soil erosion caused by wind and/or water; (ii) deterioration of the physical, chemical and biological or economic properties of soil; and (iii) long-term loss of natural vegetation (UNCCD 1994, Article 1(f)).

The UNCCD definition indicates that land degradation covers the overall negative changes in the capacity of land-based ecosystems, which include biological and water-related goods and services, as well as land-related social and economic goods and services.[21] In brief, 'land degradation' as the temporary or permanent lowering of the productive capacity of land[22] covers soil degradation and various adverse human impacts on water resources, deforestation and lowering of the productive capacity of rangelands.[23]

In the Chinese language, the term land degradation is used in both broad and narrow sense. When focusing on the overall negative changes of land-based ecosystems, 'land degradation' is used in a broad sense. For example, the PRC-GEF Capacity Building Project defines 'land degradation' as a broad term that includes soil degradation, vegetation degradation and degradation or failure of biological diversity, as well as degradation or failure of the land's value.[24] For other researches that focus on one or more particular forms of soil degradation, land degradation is often used in a narrow sense, which is a synonym of soil degradation in such a case.

---

[20]Oldeman et al. (1991).

[21]FAO (2017).

[22]UNEP (1994). Desertification, land degradation [definitions]. Desertification Control Bulletin 21.

[23]Anthony Young et al. (1994); see also FAO (1976).

[24]Hannam and Qun (2011).

3 Soil Degradation in China: Typologies, Causes and Impacts

Due to the diversity of its soil types and its vast area, China is among the most affected countries in the world in terms of the extent, intensity and economic impact of soil degradation. China is a large country with the world's largest population of about 1.4 billion. In China, only 1.3 million km^2 lands, which account for approximately 14% of the total land area, are suitable for cultivation. Rapid population growth and urbanisation, unreasonable human utilisation and influence of natural factors have caused soil degradation on 5.392 million km^2 land, accounting for about 56.2% of the total national area. Among those lands, the area of soil degradation resulting from soil erosion and water loss, desertification, soil salinisation, pasture degradation and soil pollution is 1.80, 0.334, 0.9913, 2.00 and 0.267 million km^2, respectively (Long 2013).[25] The suitable land for agricultural production is only about 1.3 million km^2, accounting for 14% of the total land area in China.

Generally, there are two categories of human-induced soil degradation processes. The first category deals with soil degradation that results from the displacement of soil material. Water erosion and wind erosion are the two major types of soil degradation under this category. The second category of soil degradation deals with internal soil physical and chemical deterioration. In this category, only 'on-site effects are recognised of soil that has been abandoned or is forced into less intensive usages'.[26] It should be noted that some types of degradation are excluded from this chapter. That is either because they are not human-introduced degradation or because they are not raised from agriculture-related activities. In detail, the following types of soil degradation are not considered in this chapter: (1) acid sulphate formation, which may occur on drainage of coastal swamps; (2) soil destruction through mining and quarrying activities or the failure to restore soil after extraction; (3) urban and industrial encroachment onto agricultural land.

3.1 *Typologies and Causes of Soil Degradation in China*

In general, the causes that lead to soil degradation are complex and can be of a different nature. Soil degradation can occur as the result of the physical (i.e., erosion, compaction), chemical (i.e., acidification, salinisation) and biological (i.e., loss of soil organic matter, loss of biodiversity) processes. Some typical natural and anthropogenic factors that alter the structure and quality of soils include deforestation and the removal of natural vegetation, agricultural activities,

[25]Long (2013).

[26]Tiziano (2016).

overgrazing, overexploitation of vegetation for domestic use and industrial activities. In China, common types of soil degradation include soil erosion, desertification, salting, sterility and pollution.[27]

3.1.1 Soil Erosion

Erosion is a process of soil degradation that occurs when soil is left exposed to rain or wind energy. Poor management of agricultural land induces soil erosion that leads to reduced productivity or, in extreme cases, to the abandonment of the land. Soil erosion leads to reduction of soil fertility and deterioration of the ecological environment.

Like many countries in the world, China is facing serious water loss and soil erosion. Globally, mild to severe soil erosion is possibly affecting about 80% of the world's agricultural land; soil erosion was estimated to reduce yields on about 16% of agricultural land around the year 2000.[28] Official reports show that soil erosion in China is much worse than the world's situation. According to the *Report on the State of the Environment in China* released by the MEP of China, by the end of 2014, the overall area of agricultural land in China was 645.74 million hectares, in which the area of arable lands was up to 135.05 million hectares.[29] Stastics also show that China's existing area of soil erosion is up to 2.949 million km^2, accounting for 31.1% of the total soil area, of which the water erosion area is about 1.293 million km^2, and the wind erosion is about 1.656 million km^2.[30] Soil erosion in the east, south and southwest regions of China can mainly be attributed to the water and wind erosion, while wind erosion that results from desertification is a major concern in the north and northwest regions.[31]

Geographical and hydrological and human factors are major attributors of soil erosion in China. Human-introduced soil degradation has generally raised from agricultural practices (deforestation, overcutting of forest areas, heavy grazing of grassland, mining, road building and other large-scale capital construction).[32] The shrinking arable land area pressures farmers to extract higher yields from their land; this in turn leads to increased soil erosion. Almost 90% of rural people living in poverty are located in areas suffering from soil erosion. In arid areas, increasing livestock numbers exacerbate the spread of deserts. Over 90% of the 1.3 million km^2 of grasslands suffer from moderate to severe degradation, while demand for meat and other livestock products is rising.[33]

[27]Cheng et al. (2006).

[28]Wood et al. (2000).

[29]China MEP (2015).

[30]MEP (2015).

[31]Zhou et al. (2008).

[32]Ibid.

[33]ADB (2002).

3.1.2 Soil Desertification

UNCCD defines 'desertification' as 'land degradation in arid, semi-arid and dry sub-humid areas resulting from various factors, including climatic variations and human activities' (UNCCD 1994, Article 1 (a)). Scientists consider desertification to be a form of soil degradation that occurs within specified geomorphic environments. Desertification as a form of land degradation occurs particularly, but not exclusively, in semi-arid areas. Just like other forms of soil degradation, it is a result of mutual actions of natural and human factors. Desertification is the result of complex interactions among various factors, including climate change and human activities. While the agricultural activities are considered, excessive reclamation, grazing and cutting are the main causes that affect the ecological equilibrium. Scientific research has shown that human-introduced desertification was thought to be the dominant factor that affected land.[34]

China has a vast distribution of deserts and desertified land. It is estimated that the area of desertified land in China is as high as 2.63 million km^2.[35] While the vulnerability to desertification is considered, stastics in 2001 showed that about 262,410 km^2 of land in China was under low risk, 239,107 km^2 was under moderate risk, 65,638 km^2 was under high risk and 72,214 km^2 was under very high risk.[36] Geographically, most of the desertification areas are found in northen and northwesten China.

Severe soil desertification may result in the loss of overall land productivity. It deteriorated China's ecological environment and reduced agricultural production. In recent years, the creation of controlled areas across the country has significantly improved the soil desertification situation of China. Study from Asian Development Bank (ADB) suggested that, by the end of 2004, the desertification area in China was up to 2.636 million km^2, taking up 27.46% of the total land of China. In addition, 79.48% of the total land was prone to desertification, which was higher than the world's average proportion, 69%.[37] By 2009, the area of desertification declined to 1.73 million km^2, accounting for 18.03% of the total land (Liu 2014).[38] Therefore, from 2004 to 2009, the overall trend of desertification in China is slowing down.

---

[34]Qi et al. (2015).

[35]CCICED (2010).

[36]Eswaran et al. (2001).

[37]Asian Development Bank (2007).

[38]Liu (2014), Xinhua News Agency, State Forestry Administration Bureau Leader on soil desertification in China (in Chinese), www.gov.cn, Jun 16, 2014. Available at: http://www.gov.cn/xinwen/2014-06/16/content_2701798.htm.

3.1.3 Soil Salinisation

As in many other parts of the world, soil salinisation is a severe problem in China. Soil salinisation refers to a process of salt accumulation in the soil. It mainly happens in areas with a dry, semi-dry or semi-humid climate or in coastal lowland areas vulnerable to being soaked and irrigated by seawater. In China, about 3.69 million hm^2 land is affected by this. Arable lands in the Huanghuaihai Plain, the west of the Northeast Plain, the Hetao area of the Yellow River, the inland area of Northwest China and the coastal areas of East China are most affected by salinity.[39] By 2010, the total area of the affected arable land was up to 0.624 million hm^2, accounting for about 7% of China's total land.[40]

Overuse of chemical fertilisers is now considered to be the major source of soil salinisation in China. Over the past decades, the use of chemical fertilisers has risen dramatically in China. Almost half of the impressive increase in grain productivity since 1980 has been attributed to inputs[41] of chemical fertilisers, which increased fivefold, meeting similar trends in the application of pesticides, fungicides and herbicides and in the use of plastic sheeting.[42] This has, on the one hand, increased the crop yield and enabled China to feed its large population with limited farmlands, but on the other hand, it intensified the risk of secondary soil salinisation.[43] It is worth noting that China is the world's largest and most inefficient fertiliser user. It has consumed 25% of the annual global supply of chemical fertilisers since 2002.[44] Every year, more than 360 kg of chemical fertiliser per hectare of land are used in China, that is, 3.3 times more than the United States and 1.6 times more than the average for EU countries. The number could be even higher in those areas where agriculture is well developed.[45]

3.1.4 Soil Sterility

Soil sterility is another significant process of soil degradation, with severe economic impact on a global scale.[46] Soil sterility means content of nutrient elements reduces from top to bottom in the soil profile, and it is often thought to be one of the most fundamental results of soil erosion and degradation. With increasing soil degradation, the content of organic substances, whole nitrogen and whole phosphorus in the soil decreases.

---

[39]CCICED (2010).

[40]Ibid.

[41]FAOSTAT (2013).

[42]Carter et al. (2012)

[43]Liu et al. (2010).

[44]Su (2006); Zhao (2013), p. 18.

[45]CCICED (2010).

[46]Lal (1997).

China has been constantly making efforts to improve its crop yields. This naturally means greater consumption of nutrients of the soil. The nutritive elements are taken away from the soil as a result of the harvest. The organic contents in China's arable land is relatively low. The average rate of organic contents in China's arable land is 1.8%, which is two times lower than that of similar soil types in Europe.[47] The *State of the Environment Reports of China* in recent years showed that over 50% of China's arable lands are short of microelements, 51% are short of phosphorus and 60% are short of potassium.[48] To cover the losses, more land would have to be converted into agricultural use and more inputs would be used to replace the reduced soil fertility, which in turn increased the burden of soil environment and led to other types of land degradation, as we discussed in this section.

3.1.5 Soil Pollution

Although there is no solid agreed definition for soil pollution, this term generally refers to human activities that add a poisonous or polluting substance to the soil. Concerns with soil contamination stem primarily from health risks caused by 'direct contact with contaminated soil, vapors from the contaminants, and from secondary contamination of water supplies within and underlying the soil'.[49]

Soil pollution in China was, for a long time, perceived as involving relatively rare environmental pollution incidents. But today, it has been widely recognised as a nationwide problem in China as an inherent consequence of industrial and farming practices. As the legacy of industries, mining areas and commercial activities in cities and suburbs, soil pollution is now severly threatening human health, food safety and the quality of natural environment in China. From 2005 to 2013, China MEP and the Ministry of Land Resources jointly conducted the first National Soil Pollution Survey in mainland China, which covered 2.4 million square miles of land across the country.[50] The Survey showed that, by the end of 2014, approximately 16.1% of the country's total lands have been polluted. About 19.4% of the total farmlands were identified as contaminated.[51] The Survey also found that some 82.8% of the polluted lands were contaminated by inorganic materials, such as cadmium, nickel and arsenic. The levels of these materials in the soil increased sharply since the 1990s. The level of cadmium had risen by 50% in the southwest and in some coastal areas. While the geographical features are considered, the overall situation of soil pollution in southern China is worse than that in the north.[52]

---

[47]CCICED (2010).

[48]Ibid.

[49]EPA (1989).

[50]China MEP (2013).

[51]China MEP (2014).

[52]Ibid.

Causes for soil pollution in China are diversified and complex, which can be historical or current and inorganic or organic.[53] According to the Survey, agricultural activities have been identified as the major sources of soil polution in China,[54] the adverse effects of which have been noticed by an increasing number of people.[55] Typical actions include the over-application of pesticides, chemical fertilisers, animal and poultry excrement and sludge; sewage irrigation; discarded residues of agricultural plastic membrane; and mulching, most of which can be classified as non- point source pollution (NSP).[56]

China is the world's main chemical pesticide producer and the world's largest chemical pesticide consumer; over-application of chemical pesticides has long been criticised in China. In 2007, China yielded some 0.173 million tons of chemical pesticide and exceeded the US and became the world's largest chemical pesticide producer.[57] According to China MEP, the total application amount of chemical pesticides in 1983 was 0.862 million tons; in 2006, pesticide consumption in China was over 1.31 million tons, and the average dosage deposited on agricultural land was 13.97 kg/ha, which is higher than that of developed countries.[58] By 2008, the application amount of chemical pesticides in China has risen up to approximately 1.672 million tons.[59]

Low efficiency of using chemical pesticides deteriorated the situation. Although the total application amount of chemical pesticides in China is extremely high, the utility rate is as low as 30%, which means approximately 70% of the pesticides directly diffused into the environment and accumulated in soils.[60] In addition, a large percentage of chemical pesticides applied in China are of high toxicity. Research showed that more than 30 types of organophosphorus pesticide with a total weight of 200,000 tons were applied in China, among which more than 80% were of severe toxicity.[61]

In the meantime, large-scale wastewater irrigation has also contributed heavily to China's soil pollution. By 1998, farmland wastewater irrigation area in China was over 36,200 km^2, which accounted for 7.33% of the overall irrigation areas and 10% of the surface water irrigation areas.[62]

---

[53] Zhao (2013), at 18.

[54] Ibid.

[55] Walter (2009) and Zhang et al. (2004).

[56] NSP generally refers pollution that comes from dispersed sources such as agricultural fields, parking lots, golf course, and etc. See William (2001), p. 1; Zhao (2013), p. 11.

[57] Zhao (2013), at 14.

[58] Conradie and Field (2000).

[59] National Bureau of Statistics of China and Ministry of Environmental Protection of China (2009).

[60] Kenneth and Lewis (2009).

[61] Ibid.

[62] China MEP (2010).

3.2 *Impacts of Soil Degradation in China*

China is among the most affected countries in the world in terms of the extent, intensity and economic impact of soil degradation. As China's economy has expanded rapidly over the past decade, soil degradation has intensified with direct and indirect impacts and has become a deep-seated concern in China.

A report from the PRC-GEF Partnership Project on Land Degradation in Dryland Ecosystems of China showed that, in the six project provinces and autonomous regions in Western China, the impacts of soil degradation of various levels include

> lowered land productivity, degraded natural ecosystems, increase in wind-blown sand and increased number of sandstorms, increased poverty and deterioration of livelihoods, reduced agricultural production, damage to transportation routes, degraded water resources, and impacts on infrastructure construction of large and medium-sized cities.[63]

3.2.1 Production Losses

Production losses in cropping systems are the primary result of a deteriorating soil environment.[64] Severe soil erosion and nutrient losses have reduced soil fertility, and the resultant degradation of the soil's physical and chemical properties has reduced land productivity.

Research shows that the total direct costs of land degradation in China were estimated at 7.7 billion USDs in 1999, which weigh 4% of GPD of the year, while indirect costs are thought to be around 31 billion USDs a year. The rate of grain yield reduction due to nutrient loss was estimated at 5% per annum from 1976 to 1989. This is equivalent to six million tons of grain valued at 700 million USDs and representing an equivalent of 30% of the imports of grain for that period.[65]

The situation seems getting worse after 2000. Research that is more recent shows that the total cost of land degradation due to land use and cover change in 2007 was up to 195.747 billion USDs, which accounts for 5.4% of China's GDP of the year.[66] The costs that have been calculated include the cost of taking action against land degradation and the cost of establishing and maintaining degraded biomes. To rehabilitate the degraded land raised from land use and land cover change between 2001 and 2009 in China, the total amount was supposed to be 255.45 billion USDs, while 30 years' time would be required. However, if no action is to be taken to rehabilitate these degraded lands, approximately 1208.08 billion USDs is deemed to lose during the same period.[67]

---

[63]Hannam and Qun (2011), at 38.

[64]Berry (2003).

[65]Huang and Roselle (1995).

[66]Deng and Li (2016).

[67]Ibid.

Besides, since soil degradation directly affects the potential land productivity, more inputs such as fertiliser and irrigation water will be needed in order to achieve the same level of production and yield, which in turn will increase the production costs.[68]

In brief, production losses due to soil degradation in cropland in China are severe. It has been commonly accepted that the situation will get worse in the next couple of years if no proper actions were taken.[69]

3.2.2 Food Safety Crisis

Soil degradation has long been threatening the quality of agricultural products of China. Frequently reported food safety incidents and heavy metal pollution cases in China have deepened the spreading fears among the Chinese people.

Among others, soil pollution is the major contributor of China's food safety crisis. Without healthy soil, pollutants may enter into the food chain and endanger the health of human beings and livestock. It has been noted that soil pollution or land contamination has greatly damaged taste and the quality of agricultural products and, in some extreme cases, made them unsuitable for eating because of chemical residue. As the major rice-growing area in the middle of China, Hunan Province suffered some of the worst soil pollution. Hunan produced 16% of the country's rice in 2012, but official investigation suggested that a large quantity of rice with excessive levels of cadmium has been sold to some southern provinces like Guangdong.[70]

Soil pollution does not raise the domestic concerns about food quality alone. In recent years, large quantity of agricultural products have been returned to China, trading cancelled and even enforced disposal because the products failed to meet the technical requirements of the importers like the US and EU.[71]

3.2.3 Food Security

In addition, land contamination, together with salinisation, desertification, erosion and radioactive pollution, has aggravated the shortage of arable lands and endangered the food security of China. China has to feed 22% of the world's population on 6.4% of the world's total land area, 7.2% of the world's arable land and 5.8% of the world's annual water resources. Food security has therefore has been a chief mission and a primary State objective in the early twenty-first century. Some traditional residential areas in China, like the *Taihu Lake* area, Yangtze Delta and

[68] Li et al. (2011).

[69] Hannam and Qun (2011).

[70] Edward (2014).

[71] The Department of Research Center of the State Council (2006).

the Pearl Delta, have been suffering from serious pollution over recent decades.[72] Meanwhile, large number of coastal areas such as the Jiaozhou Bay have been severely contaminated by heavy metals, with arsenic being the most concerning metal contaminant. Studies also showed that heavy metal such as chromium (Cr) and mercury (Hg) accumulation in soil might potentially threaten newly industrialised regions such as the Yangtze Delta and the Pearl Delta of China.[73]

Some other human-related factors, e.g., the enormous population, fast urban expansion and poor land zoning all threaten the food security of China. The population size of China is projected to peak at 1.45 billion in 2030.[74] The percentage of population residing in urban areas is projected to be 60% in 2030 and 73% in 2050.[75] In the context of fast urban expansion of China, limited arable land and water resources are being shifted to non-agricultural sectors.[76] In China, cropland has been lost at a rate of 1.45 Mha/yr since 2000.[77] Moreover, soil degradation threatens food security by lowering crop yields and increasing the need for substituting inputs like fertiliser and lime.[78] Food deficits were predicted to be 3–5%, 14–18% and 22–32% by 2030–2050 under the zero-degradation, the current degradation rate and double-degradation rate scenarios, respectively.[79] Therefore, food security is and will be constantly a primary concern of Chinese policy makers. Soil degradation now creates a strong case for government action to mitigate its impacts on food safty.

3.2.4 Other Impacts

It is commonly accepted that soil degradation has weakened the ablity of local residents to combat natural disasters that affect soil stablity. For example, towns in dryland Western China[80] are vulnerable to sandstorms; for this reason, some residents have abandoned their homes, effectively becoming 'ecological refugees'.[81] Sandstroms have been a sensitive public and political topic in China over the past decades. In the same time, the health effects of soil pollution are becoming increasingly clear and generating significant concern.

---

[72]Jin et al. (2010).

[73]Xu et al. (2007).

[74]UN (2004).

[75]UN (2007).

[76]Liao (2010).

[77]Ye and Ranst (2009).

[78]Ibid.

[79]Ibid.

[80]Western China refers to the western part of China. According to the Chinese government, Western China covers one municipality (Chongqing); six provinces (Sichuan, Guizhou, Yunnan, Shaanxi, Gansu, and Qinghai) and three autonomous regions (Tibet, Ningxia, and Xinjiang).

[81]Hannam and Qun (2011).

Expanding soil degradation also contributes to the acceleration of poverty and has negatively affected social stability. Many minority ethnic groups in Western China occupy areas affected by severe soil degradation. The increase in desertifcation has accelerated their poverty and has broadened the socioeconomic gap between the western and eastern regions of China.[82]

4 Soil Degradation and Its Implications for Environmental Law Reform of China

4.1 Legal Frameworks Governing Soil Degradation in China: An Overview

China has made great efforts to develop its legislative framework for environmental protection and natural resource conservation since the 1980s. Today, a relatively advanced environmental law system to address environmental protection and sustainable development has been established. This legislative framework encompasses various national and local laws, regulations and environmental standards. It is a hierarchy system, which comprises the following:

(1) the Constitution of China;
(2) the 1989 Environmental Protection Law (as amended in April 2014);
(3) other nationwide environmental laws promulgated by the National People's Congress and its Standing Committee;
(4) national environmental administrative regulations, orders, decisions and other normative documents with legal binding force, promulgated by the State Council and MEP (the former SEPA);
(5) local legislation, including the laws, regulations, decisions and orders promulgated by the People's Congress of provinces, autonomous regions/cities/ counties, SC-directed municipalities and relevant cities; and
(6) environmental protection standards.

Soil degradation has been addressed by laws and regulations that fall into all the above categories. In general, regulatory frameworks in relation to soil degradation control are threefold: (1) laws and regulations on soil degradation control at the central level; (2) laws and regulations on soil degradation control at the provincial and local levels; (3) policy instruments for soil degradation control at both the central and local levels.[83]

[82] World Bank (2001).

[83] More details about the legislation and policy frameworks for land degradation control of China please refer to Chapter 4 of the PRC-GEF partnership report. It systematically summarised the current laws and regulations in this regard. See Hannam and Qun (2011).

At the central level, laws and regulations in relation to soil degradation control may fall into nine categories, which include land resources, desertification prevention and control, water and soil conservation, grassland, forest, water, agriculture, protection of wild animals and plants, and environmental protection.[84]

4.2 *The Role of Environmental Legislation in Soil Degradation Control*

4.2.1 Farmland Protection

The Land Administration Law of P.R.C 1986 (as amended in 2004) is the basic law to protect and manage land resources. By adopting this Law, land-use control has served as a fundamental instrument to protect the land resources in China. Farmland is under strict protection according to this Law, which aims to maintain the national food security and provide a sound ecological environment. The conversion of agricultural land to other uses is strictly prohibited. Issues like land-use planning, protection of farmers' rights and interests and promoting intensive use of land resources have been addressed by this Law (Article 3, *Land Administration Law of P.R.C.*).

4.2.2 Desertification Prevention and Control

The Law on Prevention and Control of Desertification 2002 and the *State Council Resolution on Strengthening the Prevention and Control of Desertification* stipulate special principles, legal responsibilities and supporting policies for various departments and local governments to carry out the national objectives of desertification control. The National Program for Combating Desertification 2005–2010 specifies priority areas and national desertification investment. This programme was renewed in 2013, and more ambitious goals were set in this national agenda. The National Program for Combating Desertification (2011–2020) aims to remediate 20 million ha desertified lands in China by 2020. Combating desertification has been identified as a priority by the government to improve the eco-environment and living conditions in Western China and to coordinate sustainable development. Those requirements have been addressed by laws and policies at provincial and autonomous region levels.

[84]Ibid, at 42.

4.2.3 Soil Erosion Prevention and Control

Chinese legislature has put water and soil conservation on its agenda to protect the ecological environment. The *Water and Soil Conservation Law of P.R.C* 1991 (as amended in 2010) is the most important legislation in this respect. After its enactment, the State Council circulated the *Notice on Strengthening Water and Soil Conservation*, which aims to improve agricultural conditions, enhance economic development, reduce poverty and protect land. In the meantime, some policy documents such as the *Outline for Protecting and Monitoring National Water and Soil Conservation (2004–2015)* sets out the guiding principles, goals, strategies and countermeasures for national water and soil conservation activities. This document introduced the ecological compensation mechanism to the area of water and soil conservation.

4.2.4 Grassland and Forestry Protection

The *Grassland Law of P.R.C* 1985 (as amended in 2002) and numerous statutes, regulations, provisions and national policies in relation to the grassland utilisation, conservation and management constitute the regulatory framework for grassland protection. The *Grassland Law 2002* provides for institutional arrangements, a grassland management system, and specific measures and liabilities. Some measures such as the prevention of illegal harvesting and damage of grassland vegetation partly overlapped with the efforts on soil degradation prevention and control (Article 31).

China enacted the *Forestry Law of P.R.C* in 1984 and amended the law in 1998. This Law introduced the ecological benefit compensation system. Under this system, people who use or explore natural resource that depletes ecological functions are required to compensate those who sacrifice their interests for improving, maintaining or enhancing the ecological service function. In addition, governments at different levels have initiated some key natural forest resource protection projects. The Chinese central government funded those projects. Among others, the logging ban and standard compensation scheme were introduced under the Law. Forestry environment protection in selected riverbanks and areas affected by dust storm, desertification and soil erosion were addressed in these projects (e.g., Article 23).

4.2.5 Soil Pollution Prevention and Control

So far, no specific nationwide laws have been made to deal with the deteriorating soil pollution in China. A new law is currently on the legislative agenda of China's highest legislature, the Standing Committee of the National People's Congress.[85]

[85]Ministry of Land and Resources of P.R.C (2015).

Before the new law takes its place, a couple of existing environmental laws and regulations can be applied when soil pollution issues emerge. The following laws are thought to be relevant.

The *Environmental Protection Law of the People's Republic of China* (as amended in 2014) (EPL hereafter) defines 'land' as one element of the ecosystem. It provides that the People's Governments at all levels are liable for the protection of the agricultural environment by preventing and controlling soil pollution, desertification and other negative impacts to land; by extending the scale of comprehensive prevention and control of plant diseases and insect pests; and by promoting a rational application of chemical fertilisers, pesticides and plant growth hormones. The Central Government should provide financial support by creating a special budget for the prevention and control of 'soil pollution'.[86] The EPL also provides instructions for pollution prevention and abatement.

The *Law of the People's Republic of China on the Prevention and Control of Environmental Pollution by Solid Wastes* 1995 (as amended in 2004) provides that local governments should bear the costs of dealing with discarded installations or sites that have historically been used for the purpose of storing or treating industrial solid wastes (Article 35). This Law can be applied when the liable party cannot be reached or the liable parties can be reached but are incapable to discharge their liabilities.

Another important law, the *Land Administration Law of the People's Republic of China* 1986 (as amended in 2004), requires the People's Governments at all levels to take measures to maintain and protect irrigation and drainage facilities, as well as ameliorate the soil to increase fertility and prevent desertification, salinisation, water loss, soil erosion and pollution (Article 35).

In addition, the *Agriculture Law of the People's Republic of China* 1993 (as amended in 2002) encourages the rational use of chemical fertiliser and pesticide management to improve the quality of arable land. The Law has introduced a registration system as a way of assuring integrated agricultural environment management (Article 58). The production of pesticides, veterinary medicines, fodder and feed additives, fertilisers, seeds and farm machines, which may endanger the safety of human beings and/or livestock (Article 21), can only be conducted by acquiring a licence and registration. Similar provisions on agricultural product safety are available in the *Regulation on Pollution-Free Agricultural Product Management* (2002), under which the environmental conditions of the areas producing pollution-free agricultural products should meet specific environmental quality standards (Article 9 (1)).

Moreover, the *Rules on Land Reclamation* 2011 introduces a principle of 'those who destroyed the land should be responsible for reclamation work' to reclaim lands damaged by digging or sinking during the construction process. Finally, the

[86]To effect this proposal, the Ministry of Finance and the Ministry of Environmental Protection issued the *Measures for the Administration of Special Funds for the Prevention and Treatment of Soil Pollution* in July 2016.

Soil and Water Conservation Law of P.R.C 1991, which was amaneded in 2010, instroduced the eco-compensation system to the area of soil and water conservation at the national level (Article 31). If soil and water conservation function cannot be recovered due to the construction of projects, fees as a form of compensation shall be paid to cover the water and soil loss (Article 32).

Some administrative orders have also been promulgated by the Ministry of Environmental Protection (MEP) of China to handle some emergent issues stemming from soil pollution. Most recently, MEP promulgated the *Measures for the Soil Environment Management of Contaminated Sites* in 2016.[87] This decree took effect on 1 January, 2017. The Measures have defined the contaminated sites and liability issues regarding contaminated land remediation and the prevention of new soil pollution. This decree consists of seven chapters with 33 articles. It defines 'contaminated site' as land the soil of which has been contaminated because of (1) toxic and harmful materials, (2) abandoned materials or (3) mining (Article 2). It confirms that the polluter and/or owner who has land use rights may be liable for the land contamination; the local governments shall bear the site remediation costs if no liable parties can be reached (Article 10). Without proper remediation, the contaminated sites cannot be transferred and redeveloped (Article 27).

Finally, soil environment protection in China is highly dependent on two essential standards: the *Environmental Quality Standard for Soils 1995* and the *Environmental Quality Risk Assessment Criteria for Soil at Manufacturing Facilities 1999*. The first standard is mandatory, but it regulates agricultural land based on crop safety only. The second one is not mandatory and fails to address contamination issues for future land use. Some typical environmental standards involving soil pollution control are under preparation, which include (1) soil environment quality standards: the *Soil Environmental Quality Assessment and Remediation Standards for the Exhibition Site* (Provisional for Shanghai Expo) (HJ 350-2007); (2) testing standards: *the Soil Environmental Monitoring Technical Specifications* (HJ/T166-2004), *Groundwater Monitoring Technical Specifications* (HJ/T164-2004) and *Water Environment Monitoring Specifications* (SL219-98) issued by the Ministry of Water Resources; and (3) monitoring standards: the *Groundwater Quality Standards* (GB/T14848-93), *Soil Environmental Quality Standards* (GB15618-1995) and *Soil Environmental Quality of Industrial Enterprises* (HJ/T25-1999).

[87] China MEP (2016).

5 Addressing Challenges by Reforming the Environmental Regulatory System

5.1 Lessons Learned from the International Society

Experiences from other jurisdictions on soil degradation control may help China address soil degradation problems caused by agricultural land use. As suggested by the PRC-GEF Partnership Dryland Programme, the following experiences have been found relevant:

(1) introducing public education programmes to raise the public awareness on environmental protection and prevention and control of desertification;
(2) ensuring public participation and strong leadership at the national level on environmental issues;
(3) improving the capacity of environmental law and policy to protect ecological resources;
(4) improving the decision-making process for planning rural land use and ecological resources;
(5) introducing incentives for farmers to adopt soil conservation techniques;
(6) using a national strategy for land degradation control to prioritise funding programmes for the prevention and control of desertification.[88]

On top of this, IEM, as a comprehensive strategy and method to manage natural resources and the natural environment, is recommended as a suitable way to deal with the soil degradation problems in China.[89] As the Capacity Building Project of PRC-GEF conducted in Western China shows, benefits of applying IEM in the legal and policy processes are substantial:[90]

(1) IEM provides a scientific approach for China to fulfil its commitments to various multilateral conventions concerning environmental protection and sustainable use of natural resources. It establishes a strategic framework to manage land, water and biological resources for sustainable development.
(2) The IEM approach represents cross-cutting mechanism that accommodates multiple scientific means. It well coordinates national implementation requirements of international environmental conventions and strategies.[91]
(3) IEM is an effective means to achieve a sustainable use of natural resources and combat soil degradation. It is a sound framework having considered natural resource ownership, use of protected areas, access to resources, benefit sharing and stakeholder involvement.

[88]Hannam and Qun (2011).

[89]Ibid.

[90]Ibid. Cited by Hannam and Qun (2011), at 15.

[91]Hannam (2007).

(4) IEM provides multiple options for implementation; a country like China can apply IEM principles in national strategies and action plans, in regional plans and in policy making, land use and institutional planning. It sets solid basis for reform of institutions and organisations to support sustainable use of ecosystems and natural resources.
(5) IEM comprehensively applies administrative, market-based and societal mechanisms in natural resource management. It is an important means to alleviate poverty and enhance sustainable development.
(6) IEM is a relevant tool for planning, decision-making and evaluating ecosystem activities associated with all aspects of management, policy and law. Outcomes from the Capacity Building Project indicated that some Western China regions have made significant improvements in law and policy reforms by using this concept, which will provide useful experience for the relevant regions in China.

Instead of treating each resource in isolation, IEM offers the option of treating all elements of ecosystems as a whole to obtain multiple ecological and socioeconomic benefits. It is thought to be the most effective way of sustainable land management.[92] PRC-GEF therefore encourages the adoption of the IEM approach to restore, sustain and enhance the productive capacity of dryland ecosystems.

5.2 *Bridge Gaps by Reforming the Environmental Legal System of China*

As pointed out above, China has established a well-structured legal and regulatory framework in terms of soil environment protection. Although the Chinese Central Government and its affiliated ministries have been responsive to this emergent crisis, it should be noted that the existing legal system is far from sufficient and effective, considering its fragmented governance, existing legislative and policy framework, weak administrative organisation and capacity, and limited technical knowledge. In line with international good practices, as well as with the notion of IEM, significant improvements should be made to China's environmental legal system to address soil degradation control in China.

Firstly, existing laws and regulations need to be amended to meet the requirements of IEM. By far, there is no special legislation governing soil degradation at the national level. The *Forest Law* 1998 and *Water and Soil Conservation Law* 1991 (as amended in 2010) were thought to be 'no longer adequate in their current form to effectively manage the main problems of land degradation'.[93] In addition, with respect to the protection of wetlands, although it is partly included in the *Forestry Law* 1998, *Grassland Law* 2002 and *Land Administration Law* 1998,

---

[92] Anna et al. (2016).

[93] Hannam and Qun (2011).

assessment indicated that there are insufficient procedures to enable all ecological aspects of wetlands to be adequately protected.

Secondly, legal mechanisms regarding nature reserves; ecological compensation in soil degradation control; forest and grassland protection; EIAs; water, farmland and soil pollution; and soil conservation call for urgent reform. Take soil pollution control for example; although the most recent decree, *Measures for the Soil Environment Management of Contaminated Sites* 2016, is directly relevant to soil environment improvement and will alleviate problems of soil pollution, the decree is applied to industry sites only; farmland pollution has been excluded from its adjustment.

Thirdly, the implementation of environmental laws and regulations in China should be further enhanced. The implementation of law in China is a major concern.[94] Measures have been recommended to be taken to further develop the ideology of legal execution and to strengthen the liability provisions. Possible solutions include, as suggested by the PRC-GEF Partnership Project,[95] to increase penalties, to reduce the overlap in administrative functions and to strengthen legal education and environmental publicity.

Fourthly, special law at the national level should be made to regulate soil pollution control and contaminated site management. Experience in the context of China and the international community shows that an effective regulatory system for soil pollution control requires, *inter alia*, a systematic integration of all aspects, including assessment, remediation, spatial planning, aftercare and monitoring.[96] Meanwhile, environmental standards in relation to soil quality should be updated to meet the current situation of soil degradation. For example, the *Soil Quality Environmental Standard of China* was developed about 20 years ago; it has been criticised as being unable to tackle the current problems effectively.

Finally, more attention should be attached to financial inputs and market incentive mechanisms in combating deteriorated soil degradation originated from agricultural reasons. Financial support, subsidies, compensation and financial incentive mechanisms already exist in the aforementioned laws and regulations, as well as numerous environmental policies. Some more specialised financial instruments, e.g. credit certification and trading systems, together with preferential policy specifically targeting soil environmental protection, should be considered by China.[97] In May 2016, the State Council of China issued the *Circular of the State Council on Issuing the Action Plan for Soil Pollution Control* (*Circular of the State*

---

[94]Ma and Ortolano (2000).

[95]The capacity assessment programme of the PRC-GEF partnership looked at the relevant elements of laws and regulations in relation to soil degradation, which include: objectives and basis of the legislation; administrative management system; administrative regulation; public participation; dispute resolution; and legal liability. Ibid.

[96]Zhao (2017), p. 47.

[97]Some efforts have been made in the area of contaminated land management in recent years. The Chinese government has also created favourable tax conditions for forest products, in order to make the conversion of farmland to forested land economically sustainable; Ephraim et al. (2016).

Council on Issuing the Action Plan for Soil Pollution Control, Guo Fa [2016] No. 31), which is by far the most important administrative order in terms of soil environment protection in China. The Circular requires the central and local financial departments to enhance the support for soil pollution prevention. The central financial department shall integrate special funds for heavy metal pollution prevention and set up a special fund of soil pollution prevention for the investigation, monitoring, assessment, supervision, management, treatment and restoration of the soil environment. Carrying out the contaminated land remediation and reclamation under the Circular will facilitate soil degradation control in China. It is also a signal for the making of China's nationwide soil pollution prevention and control law.

6 Concluding Remarks

Soil degradation is a worldwide problem. It has become one of the most serious problems in China, which has profoundly affected the population's health and living conditions and arguably the sustainable development of the Chinese society. There are two interlocking complex systems involved in the soil degradation process, the natural ecosystem and the human social system, and both changes in biophysical natural ecosystem and socioeconomic conditions affect the soil degradation process. This chapter focused on the anthropogenic drivers of soil degradation in China and examined the human-introduced (especially through agricultural activities) soil degradation through soil erosion, desertification, salting, sterility and pollution.

Although the common types of soil degradation are understood adequately, there is currently a lack of accurate information on the extent of soil degradation in China. On the one hand, soil degradation is complex and varies from place to place, and over time, exact measurements of which would be difficult. In addition, huge differences among the different models concerning the estimation of the extent and intensity of soil degradation, some inherent limits and errors in the information, data and instruments used make the assessment of the amount of degraded soil and land in China a difficult task.[98] On the other hand, global data on land degradation have not been updated since late 1990s, which makes it difficult for China to seek accurate soil degradation information from international sources. At the time this chapter is being written, there is no comprehensive nationwide soil degradation/pollution database in China.[99] Therefore, maps for the purpose of outlining the status of soil degradation, especially soil pollution, should be developed by Chinese environmental authorities in a timely matter.

[98]Tiziano (2016).

[99]Ramsar Convention Secretariat (2007).

Although it is a difficult task to address challenges arising from soil degradation, China is at an advantage over the early industrialising nations because the social economic and health impacts of soil degradation are better understood and some international best practices for soil conservation and pollution control have been developed.[100] China has the opportunity to refer to the regulatory systems and methods developed in other jurisdictions and international society.[101] However, as 'finished products', these regulations and mechanisms are the result not only of lengthy processes of policy development but also of scientific, regulatory and administrative capacity building. These processes will also take time in China. It is likely for China to move faster if stronger human and financial resources were provided.

China has benefited from the programmes on soil degradation mitigation initiated by international agencies, e.g. UNDP, UNEP and the International Fund for Agricultural Development (IFAD), the regional development banks and the World Bank. With the development of international collaborations in this area, China will continually benefit from programmes of this type and in turn have its own contributions to the international society. International experiences indicate the IEM approach to be an effective way to manage soil degradation. More effort should be made by Chinese legislatures to improve the existing legislative landscape by revising and removing barriers for incorporating IEM approach into its environmental legislation. At the same time, some new laws, such as an integrated legislation at the nation level concerning the prevention and control of soil degradation and soil pollution, should be introduced in China to bridge the gaps. By doing so, it is highly possible that the impacts of soil degradation through agriculture in China will be substantially mitigated.

References

ADB (2002) PRC national strategies for soil and water conservation. Final Report (TA 3548). https://www.adb.org/sites/default/files/project-document/71709/tar-prc-33446.pdf

ADB (2007) Country environmental analysis for the People's Republic of China. Available at http://www.adb.org/Documents/Produced-Under-TA/39079/39079-PRCDPTA.pdf

[100]E.g., US adopted the Comprehensive Environmental Response, Compensation, and Liability Act (CERCLA) in 1980. CERCLA provides a Federal 'superfund' to clean up uncontrolled or abandoned hazardous-waste sites as well as accidents, spills, and other emergency releases of pollutants and contaminants into the environment. In the UK, The system for identifying and remediating statutorily defined contaminated land under Part 2A of the *Environmental Protection Act* 1990. Similar laws can be found in Germany, Canada, Australia and Japan.

[101]Although there is no internationally recognised convention on soil environment protection or soil degradation prevention and control, proposals for global soil regime have been made by some distinguished researchers. See, e.g., Boer and Hannam (2015).

Anna T et al (2016) Scaling up of sustainable land management in the Western People's Republic of China: evaluation of a 10-year partnership. Land Degrad Dev 27(2):134–144. https://doi.org/10.1002/ldr.2270

Bai ZG, Dent D (2009) Recent land degradation and improvement in China. AMBIO J Hum Environ 38(3):150–156. https://doi.org/10.1579/0044-7447-38.3.150

Berry L (2003) Land degradation in China: its extent and impact. Report submitted to the United Nations. Available at http://lada.virtualcentre.org. http://documents.worldbank.org/curated/en/537621482745818202/pdf/109976-WP-Box396323B-PUBLIC-Land-Degradation-GEF-UNCCD.pdf

Boer B, Hannam I (2015) Developing a global soil regime. Int J Rural Law Policy (1):1–13

Carter C, Zhong F, Zhu J (2012) Advances in Chinese agriculture and its global implications. Appl Econ Perspect Policy 34:1–36

CCICED (2010) Developing policies for soil environmental protection in China. Available at: http://www.cciced.net/cciceden/POLICY/rr/prr/2010/201205/P020160810466173582807.pdf

Cheng D, Cai C, Zuo C (2006) Advances in research of soil degradation by erosion. Res Soil Water Conserv 13(5):252–254

China MEP (2010) Gazette of China's Environmental Protection (2009). Available at http://jcs.mep.gov.cn/hjzl/zkgb/2009hjzkgb/201006/t20100603_190415.htm

China MEP (2013) Soil pollution and physical health. Environmental Science Publishing House, Beijing

China MEP (2014) Gazette of the national soil pollution survey. Available at http://www.zhb.gov.cn/gkml/hbb/qt/201404/W020140417558995804588.pdf

China MEP (2016) *Measures for the Soil Environment Management of Contaminated Sites* (for Trial Implementation), Order of the Ministry of Environmental Protection No. 42, December 31, 2016

Conradie EM, Field DN (2000) A rainbow over the land: a South African guide on the church and environmental justice. Western Cape Provincial Council of Churches

DeLong C, Cruse R, Wiener J (2015) The soil degradation paradox: compromising our resources when we need them the most. Sustainability 7(1):866–879. https://doi.org/10.3390/su7010866

Deng X, Li Z (2016) Economics of land degradation in China. In: Ephraim N, Mirzabaev A, Von Braun J (eds) Economics of land degradation and improvement - a global assessment for sustainable development. Springer, pp 385–399

Deng XZ, Huang JK, Rozelle S, Uchida E (2008) Growth, population and industrialization, and urban land expansion of China. J Urban Econ 63(1):96–115

Edward W (2014) One-fifth of China's farmland is polluted, state study finds. Available at: http://www.nytimes.com/2014/04/18/world/asia/one-fifth-of-chinas-farmland-is-polluted-state-report-finds.html?_r=0

Elizabeth B (2013) Global approaches to site contamination law. Springer, p 4

EPA (1989) Risk assessment guidance for superfund, vol 1 human health evaluation manual (Part A): interim final, pp 6–47

Ephraim N, Mirzabaev A, Von Braun J (2016) Economics of land degradation and improvement - a global assessment for sustainable development. Springer, p 30

Eswaran H, Lal R, Reich PF (2001) Land degradation: an overview. In: Bridges EM, Hannam ID, Oldeman LR, Pening de Vries FWT, Scherr SJ, Sompatpanit S (eds) Responses to land degradation. Proceedings of the 2nd international conference on land degradation and desertification, Khon Kaen, Thailand. Oxford Press, New Delhi, India. Avaialble at https://www.nrcs.usda.gov/wps/portal/nrcs/detail/soils/use/?cid=nrcs142p2_054028

FAO (1976) A framework for land evaluation. FAO Soils Bulletin 32. FAO, Rome

FAO (1977) Assessing soil degradation, FAO soils bulletin. FAO, Rome

FAO (2015) What is soil. FAO, Rome, Italy. Available online: http://www.fao.org/soils-portal/about/all-definitions/en/

FAO (2017) Available at http://www.fao.org/soils-portal/soil-degradation-restoration/en/

FAOSTAT (Statistics Division of the Food and Agriculture Organization of the United Nations) (2013) FAOSTAT database. Avaiable at http://faostat3.fao.org/faostat-gateway/go/to/download/Q/QC/E

Hannam I (2007) Environmental law and policy frameworks to manage land degradation in the dryland ecosystem areas of [the People's Republic of] China. J World Assoc Soil Water 2:63–74. doi: http://e-publications.une.edu.au/1959.11/5921

Hannam I, Qun D (eds) (2011) Law, policy and dryland ecosystems in the People's Republic of China. IUCN Environmental Policy and Law Paper No. 80

Houghton PD, Charman PEV (1986) Glossary of terms used in soil conservation. Soil Conservation Service, Bathurst. Available at http://www.scs.nsw.gov.au/__data/assets/pdf_file/0004/494410/Glossary_of_Terms_Used_In_Soil_Conservation.pdf

Hua M, Lv Y, Li H (2013) Complexity of ecological restoration in China. Ecol Eng 52(3):75–78. https://doi.org/10.1016/j.ecoleng.2012.12.093

Huang J, Roselle S (1995) Environment stress and grain yields in China. Am J Agric Econ 77 (4):853–864. https://doi.org/10.2307/1243808

Jiang Z (2006) To implement integrated ecosystem management to accelerate combating land degradation. In: Jiang Z (ed) Proceedings of international workshop on integrated ecosystem management, Beijing 1–2 November 2004, pp 2–6. Global Environment Facility, Asian Development Bank and People's Republic of China Forestry Publishing House

Jin Z, Cheng H, Chen L et al (2010) Concentrations and contamination trends of heavy metals in the sediment cores of Taihu Lake, East China, and their relationship with historical eutrophication. Chin J Geochem 29–33. doi:https://doi.org/10.1007/s11631-010-0033-x

Kenneth AG, Lewis TL (2009) Twenty lessons in environmental sociology. Oxford University Press

Lal R (1997) Degradation and resilience of soils. Philos Trans R Soc B 352:997–1010

Len B, Boukerrou L, Olson J (2006) Resource mobilization and the status of funding of activities related to land degradation. Available at http://documents.worldbank.org/curated/en/537621482745818202/pdf/109976-WP-Box396323B-PUBLIC-Land-Degradation-GEF-UNCCD.pdf

Li HJ, Liu ZJ, Zheng L, Lei YP (2011) Resilience analysis for agricultural systems of north China plain based on a dynamic system model. Scientia Agricola 68(1):8–17. https://doi.org/10.1590/S0103-90162011000100002

Liao Y (2010) China's food security. Chin Econ 43:103–108

Liu Y (2014) State forestry administration bureau leader on soil desertification in China (in Chinese). Xinhua News Agency, Jun 16, 2014. Available at: http://www.gov.cn/xinwen/2014-06/16/content_2701798.htm

Liu EK, Zhao BQ, Mei XR, Li XY, Li J (2010) Distribution of water-stable aggregates and organic carbon of arable soils affected by different fertilizer application. Acta Ecol Sin 30 (4):1035–1041

Long F (2013) Introduction to the resources status in China-Land resources (in Chinese). Available at: http://www.jingchengw.cn/new/20130411/4927.htm

Ma X, Ortolano L (2000) Environmental regulation in China: institutions, enforcement, and compliance. Rowman and Littlefield, Lanham (MD), USA and Oxford, UK

MEP (Ministry of Environmental Protection of the P.R.C) (2015) Report on the state of the environment in China 2015. Available at: http://www.mep.gov.cn/hjzl/zghjzkgb/lnzghjzkgb/201606/P020160602333160471955.pdf

Ministry of Land and Resources of P.R.C (2015) Soil pollution prevention and control law has entered into legislative agenda (in Chinese). Available at: http://www.mlr.gov.cn/xwdt/jrxw/201503/t20150312_1345005.htm

National Bureau of Statistics of China and Ministry of Environmental Protection of China (2009) China environment statistical yearbook 2009. China Statistics Press

Oldeman LR, Hakkeling RTA, Sombroek WG (1991) World map of the status of human-induced soil degradation: an explanatory note, second revised edition. ISRIC/UNEP. Available at: http://www.isric.org/sites/default/files/ExplanNote_1.pdf
Qi F et al (2015) What has caused desertification in China? Sci Rep 5. doi:https://doi.org/10.1038/srep15998
Ramsar Convention Secretariat (2007) Wetland inventory: a Ramsar framework for wetland inventory. Ramsar handbooks for the wise use of wetlands, vol 12, 3rd edn. Ramsar Convention Secretariat, Gland. Available at: http://www.ramsar.org/pdf/lib/lib_handbooks2006_e12.pdf
Su Y (2006) China's rural pollution problem. Economic Information, January 14, 2006. Available at http://www.chinaelections.org/NewsInfo.asp?NewsID¼45122
The Department of Research Center of the State Council (2006) Status Quo of the agricultural pollution in China and suggested countermeasures. Stud Int Technol Econ 9(4):17–20
Tiziano G (2016) Soil degradation, land scarcity and food security: reviewing a complex challenge. Sustainability 8(3):281–322. https://doi.org/10.3390/su8030281
UN (2004) World population to 2300. New York
UN (2007) World urbanization prospects: the 2007 revision. New York
UN (2012) The future we want. UN Doc A/66/L 56. Available at http://www.un.org/ga/search/view_doc.asp?symbol=A/RES/66/288&Lang=E
UNCCD (United Nations Convention to Combat Desertification) (1994) Article 1(f)
UNCED (United Nations Conference on Environment and Development) (1992) Managing fragile ecosystems: combating desertification and drought, Chapter 12 of Agenda, 21. Rio de Janerio, Brazil, 3 to 14 June 1992
UNEP (1994) Desertification, land degradation [definitions]. Desertification Control Bulletin 21
UNEP (2006) Report on millennium ecosystem assessment. UNEP, Nairobi
Walter G (2009) Catholics going green: a small-group guide for learning and living environmental justice. Ave Maria Press
William F (2001) Ritter and Adel Shirmohammadi, Agricultural nonpoint source pollution: watershed management and hydrology. Lewis Publishers, p 1
Wood S, Sebastian K, Scherr SJ (2000) Pilot analysis of global ecosystems: agroecosystems. International Food Policy Research Institute and World Resources Institute, Washington, DC. Available at: http://www.wri.org/sites/default/files/pdf/page_agroecosystems.pdf
World Bank (2001) China, overcoming rural poverty, a world bank country study. The World Bank, Washington, DC
Xu CW, Feng H, Ma HQ (2007) Assessment of metal contamination in surface sediments of Jiaozhou Bay, Qingdao, China. CLEAN Soil Air Water 35(1):62
Ye L, Ranst EV (2009) Production scenarios and the effect of soil degradation on long-term food security in China. Glob Environ Change 19:464–481. https://doi.org/10.1016/j.gloenvcha.2009.06.002
Anthony Young et al (1994) Land degradation in South Asia: its severity, causes, and effects upon the people. World Soil Resources Reports. FAO
Zhang W, Wu S, Hong J (2004) Estimation of agricultural non-point source pollution in China and the alleviating strategies I. Estimation of agricultural non-point source pollution in China in early 21 century. Sci Agric Sin 37(7):1008–1017
Zhao X (2013) Developing an appropriate contaminated land regime in China: lessons learned from the US and UK. Springer
Zhao X (2017) Land contamination legislation in China: the emerging challenges. In: Kitagawa H (ed) Environmental policy and governance in China. Springer, pp 47–67
Zhou K, Xia C, Baiping T (2008) Toward an improved legislative framework for China's land degradation control. Nat Res Forum 32(1):11–24. https://doi.org/10.1111/j.1477-8947.2008.00172.x

Soil Legislation and Policy in the Kyrgyz Republic on the Development of the Law "On Soil Fertility Protection of Agricultural Lands"

Maksatbek Anarbaev

1 Introduction

The Kyrgyz Republic is located in the heart of Central Asia with a territory of almost 200,000 km^2. It was previously part of the Soviet Union and declared its independence in 1991. By 2016, the population of the country reached over six million with rural population of 66%.[1]

The majority of the territory is part of the Tien-Shan[2] and Pamir-Alai Mountains with the popular peaks over 7000 m in elevation. The mountain ecosystems are characterized by a high level of species richness, rarity of habitats, endemism and referred to as biodiversity hotspots.[3] The mountain ranges contribute freshwater runoff, which is widely used for agricultural irrigation, household consumption, playing a crucial role for the Central Asian region.[4] At the same time, due to altitude and climatic features, these areas create environmental conditions suitable for agro-pastoralism.[5]

According to the Kyrgyz Ministry of Agriculture, by 2012, agricultural land occupied 10.6 million hectares or 53% of the country's territory where the dominant portion of the land is pastures occupying 9.147 million hectares. The arable land is around 1.2 million hectares, which means that only 6% of the total territory

[1]NatStatCom (2017).

[2]Locally called *Tengir-Too*, see also Schmidt (2013), p. 110.

[3]MEP (1998) and Olson and Dinerstein (2002).

[4]Sorg et al. (2012), p. 725.

[5]Mamytov (1987), p. 380; Kreutzmann (2011), p. 40.

M. Anarbaev (✉)
Institute of Geographic Sciences, Centre for Development Studies (ZELF), Freie Universität, Berlin, Germany
e-mail: anarbaev@zedat.fu-berlin.de

H. Ginzky et al. (eds.), *International Yearbook of Soil Law and Policy 2017*,
International Yearbook of Soil Law and Policy,
https://doi.org/10.1007/978-3-319-68885-5_5

is suitable for arable farming.[6] Agriculture is still one of most important industries of the country producing 14% of GDP and provides around 30% of employment.[7]

Moreover, in the last few decades, the ecological aspects of land management such as soil degradation and overgrazing of pastures that leads to their desertification have become a central issue in Kyrgyzstan. State authorities, as well as many other study reports, state that land degradation is high. The problem associates with diverse factors such as loss of vegetative cover and depletion of soil from overgrazing and nonrotation of crops, soil salinization, wind and water erosion, as well as necessity of improvements in the legal framework.[8] Therefore, the State policy in the agrarian sector, particularly legislation for sustainable land management, is among the priorities of the Government.

2 The Legal Framework for Land Management

In general, since independence, the Government has undertaken many important legislative initiatives that affect the agricultural sector. In this regard, the amendments in the Constitution[9] of the Kyrgyz Republic through referendum on October 17, 1998, guaranteed the introduction of private ownership. It was crucial and resulted in the adoption of the Land Code in 1999, which abolished the State monopoly on land and introduced private ownership rights to agricultural lands, except pastures that remained in State ownership. In the same year, alongside the Land Code of the Kyrgyz Republic (1999), the Government adopted several other pieces of legislation, such as the Law of the Kyrgyz Republic "About Preservation of the Environment" (1999), the Forest Code (1999), and the Law "About Peasant Farms" (1999). Later, the Law "About Management of Agricultural Lands" (2000), the Law "About Management of Agricultural Lands" (2001), the Law "About Mountain Regions" (2002), the Water Code (2004),[10] and the Law "About Cooperatives" (2004) were adopted.

This framework of land legislation has various aims, including creating the conditions for private ownership rights to land and stability of market relations in agriculture and land management. The reforms were oriented toward the development of a secondary land market. However, one of the recognized gaps in the legislation was that which considers the fundamental aspects of soil fertility, particularly the decrease in soil fertility—which means the depreciation of the principal means of agricultural production. In this regard, soil protection has

[6]Fitzherbert (2000) and Minagro (2016).

[7]NatStatCom (2016), p. 20.

[8]Sievers (2003), p. 36; UNCCD KR (2006); PALM (2011); Kerven et al. (2012); Pachova et al. (2012), p. 7; NAP UNCCD (2014); Abdurashitov (2015), p. 59; Minagro (2016).

[9]Enforced by the President Decree on 21 October 1998 UP No 322.

[10]PALM (2010), p. 4.

become the first-priority task of landowners and authorized bodies of public administration.[11]

3 Land Reform

Within the Soviet budget, the Kyrgyz Soviet Socialist Republic had formerly both directly and indirectly subsidized approximately 20% of the GDP. Later, from 1992, this financial support vanished, and the country was challenged with a budget deficit equal to two times less than what it had been before.[12] The difficult socioeconomic situation that the country faced after the dissolution of the Soviet Union forced it to implement a number of radical reforms.

During the period 1991–1998, the Kyrgyz Government implemented a wide-ranging privatization program of State-owned properties, known as "shock therapy," that passed in three phases. A dominant proportion of State-owned plants, factories, buildings, and other domestic service-providing enterprises were privatized.[13] The Government launched privatization in the agricultural sector as well, where the agricultural land and assets of State[14] and collective farms, including agricultural equipment, machinery, and livestock, were distributed among former farmworkers and other villagers[15] (including teachers, civil servants, doctors, and others working within the territory of the *kolkhozes* or *sovkhozes*[16]).

Land reform[17] allocated plots of arable land to residents of rural areas with various land use rights under a leasehold period of 49 years. Later in 1995, the Decree of the President (November 3, No UP 297) "On measures for further development and State support of land and agrarian reform in the Kyrgyz Republic" extended the land use rights to 99 years. In 1996, the Government proposed to the *Jogorku Kengesh* (the Parliament) about expediency of introduction in the Kyrgyz

[11]The rationale for the Law on Soil Fertility Protection of Agricultural Land (2012) is available at www.kenesh.kg/ru/draftlaw/download/9162/accompdoc/ky.

[12]Akaev (2000), p. 51.

[13]Akaev (2000), p. 45; Steimann (2011), p. 58; For details of privatization programs see the Decisions of the Government of the Kyrgyz Republic from 14 March 1994 No 120 and 20 March 1997 No 157 available at http://cbd.minjust.gov.kg.

[14]See the Law "About Land Reform" from 19 April 1991 No 432-XII and the Decree of the President from 13 January 1992 No UP-10 available at http://cbd.minjust.gov.kg/act/view/ru-ru/47841.

[15]See the Decision of the Government from 22 August 1994 No 632 "About approval of the provisions on the conduct of land and agrarian reform" available at http://cbd.minjust.gov.kg/act/view/ru-ru/37462?cl=ru-ru.

[16]*Kolkhoz* is namely a collective farm originating from the Russian word "*Kollektivnoe khozyaistvo*" and *Sovkhoz* is a state farm "*Sovetskoe khozyaistvo*".

[17]See the Decree of the President from 22 February 1994 No UP 23 "On measures to deepen land and agrarian reforms in the Kyrgyz Republic" available at http://cbd.minjust.gov.kg/act/view/ru-ru/47092.

Republic of private land ownership along with the State ownership[18] that resulted in an announcement of a nationwide referendum in 1998 for the amendment of the Constitution of the Kyrgyz Republic that converted this tenure into private ownership.[19] The adoption of the Land Code (1999) facilitated and accelerated the property rights relations and registration. As a result, by the end of 1999, the Government had issued around 511,000 land certificates.[20]

In other words, 530,000 families or over 2.6 million citizens of the Kyrgyz Republic became owners of 75% of agricultural arable lands. The remaining 25% of arable land has been reserved[21] under the "National Land Fund," which is nowadays named as the "Fund for Redistribution of Land."[22] These lands have been transferred to the local municipalities (Kyrg: *Aiyl Ökmötü*[23]), which use the revenue from them for their needs.[24] The pastures that represent the dominant portion of agricultural land remained in State ownership.[25]

However, the privatization of agricultural arable land has caused a number of other issues. The 470 sovkhozes and kolkhozes[26] established in Soviet times were finally dissolved by 1996, and nowadays there are 401,350 individual farming entities in agriculture.[27] The arable land was parceled out to small units, which has resulted in a small-scale farming system functioning under a market-based economy.[28] This makes an average arable landholding of 2.98 ha, but in the southern part of the Republic, due to the higher density of population, the average arable landholding is less than one hectare in area (see Fig. 1). This situation has created difficulties for the landowners to follow and implement ameliorative activities, including land rotation and preservation of soil fertility. As these lands are privately owned and managed, State authorities have limited direct influence to land use in general. Consequently, the legal mechanisms are the best approach to regulate this situation.

---

[18]See the Decision of the Government of the Kyrgyz Republic "About additional tasks for further development and state support of land and agrarian reform" from 25 March 1996 No 130 available at http://cbd.minjust.gov.kg/act/view/ru-ru/35145?cl=ru-ru.

[19]Akaev (2000), p. 30; Lim (2012), p. 51; Abdurashitov (2015), p. 57.

[20]Fitzherbert (2000); Kyrg: *Kyzyl kitep*, the "Red book", as the cover page of the document is red in color.

[21]The Law "On introduction (enforcement) of Land Code of the Kyrgyz Republic" from 2 June 1999 No 46 available at http://cbd.minjust.gov.kg/act/view/ru-ru/211.

[22]See the Decree of the President from 3 November 1995 No UP 297 "On measures for further development and State support of land and agrarian reform in the Kyrgyz Republic" available at http://cbd.minjust.gov.kg/act/view/ru-ru/47003.

[23]The Executive body of a rural municipality.

[24]Akaev (2000), p. 169.

[25]Schoch et al. (2010), p. 214; Lim (2012), p. 50.

[26]Abdurashitov (2015), p. 19.

[27]NatStatCom (2016), p. 19.

[28]Fitzherbert (2000); NAP UNCCD (2014), p. 25; Zhumanova et al. (2016), p. 507.

Fig. 1 Small-plot farming. Each strip is private arable land that belongs to individual farmers (Kyrg: *Dyikan Charba*—Peasant Farms), which replaces nowadays the former larger collective farm

With the aim of better land management, the authorities have promoted and facilitated the organization of small-plot farming practicing farmers into cooperatives (so-called *agricultural cooperatives*) to create bigger landholdings for joint agricultural farming. In 2002, the Government adopted "The State Program for Supporting of Agricultural Cooperative Movements in the Kyrgyz Republic."[29] However, in reality, this program does not work very well as an economic incentive and regulatory framework for the improvement of land relations moreover. In this regard, the ecological condition of many agricultural lands is poor.

According to official data and various field surveys, the trend is a decrease in harvest levels, occurrence of land degradation, and soil salinization, and there is an increase in abandoned land.[30] Data provided by the Ministry of Agriculture, Food Industry and Melioration (2012) indicate that from 10.647 million hectares of agricultural lands, the stony soil occupies 4.272 million hectares or 40% of the area.[31] Further, the area of land affected by saline soils is 1.333 million hectares (12.5%); land affected by water erosion and wind erosion is 5.7 (53%) and

[29]Gosudarstvennaya programma razvitiya kooperativnogo dvijeniya v Kyrgyzskoi Respublike, adopted 24 Dec 2002 No 875 http://cbd.minjust.gov.kg/act/view/ru-ru/49510?cl=ru-ru. Lost power 28 Mar 2016 No 15.

[30]Schoch et al. (2010), p. 219; Abdurashitov (2015), p. 59.

[31]Minagro (2016).

5.8 (54%) million hectares respectively. In addition, there has been a decline by three times in the use of organic fertilizers on arable land compared to that used in 1990.

The poor ecological condition of the agricultural lands, particularly State-owned pasturelands, is a major concern. For instance, from 9147 million hectares of pastures, around 3.6 million hectares (40%) are now degraded.[32]

4 Legislative Improvement

In an effort to improve the condition of pastureland and to solve various conflicts and administrative issues in pasture use, the "Law on Pastures" was introduced in 2009. With the aim of decentralization, the administrative functions over pastureland were transferred to a Pasture Committee (Kyrg: *Jaiyt Komiteti*) at the municipality level.[33] A rental system of pastures was replaced by the "fee per animal head" system called "pasture tickets" by taking into account carrying capacity of pasture.[34] Each Pasture Committee develops a management plan, which covers monitoring, soil and vegetation improvement, activities that address soil erosion, restoration of degraded lands, and improvement of infrastructure in pastureland. However, a legal framework to regulate soil protection, soil fertility, preservation of quality, and protection from degradation and other negative phenomena associated with the ownership, use, disposal of agricultural land was still necessary.

5 The Law on Soil Fertility Protection of Agricultural Lands

In 2012, the *Jogorku Kengesh* (the Parliament) of the Kyrgyz Republic passed "The Law on Soil Fertility Protection of Agricultural Lands."[35] This Law was introduced for the purpose of protecting the soil and its fertility, preservation of soil quality, and protection of soil from degradation, in relation to the use of agricultural land. In particular, it aims at regulating activities and competence in the area of soil protection and ensuring that fertility of agricultural land is maintained or improved.

[32]Liu and Watanabe (2016), p. 114; Minagro (2016).

[33]Steimann (2011); Dörre (2012); Shirasaka et al. (2016), p. 141.

[34]Lim (2012), p. 52; Zhumanova et al. (2016), p. 507; the "carrying capacity" of pasture considers an assessment of available forage resources for livestock. Based on this assessment a Pasture Committee will estimate the number of livestock that can be allocated in a specified area of pasture.

[35]This work was facilitated by an experts group within the PALM project. See PALM (2010) and Pachova et al. (2012).

With an emphasis on fertility, this Law serves as the basis of agriculture and food security of the country. In this regard, it emphasizes that soil, if managed correctly, not only retains its quality but also improves its fertility. The value of soil is determined not only by its economic significance for rural, forest, and other sectors of the economy; it is also determined by the irreplaceable environmental role of soil as the most important component of the biosphere.

The Law defines 15 key terminologies such as "soil" itself, soil fertility, soil protection, soil condition, soil quality, soil pollution, soil degradation, as well as other characteristics and properties of the soil that it regulates. In the Land Code (1999),[36] these terms were absent or not clearly defined. In this regard, the Law on Soil Fertility Protection of Agricultural Lands has introduced and improved the definition of "soil" and other soil "properties" and highlights the preservation and restoration as one of the qualities of soil, namely its fertility.

Articles 2 and 3 establish the basic principles and main directions of the State policy in the area of national soil conservation. In particular, the role of the Parliament, national administrative and local executive bodies, and the role and obligations of land owners are defined in Articles 4 to 8.

The Law on Soil Fertility Protection of Agricultural Lands establishes a special authorized body and outlines its duties and functions. According to Article 6, an authorized body has the responsibility to implement a common national soil policy, developing and implementing a strategy, target program, and actions plans of the Kyrgyz Republic in the interests of soil protection. This body can implement scientific and research activities, conduct monitoring on soil fertility, and organize activities to create among citizens awareness toward soil fertility protection. The Land Code covers some general aspects of land management, such as the aim and purposes of soil protection and evaluation, but regulates the relationship in land ownership and the registration of property rights on land. This function belongs to the Department of Cadastre and Registration of Rights to Immovable Property under the State Registration Service of the Government. Importantly, under the Law on Soil Fertility Protection of Agricultural Lands, the Government appoints the Ministry of Agriculture, Food Industry and Melioration as the responsible body to implement the "Soil Protection Policy."

6 Conclusion

The main principle of the Law on Soil Fertility Protection of Agricultural Lands is to protect soil quality and prevent its degradation, contamination, and general destruction. It regulates the activities of organizations and citizens in the area of

[36]Land Code of the Kyrgyz Republic (1999); available at http://www.gosreg.kg/index.php?option=com_content&view=article&id=89.

soil protection, as well as establishes the requirements for territories to rehabilitate contaminated soil and conduct regular monitoring and soil sampling.

However, this Law does not regulate cases of soil pollution. This is regarded as a significant omission as it is necessary to identify those that cause soil pollution, as well as the approach to estimate the cost of rehabilitation.

In order to implement the State policy on soil protection in the agricultural sector, particularly the conservation of soil fertility, in 2016 the Ministry introduced "The Concept for Soil Fertility Conservation and Improvement of the Kyrgyz Republic for 2017–2020 years," together with a detailed action plan. Finally, there is now a State body to implement the State policy on soil fertility protection and to monitor soil protection and the other important activities set out in this Law.

The public and ecological significance of the Law on Soil Fertility Protection of Agricultural Lands is high. For this reason, it is important for the Kyrgyz Republic to promote and implement this Law and continue the harmonization and development of the entire legislative system for sustainable land management.

Acknowledgements Special thanks to Prof. Ian Hannam (Australian Centre for Agriculture and Law, University of New England) and Prof. Emmanuel Kasimbazi (Environmental Law Centre, Makerere University) for their enormous help and comments on various drafts of this paper.

References

Abdurashitov A (2015) Rynochnaja transformatsija agrarnoi ekonomiki Kyrgyzskoi Respubliki [Market transformation of agricultural economics of the Kyrgyz Republic]. Kyrgyz – Russian Slavic University, Bishkek. ISBN 978-9967-19-206-5

Akaev A (2000) The transition economy through the eyes of a physicist (mathematical model of the transition economy), Bishkek. ISBN 9967-413-03-4. http://www.askarakaev.org/science/. Accessed 12 Apr 2017

Dörre A (2012) Legal arrangement and pasture-related socio-ecological challenges in Kyrgyzstan. In: Kreutzmann H (ed) Pastoral practices in high Asia, advances in Asian Human-Environmental Research. Springer, Dordrecht, pp 127–144

Fitzherbert AR (2000) Pastoral resource profile for Kyrgyzstan. http://www.fao.org/WAICENT/FAOINFO/AGRICULT/AGP/AGPC/doc/Counprof/kyrgi.htm. Accessed 22 Dec 2016

Kerven C et al (2012) Researching the future of pastoralism in Central Asia's mountains: examining development orthodoxies. Mountain Res Dev 32(3):368–377

Kreutzmann H (2011) Pastoralism in Central Asian mountain regions. In: Kreutzmann H, Abdulalishoev K, Zhaohui L, Richter J (eds) Regional workshop in Khorog and Kashgar. Pastoralism and rangeland management in mountain areas in the context of climate and global change. GIZ, Bonn, pp 38–63

Lim M (2012) Laws, institutions and transboundary pasture management in the High Pamir and Pamir-Alai Mountain Ecosystem of Central Asia. Law Environ Dev J 8(1):43. http://www.lead-journal.org/content/12043.pdf

Liu J, Watanabe T (2016) Seasonal pasture use and vegetation cover changes in the Alai Valley, Kyrgyzstan. In: Kreutzmann H, Watanabe T (eds) Mapping transition in the Pamirs, advances in Asian Human-Environmental Research. Springer, Cham. https://doi.org/10.1007/978-3-319-23198-3_8

Mamytov A (1987) Utilization and conservation of land resources of the mountains of Central Asia and southern Kazakhstan. In: Manzoor Alam S, Habeeb Kidwai A (eds) Regional imperatives in utilization and management of resources: India and the USSR. New Delhi
MEP (1998) Ministry of Environmental Protection of the Kyrgyz Republic. Biodiversity Strategy an Action Plan, Bishkek, pp 1–132
Minagro (2016) Ministry of Agriculture, Food Industry and Melioration of the Kyrgyz Republic. The concept of conservation and soil fertility improvement in the Kyrgyz Republic for 2017–2020 (in Russian). http://www.agroprod.kg/index.php?aux_page=aux4. Accessed 13 Jan 2017
NAP UNCCD (2014) The National Action Plan (NAP) and the Activity Frameworks for Implementing the UNCCD in the Kyrgyz Republic for 2015–2020. The GEF – World Bank Project "Support to UNCCD NAP alignment and reporting process". Bishkek
NatStatCom (2016) National Statistic Committee. Sel'skoe khozyaistvo Kyrgyzskoi Respubliki [The agriculture of the Kyrgyz Republic]. http://www.stat.kg/ru/. Accessed 12 Apr 2017
NatStatCom (2017) Chislennost' zanyatogo naseleniya po otraslyam ekonomiki [The number of employed population by industry]. National Statistic Committee. http://www.stat.kg/ru/. Accessed 12 Apr 2017
Olson D, Dinerstein E (2002) The Global 200: priority ecoregions for global conservation. Ann Missouri Botanical Garden 89(2):199–224
Pachova NI et al (2012) Towards sustainable land management in the Pamir-Alai Mountains. UNU-EHS PUBLICATION SERIES. Policy Brief No 5. August 2012, pp 1–46
PALM (2010) Analysis of legislative, political and institutional system of the Kyrgyz Republic. GEF/UNEP/UNU PALM Project on "Sustainable Land Management in High Pamir and Pamir-Alai Mountains – Integrated Trans-Boundary Initiative in the Central Asia"
PALM (2011) Strategy and action plan for sustainable land management in the High Pamir and Pamir Alai Mountains. GEF/UNEP/UNU PALM Project on "Sustainable Land Management in the High Pamir and Pamir-Alai Mountains – Integrated and Transboundary Initiative in Central Asia"
Schmidt M (2013) Mensch und Umwelt in Kirgistan – Politische Ökologie im postkolonialen und postsozialistischen Kontext In: Erdkundliches Wissen 153. Franz Steiner Verlag, Stuttgart
Schoch N et al (2010) Migration and animal husbandry: competing or complementary livelihood strategies. Evidence from Kyrgyzstan. Nat Resources Forum 34:211–221
Shirasaka S et al (2016) Diversity of seasonal migration of livestock in the Eastern Alai Valley, Southern Kyrgyzstan. In: Kreutzmann H, Watanabe T (eds) Mapping transition in the Pamirs, advances in Asian Human-Environmental Research. Springer, Cham. https://doi.org/10.1007/978-3-319-23198-3_8
Sievers EW (2003) The post-Soviet decline of Central Asia: sustainable development and comprehensive capital. Routledge Curzon, London, p xii–248
Sorg A et al (2012) Climate change impacts on glaciers and runoff in Tien Shan (Central Asia) Nat Climate Change 2, pp 725–731 doi:https://doi.org/10.1038/nclimate1592. Published online 29 July 2012
Steimann B (2011) Making a living in uncertainty: agro-pastoral livelihoods and institutional transformations in post-socialist rural Kyrgyzstan. University of Zurich. ISBN 3-906302-09-1. Zurich Open Repository and Archive. http://www.zora.uzh.ch
UNCCD KR (2006) UN Convention to Combat Desertification Kyrgyz Republic. The Third National Report of Kyrgyz Republic on United Nations Convention to Combat Desertification Implementation. www.unccd-prais.com/
Zhumanova M et al (2016) Farmers' decision-making and land use changes in Kyrgyz Agropastoral Systems. Mountain Res Dev 36(4):506–517

Current Status and Improvement of Institutional Base of Solving Soil Degradation and Low Agricultural Productivity Problems in Tajikistan

Murod Juraievich Ergashev and Islomkhudzha Ahmadovich Olimov

1 Introduction

Tajikistan is an agrarian country, and agriculture is the basic sector of the national economy. Seventy-two percent of the population lives in rural areas, and most people are engaged in this sector. Agricultural production is the main form of income of the rural population. Food production (for consumption and sale) and the income received from farming on average constitute 48% of the total household income.[1]

Tajikistan has limited land resources. In 1991, the area of farmland per capita was only 0.8 ha, and in recent years, it has decreased to 0.6 ha as a result of population growth. In future, forecasting the population growth, this figure may even be reduced. In the agricultural sector, there are a number of environmental problems and low labor productivity, which increases its sensitivity to climate change.

Along with the above factors, geographical location, natural and climatic conditions, low potential of the Republic, and other reasons have become the basis for recent large-scale degradation of land, which has had serious social and economic consequences. In this regard, the development of the country's economy, especially its agricultural sector, based on the concept of sustainable land management (SLM), is a time-consuming and priority task for the country and the Tajik society as a whole. In view of this, immediately after the Soviet Union collapse in the country,

[1]World Bank (2012), p. 48.

M.J. Ergashev
Soil Institute of Academy of Agriculture Science, Dushanbe, Tajikistan

I.A. Olimov (✉)
Russian-Tajik (slavonic) University, Dushanbe, Tajikistan
e-mail: islom_72@mail.ru

H. Ginzky et al. (eds.), *International Yearbook of Soil Law and Policy 2017*,
International Yearbook of Soil Law and Policy,
https://doi.org/10.1007/978-3-319-68885-5_6

reforms began in the agrarian sector. Reforming agriculture now is aimed at modernization and the sustainable development of this industry. Government programs and initiatives are being implemented, the amount of State investments is increasing, and the legal framework for land use is being substantially improved. The country has a potential to improve its environmental situation, increase productivity in the agricultural sector, and ensure food security and resilience to the effects of climate change in the coming years.

1.1 Current State of Soil Cover, Soil and Land Degradation in the Republic of Tajikistan

Summary This article reveals the state of soil cover and land resources of the republic, anthropogenic and other factors of impact on them, the degree of their degradation, as well as the causes and consequences of degradation of soil and land resources of the republic. Among the land resources including water, land, flora, pasture, arable land, and other resources, soil resources are highlighted, including research of them, which is devoted to the topic of scientific work.

Maintaining soil cover and soil fertility in satisfactory condition is important for increasing the productivity of agriculture and accordingly plays an important role in ensuring the country's food security. Intensive use of land resources without strict observance of environmental and agronomic requirements has resulted in a decline in the ecology and destabilization of the landscape zones of the Republic. Landscapes are represented by areas of mixed and deciduous forests, forest-steppe and steppe zones, and arid territories.

A large stock of organic matter is concentrated in mixed and broad-leaved forests, in which living biomass is about 45% (90% of the plants). Forests have high soil fertility. The capacity of the biomass primary productivity is very significant; broad-leaved forests are able to effectively maintain the oxygen balance of the atmosphere. In the south part of the Republic of Tajikistan, Tugai forests have almost disappeared by the beginning of the twenty-first century. Arid zones have appeared, for which small stocks of living organic substances are characterized. Dead organic matter prevails in the soils. Arid territories have always been sparsely populated with a conglomeration of the population in oases, river valleys, and foothills.

Large reserves of organic matter are maintained in the meadow steppe ecosystems with soil humus of 96%, and there is a high fertility of the land. In living biomass, 92% belongs to the herbaceous plants. Primary productivity is very significant, but the contribution to the oxygen balance is low. Plowing of lands and reclamation of drought-affected areas using irrigation systems led, first of all, to the loss of the humus layer and a sharp, up to 85%, drop in yield. Disturbance of the natural soil cover worsened the conditions of heat and moisture exchange. Arid periods in the area of arable lands intensified and increased, and dust storms appeared with wind speeds of 20–30 m/s and a decrease in relative humidity of up to 10–15%.

Destabilization of the landscape has changed the environment in the areas of agricultural activity. The total area of arable lands where it is required to carry out measures to protect the soil from erosion is 106.2 thousand hectares, and for hayfields and pastures, it is 704.8 thousand hectares. Every year, for various reasons, 4–8 thousand hectares of irrigated farmlands are not used for agricultural production. Out of the available 603.1 thousand hectares of irrigated farmland, 7.2 thousand hectares are not used.

One of the main reasons for the low level of agricultural production development in Tajikistan is the unsatisfactory state of agricultural lands, mainly due to a constant decline in fertility. There is a destruction of the productive layer, and the humus content is decreasing (Fig. 1).

The percentage of eroded and degraded soils reaches 97.9% of the total area, although it was 68% in 1971. The area of poorly eroded and degraded lands is 9.2% of the total area, the category of strongly and very strongly eroded and degraded lands is 45.1%, and medium eroded lands—43.6%. Degradation of soils sharply reduces their fertility and bio-productivity of natural lands; reduces the area of irrigated lands, forests, pastures; destroys numerous structures; and turns dry lands into corrugated surfaces. Annual damage from these dynamic processes is several million US dollars.

The processes of waterlogging and salinization of irrigated lands are widespread. The areas of irrigated land with an unsatisfactory reclamation state have increased, and land with an unacceptable depth of groundwater level against the background of salinity has decreased by 11%. According to the prevalence of saline lands in the irrigated area, Tajikistan is included in the second group of countries in the world that are characterized by at least 15% salinity where currently the area of salinity is 116.5 thousand hectares. The situation is complicated by the high predisposition of irrigated lands to resalinization, the area of which is currently 310 thousand hectares (Fig. 2).

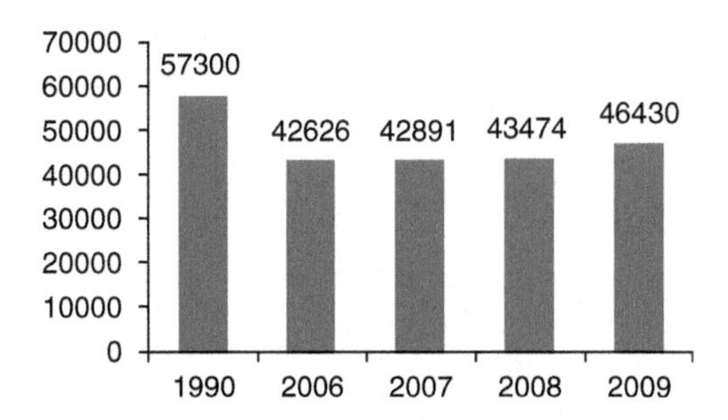

Fig. 1 Legislative system of the Republic of Tajikistan on SLM

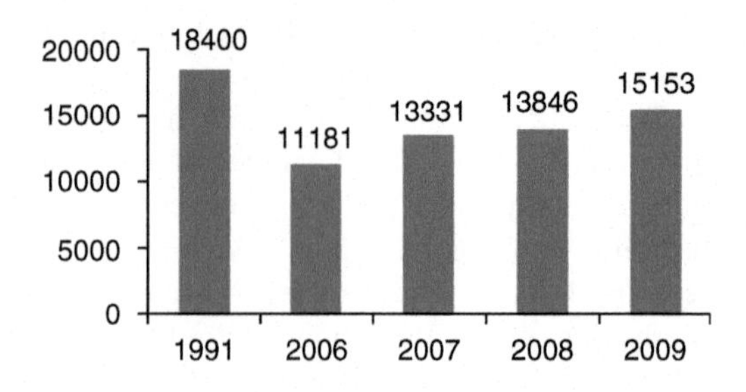

Fig. 2 Saline land areas (thousand hectares). *Source:* Environment protection/Statistical collection, 1990–2000, Dushanbe, 2014, 299p and Agriculture of the Republic of Tajikistan/Statistical collection, Dushanbe, 2014, 286p

The problem of resalinization arose in Tajikistan after 1992 due to the fact that reclamation of saline soils was abandoned. Reclamation of resalinized lands is a time-consuming and capital-intensive process. The problem of waterlogging is associated with untimely cleaning of collector-drainage networks. All of the above processes, plus the anthropogenic effect on the ecosystem, have adversely reduced the effectiveness of land management.

Since the period 2005–2011, the Ministry of Land Reclamation and Water Resources carried out works to improve and ameliorate irrigated lands on an area of 60,275 ha. With a view to implement the Decree of the Government of the Republic of Tajikistan "On Measures to Improve the Reclamation State of Irrigated Agricultural Lands for 2010-2014," it was planned to improve the ameliorative condition of 49 thousand hectares of land.

One of the threats to land degradation is the unregulated and excessive grazing of pasturelands. Summer pastures are degraded by more than 90%, and pasture productivity has decreased by 5–10 times due to changes in the composition of plants.

The area of degraded arable land in Tajikistan is 720 thousand hectares. Land degradation is mainly caused by water, wind, and irrigation erosion. The area of land subjected to desertification processes over the past decade is more than 4 million hectares, which is one third more than in 1990. Erosion processes are activated under the influence of natural phenomena such as mudflows, flooding, and landslides, especially when the slope of the land exceeds 12 degrees. Annually, about 42 thousand hectares of irrigated land undergoes different degrees of desertification.

The reasons why soil degradation occurs in and around settlements are mainly as follows:

- location and siting of settlements;
- use of steep land;
- nonobservance of good irrigation techniques;
- vegetation destruction, especially forest;
- overgrazing pasture and without rotation;
- absence of hard-walled road networks;
- improper construction of terraces.

Soil is excavated and removed from hillsides and hill slopes along highways for building materials. This practice promotes the formation of steep, precipitous slopes that have a high level of danger of ravine formation.

The pasture area is 3.7 million hectares in Tajikistan. Duration and excessive grazing worsen the condition of the grass stand; 75–90% of the grass mass has low palatability or is poorly grazed. More than 100 thousand hectares of pastures and hayfields on steep erosion-prone slopes were plowed up in 1995–1997, and as a result, the erosion of soils intensified.

With regard to soil contamination, the bulk of pollution enters the soil by atmospheric precipitation. The main sources of waste include authorized and unauthorized landfills and application of fertilizers and pesticides. Intensive cultivation of cotton (as a monoculture) practiced over a long period has caused degradation of irrigated lands. In this case, the soil is depleted and the biological processes and accumulation of nutrients are reduced. Repeated passage of agricultural machinery in a season leads to compaction of the 25–60 cm zone of the soil profile, where the bulk density exceeds 1.45. Compaction prevents root penetration and activates the rise of capillaries of saline solutions. When cultivating a permanent crop of cotton, the yield gradually decreases and the role of individual soil nutrition elements worsens. From 1991 to 2015, the production of raw cotton in all categories of farms decreased from 819.6 to 270.0 thousand tons, and the area of cotton has dropped as a result of the grant of economic freedom to farmers. Beginning 2009, by recommendation of the International Monetary Fund, the Government of the Republic of Tajikistan gave full freedom to farmers in the land use planning, i.e., independence in decision making on the breeding of different types of agricultural crops.

The application of fertilizers with agricultural production, to increase the yield of cultivated plants, especially cotton, has been associated with a steady increase in the use of artificial fertilizers. Because of the low farming standards, plants absorb violations of norms and rules, fertilizers by less than 30% the rest remains in the soil. Overuse of fertilizers leads to the accumulation of residual chemicals in the soil and polluting the atmosphere and hydrosphere. Pollution of agricultural lands with pesticides poses a real danger for the ecological systems of the Republic. These compounds, in fact, being artificial poisons, possess the ability to accumulate in plants and animal organisms and be transmitted through food chains. The accumulation of nitrates in the soils and agricultural plants has negative impacts for health.

Overuse of nitrogen, phosphorus, potassium, and other types of new fertilizers adversely affects the microorganisms in the soil, which impairs soil properties. As a result of the application of pesticides against pests (insects) and plant diseases, poisoning of the fauna also occurs, and the biologic activity of the soil is reduced.

Particular danger to soils is the disturbance caused by anthropogenic activity, especially in the extraction of minerals. As a result of open mining of mineral resources, quarries are formed, reaching depths of hundreds of meters and diameters of several kilometers. Specific loads of anthropogenic nature on the soil are quickly growing with rich mineral deposits and developing industry in the regions of our republic.

The total area of land in Tajikistan occupied by various types of waste is 1100 ha, of which 12 ha is affected by solid wastes and liquid industrial wastes and about 800 ha by mining and other wastes.[2]

Soil fertility is significantly affected by wind and water erosion. A million tons of fertile soil is lost every year from arable land through water and wind erosion. Water erosion is widespread and is the main form of land degradation. The area of land affected by these processes is 97.9% of the total area of agricultural land. Gully (linear) erosion occupies 0.4% of the total area of the Republic. Water erosion leads to flushing of the soil's fertile layer with the subsequent appearance of ravines. The high terrain of Khatlon and Sughd provinces and some Districts of Republican Subordination (districts that are directly under central rule) is especially prone to water erosion.

The process of wind (aeolian) erosion blows soil particles from the earth's surface. Subsequent destruction of these particles occurs when they interact with each other (corrosive erosion). As a result of the soil particles moving, they accumulate in local zones (*eolian accumulation*). With strong manifestations of erosion, dust storms occur when dust particles can rise to a height of 5–6 km and be blown away for distances of hundreds and thousands kilometers. Dust particles can also concentrate under the action of air masses and is deposited as sediment. According to the Academy of Sciences of the Republic of Tajikistan, for the past 30 years, the amount of dust storms in the country has increased at least ten times. In the early 1990s, they occurred only two to three times a year, but in recent years, up to 35 storms have been recorded annually. The main direction where the dust storms come from includes Afghanistan, the Middle East desert, and the Aral Sea.

2 Legislative System of the Republic of Tajikistan: Procedure of Development, Discussion, Acceptance, and Monitoring of the Implementation of Laws

Summary This section details the legislative system on sustainable land management and the process of developing, discussing, and implementing relevant laws and legislation on sustainable land management (SLM) in Tajikistan.

The Government has paid attention to environmental protection, improving the ecological situation and the rational use of natural resources by adopting around 60 laws and regulations on the environment to date. The right of legislative initiative belongs to the members of the Majlisi Milli, deputies of the Majlisi Namoyandagon, the President of the Republic of Tajikistan, and the Government of Tajikistan. Laws in the Majlisi Namoyandagon are adopted by a majority vote of

[2]http://refdb.ru/look/2550433-p2.html.

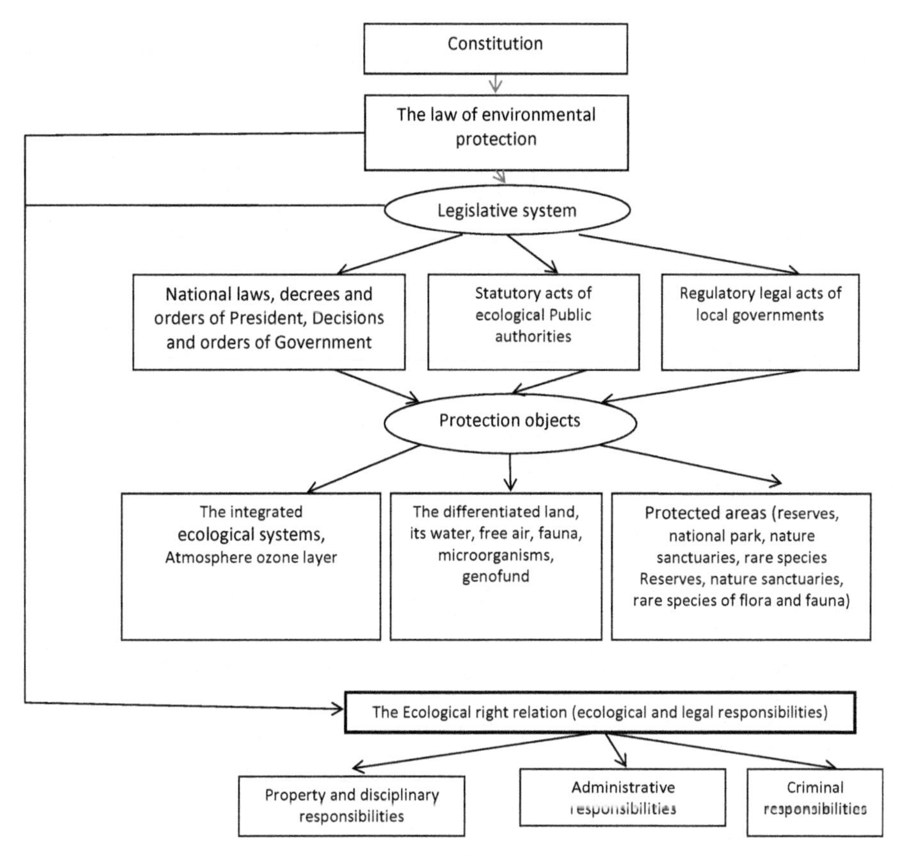

Fig. 3 Legislative system of the Republic of Tajikistan on SLM

the total number of deputies. The laws passed by the Majlisi Namoyandagon are submitted to the Majlisi Milli for approval then submitted to the President of the Republic of Tajikistan for signature and promulgation (Fig. 3).

Many laws and regulations adopted over the past two decades contribute to the goal of SLM. These legislative acts include the Land Code of the Republic of Tajikistan dated December 13, 1996, under № 327; the Forestry Code of RT dated June 24, 1993, under № 770; and others (see Table 1).

The main responsibility for land management in Tajikistan is entrusted to the State Committee for Land Management and Geodesy of the Republic of Tajikistan and the Committee for Environmental Protection under the Government of the Republic of Tajikistan. The State Committee for Land Management is responsible for regulating land ownership in rural areas and approving land tenure and is also authorized to make proposals for protecting (preserving) lands from pollution and degradation. The Committee for Environmental Protection under the Government

Table 1 List of normative legal acts on SLM in Tajikistan

Existing	Adopted and modified in the last few years
In the field of environmental protection and sustainable development 1. Forest Code of the Republic of Tajikistan, June 24, 1993, № 770 2. Law of the Republic of Tajikistan "On the Protection of Nature," August 2, 2011, № 760 3. Law of the Republic of Tajikistan "On the Protection of Atmospheric Air," February 1, 1996, № 229 4. Law of the Republic of Tajikistan "On Specially Protected Natural Territories," December 13, 1996, № 329 5. Law of RT "On Production Waste and Consumption," April 23, 2002, № 270 6. Law of RT "On Ecological Expertise" of April 7, 2003, № 379 7. Law of the Republic of Tajikistan "On Radiation Safety," June 26, 2003, № 888 8. Law of the Republic of Tajikistan "On Protection and Use of the Growing World," April 29, 2004, № 529 9. Law of the Republic of Tajikistan "On Fauna," December 19, 2007, № 433	1. The Law "On Pastures," 2013 2. The Forest Code (as amended), 2012 3. The Law "On Mountain Regions," June, 22, 2013
In the field of social and economic development 1. The Law of the Republic of Tajikistan "On Foreign Economic Activity of the Republic of Tajikistan," December 27, 1993 2. The Law of the Republic of Tajikistan "On Certification of Products and Services" of December 13, 1996, № 314 3. The Law of the Republic of Tajikistan "On Tourism," September 9, 1999, № 825 4. The Law of the Republic of Tajikistan "On Free Ecological Areas in the Republic of Tajikistan," February 18, 2004, № 1068 5. The Law of the Republic of Tajikistan "On Licensing of Certain Types of Activity," April 29, 2004, № 535 6. The Law of the Republic of Tajikistan "On Investments," April 25, 2007, № 545	
In the field of land use and agriculture 1. The Law of the Republic of Tajikistan "On Land Reform," March 14, 1992, № 604 2. Land Code of the Republic of Tajikistan, December 13, 1996, № 327 3. Law on Land Management, 5.01.2008, № 356 4. The Law of the Republic of Tajikistan "On Dekhkan Farms," March 15, 2016, № 1289 5. The Law of the Republic of Tajikistan "On Geodesy and Cartography," February 2006, № 188 6. The Law of the Republic of Tajikistan "On Seed Production," December 19, 2007, № 434	

(continued)

Table 1 (continued)

Existing	Adopted and modified in the last few years
Other legal acts 1. The Code of the Republic of Tajikistan on Administrative Offense, November 26, 2008, № 1177 2. The Labor Code of the Republic of Tajikistan, May 15, 1997, № 417 3. The Criminal Code of the Republic of Tajikistan, May 21, 1998, № 575 4. The Civil Code (part of the third), the Republic of Tajikistan, February 11, 2005, № 598 5. The Tax Code of the Republic of Tajikistan, 28.12.2013, № 901, 6. The Customs Code of the Republic of Tajikistan, November 25, 2004, № 573	

of the Republic of Tajikistan is responsible for implementing governmental programs for the control and prevention of land degradation and air and water pollution; protection and rational use of flora, fauna, fish resources; conservation of specially protected natural areas; control over the treatment and use of all types of waste; restoration and protection of forest resources; and hydrometeorology.

In order to achieve SLM, in addition to the adopted laws and legislation, Tajikistan has ratified and acceded to nine United Nations' Conventions in recent years, including the 1992 United Nations Convention on Biological Diversity, 1995 United Nations Convention to Combat Desertification, 1992 United Nations Framework Convention on Climate Change, 1983 Convention on the Conservation of Migratory Species of Wild Animals.

In December 1998, the United Nations Convention to Combat Desertification was ratified by the President's Decree under № 1144. A National Program of Actions to Combat Desertification was prepared, and a series of national workshops on problems of desertification was conducted in various regions of the country with various segments of the population, including women and youth. With the purpose of conservation and rational use of land, control of land degradation, and overcoming of desertification, in 1997 the Government decision "On State Control over the Use and Protection of Land in the Republic of Tajikistan" under № 294 was adopted.

To overcome the process of desertification and in accordance with the context of the United Nations Convention to Combat Desertification, the Government decision "On Approving the National Program of Action (NAP) to Combat Desertification" was adopted on December 30, 2001, under № 598. This program is the main strategy for Tajikistan to combat desertification. Along with the strengths in the implementation of the UN Convention to Combat Desertification, a number of existing shortcomings in this field have been identified:

- insufficient funding;
- lack of literature on the main ideas of the United Nations Convention to Combat Desertification in the national language;
- poor coordination between the governmental bodies addressing various problems of desertification;
- lack of monitoring of desertification processes;
- weak development of the technical base for combating desertification;
- noncompliance and ignorance of traditional methods to combat desertification;
- lack of rational use of natural resources by the local population.

In 1998, the Republic of Tajikistan acceded to the United Nations Framework Convention on Climate Change and on October 21, 2008, ratified the 1997 Kyoto Protocol. According to the obligations under the United Nations Framework Convention, Tajikistan prepared and submitted a National Action Plan (NAP) for mitigating climate change, which is the main strategic document on addressing climate change problems.

The concept of environmental protection and SLM has been adopted in the Republic of Tajikistan in several governmental programs, including the following:

- The Program on Improving the Condition and Rational Use of Pastures in Tajikistan for 2009–2015;
- The Program on Development of Forestry of the Republic of Tajikistan for 2006–2015, October 31, 2005;
- The State Program on the Development of Specially Protected Natural Areas for 2005–2015, March 4, 2005;
- The State Environmental Program of the Republic of Tajikistan for 2009–2019, February 27, 2009.

The State Environmental Program of the Republic of Tajikistan for 2009–2019 is the main document defining the principal directions of development of the Tajik society, maintaining a balance between natural resources and its users, organizing and coordinating relations between users of natural resources and nature itself, healthy development of society, as well as ways to restore the destroyed aspects of ecology. The implementation of measures envisaged by the Program provides an opportunity to address issues of the rational use of natural resources and environmental protection. In accordance with a Decree of the Government of Tajikistan, implementation of the Program is entrusted to all ministries, departments, enterprises, institutions, organizations, and farms, including joint ventures, regardless of ownership and subordination, whose activities are related to the use of natural resources and influence on the ecological situation. Coordination and control over the implementation of this Program is vested in the environmental authorities.

2.1 *The Current Laws and Legislative Acts on Soil Protection and Sustainable Land Management: Advantages and Disadvantages*

Summary In this section, the main laws and legislative acts on the effective management of soil and land resources are analyzed in detail; their positive aspects, as well as their shortcomings that hinder their effective implementation in the Republic, are examined.

As already noted, basic laws are adopted that promote the rational use of natural resources and protect the environment. We consider in detail the advantages and disadvantages of the existing basic laws and legislative acts that provide for soil protection.

Law on Soil Protection The "Law on Soil Protection" was adopted by Resolution of the Majlisi Namoyandagon of the Majlisi Oli of the Republic of Tajikistan on October 16, 2009, under № 555. This Law defines the main principles of State policy; the legal basis for the activities of public authorities, individuals, and legal entities toward the rational and careful use of soils, preservation of the quality and fertility of the soil, and soil protection from adverse events; it also regulates a complex of relations related to protection of soil quality and prevention of its degradation. State control on the use and protection of soils is established by the Law on Soil Protection, which monitors the following:

- soil condition during economic and other activities;
- compliance with requirements and standards in the field of soil protection;
- carrying out of activities provided for land management, city planning, environmental and other documentation containing measures to prevent soil degradation and contamination, soil restoration, and the elimination of negative processes on soils;
- reliability of information provided on the state of soil quality;
- taking of measures to address deficiencies in the field of soil protection.

This Law consists of five chapters and 21 articles. The main principles of the State policy on soil protection are as follows:

- ensuring the rational use and conservation of soil as an essential component of the natural environment;
- ensuring the application of soil protection technologies and other measures to prevent soil pollution and soil degradation during economic and other activities;
- timely conduct of surveys and activities to improve soil fertility and restore degraded soils;
- scientific validity of measures of soil protection;
- publicity, completeness, and reliability of information on the state of soil and ongoing measures to protect soil.

The Government's powers and responsibilities are provided in the Law on Soil Protection, as well as policies in this area. However, it is necessary to expand the tasks and obligations of local authorities and organizations that can prevent soil erosion and degradation. It is also necessary to introduce provisions covering the educational policy for the local population so that they know what damage will be inflicted to the soil if it is misused. The provisions of this Law do not resolve controversial issues, and that responsibility is transferred to other laws.

Land Code of the Republic of Tajikistan This Code was adopted on December 13, 1996. It consists of 19 sections and 105 articles to regulate land relations. It aims to create conditions for the rational use and protection of land, reproduction of soil fertility, preservation and improvement of the natural environment, and equal development of all forms of management.

Goals, objectives, and measures on the use of land resources are not provided in the Land Code. However, Articles 51 to 54 of Chapter 8 cover the objectives, tasks, and measures for land protection. Chapter 9 covers land management and land protection. In particular, Article 55 states that land management and protection is carried out by the authorized State body of the Republic of Tajikistan for land management and its local bodies and the authorized agency for nature protection of the Republic of Tajikistan.

It is necessary to supplement the Code with terminologies that describe the meaning of keywords, phrases, and concepts. It is required to specify in more detail specific provisions on the different levels of responsibility, including standards for their implementation. The Land Code clearly defines its direct responsibilities for meeting environmental requirements. However, there are no separate articles that provide for the functions, activities, policies, and regulations on land use, as well as the obligations to conduct research to obtain information and scientific, technical, and cultural knowledge relevant to land use and land protection. It is important to involve the public in the process of land management and use.

The Land Code includes mechanisms that make it possible to abolish land use permits for farmers, including where land use leads to land degradation. In practice, land is seized only if it is not used for two consecutive years. For less violation, the land user will pay a fine. Household plots and additional personal plots of land allocated in accordance with the established procedure represent cities and districts of local executive bodies of the Government power.

The Land Code was amended in 1997, 1999, 2001, 2004, and 2007. It is regarded that the Land Code will benefit from the following amendments:

- to define the functions and tasks of the Government, the Ministry of Agriculture, State Land Committee, and the local authorities in the field of land use and protection;
- to define the functions and tasks of primary and follow-up land users, regardless of the type of ownership;
- to define responsibility for violation of the land legislation;
- to define the procedure for pastoral land use.

Forest Code The Forest Code 1993 consists of 19 chapters and 59 articles. It regulates the management of forest resources and is aimed at creating conditions for the rational use of forests, their protection, the preservation and improvement of the natural environment, and the production of forest and agricultural products. State forestry authorities are obliged to ensure the reproduction, protection, and improvement of the sanitary condition of forests; care for them; increase the productivity of forests and fertility of forest soils; and organize forest use and records. The Forestry Agency under the Government of the Republic of Tajikistan exercises the function of development and implementation of a single State policy and legal regulation and of public administration in the sphere of forestry, forest resources, hunting, flora and fauna, and specially protected natural areas; carries out economic functions of system organization; and provides State control.

It is necessary to include further articles that provide for the goals and objectives to implement measures on the protection of forestland, education and awareness, and a commitment to cooperate in the implementation of this Code. Additional terms should be included, such as forest degradation, forestland, forest cadaster, etc.

The Code provides for the obligation of State forestry authorities. However, it is necessary to describe in detail the duties of individuals in the field of protection and rational use of forestland and specify the standards for the hierarchy of forest resource use. The Code provides for forest monitoring, which specifies the content and monitoring procedure established by the Government.

The Code requires detailed provisions for education, forest management objectives, and areas where knowledge and capacity building are required, including in general forestry education, conservation, and sustainable use of forests. Specific areas of research, the purpose of disseminating results, cooperation with other research organizations on forest degradation and desertification should also be included. Provisions to involve the public in the planning, management, and development of projects on the conservation of forestland should also be specified. The State forest cadaster is maintained at the expense of the State budget. There are no specific details on financing for forestry by a specially authorized body.

In accordance with the Forest Code, when forest users violate the established procedure for the use of forests, the Government may suspend forestry activities. Forestry disputes are considered in the order stipulated by the legislation of the Republic of Tajikistan. It is necessary to indicate the specific authority for the resolution of disputes.

Nature Conservation Law from August 2, 2011, № 760 (as amended by the Law of the Republic of Tajikistan, № 1160, of November 27, 2014): the policy of the Republic of Tajikistan in the field of environmental protection is aimed at ensuring the priority of environmental interests of the Republic. Under the Law, it will achieve this objective by taking a scientific approach for the development of economic and other activities with a careful consideration of nature, its wealth, the rational use of natural resources, and the guaranteed protection of the human right to a healthy and favorable environment. The Law regulates the use of fertilizers or pesticides, biological and chemical substances, protection from contamination of foodstuff, misuse of soil and land, and protection of land against pollution by discharge from livestock farms.

Individuals and entities are obliged to protect nature and its wealth and comply with the environmental requirements of this Law. Chapter 12 highlights ecological education and research. Public environmental organizations and associations carry out their activities in accordance with their charters, regulations, and legislation and interact with other environmental organizations based on the provisions of this Law. A link with the "Law on Public Associations and Organizations" is required. Organizations should provide tax privileges for environmental protection in this Law at the time of financing the environmental protection activities.

Legislation of the Republic of Tajikistan establishes responsibility for violations of environmental legislation. Disputes between State public organizations of the Republic of Tajikistan related to natural environment protection and the use of natural resources are resolved in court, and in case of disputes with other countries, these are resolved in accordance with the agreements[3] made between the respective countries. Changes and amendments to the Nature Conservation Law were made in 1996; on May 10, 2002; on December 2, 2002; in 2004; in 2007; and in 2015.

Law on Dekhkan Farm This Law was adopted on March 15, 2016, and defines the legal basis for the establishment and implementing activities of dehkan farms.[4] The Law consists of 45 articles, but only Article 26 provides an obligation for dehkan farm and its members to rationally and carefully use land resources, not permitting deterioration of the ecological condition of land. The activities of the farms are sufficiently defined in the Law, but there are no mechanisms that would affect the issue of rational use and protection of land. It is necessary to include articles providing goals and objectives for land management.

The Law provides the concept of development of dehkan farms. However, it is necessary to provide for the duties and responsibilities of dekhkan farms for land management and protection. It is also necessary to specify the standards and hierarchy of the land-use-planning process, the development of cooperation policies, research activities, education, and establishment of public consulting groups. There are few provisions concerning institutional responsibility and public involvement in this process. The financial mechanisms of dehkan farms are described in detail in the Law. Individuals and legal entities are held liable in accordance with

[3]Agreements between the Government of Tajikistan and Kyrgyzstan: on checkpoints through the Tajik-Kyrgyz border (31.008.2004, № 357), on the transit movement of goods by road (02.12.2005, № 463), on reconstruction, operation and maintenance of the Osh-Khorog road (13.11.1996, № 489), on free trade (11.03.2000, № 108); Agreement between the Government of Tajikistan and Uzbekistan: on cooperation in the field of veterinary (9.11.2000,№ 451); the decision to join the Agreement between Kazakhstan, Kyrgyzstan and Uzbekistan in field of environment protection and rational use of nature resources (24.11.1998, № 444).

[4]Dekhkan farms—business entities based on which the production, storage, processing and sale of agricultural products are carried out under the individual or joint activities of a group of persons, on a plot of land and property belonging to them.

the legislation of the Republic of Tajikistan for breaching this Law, and disputes that have arisen in the subjects of these relations are considered by the court.[5]

From 1992 till 2009, the Law was changed three times, and there have been both positive and negative aspects with each change. For example, this Law does not define the body that issues land use certificates, whereas the previous law clearly defined that. It is regarded that the status of a legal entity should be granted to dekhkan farms.

Law on Land Management (5.01.2008, No. 356) This Law consists of five chapters and 38 articles and regulates numerous activities related to land management, the procedure for conducting land management activities, as well as the relations in the process of land management activities[6] between the Government, local government bodies, local governments, land management authorities, land users, individuals, and legal entities that conduct land management work.

Land management shall be carried out to implement the goals specified in Article 5 of this Law based on the Decision of the Government of Tajikistan, the authorized State body for land management, local public authorities, applications (petitions) of individuals and legal entities, and court decisions. Developers of land management documentation, while carrying out their activities, are obliged to work in accordance with the requirements of this Law, other regulatory and legal acts, and legislation adopted by the authorized State land management body to regulate land management works. These bodies, when developing land management activities, must give priority to environmental protection over the economic necessity of using land.

Groups of provisions covering the overall objectives of the policies and implementation of measures to achieve the rational use of land resources are specifically formulated. The Law clearly describes the meaning of keywords or concepts that are directly related to this Law. It identifies general duties, and various articles reflect the obligations of individuals and legal entities. It is regarded that there is no need to supplement or amend the provisions of this Law at this stage.

The Government of the Republic of Tajikistan, the authorized State body for land management, and local government bodies carry out land management activities. The Law includes the functions, goals, and tasks that allow organizations and relevant administrative bodies to determine their immediate responsibilities for land management. There is no coordination with the relevant articles of the Nature Conservation Law and the Land Code.

A careful attitude to land resources and natural landscapes is envisaged. However, it is necessary to develop a mechanism to implement these actions. For example, there are no provisions to increase the level of public knowledge on land management, rational use of land, and environmental protection. Study of

[5]Disputes that have arisen between legal entities are considered in economic court, disputes that have arisen between legal entities and individuals are considered in civil court.

[6]Topographic, geodetic, and cartographic studies; agrochemical, geobotanical and other soil surveys; land inventory; compilation of special thematic maps.

the state of the land is carried out through land cadastral works, land monitoring and research to clarify various lands, their quality and assessment, as well as the development of measures to improve the state of land.

Participation and involvement of the public in the land management process should be specified in the Law. Land-use planning is carried out with a view to determine the prospects for specific and long-term land use. As a result of planning the land use and land protection in cities and districts, a special land fund is created. Land management envisages the preservation of the quality of land resources. Financing land management activities and works is made from State budget, local budgets, funds of individuals and legal entities, and other sources permitted by legislation. Individuals and legal persons are held liable for breaching this Law in accordance with the legislation of the Republic of Tajikistan. Disputes[7] that have arisen between the participants in land management are resolved by the court.

3 Current State and Perspective of Agricultural Sector Development in the Republic of Tajikistan

Summary This section describes the current state of the agricultural sector, its problems in the way of sustainable development, as well as the opportunities for its sustainable development in the near future.

The agricultural sector of the Republic's economy includes 2627 large, medium-sized, small households and organizations, including one collective farm; seven sovkhozes; 73 state farms; one leased enterprise; 1341 associations of dekhkan farms; 820 collective dekhkan farms; 25 stock companies; 68 agricultural cooperatives; 284 subsidiary farms in the enterprises and organizations; seven agricultural firms; and more than 90,000 dekhkan farms. In the reform period of 1991–2008, a share of the private sector in agricultural production increased from 9.9% up to 89.3%[8].

After the socioeconomic transformation, the new forms of management that arose, especially the dekhkan farms, still do not show the advantage of this form of farming. This is facilitated by the fact that dekhkan farms occupy a significant part of the land resource where 69% of the farms produced only 26.9% of the country's gross agricultural output in 2008.[9]

In general, the country's agriculture, despite its market transformation, is still characterized by low productivity, and this once again convincingly proves that the agrarian sphere cannot effectively develop in the market conditions without governmental support.

[7]For example, the payment of taxes at irrigated land rates, which, due to the destruction of irrigational systems, turned into pasture, granting of land to users belonging to the forest fund, etc.

[8]Agriculture of the Republic of Tajikistan. Statistical Book - Dushanbe (2009), p. 5.

[9]PPCR, Phase I, Final Report, 2011.

Agriculture totals about 25% of gross domestic product (GDP) and provides two thirds of employment of the population.[10] In this sector, there are a number of key problems caused by objective and subjective reasons, which are of a systemic nature, including the following:

- weakening of State regulation and support of the agrarian sector;
- sharp decline in the purchasing capacity of agricultural producers caused by the disparity of prices between industrial and agricultural products;
- deterioration of the material and technical base of agricultural enterprises and a low self-sufficiency of dekhkan farms and individual household plots;
- decrease in soil fertility and deterioration of the condition of the irrigation system.

A reduction in the scope of agricultural production in Tajikistan is primarily a consequence of the destruction of economic potential in this sector, the rural social sphere, and a lack of motivation to work. The number of investments in agricultural development has decreased sharply. In 1990, the share of agriculture in total was 18.3%, where at present it is 1.5%. The number of imported mineral fertilizers has decreased by 2.9 times. The technical base of agriculture has greatly degraded, and technical and electric power of labor has decreased by several times.

In some areas, the serviceability of agricultural machinery and the level of technical services provided to agriculture are in critical condition. In some dekhkan farms, there is no machinery at all.

So far, little progress has been made in improving agricultural techniques. In 2008, a permanent leasing fund was established under the budget to the amount of 8 million somoni,[11] and also "Tajikagroleasing" SUE was established,[12] which, according to the list of machinery and equipment approved by the Ministry of Agriculture of the Republic of Tajikistan, would import agricultural machinery and supply them to the agricultural enterprises under a lease arrangement. In 2008–2009, "Tajikagroleasing" SUE provided about 1000 units of agricultural machinery and, on this basis, created machine-tractor stations in some areas.[13]

Since 2008, the attitude to agriculture has changed radically in the budget and investment policies of Tajikistan. Already in 2008, the volume of investments in agriculture amounted to 166.2 million somoni against 41.2 million somoni in 2007. This represents an increase by a factor of 4. In 2008, the budget-funded support to agriculture, taking into account the "Fund for preferential lending" (without investment projects), was 206.3 million somoni against 46.6 million somoni in 2007, an increase by 64.4 times. In 2007, the amount of budget funds allocated to the development of agriculture and related industries, including investment projects

[10] PPCR, Phase I, Final Report, 2011.

[11] 907,030 US $.

[12] leasing company – State Unitary enterprise "Tajikagroleasing," gives to leasing an agricultural techniques.

[13] Asia Plus 2010, No. 6, dated February 10, p. 11.

and allocations to the "Fund for preferential lending" for soft loans, totaled to 640 million somoni. In 2010, the budget-funded support totaled to 720 million somoni or 10.9 of the consolidated budget for 2010, taking into account investment projects and deductions to replenish the "Fund for preferential lending" for soft loans.[14] This is a relatively high level of budget-funded support to the development of agrarian business sectors and to the strengthening of their material and technical base, as well as to addressing other rural socioeconomic problems.

One of the reasons for low soil fertility and, accordingly, low crop yields is a significant reduction in the use of fertilizer. The reason for this decline is that due to a shortage of energy resources and restrictions in the electric power supply, the population mainly uses organic fertilizer as dung for heating houses and cooking food. If in 1994 organic fertilizers applied in all categories of farming amounted to 800.6 thousand tons; in 2012, this figure was only 232.6 thousand tons. However, there is an insignificant increase of fertilizers used throughout the country. With the reduction in the use of organic fertilizer, the volume of mineral fertilizer used should increase several times, but due to inability of farms and households to afford such fertilizers, there is only a slight increase in the use of this type of fertilizer. In 1994, mineral fertilizers applied in all categories of farming amounted to 53.7 thousand tons, and in 2012, this figure was less at 52.8 thousand tons. Agricultural enterprises, like other entities, increase the amount of organic fertilizers used in crop production for cereals, potatoes, and vegetables, which ensures their annual increase in yield as the main types of food products to meet their own needs and sale. This is one of the reasons why the yield of vegetables and potatoes is high in Tajikistan by comparison with other "Commonwealth of Independent States" countries.

Since 1992, agricultural land has been largely redistributed from the Soviet collective and State farms to the dekhkan farms, where the latter exist in three organizational forms: individual dekhkan farms, family dekhkan farms, and partner (or collective) dekhkan farms. Individual and family dekhkan farms are the smallest (about 5 ha on average), and partner dekhkan farms have a maximum area of 100–200 ha. However, within the framework of a government program on land redistribution to individual members of households, the number of partner dekhkan farms is rapidly reducing. There was a significant expansion of the household plots with the help of "presidential lands" (redistributed according to Presidential Decrees issued in two stages, in 1995 and 1997). Despite the expansion, the area of personal plots remains very small, averaging 0.3 hectares, smaller than the average area of dekhkan farms.

There are 750,000 rural households (all with household plots) and 50,000 dehkan farms in Tajikistan.[15] A huge number of household comprises a small part of the total area. Almost 20% of arable land is now in the form of household

---

[14]Message from the President of RT, Emomali Rahmon, Majlisi Oli of the Republic of Tajikistan (24.04.2010)-Dushanbe, 2010 p 70.

[15]PPCR, Phase I, Final Report, 2011.

plots, which represents a multiple increase compared to the traditional 4–5% used in the Soviet period. The area of dekhkan farms increased from a figure that is less than half the area of arable lands in 1995 up to 65% in 2009. Thus, dekhkan farms and household plots combined totaled to 85% of arable lands in 2009 compared to 5% in 1991, while the share of agricultural enterprises (corporate farms that inherited traditional collective and state farms) decreased from 95% in 1991 to 25% in 2009. Since the number of collective dekhkan farms is also rapidly reducing, it can be concluded that land management in Tajikistan since 1991 has been effectively transformed into an individual form of ownership.

Increase of land resources in farms inevitably led to an increase of their share in agricultural production (according to the data on gross agricultural output (GAO). The share of........ agricultural enterprises (production cooperatives and traditional collective and State farms) in GAO decreased from 65% in 1995 to less than 10% in 2008. The share of subsidiary farms increased from 35% up to 65%. The remaining 25% belongs to dekhkan farms—the second component of the individual economy sector, which began to contribute to GAO after 1997. Agricultural production, as well as land management, is now fully transferred to individual ownership in Tajikistan. Since household plots produce 65% of agricultural products by less than 20% of arable land, it is obvious that they are much more productive than other types of farming.

Land resources of Tajikistan are limited by their nature, while the population, especially in rural areas, is rapidly growing. For three decades, from 1979 to 2009, the rural population doubled from 2.6 million to 5.5 million people. During the period from 1991 to 2012, the total land area remained almost unchanged, but there are significant changes in the area of arable lands (+31.3 thousand hectares), perennial plantations (+23.1 thousand hectares), and deposits [(agricultural land formerly used as arable land but not used for more than a year) (+10.3 thousand hectares)] (Table 2).

During 1990–2000 years, the total area of farmland significantly decreased (−639 thousand hectares). Other areas of reduction included irrigated arable lands (−106.2 thousand hectares), irrigated area of the total farmland (−61.3 thousand hectares), hayfields (−9.1 thousand hectares), and pastures (695.7 thousand hectares). Further, the area of arable land has significantly expanded up to approximately 850 thousand hectares since 1980, while the rural population has doubled.

Irrigated lands are mainly used for intensive cultivation (arable lands and perennial plantations) and account for 89–90% of GDP generated in agriculture and 85–87% of gross income from all agricultural activities of the Republic.[16] Productivity of irrigated arable land is five times higher than the productivity of dry lands, which indicates the high social and economic importance of irrigated lands. Ongoing transformations led to the destruction of the material and technical potential of agriculture and, in particular, of the irrigated land uses, resulting in a

[16]Begmatov and Umarov (2007), p. 69.

Table 2 The area of agricultural lands in Tajikistan (in administrative boundaries, thousand hectares)

Types of land	1991	2004	2012
Total area of land	14254.5	14255.4	14255.4
From them: irrigated	713.2	732.4	
All farmland, total	4434.1	4645.0	3795.1
From them: irrigated	653.3	604.1	592.0
From them:			
Arable land,	811.2	723.7	842.5
From them: irrigated	567.9	503.9	461.7
Perennial plantations, total	103.4	102.6	126.5
Deposit, total	19.7	23.1	31.0
Hayfields, total	26.4	22.4	17.3
Pastures, total	3473.4	3773.2	2777.70
From them: irrigated	4.3	3.2	

Source: Environment protection/Statistical collection, 1990–2000 years - Dushanbe, 2014. 299p, and Agriculture of the Republic of Tajikistan/Statistical collection. Dushanbe, 2014, 286p

sharp decline in production, for example obtaining 14–16 centers (hundred weight) of raw cotton on average from each hectare of irrigated land for 2006–2008.[17] This indicates a decrease in the land's fertility by 50% compared to the 1989–1990 period.

The yield of other crops is also low. The yield of grain in Tajikistan is lower than the average yield of grain in the CIS countries as a whole (2.1 t/ha, compared to 2.5 t/ha in average in the CIS countries). Even the yield of cotton is considerably low (1.6 t/ha compared to 2.2 t/ha in other CIS countries). However, the yield of horticultural crops (potatoes and vegetables) in Tajikistan is much higher than that in other CIS countries: potato yield in Tajikistan is 21 tons/ha, and in the CIS countries, the average is 15 tons/ha[18].

The main reason for the low efficiency of irrigated land users is the lack of material and technical facilities, lack of funds for the construction of industrial facilities, lack of good quality agricultural machinery, lack of current capital, and high cost of mineral fertilizers. Therefore, for example, in 1991, 37,054 tractors were used in agriculture,[19] and in early 2010, their total number had dropped to 14,199, a reduction of 22,855 tractors.

Since 1998, there has been an increase in agricultural production. The relative growth of the agricultural sector has made a significant contribution to the postconflict situation in Tajikistan's economy and amounted to about one third of total economic growth in 2005.[20] However, this growth is not sustainable due to a

[17] Agriculture of the Republic of Tajikistan. Statistical Book - Dushanbe (2009), p. 40.

[18] PPCR, Phase I, Final Report, 2011.

[19] Agriculture of the Republic of Tajikistan. Statistical Book - Dushanbe (2009), p. 308.

[20] UNDP (2012).

number of subjective and objective reasons, including the effects of climate change, unattractiveness of agriculture to investors, division of big dekhkan farms to small dehkan farms, which became unattractive to lending by the banks.

4 Recommendations to Improve Agriculture and the Legislative Base for Soil Protection and Sustainable Land Management

Summary This section provides an overview of the main recommendations on the effective use of soil and land resources, prevention of further degradation, application of advanced technologies for sustainable use of land resources, and use of the agricultural sector domestic opportunities in the way of its sustainable development.

To increase marketability and attractiveness, radical measures are needed for modernization and sustainable development of agriculture as a fundamental component of the national economy. The solution to this problem requires a different understanding of the role of agriculture, which is slowly changing, and its efficiency by introducing methods of the "green revolution," new technologies, restoration of land, and preventing the misappropriation of land resources through the misuse of funds allocated to support the agro-industrial complex and dekhkan farms.

It is necessary to gradually introduce "green revolution" methods and industrialization of the agro-industrial sectors. Opportunity for the growth of dekhkan farms is limited due to the small size of the farms. Their size also means that dekhkan farms are unattractive for the banking sector. There should be a natural link between the processing enterprises and the development of the dekhkan farm sector because only with such a link can raw materials be supplied. Only 11% and 15% of fruits and vegetables are processed. New agricultural processing enterprises are mostly small, and like most small firms in transition countries, they face a number of problems—obtaining finance, management, marketing, logistics, corruption practices, and technological problems.

Tajikistan is up against a serious problem of maintaining sufficient infrastructure for the development of industrial entrepreneurship. There is a need for additional funds to ensure that the quantity and quality of the infrastructure are consistent with the goals of Tajikistan's future economic growth. Important objectives of rural development policy, as a priority, include the following:

- increase in value-added enterprises in the rural economy;
- support for the development of rural entrepreneurship;
- promotion of agricultural development in dekhkan farms.

The policy of increasing value added implies an increased focus on food product processing, as well as other measures that can also potentially contribute to the development of value-added production. Enterprises for food processing in rural areas face problems associated with unreliable energy and water supply. Electricity generated by diesel-electric systems is likely to be inexpedient from a financial point of view and for the production of solar energy, but at present technical and financial constraints exist. Therefore, uninterrupted power supply through a nation-wide energy system is an important basis for the development of a food-processing subsector in rural areas and, in general, for rural entrepreneurship, as well as to ensure a reliable drinking water supply.

Currently, more favorable conditions for the development of rural entrepreneurship are being created. At the same time, the focus is on infrastructure development, as well as on expanding access to financial resources and business advisory services. Besides, classification of rural areas to the category of urban settlements will lead to the improvement of infrastructure and services, and then they can gradually grow into development centers.

Development of rural entrepreneurship will obviously require the adoption of measures for practical and technical retraining and requalification, as well as release of a significant amount of labor from agriculture. Agriculture continues to play an important role as a "safety net" of the social protection system; however, encouraging the transition from the farm sector to the sectors with higher labor capacity will become a source of rural (and urban) development in the medium and long terms.

Assistance to agriculture in dekhkan farms should focus on the dissemination of knowledge and irrigation problems to increase productivity and, therefore, to provide benefits to low-income groups in the rural community. Taking into consideration the farms' dependence on irrigation, as one of the priority tasks, it is necessary to ensure that disadvantaged people can benefit from the development of irrigation systems. To do this, it is necessary to take measures at the community level, with the participation of dekhkan farms, which will complement measures on rehabilitation and development of irrigation infrastructure.

Two priority recommendations have been identified to improve the livelihood and reduce the vulnerability of rural people:

1. expansion of household plots, small individual and family dekhkan farms;
2. diversification of agricultural products and income sources.

The first recommendation, namely, the expansion of household plots, small individual and family dekhkan farms, requires as follows:

(a) Development of land markets: amendment to the Land Code is required:

   - Issue of land use certificates as "proof of ownership" to all land users (household plots and members of dekhkan farms: administrative action only is required to complete the current process.
   - Legal guarantee that a land use certificate will allow the transfer of land and to participate in the standard market land transactions (lend-lease, sale, and purchase of land): to do this, it is necessary to make appropriate changes to the Land Code.

- Legal guarantee of tenure, i.e. land cannot be taken away from the current owner without due process and appropriate compensation: it is necessary to amend the Land Code relating to the alienation of land use rights.
- Legal maintenance of the fact that the land users, including women, will not lose their land in the case of bankruptcy: thus, it is required to make appropriate changes to the Civil Code (February 11, 2005, No. 598, the third part), Law on Bankruptcy (6.10.2008, No. 432), Law on Land Reform (14.03.1992, No. 604), Law on Dekhkan Farms (March, 2016, No. 1289). A household plot should be defined as the minimum that a rural family can retain in the case of bankruptcy.

(b) Return of unused or unproductive land to productivity can be carried out without the adoption of new legislation. This requires administrative resources only (identification and research of unused lands), as well as private or governmental investments in their improvement.

There is one aspect of the provisions on land taxation, which is worth paying attention to. Tax rates are established in accordance with the established procedure based on land quality: tax rates are higher for irrigated land and orchards and lower for pastures. Unused and unproductive land often falls under the lower limits of taxation. If they are reclaimed, the tax rate rises. Since cultivation requires investments, the Tax Code (No. 901 dated 28.12.2013) should provide tax incentives or tax deductions to farmers investing in cultivation, i.e. such farmers can be exempted from land tax on cultivated land for several years. Another feature is that higher tax rates apply to the entire cadastral registration area, even if only part of it has been cultivated and returned to the cultivation process. This represents a serious distortion that prevents farmers from investing in land cultivation, and it should be remedied.

Tax incentives can also be used to encourage the rehabilitation of worn-out pastures carried out by dekhkan farmers. This issue, as well as the advantages and disadvantages of "privatization" of pastures in comparison with pastures in common use, which was adopted in the "Law on Pastures" 2013, is not included. New decisions concerning the rights on pasture may require retroactive amendments to the Land Code and Law on Dekhkan Farms, which determine the distribution of grazing areas among the pasture users.

The second recommendation, diversification of income sources and agricultural products, consists of two different components:

1. diversification of income sources; and
2. diversification of agricultural products.

Diversification of income sources should be based on a personal initiative. As for the legislative part, it should concentrate on imbalance of a social tax.[21] Firstly, it is not entirely clear what the rural population will receive in exchange for paying a social tax: the studies show a general lack of awareness of the rights arising from

[21]Or deductions to the social insurance and social providing fund (25%), paid by employers (including dekhkan farms) from the total volume of salaries of theirs employees.

the appropriate social point of the Tax Code. Secondly, the policy of establishing a fixed social tax rate per person represents a significant burden for rural people. The monthly levied tax is much higher than the level of rural incomes. The legislature should consider an opportunity to change the Tax Code so that the social tax rate is linked to an individual or family income. The low-income population can be completely exempted from paying the social tax (their social insurance will be fully covered from the State budget), while the proportion of the population with adequate income may pay more.

Diversification of agricultural products, in general, depends on scrupulous compliance with the "mechanism on Freedom of Farming." Local authorities should not dictate to farmers how to use their land, which crops to grow, or where to sell them (including cotton). The mechanism was adopted in 2009. Significant progress has been made in its implementation. One of the tangible results of mechanism of giving of independence to dekhkan farms was a significant shift toward horticultural crops, including vegetables and fruits. Deregulation on assignment of land should help farmers to increase the production of fodder crops, which are essential for the proper care of livestock. Subsequent softening and final lifting of administrative restrictions on agricultural producers should lead to further growth in the diversification of agricultural products, reducing risk, and vulnerability.

5 Conclusion

After the Soviet Union collapse the productivity slowdown and decline in the economic activity of entities in Tajikistan have led to an aggravation of the food sovereignty problem; an increase in the import of food, which is sometimes not of high quality; and expensive goods. The main task of the transition period (1991–2010) was to develop and implement such a transformation strategy that would allow putting all capabilities of a market economy at the community service in the shortest time in order to increase a level of satisfaction of people's needs for agricultural products. Market supply for the agricultural sector is characterized by low elasticity, and private ownership of land has not yet been legislatively formalized, and entrepreneurial activity in agriculture is insignificant. This explains the inadequate level of using many effective market mechanisms in agriculture (such as shares of stock, bonds, bills, and other value documents; future contracts; lease of machinery and equipment; other schemes for resolving credit issues; etc.).

Demographic conditions, primarily the decrease in an economically active population, as well as the change of the age pattern of the rural population, have influenced the development of the rural economy. Recently, there has been a process of the villages "aging," that is, a decrease in the proportion of employable population, especially those aged under 40–50 years old. In addition, there is a reorientation of the villagers, especially young people, who do not consider the management of private household plots and other forms of rural business as a necessary "attribute" of rural livelihood.

It is necessary to preserve as much as possible the main directions of specialization in agriculture, most fully meeting the natural and economic conditions of the country. Also, it is necessary to optimize sectorial structure, taking into account the conjuncture of the world and regional markets, with orientation to food self-sufficiency and rational use of resources.

Development and implementation of measures for the sustainable development of agriculture require scientific support, including improvement of the economic mechanism for the market infrastructure formation, functioning and creation, improvement of the land use legal framework, and a study of trends in the development of the agro-industrial complex. Also, there is a need to remove barriers to the improvement of technology and practices for the manufacturing and processing of agricultural products.

It is important that the Government, through comprehensive State policies and targeted investments, create favorable conditions for the development of sustainable agriculture at the local and national levels. Civil society can bring its practical experience, orientation to the problems of the poor, practical participation, and significant investigative potential into the overall picture. Finally, entrepreneurs recognizing the complexity of global environmental and economic problems should interact with these actors and profitably use their scale, resources, investments, and technologies for the purposeful development of sustainable agriculture. In general, activities should be concentrated in the following areas:

- *Investments in agricultural productivity, especially for small dekhkan farms:*
  - increased investments in agricultural research and development, as well as in agricultural institutions; extension of experience; and infrastructure development, such as construction of roads, storage facilities, and irrigation systems;
  - investment in rural development in the fields of education (especially among girls), health, and water purification;
  - ensuring the consistent observance of labor, environmental, and effective management standards.
- *Protection and improvement of the environment while increasing agricultural productivity:*
  - implementation of policies to promote integrated management of energy, water, and land resources with full consideration of these elements;
  - implementation of policies to prioritize agricultural production diversification programs as a means to maintain soil fertility, water resources, and biodiversity.

- *Support for technological innovations acceptable and affordable for poor dekhkan farms:*
  - creation of regulatory and legal conditions that attract investment in technology and promote the sustainable development of agriculture;
  - assistance in scaling up effective pilot projects, the technologies of which enable sustainable development of agriculture;
  - support in the development of standards and guidelines for sustainable development, corporate transparency, and accountability for sustainable agriculture;
  - encouraging sustainable practices and advanced soil and water saving technologies, as well as technologies for sustainable use of all land resources in the country;
  - encouraging all companies to include sustainable agriculture as a business priority.

References

Asia Plus 2010, № 6 of February 10, p 11
Begmatov AA, Umarov DM (2007) Water economy in market conditions. Dushanbe, p 69
Internet resource: http://refdb.ru/look/2550433-p2.html
PPCR, 1st Phase, Final Report, 2011
Statistical Digest (2009) Agriculture of the Republic of Tajikistan. Statistical Digest, Dushanbe, p 308
Statistics Collection (2009) Agriculture of the Republic of Tajikistan. Statistics Collection, Dushanbe, p 5
UNDP (2012) Tajikistan: institutions and development. UNDP
World Bank (2012) Tajikistan: economic and distributive Impact of Climate Change. World Bank, p 48

Governance of Pastoral Lands

Ian Hannam

1 Introduction

Many unique challenges that pastoralists face in securing governance of land tenure include the challenges that are determined by the ecological sustainability of pastoral rangelands, particularly their soil resources. Pastoral societies are well adapted to these challenges, and they have developed customs and rules governing management and use of pastoral land that are embedded in pastoral culture. A number of initiatives and studies from recent years on pastoral governance focus on the challenges that pastoralists face, the failures of governments in securing pastoral tenure, and the emerging examples of success and progress from around the world. Among these include the World Initiative for Sustainable Pastoralism, a global partnership that gathers and reviews case studies through an extensive network.[1] Other notable global reviews include "The Land We Graze"[2] and "Governance of Rangelands – Collective Action for Sustainable Pastoralism."[3]

This chapter outlines important international environmental law principles that can inspire the achievement of sound pastoral land law and presents some of the legal and institutional elements to create or improve legislation related to tenure,

[1] https://www.iucn.org/theme/ecosystem-management/our-work/global-drylands-initiative/iucns-work-drylands/world-initiative (accessed 8 November 2016).

[2] https://cmsdata.iucn.org/downloads/land_rights_publication_english_web.pdf.

[3] https://portals.iucn.org/library/node/44904.

I. Hannam (✉)
Australian Centre for Agriculture and Law, University of New England, Armidale, NSW, Australia
e-mail: ian.hannam@ozemail.com.au

H. Ginzky et al. (eds.), *International Yearbook of Soil Law and Policy 2017*, International Yearbook of Soil Law and Policy,
https://doi.org/10.1007/978-3-319-68885-5_7

which is one of the key tools used to implement sustainable soil and agricultural policy objectives.[4] Many countries have laws in place that can support progress toward responsible governance of pastoral lands, but these laws are not always implemented. Pastoral legislation alone does not solve the problem of weak pastoral tenure, but it provides the legal basis for action. These principles and elements are consistent with the objectives of the Voluntary Guidelines on the Responsible Governance of Tenure of Land, Fisheries and Forests in the Context of National Food Security (VGGT)[5] in drawing on international and regional instruments that address human rights and tenure rights, including the 2015 United Nations Sustainable Development Goals.[6] They provide guidance on internationally accepted practices for the type of legal systems to deal with the rights to use and control pastoral land. Moreover, they contribute to improving and developing policy, legal, and organizational frameworks that regulate a range of tenure rights that exist over pastoral land. They can also strengthen the capacity and operations of implementing agencies, judicial authorities, local governments, farmer organizations, cooperatives, and small-scale owners or occupiers of pastoral land.

State efforts to improve governance of pastoral land and ensure successful implementation of the VGGT necessarily encompass a variety of commitments. This includes educating both the general public and State officials about laws that promote responsible governance of pastoral lands, and harmonizing legislation and streamlining legal and administrative procedures to establish accessible and efficient procedures can be realized in all administrative and judicial institutions at every level of pastoral land governance. There is also the need to ensure that the national judiciary and all officials responsible for adjudicating conflicts apply national laws that promote responsible governance of tenure and to create or strengthen local, culturally appropriate alternative dispute resolution bodies to ensure that pastoral land conflicts are resolved in a timely manner.

Establishing national pastoral land policy and law by following the principles and elements mentioned in this chapter will contribute to better governance and potentially improve the condition of the soil and lead to sustainable agricultural systems. This may involve introducing new laws or amending or repealing existing ones and the implementation through national law of international treaties. The provisions of these treaties would normally need to be implemented through national law and policies before they can impact individuals and groups. Even without formal implementation, these international instruments may stimulate national-level processes and indirectly affect the governance of tenure,[7] focusing more specifically on land administration, taxation, transfer of land, allocation of

[4]FAO (2016a).

[5]VGGT (2012), Windfuhr (2017) explains the relevance, reception and first experiences in implementation of the VGGT.

[6]See https://sustainabledevelopment.un.org/?menu=1300 (accessed 8 November 2016).

[7]FAO (2016b).

tenure rights, spatial planning, resolution of disputes over tenure rights, land trust, customary tenure systems, climate change, and resilience to natural disasters.

1.1 Pastoral Land

The land occupied by pastoralists is often referred to as the rangelands. Rangeland ecosystems have evolved in places of climate extremes and high climatic uncertainty and account for roughly half of all land (51%).They are challenging and unpredictable environments in which nature and society have evolved, leading to unique biological and cultural diversity. Rangeland ecosystems provide many agricultural goods and services to humanity, including soil resources, food and fiber, water supply, and sequestration of carbon. Rangelands are sometimes defined in ecological terms as "land on which the indigenous vegetation is predominantly grasses, grass-like plants, forbs or shrubs that are grazed or have the potential to be grazed, and which is used as a natural ecosystem for the production of grazing livestock and wildlife."[8] Rangelands can include natural grasslands like prairies or steppes, savannahs, wetlands, drylands and deserts, tundra and certain low forbs, and shrublands like chaparral or Mediterranean maquis. Rangelands are often found in drylands and mountains where plant growth is restricted by low rainfall, extreme cold, high altitude, steep slopes, or other factors. Globally, rangelands are based on the following land categories[9]: desert, grassland, shrubland, woodland and savannah, and tundra.

1.2 The Value of Pastoral Land

The benefits to humanity from the world's rangelands are under threat in many countries from land degradation. This lack of understanding has contributed to poorly informed interventions and policies that have sometimes exacerbated degradation. Rangelands are subject to conversion to crop cultivation, overexploitation for livestock, overextraction of woody biomass, and increased aridity due to both climate change and extraction of water. These types of degradation are driven by population growth, growing demand for food, changes in management and technologies, and a range of policy and institutional factors. Policy failures can in turn be attributed to a combination of weak resource rights and governance, weak influencing capacity of rangeland stakeholders, and insufficient or inaccurate data, information, and knowledge.[10]

---

[8]Allen et al. (2011).

[9]According to the Society for Range Management - http://rangelands.org/ (accessed 9 November 2016).

[10]Mortimore et al. (2009), Davies et al. (2015).

Evidence shows that the value of pastoralism and rangelands in most countries is greatly underestimated, and conversion of rangeland resources to other uses can have greater costs than benefits when measured across the entire system.[11] Yet despite evidence that converting rangeland to cropland is one of the most significant drivers of soil and land degradation, many countries still focus overwhelming attention to crop farming to the detriment of rangelands' health. However, it should also be noted that, in many countries, the evidence for rangeland degradation is weak, and the diagnosis of degradation, usually attributed to overgrazing, may be politically motivated and has been used to justify confiscation of pastoral land. At the same time, it has been observed that where mobility and customary institutions for local governance remain effective, rangeland degradation can be scarce.[12]

2 Pastoral Land Rights

The VGGT provide an international framework to guide policy and programs to protect and enhance the rights of pastoralist communities to lands that have been historically used for social, cultural, spiritual, and economic ends. In particular, Part 3 of the VGGT refers to the importance of "legal recognition and allocation of tenure rights and duties" for protecting the rights of indigenous peoples and other communities with customary tenure. To ensure that adequate law is implemented, the institutional, political, and social factors that contribute to weak tenure security and poor governance that can affect the agricultural potential must be addressed. Such change is dependent upon a commitment to and the establishment of fair, well-functioning, and impartial administrative and justice systems and citizens' ability to access and successfully use these unbiased, equitable systems to protect their tenure rights. "Tenure" is the way that land is held or owned by individuals, families, companies, or groups. Tenure claims may be formal, informal, customary, or religious in nature and may include ownership, use, and management rights. The strength of an individual's, family's, or group's tenure rights may hinge on national legal definitions of property rights, local social conventions, or other factors. In relation to pastoral land, land tenure rights can be described as a "bundle of rights" that may include the following freedom:

- to occupy, develop, enjoy, and draw benefits from the soil resources;
- to sell or bequeath natural resources;
- to lease or grant use rights to natural resources;
- to restrict others' access to the natural resources; and/or
- to use and manage natural resources.

[11]Davies and Hatfield (2007).

[12]Niamir-Fuller (1999).

Traditionally, pastoral land rights consisted of access to the resources required to sustain mobile livestock production such as pastures, watering points, and the movement corridors that linked together seasonal grazing areas, pastoral settlements or encampments, and markets. These customary tenure arrangements mixed aspects of common property and exclusive ownership. A descent group, a clan, or an entire ethnic group could claim common rights to an area of pastoral land. Pastoral tenure rights[13] have allowed some people to use property belonging to another for specific purposes or limited periods of time. While they have been common, they have created complex systems of rights and duties among pastoral users. In these property systems, individuals could have exclusive access to some types of resources, but they held these rights as members of social groups that were capable of defending the territorial integrity of the entire group, not by virtue of a title deed issued by a government authority. However, a variety of factors, including land conversion, privatization, conflict, population pressure, and the creation of nature reserves, have all led to a reduction of pastoral land rights in recent times.[14]

2.1 Customary Land Tenure

Customary land tenure refers to the systems that communities operate to determine ownership, possession, and access and to regulate use and transfer. Customary land tenure is a major agricultural tenure system on a worldwide scale. It is extensive in pastoral areas of Africa and Asia and also governs lands in some industrial economies, such as rural commons in Spain and territories belonging to indigenous minorities in North America and Australia.[15] Unlike introduced landholding systems, the norms of customary tenure derive from and are sustained by the community itself rather than the State or State law (statutory land tenure).[16] Although the rules that a particular local community follows are known as customary law, they are rarely binding beyond that community. Customary land tenure is as much a social system as a legal code and from the former obtains its resilience, continuity, and flexibility. Of critical importance to modern customary pastoralists is how far national law supports the land rights it delivers and the norms operated to sustain these.[17]

Customary domains are rarely homogenous. Parks, mining, timber, and agricultural concessions often create sizeable holes in customary areas. When wealthier farmers obtain formal statutory title for their homesteads, they extinguish customary title, thereby creating smaller holes in the overall community land area.

---

[13]Often referred to as 'fuzzy rights'.

[14]Behnke and Freudenberger (2013).

[15]RIRDC (2014).

[16]Another term for customary land tenure is "indigenous tenure".

[17]Alden-Wily (2012).

Customary domains are also "fuzzy" at their edges, especially where they adjoin rapidly expanding cities and towns. Chiefs may sell lands on the urban fringe to developers or have these taken. There are instances where rural communities retain control over urbanized lands. The global land rush is stimulating domestic land grabs of this kind for profit, in turn accelerating concentration and the introduction of market-based norms and placing pressure on common resources such as the soils. The greater is the value of the resources affected, the greater is the tension over norms.

2.2 Rights and Duties to Use

In the VGGT, rights and duties are established as crucial elements for governance of tenure determining if and how people, communities, and others are able to acquire rights, and associated duties, to use and control land. The elements below, from the VGGT, are of particular importance in legal frameworks for pastoralism and achieving sustainable land use[18]:

- establishment of safeguards to avoid infringing on or extinguishing tenure rights of others when recognizing or allocating tenure rights, including those not currently protected by law (7.1);
- ensuring that legal recognition and allocation of tenure rights and duties are consistent with their existing obligation under national and international laws and with due regard to voluntary commitments under applicable regional and international instruments (7.2);
- ensuring that women and men enjoy the same rights in the newly recognized tenure rights and that those rights are reflected in public records (7.4);
- recognition, respect, and protection of legitimate tenure rights of individuals and communities, including customary tenure systems (8.2);
- promotion by local or traditional institutions of effective participation and consultation of local communities, inter alia, men, women, and youth, regarding their tenure systems (9.2).
- promotion and respect of customary tenure systems to solving conflicts within communities, consistent with their existing obligations (9.11).

[18] VGGT (2012).

3 International Principles

A number of international principles have been drawn from sources relevant to the responsible governance of pastoral land and, in particular, the sustainable use of soil. These principles can be used in either of two ways in reforming or framing national pastoral law[19]:

1. as a basic underlying policy and ethical position to frame a particular legal and institutional element, which will help achieve responsible governance of pastoral tenure arrangements and sustainable use of soil;
2. as a separate, specific State legal and institutional element or integrated with another State-level element.

3.1 Principle 1: Good Governance

States should adhere to responsible governance and international human rights principles for the management of pastoral land. This reflects a growing awareness of the importance to sustainable development of transparent, accountable, honest governance, as well as a growing awareness of the effect of corruption on public morale, economic efficiency, political stability, and sustainable use in general. This principle implies the adoption of democratic and transparent decision-making procedures and financial accountability, respecting due process in procedures and observing the rule of law, and conducting public procurement in a transparent, noncorrupt manner. States should recognize the human rights and the vital role of indigenous and traditional people and their communities and other local communities in the management of pastoral land, especially the benefits of their knowledge of practices to the sustainable use of pastoral land and governance of its tenure. For this principle to be adequately implemented, States should recognize the identity, culture, and interests of indigenous and traditional people and enable their transfer of customary knowledge to achieve the sustainable use of pastoral lands.[20]

3.2 Principle 2: Sustainable Development, Integration, and Interdependence

This principle requires sovereign nations to promote healthy and sustainable pastoral land development and the resources within it that will help to improve the quality of people's life, without compromising future generations. Environmental

[19]The international principles stated here are drawn from FAO (2016)a.

[20]United Nations (1992) Principle 22; Licht et al. (2007).

protection should constitute an integral part of the development process. This particularly applies to the indigenous and other traditional communities that occupy pastoral lands.[21]

Pastoral ecosystems interact with the lithosphere, biosphere, hydrosphere, and atmosphere. They produce biomass and are a biological habitat and gene reserve. They are critical to the management of the earth's climate system. The concept of integration encompasses a commitment to adapt the environmental considerations and objectives to the core of international relations. The interdependence concept included in the Copenhagen Declaration on Social Development states that "Economic development, social development and environmental protection are interdependent and mutually reinforcing components of sustainable development."[22]

3.3 Principle 3: Intergenerational and Intragenerational Equity

This principle specifies that future generations have the right to a quality of life not less than the current generation. Elder generations must provide environmental stability for future generations that will provide them the same opportunities for development as they had. This principle has a direct relationship with the ethics of environmental order and solidarity[23] and is critical for the long-term sustainable use of soil and land.

Current generations have the right to use, enjoy, and manage land in search for a better quality of life, but these generational rights have to be addressed as collective and not as individual rights, given the fact that these future rights will exist independently of the number of people of every new generation. The UNESCO Declaration on the Responsibilities of the Present Generation Towards Future Generations asserts the necessity for establishing new, equitable, and global links of partnership and intragenerational solidarity and for promoting intergenerational solidarity for the perpetuation of humankind.[24]

[21]United Nations (1992).

[22]World Summit for Social Development (1995) http://www.un.org/documents/ga/conf166/aconf166-9.htm (accessed 27 October 2016).

[23]United Nations (1992) Principle 3 http://www.un.org/documents/ga/conf151/aconf15126-1annex1.htm (accessed 9 November 2016).

[24]UNESCO (1997) http://portal.unesco.org/en/ev.php-URL_ID=13178&URL_DO=DO_TOPIC&URL_SECTION=201.html (accessed 9 November 2016).

3.4 Principle 4: Responsibility for Transboundary Harm

Countries sharing the same pastoral system should make the effort to manage that system as a single agricultural unit. They should cooperate on the basis of equity and reciprocity, in particular through bilateral and multilateral agreements, in order to develop uniform policies and strategies covering the entire biological system of pastoral land. This involves the mutual exchange of privileges between the States or nations or, at the local level, between districts and communities.

States should cooperate to prevent the transfer to other States of any activities and substances that cause a loss of integrity of pastoral land.[25] Formal agreements should be made between respective States where seasonal movements of livestock by pastoral peoples occurring according to customary tenure involve crossing international boundaries. This responsibility obliges States immediately to notify other States of any natural disasters or other emergencies that are likely to produce harmful effects to the pastoral land of those States. States should provide prior and timely notification and information to potentially affected States on activities that may have a significant adverse transboundary effect and should consult with those States at an early stage and in good faith.[26]

3.5 Principle 5: Transparency, Public Participation, and Access to Information and Solutions

This principle ensures inclusive participation in the pastoralist context, including stronger participation by pastoralists in public decision making and access to information such as public records, cadastral maps, and a land registry. This will allow pastoralists to influence decision making related to their lands and will enable problem solving to generate legitimate, adaptive, and resilient solutions.

3.6 Principle 6: Cooperation and Common but Differentiated Responsibilities

Individual States should cooperate to conserve, protect, and restore the health and integrity of pastoral lands. In view of the different contributions to global pastoral degradation, States have common but differentiated responsibilities. Developed countries should acknowledge the responsibility that they bear in the pursuit of the sustainable use of pastoral lands in view of the pressures that their society places

---

[25] Benvenisti (2002).

[26] United Nations (1992) Principles 7, 18 and 19.

on the global environment in general and on pastoral land in particular and of the technologies and financial resources they command. The principle of cooperation has become basic in international environmental law.[27]

The world community as a whole and individual States have the responsibility to protect and conserve pastoral land in a sustainable manner and preserve its tenure for the benefit of present and future generations.[28] In particular, States should take action to prevent dangerous anthropogenic interference with the climate system. They should take measures that aim to enhance the ability of pastoral land ecosystems to adapt to climate change and restore or rehabilitate degraded pastoral ecosystems.[29]

3.7 *Principle 7: Precaution*

In order to protect the pastoral environment, the precautionary approach should be widely applied by States according to their capabilities. Where there are threats of serious or irreversible damage to pastoral land, lack of full scientific certainty shall not be used as a reason for postponing measures to prevent degradation of pastoral land.[30] This principle is central to ecosystem-based environmental management and is particularly pertinent in the context of pastoral land and management of land tenure, given the risks involved in losing its capability for many generations if inappropriate management of pastoral regimes is put into place.

3.8 *Principle 8: Prevention*

States should adopt measures to prevent damage to pastoral land. One obligation that flows from the concept of prevention is the prior assessment of potentially harmful activities. Since the failure to exercise due diligence to prevent transboundary harm can lead to international responsibility, it may be considered that a properly conducted environmental impact assessment might serve as a standard for determining whether or not due diligence was exercised. Preventive mechanisms include monitoring, notification, and exchange of information, all of which are obligations in almost all recent environmental agreements.

[27]See Preamble to UNEP (1995a) Convention on Biological Diversity; see also Sands (2003), pp 285–290.

[28]Bosselmann et al. (2008).

[29]Articles 1(5) and 3(3) UNEP (1995b) Framework Convention on Climate Change; Sustainable Development Goal 13.

[30]As adapted from Principle 15 of United Nations (1992); de Sadeleer (2002).

Prevention forms a prudent complement to the international obligation not to cause significant harm and to the polluter pays principle. Preventive measures should not depend on the appearance of pastoral land ecological problems; they anticipate damage or, where it has occurred, try to ensure it does not spread.[31]

3.9 *Principle 9: Polluter Pays Principle*

This principle is taken from the Rio Declaration on internalization of costs and has become known as the "Polluter Pays Principle" or "PPP." According to the PPP, the environmental costs of economic activities, including the cost of preventing potential harm, should be internalized rather than imposed upon society at large. Anyone whose activities cause or are likely to cause a loss of the ecological integrity of pastoral land should bear the cost of full preventive or restorative measures.[32]

3.10 *Principle 10: Access and Benefit Sharing Regarding Natural Resources*

Activities in a specific area should be carried out for the benefit of humans as a whole, irrespective of the geographical location of States and taking into consideration the interests and needs of developing States and of peoples who have not attained full independence or other self-governing status recognized by the United Nations. The authority should provide for the equitable sharing of financial and other economic benefits derived from activities in the area through any appropriate mechanism, on a nondiscriminatory basis.[33]

Pastoralists and other local communities shall have the right to access and to benefit in a sustainable manner from the common natural resources on which they rely for their livelihood and existence. The terms and modalities for exercising freedom of transit should be agreed between the States and transit States concerned through bilateral, subregional, or regional agreements. Transit States, in the exercise of their full sovereignty over their territory, should have the right to take all measures necessary to ensure that the rights and facilities provided shall in no way infringe their legitimate interests.[34]

---

[31] See Article 14, UNEP (1995a), de Sadeleer (2002).

[32] The Polluter Pays Principle occurs in a binding form in Article 2.5(b) of the Helsinki Convention on the Protection and Use of Transboundary Watercourses and International Lakes (1992) http://www.unece.org/fileadmin/DAM/env/water/pdf/watercon.pdf (accessed 27 October 2016).

[33] Hart (2008).

[34] ITPGRFA (2009), UNCLOS (1982), Hart (2008).

3.11 *Principle 11: Common Heritage and Common Concern of Humankind*

This principle is based on the establishment of a common heritage of humankind, which lies behind the existence of a common concern for the protection, preservation, and enhancement of the natural environment, proper management of the climate system, biological diversity, fauna and flora. These elements exceed the particular and immediate individual objectives of States or other actors. It represents the notion that certain global elements, regarded as beneficial to humanity, should not be unilaterally managed by States but for humankind as a whole, sharing responsibilities according to specific international agreements.

4 Pastoral Land Policy

Pastoral land policy can take many forms to achieve the objective of the sustainable land use. Moreover, the procedures and activities under pastoral land legislation can produce various materials that express a strategic or ethical position on particular aspects of pastoral land and how it relates to land tenure. Anything that promotes a course of action to control or manage any aspect of pastoral land use, particularly customary values and land tenure, could generally be considered within the genre of "pastoral land policy."[35] Some suggested areas for policy development, which can also lead to the sustainable use of soils, include maintenance of customary and traditional values and practices, involving the community in the management and protection of the pastoral environment, developing ecological standards and how they will be implemented and monitored, and policies on adaptation and mitigation of effects of climate change.

The objective for developing policy on pastoralism and land tenure, or improving existing legislation, is to promote social inclusion of pastoralists with the legal recognition of their land rights, including customary tenure and community rights. Developing new legislation and policies requires the participation of pastoralist groups, which requires support to strengthen their capabilities, for example through educational programs that will allow them to understand and better exercise their rights. More responsible governance of pastoral tenure will help to secure social and environmental sustainability and protect transhumance, including corridors for mobility and respecting the spatial and temporal use of resources. In particular, the VGGT provide specifically for the improvement of policy and legal frameworks to manage pastoral land as follows[36]:

[35]Herrera et al. (2014).

[36]VGGT (2012), pp. 7–8.

5.1. "States should provide and maintain policy, legal and organizational frameworks that promote responsible governance of tenure of land, fisheries and forests. [...]"

5.3. "States should ensure that policy, legal and organizational frameworks for tenure governance recognize and respect, in accordance with national laws, legitimate tenure rights including legitimate customary tenure rights that are not currently protected by law; and facilitate, promote and protect the exercise of tenure rights. Frameworks should reflect the social, cultural, economic and environmental significance of land, fisheries and forests. States should provide frameworks that are non-discriminatory and promote social equity and gender equality. [...]"

5.5. "States should develop relevant policies, laws and procedures through participatory processes involving all affected parties, ensuring that both men and women are included from the outset. Policies, laws and procedures should take into account the capacity to implement. [...]"

4.1 Developing a National Pastoral Land Strategy

A national pastoral land strategy is a means by which the objectives of pastoral legislation can be achieved and should outline how a pastoral land institution would manage the land tenure arrangements. The strategy should address the purpose and intent of the legislation and express an objective to achieve the sustainable use of pastoral land as a national environmental goal.[37] States should develop relevant policies, laws, and procedures through participatory processes involving all affected parties, ensuring that both men and women are included from the outset. Policies and procedures should incorporate gender-sensitive approaches, be clearly expressed in applicable languages, and be widely publicized. A State pastoral land strategy could include, for example:

- references to the objectives of national and international development and environmental strategies, policies, and treaties and linking them to the use of pastoral land and land tenure[38];
- a duty of care toward the values of indigenous and traditional communities;
- a duty to ensure that the different aspects of land tenure are properly managed;
- a duty of care to manage the effects of climate change;
- a commitment to the development of programs to achieve sustainable use of pastoral land;
- an outline of the role and benefits of community training programs for pastoral land.

[37] See UNCCD 10 Year Strategy, Secretariat of the Convention to Combat Desertification Report 8th session, Conference of the Parties Madrid 3–14 September 2007 ICCD/COP (8)/16/Add.1 23 October 2007.

[38] Adaptation, under the Frameworks of the CBD, the UNCCD and the UNFCCC, Joint Liaison Group of the Rio Conventions (2008) http://www.cbd.int/doc/publications/cc-adaptation-en.pdf (accessed 1 December 2016).

4.2 Enabling Policies

Ineffective policies can be partly attributed to poor understanding of pastoral systems by policy makers. Many States still see pastoralism as a wasteful use of land, and the national priority is often to convert pastoral lands to crop cultivation by enacting policies to acquire the necessary land. The priority challenges to pastoral tenure should focus on issues related to policy, such as improving consultation and participation mechanisms for pastoralists and developing integrated land-use planning at relevant scales.

The action areas for improving governance and strengthening human capabilities illustrate that tenure security and responsible tenure governance are not just about legal arrangements but also about relationships, processes, capabilities, and resources for governance. Policies need to be integrated in the sense that each policy supports other policies. For example, policies in support of sustainable use of soil will not necessarily lead to improvements unless other policies are implemented to create new and stable markets.

4.3 Research and Extension Systems

Research and extension systems are crucial for designing and promoting appropriate systems and practices adapted to the needs of pastoralists on issues such as rangeland management, processing, technologies that are adapted to mobility needs. Participatory research methodologies, as well as inclusion in upstream decision making about research goals, are keys to ensuring that the results of research are taken up by pastoralists.

5 Defining National Legislation Relevant to Pastoralism

There should be a clear statement of the principal purpose and intent of a legal instrument for the use of pastoral land. It may be expressed as a single purpose or a multipurpose statement. Such a statement can refer to the need for a competent pastoral land institution, the use of particular strategic approaches or mechanisms, the rights and values of pastoralist peoples, the geographic area of interest, and the setting of priorities for the management of pastoral land and land tenure.[39]

There should be in law a group of statements that express a policy and strategy that connect customary land tenure with a statutory law for pastoral land. Together, the statements should establish firm goals, targets, and standards for the general administration of pastoral land while providing for customary rights and

[39]Herrera et al. (2014).

responsibilities. The objectives may be expressed as a single or a multipurpose statement but could also comprise a number of multipurpose statements, for example, to observe the customary rights of pastoral land and that appropriate ecological standards and values and the knowledge and traditional land-use practices of pastoralist communities are fostered and protected.

5.1 Legal and Institutional Elements

The legal and institutional elements presented here are regarded as "generic" and collective; they implement many aspects of the VGGT for good governance and land tenure.[40] National law should not only recognize but also protect and promote basic tenure rights. This should include enabling tenure rights holders to secure their rights, even if these are not formalized, and ensuring the availability and accessibility of law enforcement institutions. The elements provide for the basic rights and entitlements of indigenous peoples, including participation in negotiation and decision making, the development of institutions that can represent customary rights, the maintenance and transmission of customary and traditional practices, and in this regard can form a part of statutory law for pastoral land use.

Governments can also use these elements to develop priorities for national development and sectoral policies. The assessments may also identify aspects of a legal framework that is not being properly implemented, such as legislation that allows the titling of land tenure rights held by communities that may not have been used. This approach may provide an opportunity for understanding impediments to implementation and pave the way to administrative and legislative reform. Assessments pertaining to the VGGT and the application of the following elements may expose areas of human rights concern, as well as other relevant matters in environmental law.[41] The elements outlined below can be included in a statutory law system to help protect rights and values in the use of pastoral land. Within both customary and statutory tenure systems, multiple and overlapping rights may govern the use of the same pastoral land resource. Tenure rights over common pastoral land resources, seasonal and otherwise temporary rights of access and use, as well as tenancy and sharecropping rights can all be legitimate tenure rights.

5.1.1 Relevant Institutional Framework

Creating an institutional framework for pastoralism must start from the legal recognition of the right to private and customary ownership of pastoral lands (individual, communal, cooperative, or any form of association), allowing free choice of the

[40]These principles are drawn from FAO (2016a).

[41]OHCHR (2011).

model of organization, forms of exploitation and destination within the boundaries and regulations of the law. In this regard, it is necessary to create tools that enable compliance with the government policy formulated for the sector. This may require establishing a government institution responsible for implementing a regulatory land tenure framework to resolve legal insecurity and instability for pastoralists and procure the enjoyment of their land rights. This institution should be established to monitor property processes and public records, the protection of natural resources, the acceptable use of pastoral land, and equity to access of land. It should have a broad administrative function that responds to the economic development of the country, with the recognition of rights to land tenure as its main purpose. The institution for pastoralism should have the power to regulate and implement actions related to training, promotion, and organization of pastoralist lands, in coordination with other governmental institutions.

Legislation can assist a pastoral land institution to make fair and just decisions that are consistent with land tenure arrangements. In addition, relevant legislation should contain powers that enable a pastoral land institution to take action against a person or a corporation for noncompliance with the legislation. Appropriate powers may provide for the modification or revoking of an authorization or permit, require remedial action to restore the pastoral land consistent with the land tenure, or stop an activity and require compliance with specified conditions or standards.

5.1.2 Transboundary Agreements

The VGGT establish that States should cooperate, in the framework of appropriate mechanisms and with the participation of affected parties, in addressing tenure issues that traverse national boundaries, ensuring that all actions are consistent with their existing obligations under national and international laws. As a general principle of international law, a State has a responsibility to ensure that any activities within its boundaries do not affect the integrity of pastoral land of another State,[42] and there should be appropriate procedures in national pastoral land legislation to implement the "common boundary" principle. In States where transboundary matters related to tenure rights arise, parties should work together to protect such tenure rights, livelihood, and food security of any migrating populations while on their respective territories. States should contribute to the understanding of transboundary tenure issues affecting communities and should harmonize legal standards of tenure governance, in accordance with existing obligations under national and international laws, and with due regard to traditional and indigenous rules (especially on mobility and seasonal movement), and voluntary commitments under relevant regional and international instruments. Where a State shares a common boundary with another State, or States, it should work together with

[42]United Nations (1992) Principle 2; Sands (2003).

those States to ensure that mobility corridors and seasonal routes remain accessible for pastoralist and traditional communities.[43]

Transboundary agreements negotiated at the national level should be implemented in close cooperation with local authorities and communities. Management of transhumance can be facilitated through close involvement of the local authorities on both sides of an international border.

5.1.3 Resolution of Transboundary Disputes

International law increasingly prioritizes cooperation and collaboration across boundaries. Nevertheless, States have the right to take action against another State for damage to its pastoral land arising from the transboundary effects of unsustainable land use and lack of good governance of its land tenure. The role of international law is to regulate relations and thus help to contain and avoid disputes in the first place. The substantial part of international law, therefore, does not concern dispute resolution but involves dispute and conflict avoidance.[44] It is concerned with the rights and duties of States in their relations with each other and with international organizations. The United Nations Charter is principally concerned with the preservation of world peace, including through various methods for resolving disputes peacefully.[45] These methods range from informal, non-binding, diplomatic methods through to formal and binding judicial settlement, including negotiation, inquiries, mediation and conciliation, arbitration, and judicial settlement. Where applicable, a State should establish procedures to resolve a dispute through a formal resolution process and to take legal action against another State for damage to its pastoral land arising from the transboundary effects of land use in the latter State.[46]

5.1.4 Pastoral Land Institution

The term "pastoral land institution" is used as a generic term. It can be taken to mean a single independent specialist pastoral land institution. It may also mean the "equivalent" of responsibilities and functions for the use of pastoral land found in a single specialist institution but administratively dispersed among a number of different government organizations or institutions, including customary institutions with a direct or indirect responsibility to land tenure (e.g., forestry, agriculture, and land administration). It is important that the legislation establish a duty of care and commitment to achieve the sustainable use of land and preserve the rights and

[43] Schulz (2007).

[44] Blay et al. (2005).

[45] See Article 33, United Nations Charter (1945).

[46] O'Connell (2015).

values of pastoralist peoples. This can be facilitated through well-defined responsibilities that can be spread across a number of organizations or legislative instruments. Particular "rights" and "obligations" may be established within an organizational hierarchy, and at respective levels of administration, for individuals or for specific classes of officials.

A pastoral land institution should preferably be an independent body with a broad range of functions and a dedicated budget, with the right to manage responsibilities in relation to the customary areas and other noncustomary areas of pastoral land.

5.1.5 Monitoring

States have a general obligation to monitor the condition and health of pastoral land and inform the community on a regular basis. Information should be provided to the public on a regular basis on the environmental condition of pastoral land. The results of monitoring can be used to systematically evaluate the performance of a pastoral land institution, which should also include an evaluation of the implementation of customary and traditional practices, policies, field programs, and research into the condition of pastoral land.[47] A monitoring and audit program could provide for the establishment of indicators of the ecological status of pastoral land, monitoring of the relationship between land use and land tenure, and monitoring of human issues, including poverty and customary land rights.

5.1.6 Community Participation in Pastoral Land-Use Decision Making

States should facilitate the participation of all sectors of the public in the use, management, and decision making related to pastoral land; in particular, the rights and interests of indigenous and traditional peoples must be considered. Consultation should be undertaken prior to decisions being made, and participation should be informed. The existence of opportunities for consultation and participation in decision making affecting tenure depends on multiple practical issues. However, legal frameworks can also influence those opportunities, for example through integrating legal requirements for local consultation in tenure decision-making processes. The law could make it a condition that all affected populations be consulted. Public participation can also improve the quality of law, for example by helping to ensure that legislation is tailored to the local context and to land tenure arrangements. It can increase the perceived legitimacy, the sense of ownership, and ultimately the effectiveness of legislation. It is important to recognize that indigenous people are also entitled to free, prior, and informed consent.

[47]Leake (2012).

Community participation programs should be used to enable any person to participate in the management of pastoral land, and the application of land tenure, from the local level to the State level. The links between effective participation, representation, and accountability can provide building blocks for effective participation.

5.1.7 Information for Interested Persons

Wherever a pastoral land institution proposes to act under its administrative procedures, all interested persons should be informed in a manner and with facts that will enable them to judge whether their rights, freedoms, and interests are affected, in particular those of indigenous communities. Procedures should be included to set out the manner in which the pastoral land institution will inform the public.

5.1.8 Right to Information

Any person should have a right of access to information held by the State on matters related to tenure of pastoral land and, in particular, the condition of pastoral land.[48] In order to improve security of tenure rights, the VGGT call for States to provide systems to record individual and collective tenure rights. In particular, everyone should be able to record their tenure rights and obtain information, and institutions should adopt simplified procedures and locally suitable technology to reduce the costs and time required for delivering services. Information on tenure rights should be easily available. Information on the state of the pastoral land may include data on tenure in written, visual, oral, digital, or database format.

5.1.9 Procedure to Obtain Information

Legislation should set out the procedures and circumstances under which a pastoral land institution should release information to the public, in particular where any existing or proposed action is likely to affect the ecological integrity of pastoral land. Key considerations include circumstances under which certain types of information may be protected or refused.

5.1.10 Pastoral Land Information and Knowledge

A primary responsibility of a pastoral land institution should be to collect, analyze, and record general information on pastoral land, including land-tenure-related

[48] Stec et al. (2000).

information. In particular, this applies to customary and traditional knowledge and values. Through this process, a pastoral land institution can acquire knowledge to plan and target land management operations. A pastoral land institution also has a basic responsibility to implement procedures to deter users from undertaking any act that may otherwise be undesirable, or possibly illegal.[49]

5.1.11 Mobility

Mobility remains a critical element for pastoral systems in many parts of the world. However, formulating legislation that supports the spatial and temporal aspects of natural resource use is an important challenge for pastoral land management. To support livestock mobility, a State should include procedures to:

- ensure that where customary law applies, all rights and responsibilities of customary people can be exercised accordingly and to safeguard their land tenure;
- ensure that the use of livestock mobility remains an important pastoral land management strategy;
- maintain livestock corridors and associated natural and artificial infrastructure, including water points;
- address animal health to prevent obstacles to pastoral mobility, paying particular attention to the effective control of livestock diseases (may be addressed specifically under disease control legislation) to minimize restrictions to livestock movement, bearing in mind the potential risks to mobility and herd management of veterinary cordon fences.

5.1.12 Enforcement

Enforcement can take a variety of approaches to ensure that the relevant law is complied with at a desired level or standard. Special provisions should be made with regard to customary and traditional lands. Compliance may be in the form of a direct obligation or a prescribed standard of behaviour or through a legal notice or order. Relevant laws can set out the procedures for enforcement and regulate specific activities that are inconsistent with the land tenure and not beneficial to pastoral land. Enforcement functions may include investigating offences, gathering evidence, taking remedial action, confiscating items, and initiating prosecution proceedings. Legislation would normally set out the range and limits of monetary penalty for offenses, as well as appeal provisions.[50]

Remedial actions can include civil liability. The main purpose of civil liability is to seek compensation, which is to restore the balance that existed before the

[49]Government of Australia (1989).

[50]AECEN (2015), OECD (2009).

violation occurred. It has a preventive aspect, which leads citizens to exercise caution to avoid compromising its responsibility, and a punitive aspect, of private consequences. Any person whose act or omission, voluntary or without malice, unlawfully or against good morals causes damage to another is obliged to repair the damage. The establishment of liability is to try and ensure the repair of the damage caused to the property, by trying to put things in the state they were in before the damage. For these reasons, the penalty of liability is, in principle, a type of compensation rather that repression.

5.1.13 Access to Justice

States need to procure the removal of normative, social, and economic obstacles that prevent or limit the possibility of access to justice. It refers to an effective judicial and administrative solutions and procedures available to a person (natural or legal) who is aggrieved or likely to be aggrieved by harm to the pastoral environment. The term includes not only the procedural right of appearing before an appropriate body but also the substantive right of compensation for harm done.

In many contemporary indigenous communities, dual justice systems exist. One is based on a statutory paradigm of justice, and the other is based on an indigenous paradigm. For many traditional societies, law and justice are part of a whole that prescribes a way of life. Relevant legislation should outline procedures for respective parties in legal proceedings. These should cover the following[51]:

- access to relevant information relating to breaches of the legislation, through freedom of information provisions;
- access to information regarding land tenure;
- provision of financial assistance for individuals and groups to bring civil enforcement actions;
- prosecution for breach of pastoral land legislation.

In some circumstances, a community service order may be appropriate. Such orders are sometimes used as an alternative to a fine in a variety of jurisdictions. They involve tasks carried out on a periodic basis as a contribution to a community and are often related to the nature of the offense.

5.1.14 Dispute Resolution

In addition to administrative, civil, and criminal proceedings, there should be formal procedures in legislation to resolve disputes over access to pastoral land.[52] Providing effective and legitimate ways to settle disputes between pastoralists and

[51] ACHPR (2003).

[52] Markell (2000).

between pastoralists and farmers is an important factor in protecting legitimate tenure rights and is one of the key functions of the law.

Competition over pastoral land can result in disputes over tenure rights. Disputes can take place within or between families or between individuals or communities and private companies. They can involve claims against the state and can arise over a number of issues, such as inheritance, boundaries, or transactions. States should provide access through impartial and competent judicial and administrative bodies to timely, affordable, and effective means of resolving disputes over tenure rights, including alternative means of resolving such disputes, and should provide effective remedies and a right to appeal. States should also make available mechanisms to avoid or resolve potential disputes at the preliminary stage, either within the implementing institution or externally. Moreover, multiple tenure systems may coexist in the same territory, including statutory and customary systems. Alongside formal court systems, there may be nonstate systems for adjudicating tenure conflicts, including customary systems and alternative dispute resolution mechanisms.

Methods of dispute resolution include negotiation, inquiries, mediation and conciliation, arbitration, and judicial settlement. States should also consider introducing specialized tribunals that deal solely with disputes over the use of pastoral land. Where customary or other established forms of dispute settlement exist, they should provide for fair, reliable, accessible, and nondiscriminatory ways of promptly resolving disputes over tenure rights. Mediation can be an alternative to court action to resolve disputes.

6 Conclusion

Secure tenure plays an essential role in the sustainable use of pastoral land. In a rapidly changing environment with emerging issues such as climate change, rising insecurities, land privatization, and diminishing resources, the need to strengthen responsible governance of tenure has never been more important for pastoral land. Essential elements for sustainable pastoralism such as securing customary rights and mobility have to be incorporated at various levels of natural resource management, particularly the management of its soil resources. It is also critical to strengthen specific action areas, including effective participation of pastoral communities in decision making while implementing tenure solutions. In addition, legal frameworks can support legislation relevant to pastoralism. In some countries, these types of legislation are already in place, although not always implemented.

References

ACHPR (2003) Protocol to the African Charter on Human and Peoples' Rights on the Establishment of the African Court on Human and Peoples' Rights. http://www.achpr.org/instruments/court-establishment/

AECEN (2015) Asian Environmental Compliance and Enforcement Network. http://www.aecen.org/

Alden-Wily L (2012) Rights to Resources in Crisis: Reviewing the Fate of Customary Tenure in Africa - Brief 1 of 5, Vol. 1. http://www.rightsandresources.org/documents/files/doc_4699.pdf

Allen VG, Batello C, Berretta EC, Hodgson J, Kothmann M, Li X, McIvor J et al (2011) An international terminology for grazing lands and grazing animals. Grass Forage Sci 66(1):2–28. https://doi.org/10.1111/j.1365-2494.2010.00780.x

Behnke R, Freudenberger M (2013) Pastoral Land Rights and Resource Governance, Overview and Recommendations for Managing Conflicts and Strengthening Pastoralists' Rights. http://www.usaidlandtenure.net/sites/default/files/USAID_Land_Tenure_2012_Washington_Course_Module_3_Land_Tenure_Pastoral_Land_Rights_and_Resource_Governance_Brief.pdf

Benvenisti E (2002) Transnational institutions for transboundary ecosystem management: defining the tasks and the constraints. In: Sharing transboundary resources: international law and optimal resource use. Cambridge Univeristy Press, Cambridge

Blay S, Piotrowicz R, Tsamenyi B (eds) (2005) Public international law: an Australian perspective. Oxford University Press, Oxford

Bosselmann K, Engel R, Taylor P (2008) Governance for Sustainability – Issues, Challenges, Successes. Gland. http://cmsdata.iucn.org/downloads/eplp_70_governance_for_sustainability.pdf

Davies J, Hatfield R (2007) The economics of mobile pastoralism: a global summary. Nomadic Peoples 11(1):91–116

Davies, J, Ogali C, Laban P, Metternicht G (2015) Homing in on the Range: Enabling Investments for Sustainable Land Management. Nairobi. http://cmsdata.iucn.org/downloads/technical_brief___investing_in_slm_2.pdf

de Sadeleer N (2002) Environmental principles -from political slogans to legal rules. Oxford University Press, Oxford

FAO (2016a) Improving governance of pastoral land. In: Davies J, Herrera P, Ruiz-Mirazo J, Mohamed-Katerere J, Hannam I, Nuesri E (eds) Implementing the Voluntary Guidelines on the Responsible Governance of Tenure of Land, Fisheries and Forests in the Context of National Food Security. FAO, Rome

FAO (2016b) Cotula L, Berger T, Knight R, McInerney T, Vidar M, Deupmann P, Responsible governance of tenure and the law, a guide for lawyers and other legal service providers. FAO, Rome. http://www.fao.org/3/a-i5449e.pdf

Government of Australia (1989) South Australia Pastoral Land Management and Conservation Act 1989. http://www5.austlii.edu.au/au/legis/sa/consol_act/plmaca1989384/.

Hart S (2008) Shared resources: issues of governance. IUCN Gland, Switzerland 249 pp

Helsinki Convention on the Protection and Use of Transboundary Watercourses and International Lakes (1992)

Herrera P, Davies J Manzano Baena P (2014) Governance of rangelands: collective action for sustainable pastoralism. Routledge, London

ITPGRFA (2009) International Treaty on Plant Genetic Resources for Food and Agriculture. ftp://ftp.fao.org/docrep/fao/011/i0510e/i0510e.pdf

Leake J (2012) Conclusions and a way forward. In: Squires V (ed) Rangeland stewardship in Central Asia, balancing improved livelihoods, biodiversity conservation and land protection, vol 442. Springer, London

Licht A, Goldschmidt C, Schwartz S (2007) Culture rules: the foundations of the rule of law and other norms of governance. J Comp Econ 35(4) Elsevier: 659–88. http://econpapers.repec.org/RePEc:eee:jcecon:v:35:y:2007:i:4:p:659-688

Markell DL (2000) The Role of Deterrence-Based Enforcement in a 'Reinvented' State/Federal Relationship: The Divide between Theory and Reality. http://papers.ssrn.com/abstract=1547897

Mortimore M, Anderson S, Cotula L, Davies J, Faccer K, Hesse C Morton J, Nyangena W, Skinner J, Wolfangel C (2009) Dryland opportunities: a new paradigm for people, ecosystems and development. Nairobi. https://portals.iucn.org/library/efiles/documents/2009-033.pdf

Niamir-Fuller M (1999) Managing mobility in African rangelands: the legitimization of transhumance. Intermediate Technology Publications, London

O'Connell ME (2015) Enforcement and the Success of International Environmental Law. http://www.researchgate.net/publication/254620399_Enforcement_and_the_Success_of_International_Environmental_Law

OECD (2009) Ensuring Environmental Compliance Trends and Good Practices. http://browse.oecdbookshop.org/oecd/pdfs/product/9709031e.pdf

OHCHR (2011) UN Guiding Principles on Business and Human Rights, Implementing the United Nations 'Protect, Respect and Remedy' Framework, New York and Geneva. http://www.ohchr.org/Documents/Publications/GuidingPrinciplesBusinessHR_EN.pdf

RIRDC (2014) Managing Indigenous Pastoral Lands, Module 3 Land Information. https://rirdc.infoservices.com.au/items/14-019

Sands P (2003) Principles of international environmental law. Cambridge

Schulz A (2007) Creating a legal framework for good transboundary water governance in the Zambezi and Incomati river basins. Georgetown Int Environ Law Rev 19(2):117–183

Stec S, Casey-Lefkowitz S, Jendroska J (2000) The Aarhus Convention, securing citizen's rights, through access to information, public participation and access to justice for a healthy environment, an implementation guide. New York

UNCLOS (1982) United Nations Convention of the Law of the Sea. http://www.un.org/depts/los/convention_agreements/texts/unclos/unclos_e.pdf

UNESCO (1997) Declaration on the Responsibilities of the Present Generations Towards Future Generations. 12 November 1997. http://portal.unesco.org/en/ev.php-URL_ID=13178&URL_DO=DO_TOPIC&URL_SECTION=201.html

UNEP (1995a) United Nations Convention on Biological Diversity, Nairobi

UNEP (1995b) United Nations Framework Convention on Climate Change

United Nations (1992) Report on the United Nations Conference on Environment and Development, Rio Declaration on Environment and Development. http://www.un.org/documents/ga/conf151/aconf15126-1annex1.htm

VGGT (2012) Voluntary guidelines on the responsible governance of tenure of land, fisheries and forests in the Context of National Food Security. FAO, Rome

Windfuhr M (2017) Voluntary guidelines on responsible governance of tenure of land, forests and fisheries – relevance, reception and first experiences in implementation. In: Ginzky H, Heuser I, Tianbao Q, Ruppel O, Wegerdt P (eds) International yearbook of soil law and policy 2016. Springer International Publishing

World Summit for Social Development (1995) Report of the World Summit for Social Development, A/CONF.166/9, Copenhagen Denmark

Uncertainty of Land Tenure and the Effects of Sustainability if Agriculture in the United States

Jesse J. Richardson

> Changing land tenure patterns [are] considered as important as soil conservation programs in stopping serious rates of soil erosion.[1]

1 Introduction

In simple terms, land tenure systems determine who can use what resources for how long, and under what conditions.[2] The security of land tenure plays a key role not only in the productivity of the land in agriculture but also in whether the land will be farmed sustainably. A landowner that feels uncertain or insecure as to whether they can remain on the property or whether they can make decisions relating to the use of the property will be unlikely to invest in sustainable practices that take a number of years to positively impact profits.

While a wealth of literature exists on land tenure and sustainability in developing countries, little addresses the impact of land tenure on sustainable agriculture in the United States. This chapter begins to fill this gap by addressing the issue of land tenure and sustainable agriculture in the United States context. More accurately, the chapter addresses the link between the security of land tenure and sustainable

---

This chapter is derived, in part, from Richardson, Jr., Jesse J. (2016). Land Tenure and Sustainable Agriculture, Texas A&M Law Review 3:799–826. This chapter updates the earlier article, condenses much of the material, and redirects the material to the issues of soil conservation.

[1]Parsons et al. (2010), p. 5.

[2]FAO (2002).

J.J. Richardson (✉)
West Virginia University College of Law, Land Use and Sustainable Development Law Clinic, Morgantown, WV, USA
e-mail: jesse.richardson@mail.wvu.edu

H. Ginzky et al. (eds.), *International Yearbook of Soil Law and Policy 2017*,
International Yearbook of Soil Law and Policy,
https://doi.org/10.1007/978-3-319-68885-5_8

agriculture. The question of whether certain legal issues increase uncertainty of land tenure, thereby reducing the sustainability of agricultural practices, is presented and explored.

While little exists addressing this issue generally, the existing literature focuses mainly on leasing. Given the predominance of leasing issues in both the literature and in practice, this chapter focuses on the security of land tenure in the leasing context. Two other land tenure issues will be addressed, albeit in less detail: conservation easements and heirs property. These issues were chosen based on the author's perceptions as to the predominance and importance of the issues in United States agriculture. Although one law review article addresses these three issues as connected to land tenure security and sustainability, that article was written by the author.[3]

Of the few other sources addressing the issue, one lists the following as priorities for farmland tenure in the United States, displayed in no particular order:

- farm succession;
- complexity of land transactions;
- innovative ownership models and financing;
- leases;*
- conservation easements;
- tax;
- liability;
- insurance;
- zoning;
- environment.[4]

Some of these issues are not traditionally considered as part of land tenure (environmental and tax, for example). However, the list confirms the choices of leases and conservation easements as significant issues in the United States. Another source includes heirs property as a significant land tenure issue.[5] The chapter concludes with recommendations for future research.

2 Land Tenure and Land Tenure Security

"Land tenure is the perceived institutional arrangement of rules, principles, procedures, and practices, whereby a society or community defines control over, access to, management of, exploitation of, and use of means of existence and production."[6] Stated more simply, land tenure refers to the perceived right to hold land.[7] Land

[3]Richardson (2016).

[4]Vermont Law School (2015).

[5]Parsons et al. (2010), p. 16.

[6]Dekker (2006), p. 1.

[7]Ibid.

tenure security refers to the perceived certainty of having well-defined rights to land for a defined time period.[8] Land tenure relationships may be well defined and enforceable in a formal court of law or through customary structures in a community. Alternatively, they may be relatively poorly defined with ambiguities open to exploitation.[9]

Dekker further defines land tenure as "the assurance of continuing access to land for a certain period of time."[10] Access to land does not necessarily entail control over land. Access means the ability to make use of the land, not necessarily ownership or possession.[11] Usually some decision making is included with access.[12] Control involves the "command that an individual has over a particular piece of land."[13]

For agricultural production to occur, land tenure security must exist for an extended period of time.[14] In addition, land tenure security for a long term period gives the person with access an incentive to make capital improvements and manage the land in an environmentally sustainable fashion.[15] Without land tenure security, no incentives exist to invest anything into the property that will not be paid back immediately with increased revenues. The production has a short-term vision with no implementation of long-term sustainability measures.

Each of these concepts discussed in this chapter affects land tenure security and introduces uncertainty into the agricultural operation, calling into question the future of the operation and prompting decisions that negatively impact the sustainability of the operation. Uncertainty in land tenure constitutes a major factor in determining whether conservation practices will be adopted by the operator.[16] Most of these concerns have been raised in connection with lease arrangements, but uncertainty in other types of land tenure may have similar effects. Securing a clear,

---

[8]Ibid.

[9]FAO (2002).

[10]Dekker (2006), p. 5.

[11]Ibid.

[12]Ibid.

[13]Ibid.

[14]Dekker (2006), p. 3.

[15]Dekker (2006), p. 4.

[16]Cox (2010), pp. 369, 370–371; Clearfield and Osgood (1986), http://www.ssi.nrcs.usda.gov/publications/2_Tech_Reports/T014_Adoption01Main.pdf; Arbuckle et al. (2009), pp. 73, 74; Soule et al. (2000), pp. 993, 993–94, 1003 [hereinafter Soule et al., Land tenure]; see also Arbuckle (2010), available at http://www.soc.iastate.edu/extension/ifrlp/PDF/PMR1006.pdf (noting ownership plays a role in the environmental effects of farming); Duffy et al. (rev. 2008), available at http://www.extension.iastate.edu/Publications/PM1983.pdf [hereinafter Duffy et al., Farmland ownership] (discussing length of tenure and the effect on soil conservation); Carolan (2005), pp. 387, 398 (noting there is more incentive for conservation if leases are for multiple growing seasons); cf. Lee and Stewart (1983), pp. 256, 257 (noting tenure arrangements that separate ownership from operation can hinder conservation).

predictable, and affordable land tenure arrangement is the foundation of economically viable small farming operations.[17]

This connection between land tenure security and sustainable agricultural practices is well established, even in the United States. During the 1930s, portions of the United States and Canada experienced severe drought. Combined with farming practices that failed to consider soil conservation, huge amounts of wind erosion occurred, causing the soil to turn to dust and create large dust clouds. The so-called Dust Bowl affected 400,000 km^2 in the United States and displaced tens of thousands of families.

After this catastrophic event, the national government of the United States responded with a number of public policies to stop soil erosion. These policies included programs that sought to help farmers who lease their land purchase farms and programs to help find new land for families that had lost their farms for failing to repay bank loans. "Changing land tenure patterns were considered as important as soil conservation programs in stopping serious rates of soil erosion."[18]

"How farm land is acquired, held in ownership, operated or rented has always been a matter of national interest, for just and fair conditions of tenure are recognized as essential to our national welfare. . .Obviously it is of vital concern to the nation that. . .types of land ownership and operation be developed that will be conducive to a permanent agriculture and to strong rural communities."[19] The remainder of this chapter explores the connection between leasing, conservation easements, and heirs property to sustainable agriculture generally and, more specifically, to soil conservation.

3 Leasing

3.1 Introduction

Upon declaring its independence from England, the United States rejected the feudalistic land tenure patterns of England. Private ownership became the dominant form of land ownership, with leasing a possibility as well. United States land policy often involved transferring public lands to private landowners as an incentive to settle the frontier.[20] This heritage continues to influence American views on land tenure, which tend to strongly favor land ownership.[21]

---

[17]https://casfs.ucsc.edu/about/publications/Teaching-Direct-Marketing/pdf%20downloads/Unit.9.pdf.

[18]Parsons et al. (2010), p. 5.

[19]Clark (1944).

[20]Salamon (1998), p. 160.

[21]Ibid.

In 1940, tenant farmers tilled nearly 40% of the nations' farmland. President Roosevelt's Committee on Farm Tenancy viewed absentee ownership and the prevalence of landless farm families as the cause of environmental and social ills. States responded by passing laws that favored land ownership over leasing, including a ban on long-term leases in some states and a limit on the length of leases in others. Policy responses to the Dust Bowl included a series of federal programs that sought to help tenant farmers purchase land to farm and to resettle families that had lost their farms to foreclosure.

3.2 Predominance and Nature of Leasing in the United States

Farm operators generally operated owned land until the 1950s, when farmers started to become renters.[22] Some commentators assert that part owners are becoming the norm, while others maintain that the proportion of leased land is decreasing.[23]

Three types of farm landowners have been identified.[24] "Owner-operator" refers to landowners that farm all or a part of the land they own, while "operator-landlord" refers to a farmer that leases some of the land they own to other farmers.[25] Finally, "non-operator landlords" lease land they own to farmers but do not farm themselves.[26]

Over 50% of farmed land in the United States is part of an operation that farms both owned and leased property.[27] The proportion of farmland acres farmed by tenant farmers has decreased dramatically, from 32% in 1935 to 10% in 2012.[28] Conversely, the number of acres farmed by operators that farm some owned land and some leased land increased from 25% in 1935 to 54% in 2012.[29]

Tenure varies by region, type of farm operation, size of farm, and age of operator.[30] The Midwest (54%) and Plains (57%) show the lowest rates of farmland ownership, while the West and Northeast regions rank highest, both with 71% of acres farmed by owner-operators.[31] Note, however, that regional variations exist.[32]

[22]Parsons (2012), p. 12.

[23]Id.

[24]Bigelow et al. (2016), p. 3.

[25]Ibid.

[26]Ibid.

[27]Bigelow et al. (2016), p. 5.

[28]Ibid.

[29]Ibid.

[30]Bigelow et al. (2016), pp. 6–11.

[31]Bigelow et al. (2016), p. 6.

[32]Ibid.

Cash-grain operators and cotton growers rent most of their land.[33] In particular, cotton and rice growers predominately rent land.[34] Both rice and cotton require irrigation, and water rights may be costly.[35] Relatively little land is suitable for rice production. The consequent high rental rates for rice production land and cost of water rights likely combine to make purchase of land for rice unattainable for most producers.[36] For cotton growers, expensive equipment is needed, and operators may rent additional land to gain economies of scale.[37] Livestock producers, on the other hand, lease only about 25% of their land, likely due to low rental rates for pastureland.[38]

Farm size is positively correlated with percentage of leased acreage. Small family farms lease about 31% of their acreage.[39] Forty-six percent of these producers own all of the land tilled, while 47% farm both owned and leased land.[40] Only 7% of producers in this category farm only leased land.[41]

As farm size increases, a larger proportion of acres are leased. Forty-eight percent of land in midsize, large, and very large family farms is leased.[42] Part owner operations comprise 74% each of midsize and very large family operations and 69% of large family operations.[43] As family operations grow in size, a larger proportion of the operations farm only leased land, with 8, 10, and 16% of operations falling into that category for midsized, large, and very large operations, respectively.[44]

For obvious reasons, age also correlates with proportion of leased acres. Young farmers use leasing to enter the occupation at lower cost and lower risk. Producers under age 34 comprise the group with the lowest proportion of full-owner operations (8%) and the highest proportion of full-tenant operations (27%).[45] Conversely, farm operators over the age of 65 engage in full-tenant operations only 7% of the time, while 43% of operations for this age group are characterized as full-owner operations.[46]

---

[33]Bigelow et al. (2016), p. 7.

[34]Ibid.

[35]Bigelow et al. (2016), pp. 7–8.

[36]Bigelow et al. (2016), pp. 7–8.

[37]Ibid.

[38]Ibid, at 8.

[39]Bigelow et al. (2016), p. 9.

[40]Ibid.

[41]Ibid.

[42]Bigelow et al. (2016), p. 10.

[43]Ibid.

[44]Ibid.

[45]Ibid.

[46]Ibid.

3.3 *Absentee and Nonproducer Landlords*

In general, little information is known on the landlords of farmland. In 2014, for the first time since 1999, the United States Department of Agriculture (USDA) conducted a survey that sought information on these landlords. The survey, Tenure, Ownership, and Transition of Agricultural Land (TOTAL), reveals updated and enlightening information. All states were surveyed, but the top 25 farm cash receipt states[47] were surveyed in more depth. These states represent 85% of farm cash receipts, 70% of land in agriculture, and 78% of rented land in agriculture in the United States.[48]

Most rented land is rented from nonfarmer landlords. Of the 911 million acres of farmland in the contiguous 48 states, 39% (354 million acres) is rented, with 31% (about 283 million acres) rented from nonoperator landlords and only 8% (about 70 million acres) from landlords who also farm.[49] Nonoperator landlords comprise 87% of farm landlords and own 80% of the rented farmland.[50]

Seventy percent of farm leases are fixed cash leases, which means that the tenant operator pays a predetermined amount of cash rent annually.[51] While 70% of land rented by operator landlords has been rented to the same tenant for over 3 years and 28% for over 10 years, with 81% and 41%, respectively, for nonoperator landlords, leases tend to be negotiated every year.[52] Fifty-seven percent of rented acres are rented under leases that are renegotiated annually.[53] Some suggest that this lease arrangement may give incentives to engage in long-term conservation practices, similar to a long-term lease.[54] In contrast, 28% of rented acres are rented under leases that are renegotiated every 4 or more years.[55]

Long-term leases give incentives for tenants to manage the land for future productivity.[56] Conversely, a long-term lease proves problematic in periods of market volatility or where the landlord–tenant relationship is not well established.[57] In fact, long-term leases are much more likely where the landlord–tenant relationship has existed for a number of years.[58] Short-term leases may turn into long-term leases after the landlord and tenant have built a relationship.

---

[47]Alabama, Arkansas, California, Florida, Georgia, Idaho, Illinois, Indiana, Iowa, Kansas, Kentucky, Michigan, Minnesota, Mississippi, Missouri, Nebraska, North Carolina, North Dakota, Ohio, Oklahoma, Pennsylvania, South Dakota, Texas, Washington, and Wisconsin.

[48]Bigelow et al. (2016), p. 12.

[49]Bigelow et al. (2016), p. 15.

[50]Bigelow et al. (2016), p. 17.

[51]Bigelow et al. (2016), p. 22.

[52]Bigelow et al. (2016), p. 25.

[53]Ibid.

[54]Bigelow et al. (2016), p. 27.

[55]Bigelow et al. (2016), p. 26.

[56]Bigelow et al. (2016), p. 26.

[57]Ibid.

[58]Bigelow et al. (2016), pp. 26–27.

3.4 Leasing and Sustainable Agriculture

> If the American farm owner's relationship to his farm is weak, it is practically non-existent in the case of tenants.[59]

Leasing implicates the "tenancy hypothesis"—the theory that tenants have little incentive to make long-term investments in the property since the tenant has no stake in the land beyond the term of the lease.[60] The tenancy hypothesis resembles the "impermanence syndrome" in several ways. The impermanence syndrome, generally discussed in connection with agricultural operations on the rural-urban fringe, refers to an observed reluctance of farm operators to make investments in the business due to perceived threats.[61] The perceived threat generally consists of urban expansion infringing upon farm operations.[62] This threat sometimes accompanies a short-term (usually a year-to-year) lease, where the landlord may sell the property when the urban expansion meets the farmland being rented. In the case of the impermanence syndrome, both the landlord and the tenant display a reluctance to make long-term investments in the property for agricultural purposes, feeling that the farm operation is a placeholder waiting for urban development to replace the activity.

Leased agricultural land raises sustainability concerns due to the lack of incentives by the renter to invest in long-term measures to enrich the soil and otherwise improve the property.[63] In addition, trends relating to the lease terms and the relationship of the landlord to the land and the tenant exacerbate these concerns.[64]

Tenants seem to make decisions based on a shorter term view than landlords.[65] However, Knowler and Bradshaw (2007) find mixed results as to whether landlords are more likely to engage in conservation practices. "A tenant does not put forth his best efforts when he feels the insecurity that accompanies ownership by persons who not only may be remote from the locality but may have limited interest in the farm and the farm family and may not be well informed on agricultural matters."[66]

However, the tenant's confidence in the landlord's commitment to continue the lease greatly affects the uncertainty that tenants perceive with respect to farm leases.[67] The factors impacting this confidence, or lack thereof, include the distance between the landlord's home and the leased property, the owner's connection to agriculture, and the social relationship between the landlord and tenant.[68] Although

[59]R.T. Ely and G.S. Wehrwein, Land Economics 1940, quoted in Parsons (2010), p. 48.

[60]Lichtenberg (2007), p. 294.

[61]Land Use Planning and Development Regulation Law § 13:3, fn. 4.

[62]Ibid.

[63]Cox (2010), pp. 370–371.

[64]Ibid.

[65]Soule et al. (2000); Sklenicka et al. (2015).

[66]Clark (1944).

[67]Arbuckle (2010), p. 1.

[68]Ibid.

some trends are uncertain, one clear trend emerges—landlords are less likely to live close to the rented land, and, consequently, the rate of absentee landlords is increasing. Most landlords still live within 25 miles of the farm operation.[69] However, in Iowa, for example, only 6% of farm landlords lived out of state at least part time in 1982. By 2007, the rate had increased to 21%[70] (Duffy and Smith 2009). These trends also indicate that the landlord's involvement in making decisions on the property will likely decrease as well.[71]

Another factor impacting tenure insecurity among renters involves the term, or length, of the lease. Most farm leases consist of short-term, oral agreements, often year to year.[72] An Iowa study reveals that one third of the farm leases in that state were oral leases.[73] Another Iowa study indicates an even more disturbing situation, with only 25% of farm leases in that state setting a fixed term.[74] By default, the remaining 75% of leases amount to year-to-year terms. Short-term leases increase uncertainty for the renter.[75] The renter is unlikely to make long-term investments in the land, and banks generally will not extend credit.[76] Short-term lessees also are less likely to engage in sustainable practices.[77]

Short-term leases present obstacles to enrollment in government programs encouraging sustainable practices as well.[78] Long-term programs, like the Conservation Reserve Program, require that the applicant be in control of the property for the duration of the program.[79] Without the involvement of the landlord, this requirement likely prevents tenants from participation.[80] Given the lack of incentive for short-term lessees to invest in the property, fewer sustainable practices will likely result.[81]

The tenancy hypothesis posits that tenants have little incentive to engage in long-term conservation practices since the tenant has no stake in the farm past the term of the lease.[82] Owner-operators of soybean and corn operations are more likely to use medium-term conservation practices than tenants, and use of conservation tillage as a short-term residue management is affected by the type of leasing arrangement.[83]

---

[69]USDA National Agricultural Statistics Service (NASS) (1999).

[70]Duffy and Smith (2009).

[71]Cox (2010), pp. 382–383.

[72]Parsons et al. (2010), p. 13; Grossman (2000), pp. 127–128.

[73]Duffy and Smith (2009).

[74]Duffy (2008), p. 2.

[75]Parsons et al. (2010), p. 13.

[76]Ibid, 14.

[77]Higby; Soule et al. (2000).

[78]Cox (2010), p. 387.

[79]Ibid.

[80]Ibid.

[81]Cox (2010), pp. 382–383.

[82]Lichtenberg (2007).

[83]Soule et al. (2000).

Lessees in British Columbia used “less sustainable” crop rotations than owners.[84] Insecurity of tenure negatively related to sustainable agricultural practices.[85]

To increase certainty and security in farm leases, several issues should be addressed. The term of the lease clearly provides an important indicator of security and certainty.[86] While many farm leases actually remain in effect for many years, year-to-year leases subject to renewal or cancelation annually fail to promote tenant security and certainty.[87] However, the desire of the tenant to receive very early notice of termination must be balanced with the landlord’s desire to adjust the lease terms in response to changing circumstances at least annually.[88]

Cash rents discourage long-term leases because cash rents are not adjusted yearly.[89] To address this issue, an index may be used to adjust the rent a crop share or flexible cash lease can be adopted.[90] A flexible cash lease bases the rent on the crop yield, market price for the products raised, or both.[91]

The ability to sell the farm during the lease term also inhibits long-term leases.[92] A lease provision terminates the lease upon sale, while providing for payment to the tenant in such case addresses this issue.[93] However, such a provision also reduces certainty and security for the tenant.

Landowners may balk at entering into a lease with an unfamiliar tenant.[94] Default provisions within the lease address these concerns.[95] These provisions should not reduce certainty and security for the tenant, so long as the tenant feels confident that the lease requirements can be met.

Tenants under short-term leases possess little or no incentive to make long-term improvements to the property, including enriching the soil or building structures or infrastructure, as these improvements belong to the landlord upon termination of the lease.[96] Lease provisions that provide that the landlord and tenant share the costs of such improvements increase the opportunities to make such long-term improvements.[97]

Cost-sharing provisions in the lease agreement also may give incentives for the tenant to use sustainable practices.[98] Leases that provide that landlords and tenants

[84]Fraser (2004).

[85]Carolan (2005) and Carolan et al. (2004).

[86]Cox (2011), p. 13.

[87]Cox (2011), p. 17.

[88]Ibid, 19.

[89]Ibid.

[90]Cox (2011), p. 14.

[91]Ibid.

[92]Cox (2011), p. 15.

[93]Ibid.

[94]Ibid.

[95]Ibid.

[96]Cox (2011), p. 21.

[97]Ibid.

[98]Ibid, 22.

share risk also increase the security and certainty of the tenant.[99] Crop share leases and flexible leases both accomplish this risk sharing.[100] Note, however, that lease agreements that provide for risk sharing, cost sharing, and/or profit sharing must be carefully drafted to avoid characterizing the relationship as a partnership. Partners in a partnership are personally liable for any debts or liabilities incurred by other partners in the scope of the business.

The present state of leasing, with oral, short-term leases, administered by often disinterested and absentee landlords, tends to promote short-term decision-making horizons. Consequently, sustainable agricultural measures and soil conservation are not likely to be undertaken by the tenant or the landlord. Implementing these changes will provide the tenant more certainty as to the length and scope of their land tenure. Consequently, both landlord and tenant will make decisions on a more long-term horizon, promoting sustainability and soil conservation.

4 Conservation Easements

4.1 Introduction

The Uniform Conservation Easement Act defines "conservation easement" as "a nonpossessory interest of a holder in real property imposing limitations or affirmative obligations the purposes of which include retaining or protecting natural, scenic, or open-space values of real property, assuring its availability for agricultural, forest, recreational, or open-space use, protecting natural resources, maintaining or enhancing air or water quality, or preserving the historical, architectural, archaeological, or cultural aspects of real property."[101] A conservation easement protects conservation values on a particular property by placing restrictions on the development of the property.[102]

The Internal Revenue Code provides a federal income tax deduction for the donation of conservation easements.[103] A "qualified conservation contribution" involves a contribution of a "qualified real property interest" to a "qualified organization" "exclusively for conservation purposes."[104] A qualified real property interest includes "a restriction (granted in perpetuity) on the use which can be made of the real property."[105] Qualified organizations include charitable organizations and government agencies.[106]

[99] Ibid, 23.

[100] Ibid.

[101] Unif. Conservation Easement Act § 1(1) (2007).

[102] Byers and Ponte (2005), p. 14.

[103] 26 U.S.C.A. §170(h).

[104] 26 U.S.C.A. §170(h)(1).

[105] 26 U.S.C.A. §170(h)(2)(C).

[106] 26 U.S.C.A. §170(h)(3).

The requirement that the easement be "exclusively for conservation purposes" entails perpetual restrictions.[107] The provisions categorize conservation purposes as (1) "outdoor recreation by, or education of, the general public,[108] protection of ecosystems and habitats,[109] open space preservation,[110] and historic preservation".[111] The rules include farmland and forestland under the category of "open space."[112] Most farmland conservations, consequently, use the open space rules to qualify for the federal income tax benefits.[113]

To meet the conservation purpose requirement, open space preservation must (1) be pursuant to a clearly delineated governmental policy or be "for the scenic enjoyment of the general public"[114] and (2) "yield a significant public benefit."[115] The regulation provides that preserving farmland pursuant to a flood prevention and control policy illustrates the preservation of qualifying open space pursuant to a governmental policy and provides a public benefit.[116]

The conservation purposes examples laid out in the regulations include one focused on farmland. Example 5 involves a state statute that authorizes the purchase of "agricultural land development rights" to protect a loss of open space appropriate for agriculture.[117] The donation of a conservation easement that prohibits or limits nonagricultural buildings and dwellings not used by the farm operator or employees qualifies for the federal tax benefits.[118] Neither agriculture nor farmland is defined in the code or regulations addressing conservation easements.

4.2 Conservation Easements and Land Tenure

4.2.1 Overview

Although the term "conservation easement" evokes a notion of sustainability, the term is a misnomer in practice. A conservation easement does not require any sustainable practices on the land. Instead, the easement focuses on prohibiting development. With respect to so-called agricultural conservation easements, certain

[107] 26 U.S.C.A. §170(h)(5)(A).
[108] 26 U.S.C.A. §170(h)(4)(A)(i).
[109] 26 U.S.C.A. §170(h)(4)(A)(ii).
[110] 26 U.S.C.A. §170(h)(4)(A)(iii).
[111] 26 U.S.C.A. §170(h)(4)(A)(iv).
[112] 26 U.S.C.A. §170(h)(4)(A)(iii).
[113] Gentry (2013), p. 1395.
[114] IRS Regs. §1.170A-14(d)(4)(i)(B).
[115] IRS Regs. §1.170A-14(d)(4)(i).
[116] IRS Regs. §1.170A-14(d)(4)(B).
[117] IRS Regs. §1.170A-14(f)(4) Example 5.
[118] Ibid.

standard conservation easement provisions, such as the prohibitions on land disturbances and subdivision, and the inability to amend easements, create land tenure uncertainty. This uncertainty creates the paradox that a conservation easement may reduce the implementation of sustainable practices since the landowner is unsure as to the score of their ability to make decisions on land-use practices.

Conservation easements may affect land tenure positively. By (theoretically) reducing the value of the property, the easement may make purchase of the land possible for beginning farmers or farmers ready to transition from leasing land to becoming a landowner. In practice, the price of the land may not always decrease, or may not decrease significantly. Conservation easements may not deliver all that proponents claim. The value of conservation easement properties sometimes increases due to the value as a country estate.[119] Conservation easements also fail to guarantee that land will remain in agriculture.[120]

Conservation easements may also present barriers to land tenure security. Vermont Law School (2015) identified several land tenure security issues related to conservation easement land in agricultural use:

- amending easements;
- inability to subdivide large parcels that are too big for small farms;
- inability to add infrastructure;
- lack of flexibility for diversified operations;
- lack of information on how conservation easements are created;
- easement terms that do not support farming;
- information barriers to funding a conservation easement.

An additional issue involves the often conflicting purposes of preventing development and promoting productive agriculture, although this issue may be included as "easement terms that do not support farming."[121] This chapter will discuss all of these issues briefly but will focus more attention to the issues of amending easements and the issue of easement terms that do not support farming/conflicting purposes of easements.

Most conservation easements contain a provision that bars subdivision of the property. Land trusts and other groups generally prefer to hold large parcels, believing that larger parcels promote conservation benefits more easily. The bar on subdivision appears to mainly address subdivisions to add habitable dwellings. Easements that allow or promote agricultural production could be drafted to allow subdivision, and rely on the limit on habitable dwellings to insure that additional dwellings are not added. However, if different parcels are held by different owners, pressure may exist to allow each owner a dwelling unit.

Conservation easements also generally limit impermeable surfaces, land disturbances, and structures. These limits restrict the flexibility of farm operators who

---

[119] Parsons et al. (2010), p. 18.

[120] Ibid.

[121] Gentry (2013), p. 1396.

wish to add infrastructure that would promote different farm enterprises. A more careful drafting that would allow agricultural infrastructure without impairing conservation values may aid the producer without harming conservation objectives.

A related issue involves diversified operations that may wish to engage in innovative activities that were not contemplated when the easement was placed on the property. Conservation easements are generally perpetual and, as discussed in Sect. 4.2.2, prove to be difficult to amend. More flexible drafting may aid this issue.

Lack of information on how to create conservation easements and barriers to information on funding conservation easements both point to a need for more education on conservation easements. This education must also be targeted to provide the information needed by producers. This concern also relates to another land tenure issue identified in the same report. The complexity of land transactions generally impedes land tenure in agriculture.[122] Conservation easements are extremely complex land transactions.

4.2.2 Amending Conservation Easements

> Most conservation easements are written to last in perpetuity. Any change to any conservation easement should be approached with great caution and careful scrutiny.[123]

In theory, if conditions change and the conservation easement inhibits sustainable agriculture, the conservation easement may be changed to conform to new conditions. However, in practice, changing a conservation easement proves to be extremely difficult.

The Land Trust Alliance sets out seven principles that it believes should guide the amendment of conservation easements.[124] According to the Alliance's guidelines, amendment policies should only be as flexible as necessary and amendments to easements should

1) clearly serve the public interest and be consistent with the [holder's] mission;
2) comply with all applicable laws and regulations;
3) not raise concerns about the holder's tax-exempt or charitable status;
4) not result in private inurement or impermissible private benefit;
5) be consistent with the conservation purpose(s) and intent of the easement;
6) be consistent with the intent of the donor or grantor of the of the easement and any funding agencies; and
7) have a net beneficial or neutral effect on the relevant conservation values protected by the easement.[125]

---

[122] Vermont Law School (2015), pp. 8–9.

[123] Land Trust Alliance (2007), p. 9.

[124] Land Trust Alliance (2007), p. 17.

[125] Ibid.

The Land Trust Alliance also urges that the following issues be considered:

- effect on stewardship and administration of the easement;
- engagement of stakeholders, other owners, or other involved parties;
- consideration of conflicts of interest;
- resolution of title issues;
- concerns about real property tax issues;
- acquisition of additional expert advice;
- supplementation of baseline documentation and related cost; and
- completion of required tax forms.[126]

The policy also seems to discourage relaxing restrictions on one parcel in exchange for additional or new restrictions on a different parcel.[127]

The report also cites concerns about such bargains violating applicable law and lack of court review.[128] For example, IRS regulations provide that the original deduction taken by the donor remains unaffected so long as the termination results from an "unexpected change" that "makes impossible or impractical the continued use of the property for conservation purposes."[129] The termination must occur in a judicial proceeding, and the portion of the funds resulting from any subsequent sale or disposition of the property must be allocated to the holder of the easement and must be used in a manner that as closely as possible conforms to the conservation purpose of the original conservation easement.[130]

The guidance recommends a written amendment policy to facilitate application of the important principles.[131] The policy should consider the relevant tax provisions, including private inurement and private benefit prohibitions, state conservation easement enabling statutes, and state law governing charitable organizations.[132]

These difficulties in adapting conservation easements to changing conditions create what the author deems the "perpetuity syndrome." Rules are included in a document that likely will not change, locking the landowner into particular forms and agriculture, limiting choices, and stifling innovation. Whether certain activities are permitted must be subjected to a case-by-case analysis under Internal Revenue rules, creating uncertainty and destabilizing land tenure.

---

[126]Ibid, at 18.

[127]Ibid.

[128]Ibid.

[129]26 C.F.R. § 1.170A-14(g)(6)(i) (2007).

[130]Ibid.

[131]Land Trust Alliance (2007), p. 21.

[132]Ibid, 23–32.

4.2.3 Terms That Do Not Support Agriculture/Conflicting Easement Objectives

Most of the concerns relating to negative impacts of conservation easements on land tenure security of agricultural producers may be traced to a seeming contradiction between two major objectives of so-called agricultural conservation easements. The Internal Revenue Code and the regulations promulgated thereunder focus on conservation purposes, and agricultural production proves to be a mismatch for the requirements.

The "exclusively for conservation purposes" requirement must be considered when looking at activities on the farm. The regulations provide that the deduction will not be allowed if an inconsistent use contravenes a "significant conservation benefit," even where the conservation easements promote other significant conservation benefits.[133] The example used in the regulations explains that if a conservation easement on farmland promotes flood prevention and control, the deduction will nonetheless be denied if use of pesticides on the farm destroys a significant ecosystem.[134]

For the agricultural landowners, the lack of guidance on this point makes decision making difficult, if not impossible. As the cases described below detail, agricultural conservation easements, as a practical matter, present two, often conflicting, purposes. First, the easement promotes agricultural operations on the property. Second, and most important for the federal income tax deduction, the easement promotes some qualified conservation purpose, often "open space." Increasingly, courts have had to decide which purpose prevails when the purposes conflict.

In practice, this conflict most often occurs where the farm operator wishes to engage in an agritourism activity, place wind turbines or solar panels on the property, or engage in processing or sales activities. One commentator, referring to these activities as "rural enterprises," notes two approaches to deal with these activities.[135] One approach allows these activities so long as enterprise is subordinate to the farm business, while the other approach allows the activities within the area set aside in the easement for farm buildings.[136] A third approach, used by the Virginia Outdoors Foundation, a state agency, involves specifically setting out activities allowed in the "farmstead" (farm building area), allowing certain dwellings and farm structures, processing and sale of products "produced or partially produced" on the property, and alternative energy structures "scaled to provide electrical energy or pump water for permitted dwellings, structures, and activities on the property."[137] The template sets out "[E]xcess power generated incidentally"

---

[133]IRS Regs. §1.170A-14(e)(2).

[134]Ibid.

[135]Anderson and Cosgrove (1998), pp. 11–12.

[136]Ibid.

[137]Virginia Outdoors Foundation, paragraph 3. (i).

by the alternative energy structures.[138] Other miscellaneous buildings and structures are also allowed in the farmstead area so long as the structures are consistent with the conservation purposes and associated with allowed activities. Another paragraph specifically sets out allowable activities outside of the farmstead area.[139]

Three published court opinions illustrate the tension between agricultural production and conservation with respect to agricultural conservation easements. First, a Kentucky case[140] requires the court to resolve conflicting conservation easement provisions, one that allows commercial agriculture and another that prohibits alteration of the topography of the property. Section 1 of the easement provided that "the purpose of th[e] Easement [is] to assure that the Protected Property will be retained forever substantially undisturbed in its natural condition and to prevent any use ... that will significantly impair or interfere with the Conservation Values of the Protected Property."[141] The landowner filled sinkholes on the property, claiming that the filling was necessary to facilitate agricultural production on the property.[142] The United States District Court for the Eastern District of Kentucky granted summary judgment to The Nature Conservancy, reasoning that although the easement allowed agricultural activities like plowing, the prohibition against alteration of the topography prevents the filling of the sinkhole.[143] The Sixth Circuit Court of Appeals affirmed the reasoning of the district court and the court's granting of an injunction. The court also affirmed the granting of attorneys' fees to The Nature Conservancy of over $80,000, and costs of over $18,000.[144]

Second, a Maryland case,[145] involved a citizen challenge to whether a conservation easement permitted an approximately 10,000 square foot creamery that would market locally produced organic dairy products to the public; a 1500 square foot farmers market; and a parking lot. The scenario points out the inability to add infrastructure and to diversify the operation.

A community association and adjacent landowners filed suit, alleging that the proposed creamery would violate the provisions of the easement.[146]

The easement stated that "the parties [intend] that the said land be preserved solely for agricultural use..."[147] The covenants and restrictions included the following provisions:

> [A](1)(a) Except as otherwise provided in this instrument, the above described land may not be used for any commercial, industrial, or residential purpose....

---

[138]Ibid, paragraph 3. (i)(d).

[139]Ibid, a portion of paragraph 4. (i).

[140]The Nature Conservancy, Inc. v. Sims (2012).

[141]Sims, 674.

[142]Sims, 675.

[143]Ibid.

[144]Sims, 676–677.

[145]Long Green Valley Ass'n v. Bellevale Farms, Inc. (2013).

[146]Ibid, 296.

[147]Ibid, 300.

> [A](3) The Grantor reserves the right to use the above described land for any farm use, and to carry on all normal farming practices . . . including any operation directly relating to the processing, storage, or sale of farm, agricultural or woodland products produced on the said above described land. . ..[148]

The circuit court dismissed the case, finding that the plaintiffs lacked standing.[149] Plaintiffs appealed, arguing that the conservation easement creates a charitable trust, and, therefore, interested members of the public may enforce the easement's provisions.[150]

The Court of Appeals of Maryland affirmed.[151] Although Bellevale Farms prevailed in this lawsuit, the costs of litigation threatened the continuation of the farm operation.[152] Many operations would not have survived this series of lawsuits. The conservation easement made innovations in the operation uncertain, and costly, even though the operation focuses on grass-fed dairy and organic production, types of production thought to be sustainable.

Finally, a recent Virginia Supreme Court case[153] involved construction of a building on the easement property.[154] The building would be used for a creamery and bakery (using milk and wheat raised on the adjacent property), storage of aging wine (produced from grapes grown on both properties), and a tasting room.[155] The plans included sale of wine, cheese, and bakery products produced on the site, and the property would be open to the public.[156] The land trust filed suit for declaratory judgment, claiming that the construction of the building and the intended uses violated the conservation easement.[157] After a five-day trial, the trial court ruled in favor of White Cloud, with the exception of some rulings on affirmative defenses not relevant given the ultimate outcome of the case.[158] The land trust appealed, claiming as error, inter alia, that trial court's application of the common law strict construction principle for restrictive covenants to a conservation easement.[159]

Several provisions of the easement appear to conflict with each other. Section 1.1 sets out the purpose of the easement. "It is the purpose of this Easement to assure that the Protected Property will be retained *in perpetuity* predominantly in

---

[148]Ibid, 301.

[149]Id, 305.

[150]Ibid.

[151]Ibid, 321.

[152]"The costs of going organic were dwarfed by lawyer fees. Paying lawyer bills put the grain bill out of reach." Rachel Gilker, "Grazing Profitably With a Cherry on Top", On Pasture, http://onpasture.com/2014/03/24/grazing-profitably-with-a-cherry-on-top/ (March 24, 2014).

[153]Wetlands America Trust, Inc. v. White Cloud Nine Ventures, L.P. (2016).

[154]Ibid.

[155]Ibid.

[156]Ibid.

[157]Ibid.

[158]Ibid, 2.

[159]Ibid.

its natural, scenic, and open condition, as evidenced by the [Baseline Document] Report [BDR], for conservation purposes as well as permitted agricultural pursuits, and to prevent any use of the Protected Property which will impair significantly or interfere with the conservation values of the Protected Property, its wildlife habitat, natural resources or associated ecosystem."[160] Section 3.3(A)(iv) of the easement states: "No permanent or temporary building or structure shall be built or maintained on the entirety of the Protected Property other than ... farm buildings or structures. . . ."[161] The easement fails to define either term.[162]

On the other hand, Section 3.1 of the easement allows "industrial" and "commercial" agriculture.[163] Additionally, Section 4.1 provides that "changes in agricultural technologies, including accepted farm and forestry management practices may result in an evolution of agricultural activities on the Protected Property."[164]

The Virginia Supreme Court, based on provisions of the easement allowing farm buildings and industrial and commercial activities, rejected the land trust's challenge.[165] In doing so, the court acknowledged an "inherent tension between the 'conservation purposes' and the expressly 'permitted agricultural pursuits'"; the easement requires a balance between the two.[166]

The three cases, taken together, illustrate the issues raised by placing conservation easements on agricultural land. Allowing agricultural practices creates a tension with the provisions of the Internal Revenue Code and often results in ambiguous and conflicting provisions within the conservation easement. As agricultural practices evolve, and market conditions change, the uncertainty caused by conservation easements increases.

All of these restrictions and requirements limit the choices of farm operators, even if the operator owns the property. In the ever-changing environment of agriculture, these constrained choices may threaten the second prong of sustainable agriculture, economic viability of the operation.

4.2.4 Conservation Easements and Sustainable Agriculture Conclusions

A confluence of factors combines to cause conservation easements to increase land tenure uncertainty. First, the landowner will be uncertain as to which land practices may be engaged in without violating the easement provisions. This uncertainty may cause the landowner to eschew sustainable land practices. Second, conservation

[160]Ibid, 9.
[161]Ibid, 6.
[162]Ibid.
[163]Ibid.
[164]Ibid.
[165]Ibid.
[166]Ibid.

easements limit the flexibility and adaptability of land practices. Therefore, landowners often will be prohibited from changing practices to meet new conditions. With climate change and other environmental change, this issue may prevent soil conservation and other sustainable land practices. Finally, an inherent conflict exists between the conservation values sought to be promoted by conservation easements and production agriculture. Consequently, landowners may have to maintain standard practices due to an inability to switch to more innovative conservation practices that would maintain soils and promote sustainability.

5 Heirs Property

5.1 Introduction

The terms "heirs property," "heirs' property," and "land in heirs" all describe a form of ownership where at least some of the owners have acquired the property through inheritance.[167] Often numerous and related owners hold the property as tenants in common.[168] Heirs property proves particularly prevalent in poor African American and Native American communities,[169] as well as in low-income areas of Appalachia.[170]

Two concerns arise from heirs property: the vulnerability (or displacement) concern and the wealth (or efficiency) concern.[171] The vulnerability concern refers to the fear of being forcibly dispossessed from the property through a partition sale initiated by another cotenant, whether a family member or a third party.[172] "[I]t is often the case that an unscrupulous real estate speculator purchases a very small interest in a family-owned tenancy-in-common property with the sole purpose of seeking a court-ordered partition by sale."[173] Note that this concern relates directly to insecure tenure.

The "wealth concern" refers to the diminished ability of cotenants to use the land, whether to build a home, for recreation, for business, as collateral for a loan, or for other reasons, due to the nature of tenants in common property.[174] Many uses of the heirs property, like timbering or mineral mining, require unanimous consent of

---

[167]Deaton (2012), pp. 615–616.

[168]Ibid.

[169]Deaton et al. (2009), pp. 2344–2345.

[170]Ibid, 2345.

[171]Deaton et al. (2009), p. 2345; Deaton, p. 617.

[172]Ibid.

[173]UNIF. PARTITION OF HEIRS PROP. ACT, Prefatory Note, at 2. Little or no empirical evidence exists to either confirm or disprove these claims.

[174]Deaton et al. (2009), p. 2346.

all tenants.[175] This concern directly relates to the second prong of sustainable agriculture, an economically viable operation.

Included in the wealth concern is the fact that cotenants of heirs property find it difficult or impossible to borrow money against the property or lease the property.[176] Further, "heirs property owners are not able to bid competitively at the partition sale auction because they are unable to secure any financing to make an effective bid and because they are cash poor."[177] Heirs property often coincides with land loss in rural communities.[178]

5.2 Tenancy in Common

Tenancy in common involves joint ownership of property where each of the co-owners holds an undivided fractional interest in the property.[179] The shares of each co-owner need not be equal.[180] Each co-owner holds the right to occupy the entire property, in common with the other co-owners.[181] In other words, the co-owners possess separate freeholds but undivided possession.[182] The single unity of possession characterizes tenancy in common.[183]

The share of a deceased co-owner generally passes to that co-owner's heirs.[184] If the shares of each co-owner are equal, survivorship may attach if clearly expressed.[185] A co-owner may transfer their interest in the property, which does not destroy the tenancy but merely transfers the interest to a new co-owner.[186]

Any co-owner may take possession of the entire property, but not the exclusion of the other co-owners.[187] Any co-owner of the property also holds the right to request partition of the property or a division of the property among the co-owners.[188]

---

[175]Deaton et al. (2009), p. 2346; Parsons et al. (2010), p. 16.

[176]Baucells and Lippman (2001).

[177]UNIF. PARTITION OF HEIRS PROP. ACT, Prefatory Note, at 8.

[178]Parsons et al. (2010), p. 16.

[179]Deaton et al. (2009), p. 2344.

[180]20 Am. Jur. 2d Cotenancy and Joint Ownership § 32.

[181]20 Am. Jur. 2d Cotenancy and Joint Ownership § 31.

[182]20 Am. Jur. 2d Cotenancy and Joint Ownership § 31.

[183]20 Am. Jur. 2d Cotenancy and Joint Ownership § 32.

[184]20 Am. Jur. 2d Cotenancy and Joint Ownership § 31.

[185]Ibid.

[186]20 Am. Jur. 2d Cotenancy and Joint Ownership § 39.

[187]20 Am. Jur. 2d Cotenancy and Joint Ownership § 40.

[188]Ibid.

Any and all co-owners have the right to occupy and utilize the entire property but cannot exclude other co-owners from exercising the same right.[189] No co-owner may, however, appropriate any portion of the property for their exclusive use.[190]

With respect to timber, for example, any co-owner may timber the property in a customary manner.[191] However, if the timbering exceeds the co-owner's proportionate share, other co-owners may file a petition for waste.[192] All co-owners must consent to a timbering of the property as a whole.[193]

5.3 *Heirs Property and Sustainable Agriculture*

The effect of this inability of any of the cotenants to use the land without agreement of the other cotenants is an instance of what has been termed as the tragedy of the anticommons.[194] The tragedy of the anticommons results in the underutilization of the resources, or "waste."[195]

The prefatory note to the Uniform Act discusses owners of heirs property in Maine that are "unable to manage their property in a rational way because some passive or uncooperative cotenants either do not contribute to their share of the expenses needed to maintain ownership of the property or refuse to give the needed consent to plans that their more active fellow cotenants formulate to improve the management, stability, and utilization of the property."[196] "[A]ll across the country. . .[owners of heirs property] find themselves locked into a dysfunctional common ownership arrangement because there are no legal mechanisms to consolidate title to such property among family members who have been active and responsible owners."[197]

In the familiar tragedy of the commons, multiple owners each hold the right of inclusion, or the right to use the property.[198] This property regime often leads to overuse of the property.[199] By contrast, in the tragedy of the anticommons, by definition, multiple owners are each endowed with the right to exclude others from a scarce resource, and no one has an effective privilege of use.[200] When there are

[189]20 Am. Jur. 2d Cotenancy and Joint Ownership § 41.

[190]Ibid.

[191]20 Am. Jur. 2d Cotenancy and Joint Ownership § 43.

[192]20 Am. Jur. 2d Cotenancy and Joint Ownership §§ 43, 44.

[193]20 Am. Jur. 2d Cotenancy and Joint Ownership § 43.

[194]Heller (1998); Deaton et al. (2009), p. 2347; Deaton (2007a), pp. 927–928.

[195]Ibid.

[196]UNIF. PARTITION OF HEIRS PROP. ACT, Prefatory Note, at 7.

[197]Ibid.

[198]Heller (1998), p. 677.

[199]Ibid.

[200]Heller (1998), p. 668.

too many owners holding rights of exclusion, the resource is prone to underuse—a tragedy of the anticommons.[201] Heirs property often lays unused, with no heir able to muster a consensus on how to use the property. After a time of neglect, adapting the property for reuse may take significant time and resources. Land and buildings unused in productive activity and allowed to lay neglected for long periods of time "often require a significant investment of labor and money to bring back into productive and profitable use."[202] That resources cannot be leveraged to invest constrains economic development.[203]

The tragedy of the anticommons proves especially harmful with respect to "working lands," like agricultural and forestal lands. No owner can effectively exercise control over the property, so the land cannot be profitably farmed or timbered. Likewise, no owner will be willing to lime the soil, repair fences, or make other investments toward sustainable agriculture. The property sits idle, growing up in weeds.

One may argue that underuse of land may promote soil conservation and sustainability since the land will lie unused. Although unused or underutilized land may, by happenstance, preserve soils in some cases, the uncertainty also prevents active soil improvement measures and other sustainability practices. For example, since the landowners are usually unable to reach agreement, the land may not be dedicated to government programs to protect and improve soils, such as the Wetlands Reserve Program or the Conserve Reserve Program. Therefore, although passive neglect may prevent active degradation of soils, proactive measures are prevented. While "doing nothing" may promote soil conservation more than active degradation, positive soil improvement measures are impossible to institute.

6 Conclusions

Anything that introduces an element of uncertainty into the stability and long-term nature of agricultural land tenure impacts productivity and sustainability of agricultural practices, thereby impacting soil health. This chapter examined three of the primary sources of uncertain land tenure in United States: leasing, conservation easements, and heirs property. Leasing has been examined in some depth in the United States context and clearly introduces uncertainty. Long-term leases should be encouraged to make the tenants' land tenure more secure. Education of nonproducer landlords would also increase land tenure security.

Conservation easements are generally viewed as increasing the likelihood of sustainable practices. However, in the agricultural context, easement provisions often conflict with farming goals, creating land tenure insecurity. More study

[201]Heller (1998), pp. 674–675.

[202]Higby et al. (2006), p. 73.

[203]Deaton (2007b), p. 929.

should be devoted to crafting easement provisions that protect conservation values but also promote secure agricultural land tenure. Decoupling agricultural easements from the provisions of the Internal Revenue Code would advance this effort.

Finally, many parts of the country see a predominance of fractionated land interests, known as heirs property. The fractionated ownership creates land tenure insecurity and shackles the owners as to both agricultural practices and sustainable practices. Policies to encourage consolidation of property interests, while protecting vulnerable landowners, should be pursued.

References

Anderson J, Cosgrove J (1998) Agricultural easements: allowing a working landscape to work, exchange. Land Trust Alliance, Washington, DC, Fall, pp 11–12

Arbuckle JG Jr (2010) Iowa State Univ. Extension, rented land in Iowa: social and environmental dimensions 1. Available at http://www.soc.iastate.edu/extension/ifrlp/PDF/PMR1006.pdf

Arbuckle JG Jr et al (2009) Non-operator landowner interest in agroforestry practices in two Missouri watersheds. Agroforest Syst 75:73, 74

Baucells M, Lippman SA (2001) Justice delayed is justice denied: a cooperative game theoretic analysis of hold-up in co-ownership. Cardozo Law Rev 22:1191

Bigelow D, Borchers A, Hubbs T (2016) U.S. farmland ownership, tenure, and transfer, economic research service. Economic Information Bulletin Number 161

Byers E, Ponte KM (2005) The conservation easement handbook 14, 2nd edn

Carolan MS (2005) Barriers to the adoption of sustainable agriculture on rented land: an examination of contesting social fields. Rural Soc 70:387, 398

Carolan MS, Mayerfeld D, Bell MM, Exner R (2004) Rented Land: Barriers to Sustainable Agriculture. Journal of Soil and Water Conservation 59(4): 70A–75A

Clark N (1944) Improving farm tenure in the Midwest. Agricultural Experiment Station. University of Illinois. Bulletin #502

Clearfield F, Osgood BT (1986) Soil Conservation Serv., sociological aspects of the adoption of conservation practices 9

Cox E (2010) A lease-based approach to sustainable farming, Part I: farm tenancy trends and the outlook for sustainability on rented land. Drake J Agric Law 15:369, 370–371

Cox E (2011) A lease-based approach to sustainable farming, Part II: farm tenancy trends and the outlook for sustainability on rented lands. Drake J Agric Law 16:5, 13

Deaton BJ (2007a) A review and assessment of the Heirs' property issue in the United States. XLVI J Econ Issues 927, 927–928

Deaton BJ (2007b) Intestate succession and Heir property: implications for future research on the persistence of poverty in Central Appalachia. J Econ Issues 41(4):927–942

Deaton BJ (2012) A review and assessment of the Heirs' property issue in the United States. XLVI J Econ Issues 615, 615–616

Deaton BJ, Baxter J, Bratt CS (2009) Examining the consequences and character of 'Heir Property'. Ecol Econ 68:2344–2345

Dekker H (2006) In pursuit of land tenure security. Amsterdam University Press, Amsterdam, NLD, p 1

Duffy M (2008) Iowa State Univ. Extension, Survey of Iowa leasing practices, 2007. AG Decision Maker, Sept. 2008, p 2. http://www.extension.iastate.edu/Publications/FM1811.pdf

Duffy M, Smith D (2009) Farmland ownership and tenure in Iowa, 2007. Iowa State University Department of Economics

Duffy M et al (2008) Iowa State Univ. Extension, farmland ownership and tenure in Iowa 2007, at 18 (rev. 2008)
FAO (2002) Land tenure and rural development, FAO land tenure series 3. Food and Agriculture Organization of the United Nations Rome. http://www.fao.org/docrep/005/y4307e/y4307e00.htm
Fraser EDG (2004) Land Tenure and Agricultural Management: Soil Conservation on Rented and Owned Fields in Southwest British Columbia. Agriculture & Human Values 21:73–79.
Gentry PM (2013) Note, applying the private benefit doctrine to farmland conservation easements. Duke Law J 62:1387, 1395
Grossman MR (2000) Leasehold interests and the separation of ownership and control in U.S. farmland. In: Geisler C, Daneker G (eds) Property and values: alternatives to public and private ownership, pp 119, 127–128
Heller M (1998) Tragedy of the anticommons, property in the transition from Marx to markets. Harv Law Rev 111:621
Higby AM et al (2006) In: Smith M (ed) A legal guide to the business of farming in Vermont 73. Available at http://www.uvm.edu/farmtransfer/?Page=legalguide.html
Knowler D, Bradshaw B (2007) Farmers' adoption of conservation agriculture: A review and synthesis of recent research. Food Policy 32: 25–48.
Land Trust Alliance (2007) Amending conservation easements: evolving practices and legal principles. Research Report, 9 (August 2007)
Lee LK, Stewart WH (1983) Landownership and the adoption of minimum tillage. Am J Agric Econ 65:256, 257
Lichtenberg E (2007) Tenants, landlords, and soil conservation. Am J Agric Econ 89(2):294–307
Long Green Valley Ass'n v. Bellevale Farms, Inc., 432 Md. 292, 68 A.3d 843 (2013)
Parsons R et al (2010) The farmlasts project: farm land access, succession, tenure, and stewardship 12. Available at http://www.uvm.edu/farmlasts/FarmLASTSResearchReport.pdf
Richardson JJ Jr (2016) Land tenure and sustainable agriculture. Texas A&M Law Rev 3:799–826
Salamon S (1998) Cultural dimensions of land tenure in the United States. In: Jacob HM (ed) Who owns America: social conflict over property rights. University of Wisconsin Press, Madison, WI, pp 159–181
Sklenicka P et al (2015) Owner or tenant: Who adopts better soil conservation practices? Food Policy 47: 253–261.
Soule MJ, et al (2000) Land tenure and the adoption of conservation practices. Am J Agric Econ 82: 993–1005
The Nature Conservancy, Inc. v. Sims, 680 F.3d 672 (2012)
Unif. Conservation Easement Act § 1(1) (amended 2007). Available at http://www.uniformlaws.org/shared/docs/conservation_easement/ucea_final_81%20with%2007amends.pdf
USDA National Agricultural Statistics Service (NASS) (1999) 1997 census of agriculture: agricultural economics and land ownership survey (1999). USDA, Washington, DC. Available at http://www.agcensus.usda.gov/Publications/1997/Agricultural_Economics_and_Land_Ownership/
Vermont Law School (2015) Land tenure convening report. Center for Agriculture and Food Systems (March 2015)
Virginia Outdoors Foundation, Intensive Agriculture Easement Template. Available at http://www.virginiaoutdoorsfoundation.org/protect/donating-an-open-space-easement-to-vof/documents-for-easement-donors/
Wetlands America Trust, Inc. v. White Cloud Nine Ventures, L.P., 2016 WL 550339 (2016)

Soil Data Needs for Sustainable Agriculture

Luca Montanarella and Panos Panagos

1 Introduction

The Sustainable Development Goals (SDG) approved by the United Nations in 2015 include an explicit call towards ending hunger, achieving food security and promoting sustainable agriculture by 2030. SDG Goal 2 seeks sustainable solutions to end hunger in all its forms by 2030 and to achieve food security. The aim is to ensure that everyone everywhere has enough good quality food to lead a healthy life. Achieving this goal will require better access to food and the widespread promotion of sustainable agriculture. This entails improving the productivity and incomes of small-scale farmers by promoting equal access to land, technology and markets, sustainable food production systems and resilient agricultural practices. It also requires increased investments through international cooperation to bolster the productive capacity of agriculture in developing countries.

Soils play a key role in achieving this goal. Especially, target 2.4 aims towards ensuring sustainable food production systems by 2030 through the implementation of resilient agricultural practices that increase productivity and production; help maintain ecosystems; strengthen capacity for adaptation to climate change, extreme weather, drought, flooding and other disasters; and progressively improve land and soil quality.

Soil quality is indeed the key to achieve sustainable agriculture. Soils in good health produce enough food in a sustainable way since their inherent fertility limits the need for excessive fertiliser inputs. Assessing soil quality is a complex exercise requiring a range of data and information. Within the OECD countries, there is regular reporting on soil quality. For the European Union (EU), EUROSTAT, the

L. Montanarella (✉) • P. Panagos
European Commission, Joint Research Centre, Ispra, VA, Italy
e-mail: luca.montanarella@ec.europa.eu; panos.panagos@ec.europa.eu

H. Ginzky et al. (eds.), *International Yearbook of Soil Law and Policy 2017*,
International Yearbook of Soil Law and Policy,
https://doi.org/10.1007/978-3-319-68885-5_9

EU's official statistical service, reports on soil quality[1] and soil erosion[2] regularly in the framework of the agri-environmental reporting data flow.

A large amount of research has been performed in order to further refine the soil quality indicators relevant for the EU. Since the adoption of the EU Soil Thematic Strategy, the EU has recognised the multifunctionality of soils. Soils are not only relevant as basis for food and biomass production, but they perform a number of other very relevant functions as well, as recognised in article 1 of the proposed Soil Framework Directive by the European Commission[3]:

(a) biomass production, including in agriculture and forestry;
(b) storing, filtering and transforming nutrients, substances and water;
(c) biodiversity promotion, including habitats, species and genes;
(d) providing the physical and cultural environment for humans and human activities;
(e) harbouring raw materials;
(f) acting as a carbon pool;
(g) serving as an archive of geological and archaeological heritage.

These functions need to be protected for future generations, hence the relevance for sustainable development and the achievement of the Sustainable Development Goals. Sustainable development cannot be achieved without protecting the soil functions that enable us to live on this planet.

For those two indicators (soil quality, soil erosion), the EU proposed a scheme for monitoring the change per country and region. Besides comparison between different periods, it is also important to set specific objectives that are contributing to sustainable agriculture. The development of models and the subsequent production of the best estimated data are very important both for estimating trends of the indicators and measuring the achievements of goals.

2 Sustainable Agriculture

In simplest terms, sustainable agriculture is the production of food, fiber or other plant or animal products using farming techniques that protect the environment, public health, human communities and animal welfare.[4,5] This form of agriculture enables us to produce healthy food without compromising future generations' ability to do the same. However, there are many agricultural systems and techniques

---

[1]http://ec.europa.eu/eurostat/statistics-explained/index.php/Agri-environmental_indicator_-_soil_quality (Eurostat 2017a).

[2]http://ec.europa.eu/eurostat/statistics-explained/index.php/Agri-environmental_indicator_-_soil_erosion (Eurostat 2017b).

[3]European Commission (2006).

[4]Gold (2016).

[5]Foley et al. (2011).

that claim to be sustainable: organic farming, conservation agriculture, agroecology, agroforestry, sustainable intensification, precision farming, etc. Agreed criteria to define sustainable agriculture are therefore needed, as are related indicators in order to assess its degree of sustainability in an objective way.

Within the SDGs, Goal 2 specifically promotes sustainable agriculture to end hunger and all forms of malnutrition by 2030. This will require sustainable food production systems and resilient agricultural practices; equal access to land, technology and markets; and international cooperation on investments in infrastructure and technology to boost agricultural productivity. Especially, target 2.4 calls for the following: 'By 2030, ensure sustainable food production systems and implement resilient agricultural practices that increase productivity and production, that help maintain ecosystems, that strengthen capacity for adaptation to climate change, extreme weather, drought, flooding and other disasters and that progressively improve land and soil quality.' The indicator selected for measuring success is the proportion of agricultural area under productive and sustainable agriculture. Thus, the definition of sustainable agriculture adopted to measure progress with the agreed indicator will be crucially important.

There has been considerable discussion over the past 30 years about how to define 'sustainable agriculture'. Sustainability was often understood primarily within the context of its environmental dimension. Now, it is well established that sustainability needs to be considered in terms of its social, environmental and economic dimensions. Both soil quality and soil erosion indicators have been operationalised accordingly to capture this multidimensional nature. Besides the aggregated statistics per region covering the environmental aspect, both indicators are addressing also economic dimensions (e.g., cost of soil erosion in GDP terms) and societal aspects (e.g., off-site effects of soil erosion).

Challenges to sustainable agriculture vary within and across countries, as well as by region, and are affected by socio-economic and biophysical conditions. By defining sustainability across its three dimensions, countries can select those metrics within their measurement instrument that best capture the priorities most relevant to them. One important metric is soil quality, as emphasised in target 2.4, which explicitly calls for an improvement of soil quality by 2030.

3 Soil Quality Data

In the EU, 'good' soil is not only defined by its capacity to produce food and fibre (agricultural soil quality), but it also needs to be assessed for the other soil functions identified in the EU Soil Thematic Strategy (see above). This is a very different view on soil quality compared to that held by the USA and other OECD countries, where soil quality is essentially defined by the soils' agricultural productivity. The multifunctional vision of soil quality requires the adoption of a complex system of indicators responding to each of the seven soil functions.

The composite indicator, the agri-environmental soil quality index, consists of four sub-indicators of similar weight, which have relevance either to the agricultural and/or to the environmental performance of soil:

1. Productivity index has relevance to the agricultural policy field and measures the capacity of soil for biomass production (Figs. 1 and 2).[6] The level on which the soil is delivering its biomass production service is evaluated on the basis of soil properties (e.g., soil organic carbon, soil texture, pH, etc.) at prevailing climatic and topographical conditions. Since productivity is a result of the interaction of soil and climatic and topographical conditions, these factors need to be assessed in their complexity. Productivity differences between similar soils under intensive rainfed crop systems vary with the changing availability of precipitation and differences in temperature regimes. For instance, in a temperate sub-oceanic climate, the rather stable thermal regime and balanced water availability due to medium to high amounts of precipitation not only secure water availability to plants in most soil types throughout the growing period, but they also facilitate decomposition and weathering throughout the year. These processes are limited under temperate continental or semi-arid Mediterranean climates due to cold and/or arid periods in most years. With the increasing aridity of many climates, the importance of soils' physical and chemical properties in supplying water and nutrients to plants is gaining increasing significance. Based on this principle of soil productivity processes, ranking of inherent productivity of soil has been performed for eight major climatic areas.
2. Fertiliser response rate has relevance to the agri-environmental policy field and measures the input need to attain optimal productivity. Soil's productivity is only partly due to its inherent fertility; it is also due to management, mainly nutrient inputs. Thus, the effect of fertilisation was considered in this module. While acknowledging the importance of the applied technology of soil use on the actual productivity of soil on a detailed scale, such distinctions were regarded as unnecessary to directly incorporate into our continental scale study. The goal of our study was solely to determine soil productivity—i.e., the capacity of soil to supply nutrients, water and a rooting medium for plants—in a comparative manner and not to assess the effects of management differences. It is recognised that in the case of the realisation of biomass productive capacity, technological advancement plays a role if we compare regions of contemporary Europe on a continental scale. However, the efficiency of input use, consequently the selection of the most appropriate techniques and amount of input, is also determined by biophysical conditions to a great extent. For example, large regional crop yield differences within France are due to biophysical differences rather than differences in available technology, capital or other

[6]Tóth et al. (2013).

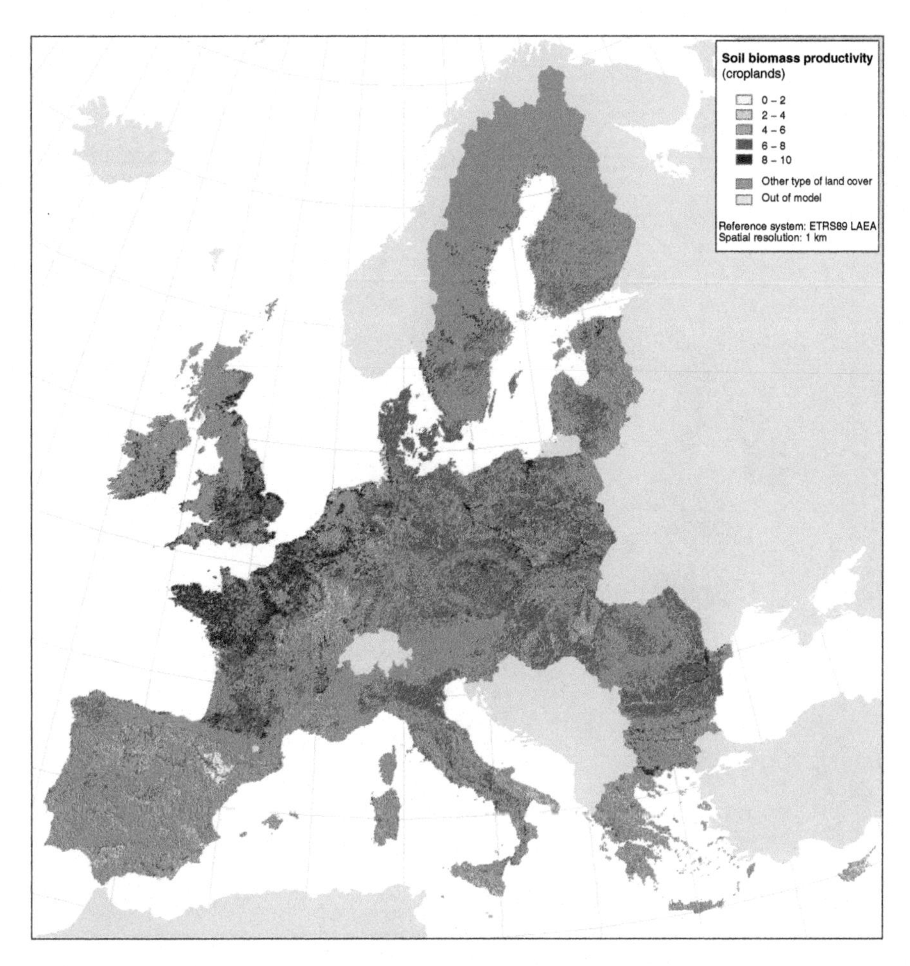

Fig. 1 Soil biomass productivity of croplands in the EU (expressed in relative terms with indices without measurement units), 2006, EU-27. Source: Joint Research Centre, European Commission

socio-economic conditions. Therefore, to evaluate the effect of management, we only considered the influence of fertilisation. To do this, we assigned a fertiliser response score to each soil unit in the seven climatic zones. Soils with the largest estimated relative fertility increase received the maximum of 8 points and soils demonstrating little influence from fertilisation received 1 point. As an example, an Albic Arenosol in the sub-oceanic climate has among the highest relative productivity increase due to proper fertilisation, while fertilisation has little effect on the crop productivity of Calcaric Rendzinas in the semi-arid Mediterranean. To calculate soil productivity for the cropland land-use type (Fig. 1), the inherent soil productivity and the fertiliser response scores were aggregated (Fig. 3). The inherit soil productivity is based on spatially weighted average productivity scores that were calculated for the soil mapping units (SMUs) of the

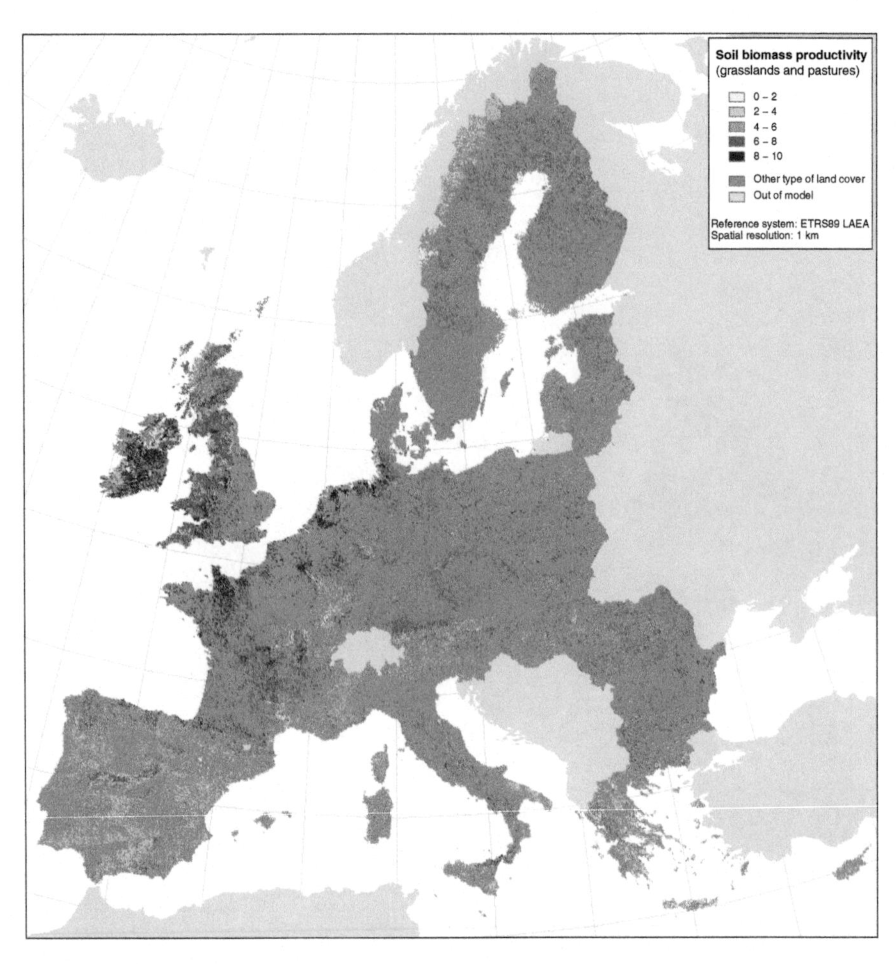

Fig. 2 Soil biomass productivity of grasslands in the EU (expressed in relative terms with indices without measurement units), 2006, EU-27. Source: Joint Research Centre, European Commission

Geographical Soil Database of Europe (SGDBE).[7] In order to avoid the bias originating from the evaluation of non-cropland soils, only those soil typological units (STUs) that have cultivated land as their primary or secondary land-use type were considered in the SGDBE. Figure 3 shows the fertiliser response rate of croplands in the European Union. The fertiliser response rate corresponds to the potential yield increase from extensive cropland management on intensive rainfed cultivated soils.

[7]King et al. (1994).

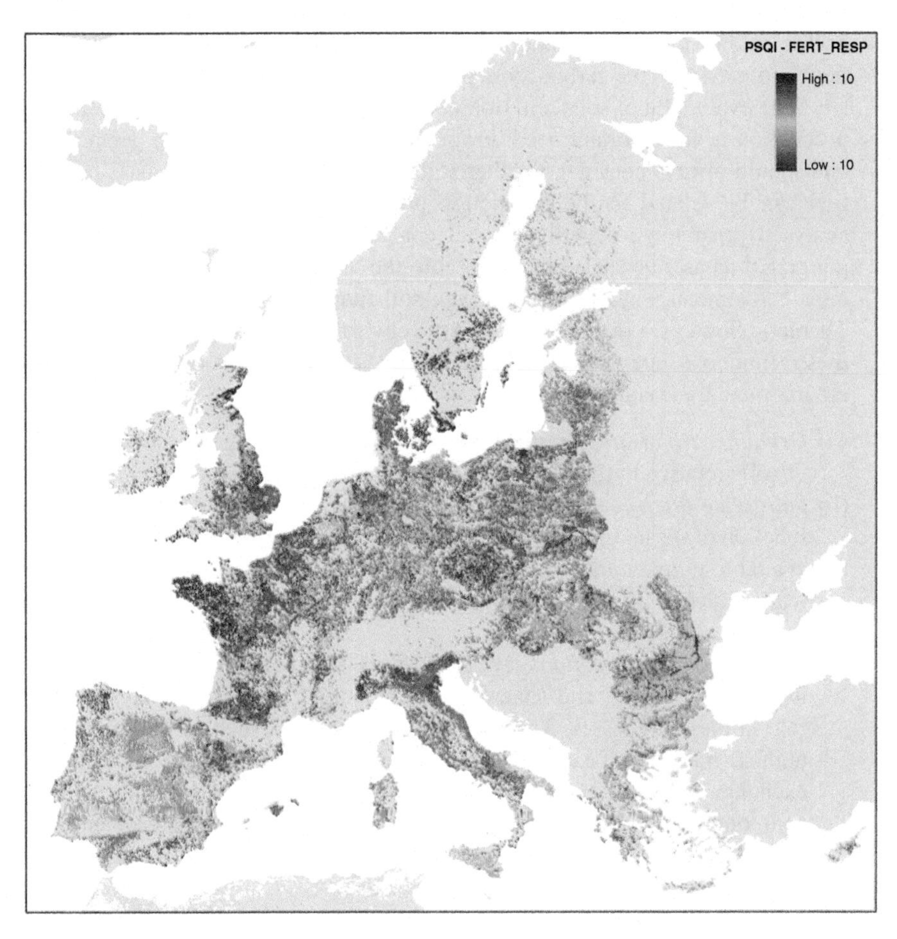

Fig. 3 Fertiliser response rate of cropland soils (1 = low, 10 = high), 2006, EU-27. Source: Joint Research Centre, European Commission

3. Production stability index has relevance to the agricultural policy field and measures the soil's response to climatic variability. In the agri-environmental soil quality domain, soil response properties are characterised by water and nutrient dynamics. Annual variations of yield depend on complex weather conditions, including precipitation and temperature regimes. The soil water element of productivity interacts with the dynamics of nutrient availability. The water regime of soil types is reflected in the variability of yields over the years. This variety may differ to a great extent among soil classes. In the case of crop production, the stability of yields is desirable. Therefore, soil characteristics responsible for higher variability should be considered as indicators of negative response properties and consequently lower their index score for the production stability sub-indicator.

4. Soil environmental service index has relevance to the agri-environmental policy field and measures the carbon storage, filtering, transforming and soil biodiversity. The evaluation of soils' environmental services independent of agricultural production is an important need. From the agricultural production perspective, one should always consider whether soil functions other than biomass production can be related to the production potential or whether they have to be analysed separately. Nevertheless, all analyses should be based on the same integrated dataset because this can secure the most accurate information in the most economical way. There are seven soil functions—as defined by the Soil Thematic Strategy—that contribute to the environmental quality of soil. Four of these functions are recommended to be evaluated in relation to agri-environmental measures:

   (a) *Organic carbon storage*: this indicator expresses the organic carbon content of soils relative to the theoretical maximum amount they can hold.
   (b) *Substance filtering*: substance filtering capacity of any porous media is a function of the textural and colloid properties. There are a number of models available to quantitatively estimate the filtering capacity of European soils (e.g., EuroPEARL for pesticides[8]). These models can be applied for the assessment.
   (c) *Substance transformation*: several patterns of material transformation exist in the soil. Aerobic and anaerobic biological decomposition plays a major role in these processes. Results of redox processes, such as stabilisation of humus, may also bear great importance for specific functions. Material exchange between soil components, as well as between soil and non-soil components, is another major process of which the magnitude is determined by the substance transformation capacity of soil. Soils need to be assessed by the availability and strength of these processes.
   (d) *Biodiversity and biological activity*: in order to answer agri-environmental policy questions, the role of soil biodiversity needs to be addressed in terms of threshold values of influences from impacts and interpretation of the functioning of soil biodiversity, among other factors. The recently published Global Atlas of Soil Biodiversity[9] includes maps of the major threats to soil biodiversity (pressures), which provide a useful indication of areas at risk of soil biodiversity decline. Land-use change, organic matter decline and intensive human exploitation are among the major pressures on soil biodiversity activities.[10]

The final assessment of soil quality will therefore consist of a complex set of indicators representing the quality of soils for each of its functions. The second agri-environmental indicator identified by the OECD as relevant for the assessment of

[8]Tiktak et al. (2004).

[9]Orgiazzi et al. (2016a).

[10]Orgiazzi et al. (2016b).

sustainable agriculture is soil erosion.[11] OECD is requesting now the soil erosion data as described in the agro-environmental indicators.[12]

4 Soil Erosion

Erosion can be defined as the wearing away of the land surface by physical forces, such as rainfall, flowing water, wind, ice, temperature change, gravity or other natural or anthropogenic agents that abrade, detach and remove soil or geological material from one point on the earth's surface to be deposited elsewhere. When used in the context of pressures on soil, erosion refers to accelerated loss of soil as a result of anthropogenic activity, in excess of accepted rates of natural soil formation.[13] It is important to have the best estimates on soil erosion and to identify hot spots (areas with high erosion rates >10 t ha^{-1} $year^{-1}$) where agro-environmental practices are necessary to reduce erosion.

4.1 Water Erosion

Soil erosion by water is one of the major threats to soils in the European Union, which has a negative impact on ecosystem services, crop production, drinking water and carbon stocks. The European Commission's Soil Thematic Strategy identified soil erosion as a relevant issue for the European Union and proposed an approach to monitoring it. A recently published paper[14] presents the application of a modified version of the Revised Universal Soil Loss Equation (RUSLE) model (RUSLE2015)[15] to estimate soil loss in Europe for the reference year 2010, within which the input factors (rainfall erosivity, soil erodibility, cover management, topography, support practices) are modelled with the most recently available pan-European datasets. While RUSLE has been used before in Europe, RUSLE2015 improves the quality of estimation by introducing updated (2010), high-resolution (100 m), peer-reviewed input layers. The mean soil loss rate in the European Union's erosion-prone lands (agricultural, forests and semi-natural areas) was found to be 2.46 t ha^{-1} $year^{-1}$, resulting in a total soil loss of 970 Mt annually. The new soil erosion map of Europe (Fig. 4) is at the highest possible resolution of

[11]OECD (2017) http://www.oecd.org/greengrowth/sustainable-agriculture/agri-environmentalindicators.htm.

[12]http://ec.europa.eu/eurostat/statistics-explained/index.php/Agri-environmental_indicator_-_soil_erosion.

[13]Huber et al. (2008).

[14]Panagos et al. (2015a).

[15]http://esdac.jrc.ec.europa.eu/themes/rusle2015.

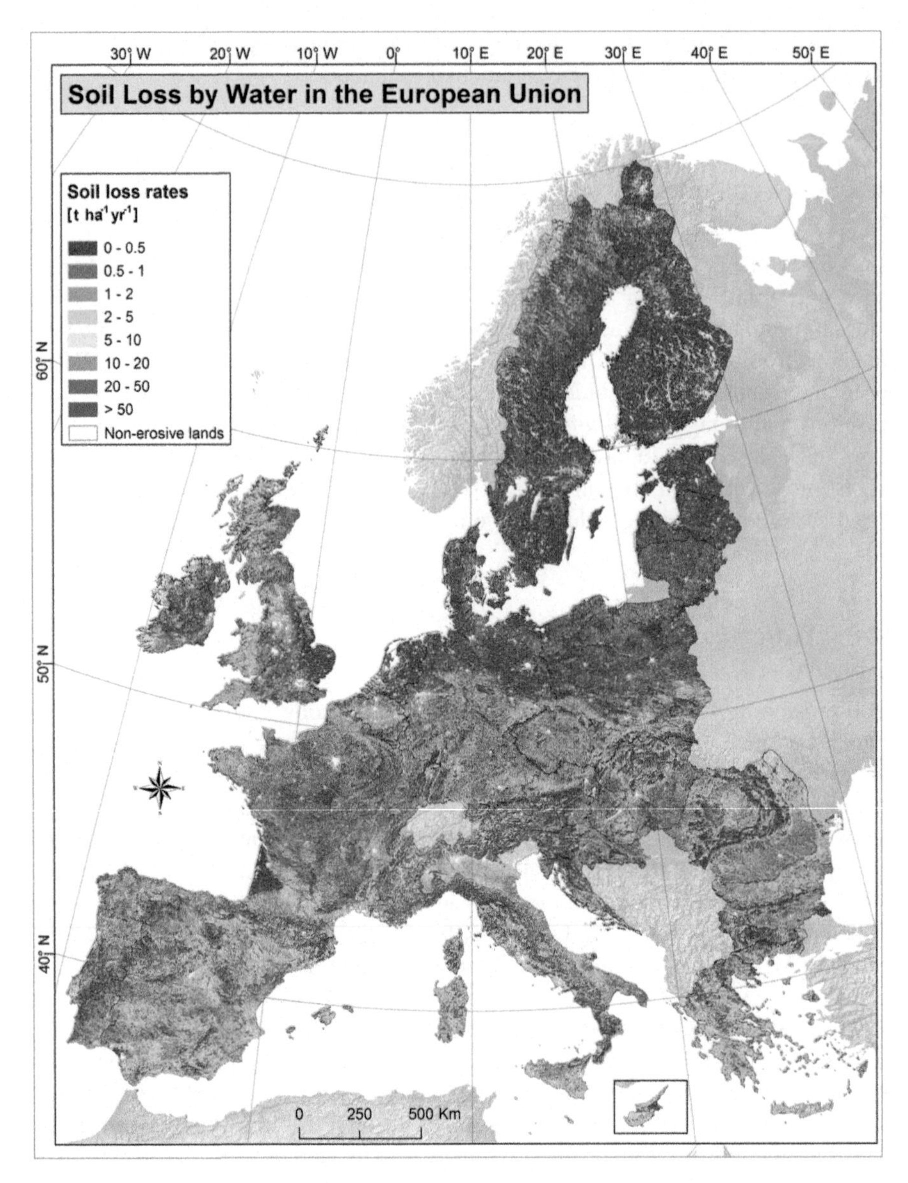

Fig. 4 Soil loss by water erosion in the European Union. Source: Joint Research Centre, European Commission

100 m and is the most detailed assessment yet of soil erosion by water for the European Union.

A major benefit of RUSLE2015 is that it can incorporate the effects of policy scenarios based on land-use changes and support practices. The impact of the Good Agricultural and Environmental Condition (GAEC) requirements of the Common

Agricultural Policy (CAP)[16] and the EU's guidelines for soil protection[17] can be grouped under land management (reduced/no till, plant residues, cover crops) and support practices (contour farming, maintenance of stone walls and grass margins). These policy interventions (GAEC, (see footnote 15) Soil Thematic Strategy (see footnote 16)) have reduced the soil loss rate by 9.5% on average in Europe over the past decade and by 20% in particular on arable lands.

Special attention should be given to the 6.7% of the total agricultural land in the European Union estimated to suffer from high erosion (soil loss rates of more than 11 t ha^{-1} $year^{-1}$). A correspondence article, 'Common Agricultural Policy: Tackling soil loss across Europe',[18] published in Nature (October 2015), underlined the effect of Common Agricultural Policy in reducing soil erosion on European agricultural land. However, further economic and political action should rebrand the value of soil as providing myriad ecosystem services, increase the income of rural landowners, foster young farmers and organise regional services for licensing land-use changes.[19] This action can be reinforced by an economic evaluation of soil erosion, taking into account the costs for the society (off-site effects).

Soil erosion is one of the indicators described within the general context of the Common Agricultural Policy (CAP), through which policy measures are designed, planned and implemented. Those indicators form part of the monitoring and evaluation framework for the CAP 2014–2020 and are used in rural development programmes for a comprehensive overall description of the current situation of each programming area (Member State or regional level).

To this end, the RUSLE2015 model has been applied with the latest available CORINE land cover version, CORINE2012.[20] It focused on agricultural areas at risk of soil erosion by water throughout each Member State of the EU-28. Table 1 presents both the results of the model for 'Estimated agricultural area affected by moderate to severe water erosion (>11 t ha^{-1} $year^{-1}$)' and 'Share of estimated agricultural area affected by moderate to severe water erosion'.

Regarding the area affected, around 6.7% of the EU-28 total agricultural area was estimated to suffer from moderate to severe erosion in 2012. This share was higher in the old EU Member States, known as the EU-15 (7.7%) than in the new EU Member States (joined after 2004), known as the EU-13 (4.3%). Cultivated land (arable and permanent cropland) was estimated to be more affected (7.5%) than permanent grasslands and pasture (4.2%). The share of agricultural land estimated to suffer from moderate to severe erosion was highest in Slovenia (42.4%), Italy

[16]https://ec.europa.eu/agriculture/direct-support/cross-compliance_en.

[17]http://ec.europa.eu/environment/soil/three_en.htm.

[18]Panagos et al. (2015b).

[19]Panagos et al. (2016).

[20]CLC (2012).

Table 1 Soil erosion by water indicator

Country	Estimated agricultural area (1000 ha) affected by moderate to severe water erosion			Share (%) of estimated agricultural area affected by moderate to severe water erosion		
	Total agricultural area	Arable and permanent crop area	Permanent meadows and pasture	Total agricultural area	Arable and permanent crop area	Permanent meadows and pasture
AT	690.6	243.7	446.9	21.0	12.2	34.3
BE	6.9	6.5	0.4	0.4	0.5	0.1
BG	204.7	191.6	13.1	3.3	3.6	1.6
CY	33.5	33.4	0.1	7.2	7.6	0.4
CZ	65.7	63.2	2.5	1.5	1.7	0.3
DE	286.9	242.7	44.2	1.4	1.7	0.7
DK	0.1	0.1	0.0	0.0	0.0	0.0
EE	0.1	0.1	0.0	0.0	0.0	0.0
EL	657.9	607.4	50.5	10.7	12.1	4.4
ES	2633.1	2381.2	251.9	9.6	10.5	5.3
FI	0.1	0.1	0.0	0.0	0.0	0.0
FR	973.3	679.5	293.8	2.9	2.8	3.0
HR	238.7	183.2	55.5	9.4	9.2	10.4
HU	166.3	162.4	3.9	2.6	3.0	0.4
IE	14.7	6.7	8.0	0.3	0.8	0.2
IT	5574.1	5043.6	530.6	32.7	33.0	29.4
LT	0.6	0.6	0.0	0.0	0.0	0.0
LU	4.7	4.5	0.2	3.4	4.5	0.5
LV	0.2	0.2	0.0	0.0	0.0	0.0
MT	1.5	1.5	0.0	9.6	9.6	0.0
NL	0.1	0.1	0.0	0.0	0.0	0.0
PL	258.0	257.0	1.0	1.4	1.6	0.0
PT	231.8	229.9	1.9	5.4	5.6	1.1
RO	1373.2	1248.0	125.2	9.7	11.2	4.1
SE	13.2	12.3	0.9	0.3	0.3	0.2
SI	306.9	242.4	64.4	42.4	41.2	47.4
SK	158.9	152.1	6.8	6.8	7.4	2.4
UK	241.2	31.2	210.0	1.6	0.5	2.5
EU-28	**14137.2**	**12025.5**	**2111.8**	**6.7**	**7.5**	**4.2**

(32.7%) and Austria (20.9%), followed by Greece (10.7%), Romania (9.7%), Spain (9.6%) and Croatia (9.4%). In contrast, the lowest share of estimated agricultural area affected by moderate to severe water erosion (< 1%) was found in Belgium, Denmark, Finland, Ireland, Sweden, the Netherlands and the Baltic states.

4.2 Wind Erosion

Field measurements and observations have shown that wind erosion is a threat for numerous arable lands in the EU. Wind erosion affects both the semi-arid areas of the Mediterranean region as well as the temperate climate areas of the northern European countries. Yet there is still a lack of knowledge and thus limited understanding about where, when and how heavily wind erosion is affecting European arable lands.[21] Currently, the challenge is to integrate the insights gained by recent pan-European assessments, local measurements, observations and field-scale model exercises into a new generation of regional-scale wind erosion models. This is an important step to making the complex matter of wind erosion dynamics more tangible for decision-makers and supporting further research on a field-scale level.

A geographic information system version of the Revised Wind Erosion Equation (GIS-RWEQ) was developed to (1) move a step forward in large-scale wind erosion modelling, (2) evaluate the soil loss potential due to wind erosion on arable land in the EU and (3) provide a useful tool to support field-based observations of wind erosion. The model was designed to predict the daily soil loss potential at a ca. 1 km^2 spatial resolution. The soil loss potential due to wind erosion (Fig. 5) was estimated by the GIS-RWEQ for the 1.17 million cells at 1 km spatial resolution into which the arable land of the EU-28 was subdivided.[22] The average annual soil loss predicted by the GIS-RWEQ for EU arable land totalled 0.53 Mg ha^{-1} $year^{-1}$, with the second quantile and the fourth quantile equal to 0.3 and 1.9 Mg ha^{-1} $year^{-1}$, respectively. The cross validation shows a high consistency with local measurements reported in literature.

The highest annual soil loss rate by wind erosion in agricultural lands was noticed in Denmark (3 Mg ha^{-1} $year^{-1}$), the Netherlands (2.6 Mg ha^{-1} $year^{-1}$), Bulgaria (1.8 Mg ha^{-1} $year^{-1}$) and, to a lesser extent, also the UK (1 Mg ha^{-1} $year^{-1}$) and Romania (0.95 Mg ha^{-1} $year^{-1}$). In northern Europe, the locations most susceptible to wind erosion were found along the North Sea coasts of Denmark, the UK, the Netherlands, Germany, France and Belgium. Noticeable soil loss rates were also predicted along the coast of the Baltic Sea, especially in the western sector; the Czech Republic; Slovakia; and Hungary. In the Mediterranean area, higher erosion rates occurred in a zonal distribution. Here, higher soil loss rates were located in the Spanish regions of Aragón, Castilla y Leon; the Italian regions of Apulia, Tuscany and Sardinia; the Provence region in France; and the Greek regions of Central and Eastern Macedonia and Thrace. In Eastern Europe, high erosion rates appeared in the Romanian and Bulgarian lowlands surrounding the Carpathian Mountains and along the Black Sea coastline.

[21] Borrelli et al. (2016).

[22] Borrelli et al. (2017).

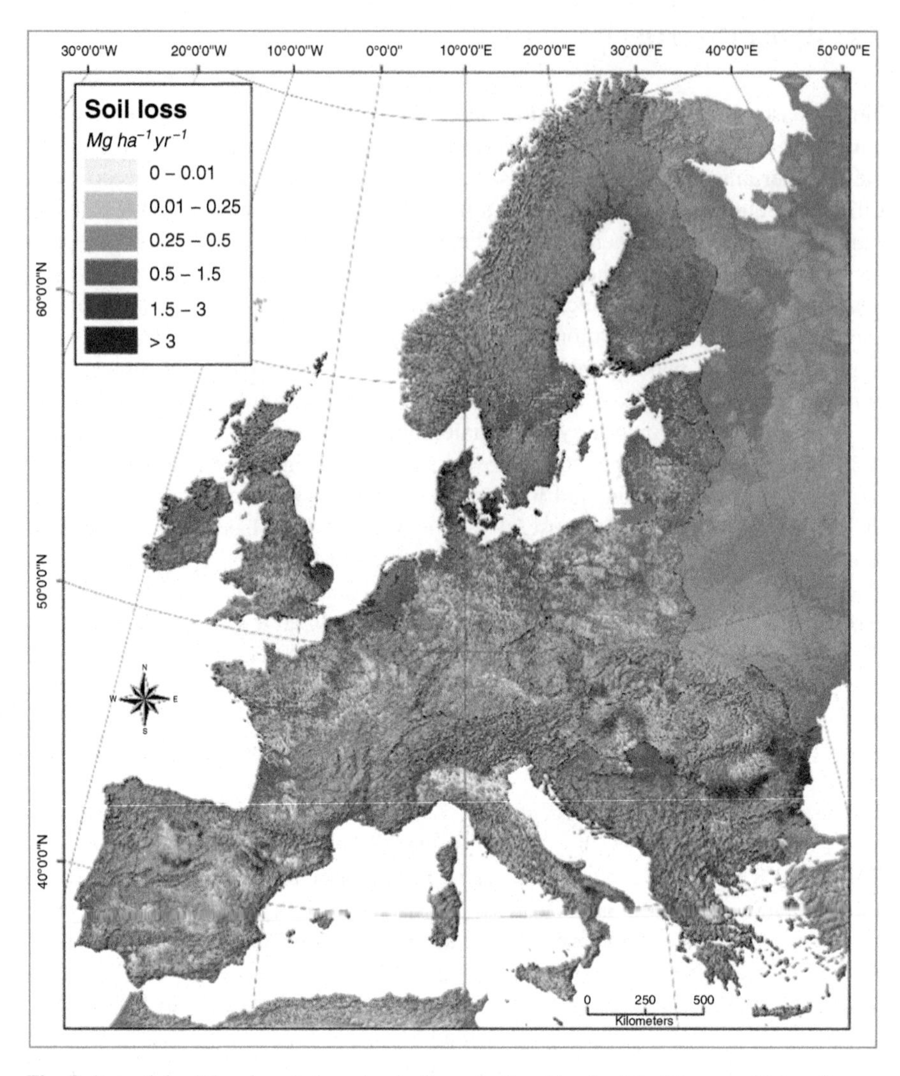

Fig. 5 Potential soil loss by wind erosion in the agricultural lands of the European Union. Source: Joint Research Centre, European Commission

5 Data Availability

The soil quality maps and the soil erosion datasets are available in the European Soil Data Centre (ESDAC).[23] The data are frequently updated based on model runs, and they are easily accessible to both the scientific community and the public. The

[23]http://esdac.jrc.ec.europa.eu.

soil erosion data are not available only as aggregated statistics, but all the input layers and the high-resolution outputs are also available.

6 Conclusions

Sustainable agriculture implies the maintenance of sufficient soil quality and limiting soil erosion. The high soil organic carbon and the importance of biodiversity activity (incorporated in soil quality) are also important factors mitigating soil erosion. For Europe, this means protecting the full functionality of soils for their seven functions identified by the EU Soil Thematic Strategy, as well as limiting erosion by water and wind below 10 t ha^{-1} $year^{-1}$. Achieving such a target by 2030 requires stringent soil protection policies and efficient monitoring systems for detecting changes in soil condition and properties. Unfortunately, the withdrawal of the proposed EU Soil Framework Directive by the European Commission in 2014 leaves the EU without a legal basis for implementing effective soil protection strategies. Voluntary instruments may adequately replace such a legally binding obligation, but it will depend on Member States' national implementation of such non-binding agreements. Great expectations have been raised by the very successful UN International Year of Soils 2015 and the recent final approval by the FAO Council of the Voluntary Guidelines for Sustainable Soil Management (VGSSM).[24] These guidelines provide the necessary basis for national governments to move towards ensuring sustainable soil management. In 2030, we will take stock of the achievements and will be in a position to assess the actual success or failure of the agreed SDGs, especially Goal 2, aiming towards ending hunger and assuring healthy food for all. Without increasing understanding as to what is sustainable agriculture and how it can be measured, this will be a difficult goal to achieve in such a relatively short time frame.

References

Borrelli P, Panagos P, Ballabio C, Lugato E, Weynants M, Montanarella L (2016) Towards a Pan-European assessment of land susceptibility to wind erosion. Land Degrad Dev 27 (4):1093–1105

Borrelli P, Lugato E, Montanarella L, Panagos P (2017) A new assessment of soil loss due to wind erosion in European agricultural soils using a quantitative spatially distributed modelling approach. Land Degrad Dev 28:335–344

CAP, Good Agricultural and Environmental Conditions (GAEC) (2017) Available at: https://ec.europa.eu/agriculture/direct-support/cross-compliance_en. Accessed 15 May 2017

CLC (2012) CORINE Land Cover 2012. Available at: http://land.copernicus.eu/pan-european/corine-land-cover/clc-2012. Accessed 20 Dec 2016

[24] FAO (2016).

ESDAC (2012) European Soil Data Centre. Available at: http://esdac.jrc.ec.europa.eu. Accessed 10 May 2017

European Commission (2006) Proposal for a Directive of the European Parliament and of the Council establishing a framework for the protection of soil and amending Directive 2004/35/ EC. COM(2006) 232 final, Brussels, 22 Sept 2006

Eurostat (2017a) Agri-environmental indicator – soil quality. Available at: http://ec.europa.eu/ eurostat/statistics-explained/index.php/Agri-environmental_indicator_-_soil_quality. Accessed 7 Feb 2017

Eurostat (2017b). Agri-environmental indicator – soil erosion. Available at: http://ec.europa.eu/ eurostat/statistics-explained/index.php/Agri-environmental_indicator_-_soil_erosion. Accessed 15 May 2017

FAO (2016) Voluntary Guidelines for Sustainable Soil Management (VGSSM). http://www.fao. org/3/a-bl813e.pdf. Accessed 7 Apr 2017

Foley JA, Ramankutty N, Brauman KA, Cassidy ES, Gerber JS, Johnston M, Mueller ND, O'Connell C, Ray DK, West PC, Balzer C, Bennett EM, Carpenter SR, Hill J, Monfreda C, Polasky S, Rockström J, Sheehan J, Siebert S, Tilman D, Zaks DPM (2011) Solutions for a cultivated planet. Nature 478(7369):337–342

Gold MV (2016) Sustainable agriculture: definitions and terms. Sustainable agriculture and food supply: scientific, economic, and policy enhancements. Apple Academic Press, pp 3–16

Huber S, Prokop G, Arrouays D, Banko G, Bispo A, Jones RJA, Kibblewhite MG, Lexer W, Möller A, Rickson RJ, Shishkov T, Stephens M, Toth G, Van den Akker JJH, Varallyay G, Verheijen FGA, Jones AR (eds) (2008) Environmental assessment of soil for monitoring: volume I indicators and criteria. Office for the Official Publication of the European Communities. Luxembourg, EUR 23490 EN/1

King D, Daroussin J, Tavernier R (1994) Development of a soil geographic database from the soil map of the European Communities. Catena 21(1):37–56

OECD (2017) Agri-environmental indicators. Available at: http://www.oecd.org/greengrowth/ sustainable-agriculture/agri-environmentalindicators.htm. Accessed 15 May 2017

Orgiazzi A, Bardgett RD, Barrios E, Behan-Pelletier V, Briones MJ, Chotte JL, De Deyn GB, Eggleton P, Fierer N, Fraser T, Hedlund K (2016a) Global soil biodiversity Atlas. Publications Office of the European Union, Luxembourg, 176pp

Orgiazzi A, Panagos P, Yigini Y, Dunbar MB, Gardi C, Montanarella L, Ballabio C (2016b) A knowledge-based approach to estimating the magnitude and spatial patterns of potential threats to soil biodiversity. Sci Total Environ 545–546:11–20

Panagos P, Borrelli P, Poesen J, Ballabio C, Lugato E, Meusburger K, Montanarella L, Alewell C (2015a) The new assessment of soil loss by water erosion in Europe. Environ Sci Policy 54:438–447

Panagos P, Borrelli P, Robinson DA (2015b) Common agricultural policy: tackling soil loss across Europe. Nature 526:195

Panagos P, Imeson A, Meusburger K, Borrelli P, Poesen J, Alewell C (2016) Soil conservation in Europe: wish or reality? Land Degrad Dev 27(6):1547–1551

Tiktak A, de Nie DS, Piñeros Garcet JD, Jones A, Vanclooster M (2004) Assessment of the pesticide leaching risk at the Pan-European level. The EuroPEARL approach. J Hydrol 289:222–238

Tóth G, Gardi C, Bódis K, Ivits É, Aksoy E, Jones A, Jeffrey S, Petursdottir T, Montanarella L (2013) Continental-scale assessment of provisioning soil functions in Europe. Ecol Process 2 (1):32

Import Regulations and Certification as a Means to Enforce Sustainable Agriculture Abroad

Andrea Schmeichel

1 Introduction

Climate change is one of the greatest challenges of our time. It is also an issue where it became starkly clear that environmental values compete not only with socio-economic issues or trade interests but also with other environmental goals. In recent years, the conflict between environmental protection and climate change mitigation reached boiling point regarding imports of palm oil. At first, bioenergy—including biodiesel from palm oil—was seen as an opportunity to curb carbon emissions. However, it quickly became apparent that the side effects outweighed the benefits if the market was left unregulated.[1] Intensive cultivation, land conversions through increased demand for cultivated land and the emissions caused by production and transport chains annihilated emission reductions through the use of bioenergy. Climate change mitigation was threatening the exact resources it was trying to protect.

In the European Union, a first attempt at integrating limits to climate change policy into legislative measures aiming at mitigating climate change was made by the provisions for sustainability for biofuels and bioliquids in Arts. 17 to 19 Directive 2009/28 of the European Parliament and of the Council of 23 April 2009 on the promotion of the use of energy from renewable sources (RED).[2]

Following this introduction (Sect. 1), the article first defines the notion of bioenergy and sets out the conflict between climate change mitigation and other issues (Sect. 2). It then turns to the substantial requirements in the sustainability

[1]OECD (2007).

[2]OJ L 140, 5.6.2009, pp. 16–62.

A. Schmeichel (✉)
Arnecke Sibeth, Berlin, Germany
e-mail: aschmeichel@arneckesibeth.com

H. Ginzky et al. (eds.), *International Yearbook of Soil Law and Policy 2017*,
International Yearbook of Soil Law and Policy,
https://doi.org/10.1007/978-3-319-68885-5_10

criteria, detailing the requirements on GHG emission balance and biodiversity and pointing out room for improvement (Sect. 3). The article then describes the compliance mechanism and assesses certification systems, as well as the impact of the sustainability criteria as a meta-standard (Sect. 4). The article concludes that the sustainability criteria should be considered a first step that will provide experiences for further development. They may also serve as a blueprint for other sectors (Sect. 5).

2 Conflicting Goals for the Use of Bioenergy

2.1 *Biomass for Energy*

The term biomass is used to describe biological material derived from living or recently living organisms. It incorporates a variety of biodegradable products, from agriculture, forestry and related industries, including fisheries. Biomass is a versatile material with multiple uses. It can be used as food or feed, as building material, for clothing, for bioplastics; the options are seemingly endless. Biomass is also an important source of energy, so-called bioenergy. Traditionally, it has been used for energy through simple combustion, i.e. by burning as a source of heating and for cooking. More recently, more modern forms of bioenergy have emerged and expanded rapidly. It has been used as biogas, biofuel or other bioliquids, in particular for transportation and also for heating and cooling.

The increasing production and use of bioenergy has socio-economic and environmental effects, both good and bad.[3]

2.2 *Socio-Economic Effects*

2.2.1 Bioenergy for Socio-Economic Development

The promotion of bioenergy is seen not only as a factor for climate change mitigation and the security of energy supply but also in ‘promoting technological development and the introduction of new agricultural processes and providing opportunities for employment and regional development, especially in rural and isolated areas’ (recital 1 RED). Thus, biofuels have the potential to contribute to rural development, generate employment, create trade and investment opportunities and thereby increase economic prosperity. This is particularly the case in countries with climates suitable for agriculture and cheap labour, often developing and newly developed countries.

---

[3]Worldwatch Institute (2007). Rosillo-Calle (2007), pp. 1–26.

2.2.2 Adverse Effects

However, an increasing use of bioenergy may also lead to negative socio-economic impacts, both specific to bioenergy and related to agriculture in general.[4]

As arable land and yields of biomass are limited, the use of the available biomass creates competition between the different uses, notably between the use of biomass for energy and its use for food and feed. Land-use changes for bioenergy mean that arable land cultivated for food and feed is no longer available for that purpose. This direct competition for space may cause shortages in the supply of food and feed. Even where food availability is not strongly affected, the prices payable may rise. Bioenergy thereby affects not only the availability but also the affordability of food and feed.[5]

Moreover, the production of biomass for energy has to be put into the wider context of agriculture. Increasing production of bioenergy will not only intensify but also extensify agriculture. This leads to a higher demand for land, causing conflicts with other users, potentially culminating in displacement, land grabbing and violent conflict.[6] Moreover, other socio-economic aspects come into play. A major issue is the working conditions of agricultural workers.[7]

2.3 *Environmental Impacts*

2.3.1 Bioenergy for Climate Mitigation

The use of renewable energy offers an opportunity to replace fossil fuels, thus diversifying energy sources and reducing the dependence on fossil fuels. Accordingly, 20% of the gross final energy consumption (electricity, heating and cooling, and transport) shall be satisfied by renewable energy in 2020 in the EU, and 10% in the transport sector (Arts. 3 (1) and 3 (4) RED). Different support schemes have been set up by the Member States to achieve this goal, ranging from quotas, through auctions to feed-in tariffs.

Of the renewable sources of energy, bioenergy has been considered particularly promising. It is easily stored and is not dependent on a grid—unlike other sources of renewable energy like wind or solar power. Moreover, it is a versatile and flexible source of energy as it can be made available in all aggregate states, solid, liquid and gaseous. Currently, three quarters of heating and cooling, and through combined heat and power (CHP) also a portion of electricity from renewable energy, is

[4]Jumbe et al. (2009), pp. 4980–4986; Zhang (2008), pp. 3905–3924.

[5]On the issue of conflicts with food, see FAO (2008).

[6]Deininger et al. (2011), 13 foll., 163.

[7]Goldemberg et al. (2008), p. 2095.

satisfied by energy from biomass. In the transport sector, first-generation biofuels, biodiesel and bioethanol are at this time the only market-ready fuels.[8]

2.3.2 Adverse Environmental Impact

At first glance, bioenergy appears carbon neutral, as carbon stored in plants is released during combustion. However, the balance looks different when taking into account the entire life cycle of the plant, i.e. all steps of the production process from preparing the soil, planting, processing, transporting to combustion (a so-called life cycle assessment). Consequently, the climate change mitigation effect of bioenergy is mainly seen in the substitution of fossil fuels and energy-intensive materials such as concrete, steel, aluminium and plastics.[9]

An increasing production of bioenergy can also contribute to degradation of farmed land, for example through the following:

- Large-scale monocultures are prone to increased erosion.
- Intensive irrigation increases the risk of desertification and the loss of biodiverse peatland.
- Increasing use of chemicals such as fertilisers to increase yields may lead to eutrophication on the farmed land and on adjoining areas.
- Besides effects on water quality, more intensive farming may also require irrigation, thus withholding groundwater from natural or near natural ecosystems.
- New crops, including genetically modified organisms or potentially invasive alien species, may spread across nearby natural or near natural habitats.

Besides degradation of already farmed land, an increasing production of raw materials for bioenergy also increases the pressure on natural or near natural habitats. Land-use changes may be direct, where natural or near-natural habitats are converted to produce bioenergy feedstock, or indirect, where previous uses of a plot of land, for example agriculture for food, are displaced into natural or near-natural habitats.[10]

2.4 *The Export of Conflicts Through the Import of Bioenergy*

These impacts of the EU climate change mitigation policy do not only affect the Member States. They also impact on other countries that were not involved in the drafting of these policies.

[8]See European Commission (2011), p. 11.

[9]IPCC (2007), pp. 282, 629.

[10]van Stappen et al. (2011), pp. 4824–4834: 4829.

The European Commission estimated in 2007 that, if a 10% biofuel target were introduced, about 43 megatons of oil equivalent (Mtoe) of biofuel would be necessary to reach the quota.[11] The quota of 20% renewables and 10% specifically for transport is not designed to be met by domestic production alone. The RED states outright that while the targets could be met with domestic production, a combination with imports would be both 'likely and desirable' (recital 16). While estimates differ on how much biomass will be imported for bioenergy alone, the Member States have estimated in their National Renewable Energy Action Plans (NREAP) to import a total of 44% bioethanol (3.1 Mtoe) and 36% biodiesel (7.7 Mtoe) or associated feedstocks in 2020.[12] International trade in biofuels has risen from practically zero to 120–130 petajoules (PJ) in the decade leading up to 2009, mainly directed to the lucrative markets in the EU or the USA.[13] It has been estimated that without the renewable energy targets, exports of palm oil from Indonesia and Malaysia would be 25% lower.[14]

There is a good chance that imports will exceed what is necessary to meet targets that cannot be fulfilled by local production as imported biomass for energy may be economically more attractive than local produce. The production of bioenergy or its raw materials is particularly attractive in the tropics as land is cheap and yields are high.[15] Many states are expected to develop their export potential, in particular the South East Asian states, including Malaysia, Thailand, Indonesia and the Philippines, and a number of African and South American countries.[16]

The requirement for imports is particularly problematic because the environmental side effects of a European Union policy are quasi-exported at the same rate as agricultural produce is imported into the EU. These exporting countries very often still have abundant natural and near-natural habitats, such as the rain forests of Indonesia. At the same time, exporting countries are often developing or newly developed countries where export is seen as a motor for economic development and environmental or social standards may be low and/or hard to enforce. Land-use rights may not be determined securely, threatening to displace indigenous population. Another issue to be tackled is corruption, which is often high in these areas.

---

[11]European Commission (2007).

[12]Bowyer (2011), p. 8.

[13]Lamers et al. (2011), p. 2672.

[14]Fonseca et al. (2010), p. 40.

[15]OECD (2007), pp. 12 foll.

[16]Worldwatch Institute (2007), pp. 9, 141.

3 The Sustainability Criteria for Biofuels and Bioliquids

The Renewable Energy Directive 2009/28/EC (RED) addresses at the European level for the first time the promotion of both renewable energy in general and bioenergy in particular and defines the limits of this promotion for the case of some forms of bioenergy in order to address its potential side effects. The sustainability criteria prescribe a certain greenhouse gas balance, the observation of certain rules of good agricultural practice, and exclude certain ecosystems from cultivation or extraction of biomass, referred to as 'no-go areas'.

3.1 Scope and Legal Consequences

The scope of application of the sustainability criteria is determined in two ways. Firstly, the criteria cover biofuels and bioliquids (Art. 2 h), i) RED), irrespective of the origin of the raw materials. Secondly, the criteria are a condition for the eligibility of economic operators for certain national support schemes, financial support for the consumption (Art. 17 (1) b), c) RED) and, importantly, to be counted towards the Member States' national renewable energy targets (Art. 17 (1) a) RED).

3.1.1 Biofuels and Bioliquids

The EU sustainability criteria apply to liquid biomass only, i.e. biofuels and bioliquids such as biodiesel (Art. 2 h), g) RED). Solid fuels are not covered, and gaseous fuels only when they are used in the transport sector (see Art. 2 g) RED). The sustainability criteria apply irrespective of the country of origin of the feed-stock. Accordingly, the criteria apply to goods grown or produced within the EU and to imported goods in free circulation (Art. 24 TEC, now Art. 29 TFEU).

The scope of the sustainability criteria is very limited, covering only liquid biomass. Thus, the majority of bioenergy or renewable energy need not be produced according to sustainability criteria, even though the environmental and socio-economic impact of those energy sources can also be considerable. Equally, other uses such as for food and feed or as a building material are not covered by the sustainability criteria. The positive impact of the sustainability criteria therefore has to rely strongly on dispersion effects, expanding environmental provisions into other areas.

3.1.2 The Twofold Incentive for Compliance

The sustainability criteria do not prohibit the production or import of 'unsustainable' produce but only discourage its consumption. Compliance with the sustainability criteria is not mandatory but is introduced as an incentive to both the EU Member States and economic operators: Member States can only count biofuels and bioliquids towards the achievement of their national overall target for renewable energy if these comply with the sustainability criteria (Art. 17 (1) - sub-para. 1 a) RED). Economic operators can only benefit from selected national support schemes for biofuels and bioliquids, such as quotas or tax exemptions, if they use compliant produce (Art. 17 (1) sub-para. 1 b), c) RED). Thus, the sustainability criteria do not directly regulate the behaviour in the way that a public authority exercises control over certain activities in the public interest.[17] Instead, they provide an incentive for operators to comply.

3.1.3 Analysis

The sustainability criteria rely on market mechanisms, by enabling biomass that is considered compliant by a reliable compliance control mechanism to be sold at a premium price, and thereby providing another incentive to comply (recital 76 RED). Thus, the sanctioning is softer than, for example, in the forestry sector (i.e., for solid biomass), where Regulation 995/2010 prohibits illegally harvested timber and derived products—both imported and domestically produced—from being placed on the EU market.[18]

Considering the limited scope of application, the sustainability criteria cannot avoid leakage, i.e. the displacement effects of 'dirty' biomass to other uses. However, the initial limitation of the scope appears to be guided by reason. The scope of application reflects an area where problems where considered particularly prominent. Moreover, first steps have already been taken in expanding the concept of limiting environmental policies for environmental concerns to other concerned forms of bioenergy.

3.2 Greenhouse Gas Emission Balance

In order to contribute to the mitigation of climate change, biofuels and bioliquids have to reduce GHG emissions, i.e. show a positive carbon balance. This is to be ensured both through a GHG emission reduction quota and the exclusion of cultivation of so-called carbon sinks.

[17]See the definition of regulation by Selznick (1985), pp. 363–364.

[18]European Parliament and Council, *Regulation (EU) n. 995/2010 laying down the obligations of operators who place timber and timber products on the market*, OJ [2010] L 295/23.

3.2.1 Maximum Emission Level

Article 17 (2) RED introduces a progressive reduction target for biofuels and bioliquids compared to fossil fuels. By 2017, the GHG balance of biofuels and bioliquids have to be 35%, from 2017 50% and from 2018 60% lower for production installations in which production started in 2017 or after (Art. 17 (2) sub-paras. 1, 2 RED). However, production installations operating before 23 January 2008, the date of the Commission proposal, only had to comply from 1 April 2013, long after the transposition period ending 5 December 2010 (Art. 17 (2) sub-paras. 4, 27 (1) RED).

3.2.2 Life Cycle Assessment

In calculating the GHG emission balance, all emissions from cultivation, production and use of biofuels and bioliquids shall be taken into account (Life cycle assessment, Art. 17 (2) sub-paras. 3, 19 (1), Annex V RED). Article 19 (1), Annex V RED, provides for different calculation pathways. Actual values can be used according to the calculation methodology foreseen in part C Annex V RED, or default values can be used (Annex V A, B RED), with aggregated default values provided where no values exist for the entire production chain (Annex V D, E RED). Default values are derived from typical values, the methodology used for the calculation of actual values, as well as a scientific data set (Art. 2 n) RED). However, due to the complexity of the production pathways, the number of default values is limited. The European Commission provides limited guidance in its Decision 2010/335/EC on guidelines for the calculation of land carbon stocks.[19]

3.2.3 Analysis

The emission reduction target appears limited, considering much more stringent suggestions in the legislative process. However, the benchmark is subject to reform and is true to the idea that the criteria are essentially a trial version.

Currently, the LCA methodology is static, both temporally and spatially, and is limited to emissions rather than general pollution, restricting its ability to assess dynamic interactions between land surface and atmosphere.[20] Major issues and uncertainties were not included in the beginning, such as the release of land carbon stocks through indirect land-use changes or emissions from agriculture.

The numerous reporting and updating requirements surrounding the required GHG emission balance, the calculation method and the default values show that the requirements on GHG emissions in the sustainability criteria are not considered

[19]OJ [2010] L 151/19.

[20]Bright et al. (2012), pp. 2–11.

final. This reflects that these provisions were introduced before the issues where fully and comprehensively understood and resolved at a scientific and technological level. It remains to be seen how they are expanded and adapted to scientific and technological progress. Accordingly, the EU has already amended RED with Directive 2015/1513/EU on the reduction of indirect land-use change for biofuels and bioliquids.[21]

3.3 Direct Land-Use Changes: The Introduction of 'No-Go Areas'

No-go areas have been introduced for two reasons. Areas are excluded from cultivation or extraction of biomass if they are considered to have a high carbon storage capacity or a high level of biodiversity.

3.3.1 The Exclusion of Carbon Sinks

Biomass for biofuels and bioliquids may not be extracted from or cultivated on land that qualified as a carbon sink in January 2008 and has since lost that status (Art. 17 (4) RED). These sinks are repositories to which CO_2 is deposited through natural processes, rather than a resource with intangible, non-economic benefits.[22] Even though carbon sinks do not expunge emissions, carbon is stored, sometimes for a considerable amount of time. their destruction has substantial influence on the climate as this carbon is rapidly released. Article 17 (4) RED enumerates specific sinks:

- wetlands, which are covered by or saturated with water for at least a significant part of the year (a);
- continuously forested areas of a certain size (1 ha), tree height (5 m) and canopy cover (over 30%) (b);
- Wooded land where the canopy cover is between 10% and 30%; the land in question is excluded if the conversion would result in exceeding the GHG emission saving requirements of Art. 17 (2) RED, currently 35% (c).

[21] OJ [2015] L 239/1.

[22] Aguirre (2009), p. 206.

3.3.2 The Exclusion of Biodiverse Areas

Besides these carbon emission reductions, certain areas presumed of high biodiversity are excluded from cultivation (Art. 17 (3) RED). This applies to areas that have held that status in or after January 2008, irrespective of their current status.

Under Art. 17 (3) RED, land with high biodiversity value includes:

- primary forests and other wooded land of native species where there is no visible indication of human activity and the ecological processes are not significantly disturbed (a);
- areas designated for nature protection under national law or according to international agreements, intergovernmental organisations or the IUCN, 'unless evidence is provided that the production of that raw material did not interfere with those nature protection purposes' (b);
- highly biodiverse natural and non-natural grassland; the latter may be cultivated where harvesting of the raw material is necessary to preserve the grassland status (c).

Moreover, Art. 17 (5) RED also excludes areas that were peatland in 2008 (unless cultivation does not involve drainage of previously undrained soil). Peatland is an example of where carbon sinks and high biodiversity coincide.

3.3.3 Analysis

The area exclusions have the potential to contribute to the protection and conservation of the enumerated areas. Especially in the forestry sector, the reference to primary forests may reinforce the protection status in real terms of these areas. However, beyond primary forests, the RED definition of forests only makes quantitative rather than qualitative requirements.

The exclusions of certain ecosystems are, however, far from providing a hard and solid framework on the protection of highly biodiverse ecosystems and carbon sinks. Further definitions of the broadly enumerated ecosystems are lacking, and definitions differ in various jurisdictions—even though this has been partly remedied by Commission Regulation (EU) No 1307/2014 on defining the criteria and geographic ranges of highly biodiverse grassland.[23] Moreover, the list includes exceptions and, in some cases, management practices short of exclusions. Every 'soft spot' and exception contained in the sustainability criteria limits their ability to reinforce national standards in resource states with low enforcement rates. This is even more the case since the sustainability criteria do not establish any procedural safeguards on the determination of the areas in question.

---

[23] OJ [2014] L 351/3.

The sustainability criteria are a negative list, excluding certain ecosystems from cultivation. Instead, a positive list could have been drafted, limiting the cultivation of biomass for energy to degraded land. Several issues are not addressed:

- Biodiversity and carbon storage beyond the described ecosystems is not taken into account.
- Apart from canopy cover, no spatial requirements are made, ignoring the requirement of species for a certain size habitat or certain size population to maintain genetic diversity. In order to influence the quality of air and water, the availability of ground water, climate regulation, sequestration of carbon, nutrient cycling and other ecosystem services, those ecosystems need to have a certain size, depending on local circumstances.
- The sustainability criteria do not contain management plans for the improvement of the biodiversity value of the ecosystem in question. Moreover, no rules are introduced on buffer zones or corridors enabling the interlinking of habitats.
- No quantitative restrictions apply on how much land may be used for bioenergy or agriculture.
- Indirect land-use changes cannot be depicted by the sustainability criteria; their inclusion in a life cycle assessment is challenging (see also Art. 19 (6) RED, Directive 2015/1513/EU).

Another issue is the implementation of the no-go areas. It requires accurate maps of the area status in 2008 that can be brought in line with the—broad—requirements of the sustainability criteria. Such surveys may not exist or may be incomplete or follow different terminology. It may also be difficult to obtain reliable information.

Nonetheless, the approach appears to be pragmatic, covering at least areas with above-average biodiversity and carbon storage capacity. Compared to requiring individual impact assessments, the exclusion of specific ecosystems appears to be easier to administer for economic operators. In pursuing an ecosystem approach, the RED seeks to make a compromise between the pursued goals and practicality in achieving them, factoring in administrative burdens.

3.4 *Unresolved Issues*

3.4.1 Agricultural Standards

Article 17 (6) sub-para. (1) RED provides that agricultural raw materials shall fulfil the environmental standards under Regulation (EC) n. 73/2009 under the Common Agricultural Policy (CAP), which establishes minimum requirements for good agricultural practice.[24] Notably, the compliance mechanism construed for the sustainability criteria does not encompass compliance with these agricultural

[24]OJ [2009] L 30/16.

standards (see Art. 18 (1) RED). Whether the requirements of cross-compliance are observed is determined solely according to the rules of CAP. However, the rules on good agricultural practice only apply to raw materials cultivated within the European Union and not to feedstock imported from third countries. Accordingly, the impacts of agriculture are not addressed where the feedstock for biofuels and bioliquids has been produced in third countries, beyond their impact on the GHG emission balance through N_2O emissions (see above Ch. 6 II A 1, B 2). Accordingly, the impacts of increasing bioenergy production both on agricultural practices on cultivated land and on surrounding natural or near natural ecosystems are not addressed for imported biomass.

Even though biomass for energy can be used to enhance soils, unregulated cultivation also poses great risks, including the increasing cultivation of monocultures, with attendant negative effects on soil fertility, erosion and carbon storage capacity. Moreover, they may require the use of chemicals to make up for vulnerability to disease, as well as an increased use of chemicals such as fertilisers and pesticides. A particular concern is that the use of chemicals for the cultivation of bioenergy is not limited through the food safety regulations that govern the use of biomass for food and feed and is therefore more likely to increase.

More intensive cultivation of biomass for energy will also affect biodiversity beyond arable land, for example as irrigation requirements affect the availability of fresh water and cause salination, and the use of fertilisers negatively affects water quality. Accordingly, the cultivation of feedstocks for bioenergy can also negatively affect ecosystem services. No provisions are made within RED for agricultural practices such as crop rotation or the use of agrochemicals such as fertilisers, pesticides, fungicides or herbicides, which also have an impact on soil and water quality. Accordingly, an area that is crucial for achieving a high level of protection, and where the national standards of resource states are usually not very high or are not enforced, is not addressed. All these factors make clear that regulation of agricultural standards would have been desirable for biodiversity protection. This is especially true where domestic environmental standards are low or not enforced, as in the potential exporting countries of biomass, mainly developing and newly industrialised countries. An opportunity has been missed to reinforce environmental standards in the agricultural sector, at least for imports.

3.4.2 Socio-Economic Impacts

Socio-economic issues do not form part of the substantial criteria but are only introduced as reporting requirements (Art. 17 (7) RED).

The Commission has to present bi-annual reports on the impact of the EU biofuel policy on food security and broader development issues. However, the Commission is not required to propose reform on these issues. Explicit provisions on the inclusion of socio-economic impacts of bioenergy are only made for competition with food: Commission reports have to take into account the impact of the use of biomass for energy on food prices when reviewing the renewable energy targets

(Art. 23 (8) sub-para. 1 c) RED).[25] Other socio-economic issues such as land rights and working conditions are not mentioned explicitly as areas to be included in the sustainability criteria in the future.

With the increasing pressure on land, land seizures—which displace a local (indigenous) population to the benefit of large investors—are a serious issue, in particular where the property status was unclear or based on customary rights. Small-scale growers are under severe pressure. Rising rents and land grabbing may also push displaced persons to cultivation into no-go areas.[26] Finally, working conditions and labour standards, which were already criticised prior to the biofuel boom, are likely to be perpetuated.[27]

3.5 Conclusion

RED recognises the potential risk of environmental and social side effects caused by the renewable energy targets that it introduces, and introduces a set of sustainability criteria for biofuels and other bioliquids to help mitigate that risk. This is a direct attempt to balance climate change mitigation with mitigation of other harms. The scope of the sustainability criteria is clearly delimited: it applies to biofuels and bioliquids as far as national targets for renewable energy, renewable obligations and the financial support of the consumption of biofuels and bioliquids are concerned. Moreover, the sustainability criteria focus on certain environmental impacts: a specific GHG emission balance, as well as the exclusion of certain ecosystems considered particularly high in biodiversity or carbon storage capacity.

However, the focus of the sustainability criteria is strongly on the environmental side effects. Socio-economic considerations are mentioned but do not form part of the criteria themselves. The requirements on the Commission for reforms beyond environmental concerns remain general and show that socio-economic concerns, besides the conflicts with food, are not a strong focus of the further evolution of the criteria. At the moment, these issues, which are particularly important to developing countries, are not taken into account sufficiently.

Moreover, an important sector—agriculture—has been completely excluded for imports from third countries. Even for domestic biomass, the agricultural aspects of biomass for energy remain governed by the Common Agricultural Policy without bridges built between these areas of law.

---

[25]Cp. European Commission, Renewable Energy Progress Report 2017, COM (2017), 57.

[26]German et al. (2011), Art. 29: 4 foll.

[27]Oxfam UK (2007), p. 4.

4 Compliance

4.1 The Recognition Mechanism

Besides the stringency of the sustainability criteria, the mechanisms in place to ensure compliance with these requirements are crucial: they must establish a transparent and impartial evaluation process, balancing the complexity of the issues against the practicality of compliance control.

Compliance with the sustainability criteria is, however, not ensured directly. Instead, they operate as a meta-standard. Certification systems are benchmarked against the sustainability criteria by the European Commission or an individual Member State. Produce certified by independent certification bodies based on independent audits under that certification system is then considered compliant with the sustainability criteria (see Art. 18 (3) to (5) RED). Thereby, compliance control is delegated to private operators within a framework set by RED and the Member States.

Substantially, certification systems have to observe the sustainability criteria. Certification systems can also be recognised for only partial compliance. The recognition decision can be extended to measures taken beyond the scope of the sustainability criteria for the conservation of areas that provide for soil, water and air protection, indirect land-use changes, the restoration of degraded land, the avoidance of excessive water consumption in areas where water is scarce and socio-economic issues (Art. 17 (7) RED).

Procedurally and organisationally, voluntary schemes must meet 'adequate standards of reliability, transparency and independent auditing', i.e. rely on a privately organised compliance control mechanism (Art. 18 (5) s. 1 RED). These are, at least for voluntary certification systems, further detailed in a non-binding Commission Communication.[28] Voluntary schemes require an auditable evidence system and acceptance of responsibility for auditing, and data archival for 5 years. Auditing is required to take place before the economic operator may participate in the scheme. Group auditing is permitted, in particular for small operators and cooperatives. Moreover, the voluntary scheme should arrange for regular—at least annual—retrospective auditing by an independent external auditor with the appropriate auditing capabilities, in particular relating to the sustainability criteria. Verifiers can show compliance with these requirements if they follow the International Standardisation Organisation's (ISO) standard ISO 19011 establishing guidelines for quality and/or environmental management systems auditing or partial compliance through other international standards. The sustainability criteria make no provisions on the governance structure of certification systems or accreditation of certification bodies employing the auditors. The accreditation of certification bodies, crucial for the independence of the auditing,[29] is not addressed in Art.

[28] Commission Communication on the practical implementation of the EU biofuels and bioliquids sustainability scheme and counting rules for biofuels, OJ [2010] C 160/8.

[29] Deaton (2004), p. 618.

18 (4) RED. There are no monitoring requirements once the recognition has been granted, and this recognition is valid for 5 years. No specific provisions are taken to avoid corruption. It is therefore unclear if the recognition process in its current form is suitable to ensure compliance with the sustainability criteria.

4.2 The Recognised Certification Systems

A variety of different certification systems exist touching on bioenergy, ranging from general agricultural standards to specific biofuel standards, from company initiatives to multi-stakeholder initiatives. The Commission has recognised a number of voluntary schemes, such as the following:

- Bonsucro EU (formerly known as the Better Sugarcane Initiative);
- Greenergy (Greenergy Brazilian Bioethanol verification programme);
- ISCC (International Sustainability and Carbon Certification);
- RBSA (Abengoa RED Bioenergy Sustainability Assurance);
- RSB EU RED (Roundtable of Sustainable Biofuels EU RED);
- RTRS EU RED (Round Table on Responsible Soy EU RED);
- 2BSvs (Biomass Biofuels voluntary scheme).

All of these standards either were developed explicitly to demonstrate compliance with the RED sustainability criteria or have been amended to comply.

Overall, the standards in multi-stakeholder initiatives cover a wider scope and are more detailed than company initiatives. However, variations also exist between different multi-stakeholder initiatives.

The objectives of the standards concerning various issues differ, from avoiding negative impact to improving quality and quantity through restoration or promotion. Similarly, the various schemes choose different mixes of management (process) requirements and minimum targets, although the focus remains on process rather than target requirements. Notably, minimum requirements are often implied through management practices.

Leaving aside GHG emissions, all voluntary schemes address broader environmental and social impacts. However, the variety among the different standards is considerable. Overall, certification systems organised as multi-stakeholder initiatives tend to have higher requirements.

The different certification schemes address environmental and socio-economic standards, in particular environmental impact assessment, environmental management and biodiversity, and abiotic factors such as soil, water and air, as well as human rights, labour conditions, land and land-use rights. However, the focus of the different schemes, as well as the regulatory density of each issue, differs across the schemes.[30]

[30] van Dam et al. (2010), p. 2457.

The certification process is undertaken by certification bodies that are accredited to the certification system by an independent accreditation body. Notably, they do not certify a product, but an economic operator that may then trade certified produce. The requirements on the certification body's auditors are also very varied across the different initiatives, although reference is usually made to ISO 19011. ISO 19011 contains procedural rules on the conduct of the audit, as well as requirements on the expertise and competence of auditors. The standard has recently been reformed and is available in a 2011 version.[31] The schemes developed for recognition by the Commission show a lower level of regulation of the certification process, only making very general requirements on audits and the governance of certification bodies overall. The schemes already in existence prior to the drafting of RED in many respects go considerably further, introducing safeguards for the certification of operators coinciding with product claims, introducing detailed procedural rules for auditing, on independent decision-making, dispute resolution or stakeholder participation.

4.3 Analysis

Where state law refers to certification systems or other transnational law, it significantly alters the character of the scheme. It appears that the recognition mechanism dramatically increased the interest of economic operators in gaining certification. This also alters the character of the certification systems: voluntary certification systems are based on the notion of self-regulation, creating market incentives for good environmental management rather than command and control instruments. These voluntary standards must now fit into hard regulatory regimes.[32]

The use of a meta-standard has great potential but also faces some challenges. Voluntary standards could partly compensate for compliance deficits at national standards, especially in exporting countries where low standards and/or low enforcement rates hinder environmental protection. However, there is also some evidence that certification is most successful where a state governance framework guarantees the enforcement of the relevant law.[33] Nonetheless, certification has been found to contribute indirectly to sustainable management standards by reaching an agreement on good practice, which is then fed into national legislation.[34] Where public capacities are not yet available, certification systems have sometimes been found to produce more credible and effective governance

[31]ISO, *ISO 19011 – Guidelines for quality and/or environmental management systems auditing* [2011].

[32]Matus (2010), p. 4.

[33]Guénéau and Tozzi (2008), pp. 550–562.

[34]McDermott et al. (2007), pp. 47–70.

structures, but at the risk of legitimizing existing practices where standards are too low or are not enforced.[35]

Certification cannot replace legislation but has to operate side by side with greater regulatory efforts: certification is undertaken at operation level. It is not suitable to address macroeconomic issues such as competition with food or indirect land-use changes.

Nonetheless, the level of regulation achieved by these voluntary certification schemes shows that there is a great potential for self-regulation. However, the standards often focus on the implementation of management practices rather than impact assessment. Therefore, compliance with a certification system's principles and criteria may not give an indication of whether the specific goal of the management practice has been achieved.

Moreover, the variety in scope, definition, individual standards and density of regulation in the certification schemes creates a very opaque market for consumers and also the economic operators themselves. Where standards are not harmonised, the risk of 'standards shopping' leads to market distortion and loss of credibility.[36] Considering the uncertainty caused by the diversity of standards, efforts for cross-harmonisation are made by some certification systems, considering certain certifications as equivalent for the certification under their own, for example where the latter is recognised by the Commission. However, the RED criteria are much too limited in scope of application (i.e., only biofuels and bioliquids and biomass for their production). Most certification systems address more issues than the RED sustainability criteria, in particular on socio-economic issues and also on environmental matters. Where RED only makes general provisions on auditing, certification systems typically address the entire certification process and quite often include the assessment of the operation after certification.

As such, the RED criteria lead to worries of cementing—and legitimising—lower standards.[37] In practice, the sustainability criteria do not serve as a benchmark for certification systems, as the recognised schemes have drafted specific EU versions that apply to produce traded in Europe, contributing to the proliferation of standards. The European Court of Auditors has criticised the recognition mechanisms, as some of the voluntary schemes have been found not compliant in reality both on standards and governance. Another major issue for the Court was that the functioning of a recognised scheme is not supervised.[38]

Moreover, there is evidence that certification systems are being developed to comply with, but not to exceed, the minimum requirements set for Commission recognition. Thereby, there is competitive pressure between high- and low-level standards, which are recognised as equivalent for the purpose of RED but in fact

[35] Cashore et al. (2005), pp. 53–69.

[36] van Dam et al. (2010), p. 2452.

[37] German and Schoneveld (2011), p. 28.

[38] European Court of Auditors (2016).

demand significantly different levels of environmental and socio-economic protection.

5 Conclusion

5.1 *The Impact of the Sustainability Criteria*

The sustainability criteria in Art. 17 to 19 RED seek to further the environmental integrity of the use of renewable energy. The change from fossil fuels to renewable energy is being undertaken with a view to climate change mitigation and also energy security and economic development. Thus, the sustainability criteria are at the nexus of environmental protection and free trade and development.

Overall, the sustainability criteria appear to be a blueprint, launched in a limited sector to test the waters. Their scope is limited, the substantial criteria open to adaptation, the compliance mechanism limited. This is even more true for imported produce where, for example, agricultural standards do not apply. The sustainability criteria partly lag behind the voluntary certification systems when benchmarked against them, which include management practices and binding socio-economic criteria. Thereby, the sustainability criteria only address a small fragment of the conflicts surrounding bioenergy. However, issues like food safety or land grabbing are critically important in developing and newly developed countries.

The scope of the sustainability criteria is limited both geographically and substantively, which raises the issue of the redirection of trade streams and production paths (leakage). Thus, 'non-compliant' biofuels and bioliquids can easily be traded outside the EU. Exporting states can choose different markets, and non compliant liquid biomass cultivated in the EU can be exported. For example, the rising industrial powers in South East Asia such as China are already becoming more and more important trading partners for resource-exporting countries.

Moreover, the focus on biofuels and bioliquids distorts the perception of the conflict between biodiversity protection and climate change mitigation or development interests. The focus on (liquid) biomass for energy creates the illusion that bioenergy is the single factor putting pressure on the environment. Other driving forces such as food and feed supply, population growth, etc. are important factors in biomass but are governed by their respective sectoral regimes. The focus on bioenergy negatively affects the image and reputation of renewable energy generally, for which there are no alternatives in combating climate change and ensuring energy security. Moreover, other conflicts surrounding agriculture are sidelined by the sustainability criteria in their current form, forgoing the potential for a more comprehensive solution.

5.2 The Benefit of Certification Systems in the Case of Imports

The meta-standard approach is well suited to further the application of environmental standards (equivalent to the sustainability criteria) beyond the territory of the EU Member States. Public authorities do not commonly operate uninvited in other countries' jurisdictions, following the non-intervention principle. This constraint does not apply to private entities; particularly in former colonies, a privately organised conformity assessment does not have the same stigma as inviting other countries' public authorities onto their territory would. Furthermore, unlike national public authorities, voluntary schemes can operate across national borders with relative ease. In relying on voluntary schemes, the EU and Member States can take advantage of the accumulated expertise and knowledge of local conditions of these certification systems. Moreover, the meta-standard approach allows for principles and criteria to be adapted to different regional or local circumstances. In this context, the general provisions on the no-go areas appear appropriate as they can encompass numerous specialised or country-specific principles and criteria in certification systems with various scopes.

However, it can be questioned whether the referral to voluntary certification systems is a sufficient tool in assessing compliance with the sustainability criteria. There is great variation between the different certification systems. The stringency of the principles and criteria, the degree of management requirements or hard targets, as well as the provisions on the governance of certification systems, certification bodies and the independence of the auditing process differ considerably. While the level of available data is low, first indications show that compliance of certified operators needs improvement. Moreover, where the focus is on management practices, as is the case with many schemes, no measure is available to what degree these practices contribute to achieving the objective of the certification scheme. Accordingly, the suitability of certification systems to assess compliance has been called into question—in particular, in countries with low public governance standards, high corruption and low environmental standards. State regulation may also be required for another reason: certification systems only operate at the level of individual operations and cannot address indirect and global impacts such as food security. Consequently, the meta-standard making reference to certification systems may be a necessary tool in assessing conformity with standards equivalent to the sustainability criteria, in particular in cross-border compliance assessment, but does not appear sufficient to solve the issues surrounding biofuels and bioliquids, in particular in developing and newly industrialised countries.

5.3 *The Way Ahead*

The sustainability criteria can be seen as providing a balance that allows Member States to achieve their renewable energy targets and to limit environmental side effects on the way. Considering the focus on the expansion of renewable energy—if fossil fuels are to be completely replaced, much higher targets will prove necessary—the balance is likely to lean towards the increased use of bioenergy. Alternatively, the focus might move to other sources of renewable energy, which have or are perceived to have less environmental or socio-economic impacts.

The sustainability criteria have a very narrow scope of application, which does not inhibit leakage. Binding criteria have only been established in areas that are relatively uncontroversial. No-go areas are only declared where there is no doubt about the biodiversity value. Default values for the calculation of the GHG emission balance have only been provided rarely, where adequate data are available. Areas with more uncertainty, such as indirect land-use changes or emissions from agriculture, are only subject to reporting requirements. This reinforces that further research is necessary on scientific issues such as GHG balancing, and also on the socio-economic issues, such as the competition with food and feed. Accordingly, the criteria are accompanied by numerous periodic reporting and review requirements. Moreover, more research needs to be conducted into the impact of environmental provisions, not only of the sustainability criteria but also of the certification systems to which they refer for compliance control. Notably, the certification decision only describes the compliance with the requirements of the certification system. It does not assess whether the objectives of the certification system are met. Accordingly, further research is necessary on the impact of legislation and certification systems in practice. Currently, the level of data is very low; the research on the compliance with and the impact of certification systems is only just beginning. Only when the evidence is clearer can an informed choice be made on the most effective instrument for achieving environmental integrity. Thus, the current Art. 17 to 19 RED may represent a test balloon for more stringent criteria. Moreover, experience from bioenergy certification can be drawn on for parallel problems in wider areas. A prime contender appears to be agriculture in more general, where similar issues exist and certification systems are already in place.

In the long term, the pressure on biodiversity from bioenergy and biofuels in particular is likely to decrease with the development of more-efficient biofuels, the so-called second-generation biofuels. More scientific research is required to bring these fuels to market. Through these efficiency gains, the further development of e-mobility and the adaptation of the electricity grid and storage capacities for renewable energy demands, the importance of biomass as an energy source is likely to decrease. Amended sustainability criteria for biofuels and bioliquids can provide useful controls during the current transitional period and will serve as a starting point in addressing the environmental side effects of these other renewable energy sources.

References

Aguirre GJ (2009) Why cutting down trees is part of the problem, but planting trees isn't always part of the solution – how concepualizing forests as sinks can work against Kyoto. Or Rev Int Law 11:205–224

Bowyer (2011) Anticipated indirect land use change associated with expanded use of biofuels and bioliquids in the EU – an analysis of the National Renewable Energy Action Plans. Institute for European Environmental Policy, London

Bright RM, Cherubini F, Strømman AH (2012) Climate impacts of bioenergy: inclusion of carbon cycle and albedo dynamics in life cycle impact assessment. Environ Impact Assess Rev 36:2–11

Cashore B et al (2005) Private or self-regulation? A comparative study of forest certification choices in Canada, the United States and Germany. Forest Policy Econ 7:53–69

Deaton BJ (2004) A theoretical framework for examining the role of third-party certifiers. Food Control 15:615–619

Deininger K et al (2011) Rising global interest in farmland: can it yield sustainable and equitable benefits? World Bank, Washington. 13 foll., 163

European Commission, Renewable Energy Progress Report 2017, COM (2017), 57

European Commission Energy Road Map 2050 COM 2011 885 final

European Court of Auditors, The EU system for the certification of sustainable biofuels, special report no. 18, 2016

Fonseca MB et al (2010) Impacts of the EU biofuel target on agricultural markets and land use. JRC, Seville

Food and Agriculture Organisation (FAO) (2008) Soaring food prices: facts, perspectives, impacts and actions required. FAO, Rome

German L, Schoneveld GC (2011) Social sustainability of EU-approved certification schemes for biofuels. CIFOR, Bogor

German L, Schoneveld GC, Pacheco P (2011) Local social and environmental impacts of biofuels: global comparative assessment and implications for governance. Ecol Soc 16:Art. 29; 4 foll

Goldemberg J, Coelho ST, Guardabassi P (2008) The sustainability of ethanol production from sugarcane. Energy Policy 36:2086–2097

Guénéau S, Tozzi P (2008) Towards the privatisation of global forest governance? Int Forestry Rev 10:550–562

IPCC (2007) Climate change 2007 – mitigation of climate change – Working Group III contribution to the Fourth Assessment Report Cambridge. Cambridge University Press, New York

Jumbe CBL, Msiska FBM, Madjera M (2009) Biofuels development in sub-Saharan Africa: are the policies conducive? Energy Policy 37(11):4980–4986

Lamers P et al (2011) International bioenergy trade – a review of past developments in the liquid biofuel market. Renew Sustain Energy Rev 15:2655–2676

Matus K (2010) Assessing challenges for implementation of biofuels sustainability criteria – workshop report. UNEP, Inter-American Development Bank, Washington

McDermott CL, Noah E, Cashore B (2007) Differences that 'matter'? A framework for comparing environmental certification standards and government policies. J Environ Policy Plan 10:47–70

OECD – Round Table on Sustainable Development (2007) Biofules: is the cure worse than the disease? OECD, Paris

Oxfam UK (2007) Bio-fuelling poverty – why the EU renewable-fuel target may be disastrous for poor people. London

Rosillo-Calle F (2007) Overview of bioenergy. Introduction. In: Rosillo-Calle F et al (eds) The biomass assessment handbook: bioenergy for a sustainable environment. Earthscan, London, pp 1–26

Selznick P (1985) Focussing organizational research on regulation Chap. 11. In: Noll RG (ed) Regulatory policy and the social sciences. University of California Press, Berkeley, pp 363–364 van Dam J, Junginger M, Faaij APC (2010) From the global efforts on certification of bioenergy towards an integrated approach based on sustainable land use planning. Renew Sustain Energy Rev 14:2445–2472

van Stappen F, Brose I, Schenkel Y (2011) Direct and indirect land use changes issues in European sustainability initiatives: state-of-the-art, open issues and future developments. Biomass Bioenerg 35(12):4824–4834: 4829

Worldwatch Institute (2007) Biofuels for transport – global potential and implications for sustainable energy and agriculture. Earthscan, London

Zhang Z (2008) Asian energy and environmental policy: promoting growth while preserving the environment. Energy Policy 36(10):3905–3924

Part III
Recent Developments of Soil Regulation at International Level

Implementing Land Degradation Neutrality (SDG 15.3) at National Level: General Approach, Indicator Selection and Experiences from Germany

Stephanie Wunder, Timo Kaphengst, and Ana Frelih-Larsen

1 Introduction

The continuing degradation of land and soils is a severe threat to the provision of ecosystem services and economic development globally. The pressures on land are increasing due to urbanisation, population growth and rising demands for food, feed, fuel and fibre. Halting land degradation is therefore a prerequisite for sustainable development.

In recent years, the concept of 'Land Degradation Neutrality' (LDN) has received increasing attention at the international policy level. Rather than focusing only on halting the loss of healthy and fertile land, 'neutrality' requires also active reversal of degradation by restoring land in order to counterbalance losses that cannot be avoided. With the adoption of the 17 UN Sustainable Development Goals (SDGs), which apply to both developing and developed countries, 'Land Degradation Neutrality' (target 15.3) is now also a part of the 2030 UN Agenda for Sustainable Development. Target 15.3 aims to achieve land degradation neutrality by 2030, and it is now a responsibility of all countries to implement this goal. A review of the past 2 years after the adoption of the SDGs in general and SDG 15.3 in particular has shown that the implementation needs of LDN has served as a 'window of opportunity' for many countries to strengthen policies for sustainable use of land and soils.

Since 'Land Degradation Neutrality' remains a concept that leaves ample room for interpretation, its implementation at national levels requires that the meaning of land degradation within the specific context of each country is defined and ways for how land degradation neutrality can be achieved are outlined. The first attempts to

S. Wunder (✉) • T. Kaphengst • A. Frelih-Larsen
Ecologic Institute, Berlin, Germany
e-mail: stephanie.wunder@ecologic.eu

H. Ginzky et al. (eds.), *International Yearbook of Soil Law and Policy 2017*,
International Yearbook of Soil Law and Policy,
https://doi.org/10.1007/978-3-319-68885-5_11

do so have been undertaken by the UNCCD, e.g. through the LDN 'Target Setting Program', such that by now more than 100 countries have joined. In September 2016, and again in more detail in February 2017, the UNCCD Science Policy Interface also published a 'Conceptual Framework' intended to support the processes to deliver the 15.3 goal.[1]

However, as of April 2017, national implementation is still in the early stages, taking place in only a few countries and mainly based on the first steps of target setting. Ultimately, to implement LDN nationally, one or a set of indicators have to be agreed on, a monitoring system needs to be established and appropriate instruments implemented in order to steer land use and land management into a sustainable direction.

Against this background, this chapter first reviews the concept, the main elements and the current international framework for assessing the progress towards SDG target 15.3, which aims to achieve land degradation neutrality (LDN). It then proposes a stepwise approach to further tailor national requirements with the overall approach of LDN implementation. As the concept of LDN and the internationally agreed implementation framework leaves room for national interpretation of the concept, this chapter then draws on Germany to illustrate how the implementation process has been kick-started. Finally, this chapter outlines how an indicator that combines the assessment of land-use changes (LUC) with the hemeroby (naturalness) can be used to serve as a proxy indicator for LDN.

This chapter builds on the insights of the German research project 'Implementing the Sustainable Development Goals on Soils in Germany', which was carried out by the Ecologic Institute on behalf of the German Environment Agency and the German Federal Environment Ministry. The project was finalised in July 2017 and included more than 50 interviews and three expert workshops—these all served as an important input for this chapter.

2 Concept and Main Elements of Land Degradation and Land Degradation Neutrality

The concept of 'land degradation neutrality' is ambitious as it not only seeks to avoid and reduce degradation of land but also combines this objective with measures to reverse degradation in order to arrive at neutrality/no net loss. Because there is quite some room for interpretation when it comes to national implementation, arriving at a common understanding of the key terms 'Land Degradation' and 'Land Degradation Neutrality' is a starting point for implementing SDG 15.3 and requires an evaluation of the currently used terminology and concepts. Recently, some work has been done to further specify a conceptual approach towards the implementation of LDN. In particular, the UNCCD has been very active in further

---

[1]Orr et al. (2017).

defining the concept through its 'conceptual framework'[2] and through the UNCCD LDN Target-Setting Programme. These have helped many countries to start implementing the LDN targets in their national policies.

Before looking at how the concepts have been defined, it is important to point out the distinction between 'land' and 'soil'—terms that are often used interchangeably. In July 2015, the 12th Conference of the Parties to the UN Convention to Combat Desertification (UNCCD) agreed on a differentiation between the terms. According to the UNCCD, there are overlaps between land and soil, but they do not denote the same thing: 'while soil constitutes one of the most essential natural resources of our planet, the land comprises a multifunctional ecological system, whose natural capital, soil and biodiversity, interacting with water and atmosphere, generate the flow of ecosystem services that support human well-being by securing the life and livelihood of individuals and communities'.[3] Land therefore comprises soil but also consists of many more dimensions and interactions with vegetation.[4]

2.1 *What Is 'Land Degradation'?*

The subject of land degradation has long been the object of scientific and political debate, for example in connection with the subject of desertification, deforestation, soil erosion or certain management approaches such as 'sustainable land management'. With the discussion about a 'land degradation neutral world' in the context of the SDGs, the term 'land degradation' has acquired new and stronger political weight.

The UNCCD has played a major role in establishing the definition and character of the term 'land degradation' at the international level, referring to Article 1 of the text of the UNCCD.

Parts of the UNCCD definition also became part of the official definition for the implementation of Goal 15.3:

> Land degradation is the reduction or loss of the biological or economic productivity and complexity of rainfed cropland, irrigated cropland, or range, pasture, forest and woodlands resulting from land uses or from a process or combination of processes arising from human activities (IAEG-SDGs 2016).[5]

[2]Orr et al. (2017).

[3]UNCCD (2015), para. 22).

[4]Stavi and Lal (2015), see also Article 1 UNCCD (2015).

[5]IAEG-SDGs (2016).

At the core of this definition is the production of functions and economic value of land for agriculture and forestry, which are promoted by human activities and negatively manifest themselves in the form of measurable phenomena such as erosion, soil quality loss and loss of vegetation.

In contrast to the UNCCD definition, other sources such as the IPBES and the Global Environmental Facility (GEF) have a broader and less specific interpretation of land degradation and see it more as the loss or reduction of ecosystem services and functions and do not differentiate between human activities or natural degradation processes. All definitions share the understanding that land degradation includes both the absolute and the relative/partial loss of ecosystem functions.

Despite the attempts to clearly define land degradation, it remains a fuzzy concept or, as Caspari, van Lynden and Bai (2015) put it, a 'blurred entity',[6] because land degradation has various dimensions, occurs at multiple scales, involves a great variety of actors and above all is value laden. This has to be kept in mind when discussing land degradation neutrality in detail.

2.2 *What Is 'Land Degradation* Neutrality*'?*

Land degradation neutrality was first introduced as a concept in the run-up to the Rio+20 conference. In this context, the UNCCD Secretariat published a policy brief on a potential goal of 'zero net land degradation'.[7] In the outcome document of Rio +20 entitled 'The future we want', the heads of state and government 'recognize the need for urgent action to reverse land degradation. In view of this we will strive to achieve a land degradation neutral world in the context of sustainable development.'[8]

As mentioned above, the term is included as target 15.3 under Goal 15 of the Sustainable Development Goals:

> Goal 15. Sustainably manage forests, combat desertification, halt and reverse land degradation, and halt biodiversity loss.[9]

Target 15.3 states:

> By 2030, combat desertification, restore degraded land and soil, including land affected by desertification, drought and floods, and strive to achieve a land degradation-neutral world.[10]

The definition used in the context of the SDG implementation at the international level makes clear reference to the third decision of the 12th COP of the UNCCD (Dec 3/COP12). This decision was provided by the UNCCDs Intergovernmental

[6]Caspari et al. (2015).

[7]UNCCD Secretariat (2012).

[8]UN General Assembly (2015).

[9]UN General Assembly (2015).

[10]UNDESA (2016).

Working Group (IWG) and represents the general political as well as scientific consensus on the term:

> Land degradation neutrality is a state whereby the amount and quality of land resources necessary to support ecosystem functions and services and enhance food security remain stable or increase within specified temporal and spatial scales and ecosystems (IAEG-SDGs 2016).[11]

The core and innovative part of this definition is that in order to 'remain stable or increase' efforts for land restoration, rehabilitation and sustainable management are necessary if degradation processes cannot be avoided.

Nonetheless, due to the many ambiguous terms in the definition and inherent conflict of objectives, the definition of land degradation neutrality needs further specification, technical guidance and adaptation to the regional context in order to be implemented. This includes, for example, the following issues:

- *Baseline*: In order to assess if land resources remain stable or increase a baseline needs to be set. Given the shortcomings of global available data, practical solutions need to be found.
- *Ecosystem functions*: the definition broadly refers to ecosystem functions and services. However, there are often trade-offs between provisioning, regulating and cultural services that need to be taken into account (e.g., intensifying agricultural food production increases provisioning services but often decreases regulating services).
- *Balancing quality and quantity of land degradation and restoration*: both the 'amount and quality of land resources' should remain stable or increase. However, questions remain regarding how degradation and restoration can be balanced: within or across ecosystem types? Within what spatial scale and time frame? How to deal with the different quality of land degradation processes (i.e., complete loss versus certain reduction in services and functions)?

In an effort to make the land degradation neutrality concept more concrete, the SPI of the UNCCD defined a number of principles[12] to be followed by all countries that choose to pursue LDN. They state that 'there is flexibility in the application of many principles but the fundamental structure and approach of the framework are fixed, to ensure consistency and scientific rigour'.[13]

---

[11] IAEG-SDGs (2016).

[12] 19 principles that include requirements such as to "protect rights of land users" and "respect national sovereignty", but also more specific principles such as to "apply response hierarchy: avoid, reduce, reverse land degradation" and "counterbalance like for like".

[13] UNCCD/Science Policy Interface (2016), Orr et al. (2017).

2.3 International Agreements on Indicator Selection and Monitoring of LDN

In March 2016, half a year after the adoption of the SDGs, an agreement on indicators for the SDGs was achieved. For SDG 15.3, the agreed indicator is the 'proportion of land that is degraded over total land'. It is defined as the amount of land area that is degraded. The measurement unit for indicator 15.3.1 is the spatial extent (hectares or km^2) expressed as the proportion (percentage) of land that is degraded over total land area.[14]

The minimum set of indicators recommended (but not compulsory) for tracking progress towards LDN against a baseline is as follows:

- **land cover**;
- **land productivity** (metric: net primary productivity);
- **carbon stocks above and below ground** (metric: soil organic carbon).

These indicators are part of a set of six progress indicators used by the UNCCD to track progress in the implementation of the Convention through national reporting. They have also been included as suggested indicators for the implementation of target 15.3 and provide a common ground for monitoring and reporting progress towards SDG target 15.3.

The rationale behind the three indicators is spelled out in the report of the Global Mechanism of the UNCCD 'Building Blocks for LDN Target Setting'[15]:

- Land cover and land-cover change has multiple applications for evaluating progress towards various SDG targets and gives a first indication of land degradation and provides a first indication of a reduction or increase in vegetation, habitat fragmentation and land conversion.
- Land productivity points to long-term changes in the health and productive capacity of the land.
- Soil organic carbon denotes overall soil quality.[16] On seasonal to decadal timescales, carbon stocks of natural and managed systems may be explained largely by changes in plant biomass (fast variable). However, over the long term, soil organic carbon stocks (slow variable) is also a relevant indicator of the overall functioning of the ecosystem, its adaptive capacity and resilience to perturbations (e.g., floods, drought).

However, while these sub-indicators are a recommended starting point, they do need a national adjustment. In the accompanying document to the SDG indicators provided by the UN Statistical Commission (UNSTAT) and the Inter-Agency and Expert Group on Sustainable Development Goal Indicators (IAEG) with support

[14]IAEG-SDGs (2016).

[15]Global Mechanism of the UNCCD (2016a).

[16]Global Mechanism of the UNCCD (2016a).

from the UNCCD secretariat, it is therefore stated that 'As the sub-indicators will never fully capture the complexity of land degradation processes, there will always be a need for other relevant national or sub-national indicators' and 'further work is needed to provide a standardized approach and "good practice guidance" to derive the sub-indicators and help build monitoring and reporting capacities at the national, regional and global levels'.[17]

The current approach for measuring and assessing LDN, as described by the Inter-Agency and Expert Group on Sustainable Development Goal Indicators,[18] argues for three steps and the adoption of a mixed-method approach, which makes use of multiple sources of information (e.g., Earth observation, geo-spatial information, biophysical variables):

(a) **Set baselines** to determine the initial status of the sub-indicators in absolute values. This would include (1) the preparation of base land cover information that builds on standard land cover ontology (e.g., LCCS/LCML), (2) the establishment of a baseline for land productivity (e.g., NPP/NDVI) and (3) the establishment of a baseline for carbon stocks, above and below ground, with an emphasis on soil organic carbon below ground and building on the IPCC's work on carbon above ground.
(b) **Detect change** in each of the sub-indicators, including the identification of areas subject to change and their validation or evaluation by a participatory national inventory of land degradation, particularly where change in two or three of the sub-indicators coincide or overlap spatially. When contextualised with information at the national and sub-national levels, areas with declining productivity and carbon stocks may be considered degraded, while areas with increasing productivity and carbon stocks may be considered improving. The definition of adverse or desirable land cover changes is highly contextual and needs to take into account local ecological and socio-economic circumstances that require in-situ validation.
(c) **Derive the indicator** by summing all those areas subject to change, whose conditions are considered negative by national authorities (i.e., land degradation) while using 'good practice guidance' in their measurement and evaluation of changes within each sub-indicator and their combination.

While the document of the IAEG suggests that 'areas with declining productivity and carbon stocks may be considered degraded' (suggesting that if two of the three indicators show a negative trend, land is to be considered as degraded), more recent documents provided by the UNCCD refer to a 'one-out, all-out' approach. This means that 'if any of the three indicator metrics shows significant negative change, it is considered as a loss (and conversely, if at least one indicator/metric shows a significant positive change and none show a significant negative change it is considered a gain)'.[19]

---

[17] IAEG-SDGs (2016).

[18] IAEG-SDGs (2016).

[19] UNCCD/Science Policy Interface (2016).

However, since the application of all three indicators is not compulsory for the implementation of LDN, what indicators countries choose to complement the proposed set of indicators will only be revealed in the next few years. Yet a largely common application of these indicators would bear the advantage that reporting on land degradation from the various countries could be compared at international level.

3 Steps and Guiding Questions to Implement LDN on a National Level

Due to the open questions derived from the incomplete definition of LDN (see Sect. 2), many aspects of the concept need to be concretised in order to guide countries in applying the concept in practice.

Moreover, the SDG targets are defined in the 2030 Agenda as 'aspirational and global', with each government tailoring its own national targets and indicators 'guided by the global level of ambition but taking into account national circumstances'. So there is not only the room but also the intention to tailor the targets and their monitoring according to national needs.

Due to the complexity of land degradation, countries will need guidance for the LDN implementation in their national policies. This chapter will outline a stepwise approach for advancing the implementation of the LDN concept. The approach is divided into seven key strategic steps:

1. Define and tailor LDN in the national context.
2. Define suitable indicators.
3. Define baseline and set targets.
4. Specify the spatial dimension.
5. Determine compensation mechanisms.
6. Set up and maintain monitoring system.
7. Improve enabling environment.

The activities of the UNCCD and the Target-Setting Programme are a valuable source of learning for implementing LDN as many activities already started before the SDGs were adopted. For example, a summary about the experiences made and lessons learned with regard to the 14 UNCCD LDN pilot projects[20] and the summary on 'Building blocks for LDN Target Setting'[21] are very useful inputs for designing national policies. Another recent publication of the UNCCD Science Policy Interface provides a Scientific Conceptual Framework for Land Degradation

[20]Global Mechanism of the UNCCD (2016b).

[21]Global Mechanism of the UNCCD (2016a).

Neutrality, including LDN principles and suggestions for monitoring and effective implementation.[22]

Accordingly, the list of key steps and guiding questions builds on recent activities, experiences and publications of the UNCCD but also goes beyond to include new thoughts on further aspects to be considered.

3.1 Define and Tailor LDN in the National Context

Different biophysical and climatic preconditions, and also varying economic developments in different countries, result in a wide spectrum of soil threats, drivers of land degradation and trends in land degradation. Some soil threats are less relevant in some countries, such as salinisation in Germany or landslides in the Netherlands, but more relevant in others. An important starting point to better understand the national context and particular needs is to carry out an assessment of historic and ongoing land degradation trends and to identify the relevant types of land degradation.

The first step therefore involves a thorough analysis of the biophysical environment and socio-economic aspects[23] of the country. Building on the key drivers of land degradation, scenarios that forecast gains and losses if current land-use trends continue and planned projects and policies are implemented should be compiled showing different alternatives of future development.[24] In parallel, efforts to work on a long-term perspective of land use, i.e. 'the progressive development of a manageable final landscape', should be undertaken.[25]

The scenarios can help to make decisions on national priorities and goals in the context of land degradation, for example if LDN should particularly enhance the productivity of the land for agriculture or if natural soil functions should be restored.

Multi-stakeholder engagement needs to be secured from the very beginning of such target and priority-setting process.[26] Experiences made in the 14 UNCCD pilot projects showed that involving existing institutions that have already dealt with LDN-related issues in the past proved more effective for most countries than creating completely new consultation bodies.[27, 28] The assessment also showed

[22]UNCCD/Science Policy Interface (2016), Orr et al. (2017).

[23]Akthar-Schuster et al. (2016b).

[24]Global Mechanism of the UNCCD (2016a).

[25]Akhtar-Schuster et al. (2016b).

[26]Global Mechanism of the UNCCD (2016a).

[27]Global Mechanism of the UNCCD (2016b).

[28]The experience of pilot countries also suggests that working groups should at least meet twice a year with the sole purpose of reviewing the LDN target setting process. The process should also involve the Ministry of Finance and Planning early in the process (Global Mechanism of the UNCCD 2016b).

that a high level political commitment (e.g., from ministries) is a key factor for successful LDN target setting and implementation.[29]

3.2 *Define Suitable Indicators*

Suitable indicators have to be developed considering both the national priorities of LDN (see 2.1.) as well as data availability and representation of all sectors (e.g., the indicators' relevance for agricultural, forestry and urban land).

In many of the 14 UNCCD pilot countries, the choice of the three proposed indicators (see 2.3) provided 'a simple, practical way to consistently and uniformly assess the extent of land degradation. (. . .) Some countries used global data as the main source, while others used national data in combination with – or as an alternative to – global data, according to their needs and capacities. In isolated cases, significant differences between global and national data were found.'

However, that does not automatically mean that the three proposed indicators are suitable for measuring LDN in all countries. Priorities in soil and land degradation trends, as well as existing national reporting activities, could direct attention onto other indicators or monitoring concepts. That also involves the question of how an aggregated indicator could be developed to reduce efforts in data collection and aggregation of different indicators in an overall LDN balance. When developing such an indicator, it needs to be kept in mind that every aggregation step bears a potential loss or distortion of information. This should be considered when choosing a methodology.[30] Section 5 of this chapter provides a suggestion for an integrated/proxy indicator for LDN based on land-use change assessments combined with the hemeroby concept.

3.3 *Define Baseline and Set Targets*

SDG 15.3 sets the year 2030 as the target date to *aim for* global LDN. Strictly speaking, LDN does not have to be *achieved* by 2030 as the goal formulates 'to *strive to achieve*' by 2030. The time factor has two aspects: first, it depends on the baseline year that the LDN target refers to. Second, it needs to be settled at which point the intended target state should be achieved.

From the text of SDG 15.3 and the UNCCD definitions, it remains uncertain which reference state should apply to neutrality, i.e., from which original condition soil should not be more degraded by 2030. The UNCCD definition merely states that there is a reference period, but it is not specified further: 'the amount and

[29]Global Mechanism of the UNCCD (2016a).

[30]Feldwisch et al. (2006).

quality of land resources ... remain stable or increase within specified temporal ... scales and ecosystems'.

Consequently, countries have to define their own 'baseline'. Given the ambiguity of the LDN definition, they have room to set different levels of ambition, depending if they want to achieve 'stable' conditions of land resources or even want to achieve an improvement of the current situation. Earlier reference dates (for example the year 2000) are usually more ambitious than later ones (e.g., 2016) as land degradation has increased almost everywhere in the past decades. If a reference year is even set in the future (e.g., 2020), land-use changes that are implemented prior to that year will not be considered, which could have harsh consequences regarding further land degradation (see, for example, political discussions about the relevance of reporting dates for the conversion of grasslands and the removal of landscape elements in NABU and DVL 2014[31]).

Once the baseline is set, a clear understanding of the intended target state in 2030 needs to be developed. It has to be taken into account that compensatory or restoring measures can vary significantly in terms of their temporal effect. In other words, soil decontamination and restoration processes usually need markedly more time than soil degradation processes. The question therefore is how different effects and timespans between degradation, regeneration and restoration should be balanced against the target state in 2030. This leads to further questions such as: should degradation already encompass small or early decreases in soil function[32] or only threats that can lead to a massive loss of soil functions/soil-related ecosystem functions?[33] What is a 'significant' change in soil quality, e.g., is an increase from 47 to 50 tC/ha already 'significant', or should the 'threshold' be higher or lower?[34] Such questions must also be regarded in consideration of national experiences and existing assessment frameworks.

During the UNCCD pilot project, most countries also analysed the financial feasibility of the measures required to meet the proposed targets. Some set several targets with different levels of ambition, according to their respective capacities and potential financing opportunities.[35]

3.4 *Specify the Spatial Dimension*

For the implementation of land degradation neutrality, the spatial scale for balancing degradation against restoration and regeneration must be determined. Depending on the national context, it could be beneficial to separate the country in

[31] Cf. NABU and DVL (2014).

[32] Akhtar-Schuster et al. (2016a).

[33] König (2016).

[34] UNCCD/SPI (2016).

[35] Global Mechanism of the UNCCD (2016b).

different regions with different geographical conditions (e.g., mountainous areas and plains) or administrative units (e.g., provinces or federal states). The International Union for Conservation of Nature and Natural Resources[36] argues for a stronger orientation on ecological parameters and proposes the ecosystem level for balancing land degradation. Similarly, the Conceptual Framework of the UNCCD/SPI[37] argues for a 'landscape level' (such as catchment areas). Furthermore, it argues for counterbalancing 'like for like', which means that compensation measure could apply only within the same land (use) category. It has to be noted that the smaller is the scale of balancing, the larger are the efforts of aggregating results in order to track degradation neutrality at national scale. Such a decision also involves the question of which administrative unit or body is responsible for the data collection and analysis.

Many experts who were interviewed within the German research project 'Implementing the Sustainable Development Goals on Soils in Germany' raised the point to even include 'extra-territorial effects' in the national balance for LDN, given that the 'virtual net import of land' through the consumption of imported goods puts pressure on land resources in other countries.

3.5 Determine Compensation Mechanisms

Along with considering the spatial-temporal dimension of the LDN concept, compensation mechanisms need to be established. However, beforehand it must be clarified if and to which extent land and soil degradation is taking place. For this, information about the state and management of the concerned areas needs to be compiled and analysed. Two possible approaches for the determination of degradation are as follows:

- *Benchmarks*: reaching a fixed benchmark (e.g., concentration of pollutants, amount of soil organic carbon) would give insight into whether a soil area is degraded or recovered. However, due to the very complex characteristics of soils, it is nearly impossible to deduce such an absolute benchmark from a scientific perspective (except for some pollutant concentrations). Furthermore, benchmarks are usually defined in terms of impairments of soil functions—and for many impacts, it is almost impossible to set fixed benchmarks for degradation as it is usually a gradual process. Finally, all impacts that are severe but do not reach the set benchmark would not be considered in LDN accounting or considered in the net balancing and would thus not have to be compensated.
- *Dynamics of change*: the benefit of observation and assessment of the dynamics of improvement and degradation is that they make it possible to gain a quick

[36] IUCN (2015).
[37] UNCCD/SPI (2016).

overview, as long as indicators are available. However, the question remains at which point compensation measures are to be undertaken when the initial condition of the considered areas varies significantly.

As outlined above, degradation must be counterbalanced against compensation measures. The UNCCD Secretariat differentiates between three main measures that can prevent or reverse land degradation:[38]

- *Natural regeneration*: this involves avoiding and reducing anthropogenic impacts (for a set period of time) on the degraded area to ensure regeneration through natural processes, e.g., laying fallow. However, by solely utilising regeneration, degraded ecosystems have little chance to reach their original state.[39]
- *Improved land-use practices* (sustainable land management): sustainable land management (SLM) can lead to an improvement of soil quality and thereafter stabilises the state of the soil.[40] The FAO defines SLM as 'the use of land resources, including soils, water, animals and plants, for the production of goods to meet changing human needs, while simultaneously ensuring the long-term productive potential of these resources and the maintenance of their environmental functions'. SLM represents regionally adapted land-use systems that often make sense only at local level. In April 2014, the UNCCD recommended the World Overview of Conservation Approaches and Technologies (WOCAT) as an important SLM database. More than 470 technologies and 235 approaches for SLM are presented there. The UNCCD identifies several SLM practices, such as mulching, zero tillage, green manuring and water harvesting.[41]

  In contrast to SLM, landscape management is a concept with a broader scope.[42] Unlike SLM, the Landscape Management Approach and the interlinked monitoring evaluation scale focus on the landscape level and thus on the interdependence of ecosystems. This management approach embraces important functional interlinkages.
- *Restoration* (human activity to restore the natural basis of an ecosystem): active restoration of ecosystems is necessary when the degree of degradation is too high to utilise the land productively and natural regeneration is not practical or too slow. The Convention on Biological Diversity defines the restoration of ecosystems as an active management process to restore a degraded, damaged or destroyed ecosystem for the conservation of ecosystem resilience and biodiversity. However, this definition does not specify at which state an ecosystem is restored, so this has to be reassessed separately for each ecosystem. A variety of

[38]UNCCD (2012).

[39]Ngo (2015).

[40]Gnacadja (2012).

[41]UNCCD/SPI (2015).

[42]Dernier et al. (2015).

ecological factors play a role in the composition of species and functional groups that define the stability or resilience of ecosystems.[43] Restoration of ecosystems demands a lot of time, as well as physical and financial expenditures. However, this approach is worthwhile if long-term effects and cost reduction are considered.[44] In the global analyses by 'Economics of Land Degradation', it was stated that the financial benefits of investing in the restoration of ecosystems is up to five times higher in many regions than the associated costs for a period of 30 years.[45]

For the practical implementation, the UNCCD proposes a clear **hierarchy** between these compensation measures:[46]

- Interferences with ecosystems should be **avoided as a first priority**.
- If this is impossible, **negative impacts should be reduced**.
- If both are impossible, **negative impacts should be compensated** (in another location).

Furthermore, a compensation principle should be implemented that focuses on the ecosystem-based interlinkages between degradation and restoration:

1. More area should be restored than degraded (especially due to the time lag between rehabilitation and the uncertain effects of taken measures).
2. Compensation measures should be applied in similar ecosystems (i.e., the same ecosystem type).
3. Compensation measures should be in situ or as close to the area of degradation as possible.[47]

3.6 Set Up and Maintain LDN Monitoring System

To monitor and evaluate LDN achievements, a centralised land management/land degradation monitoring and evaluation information system must be established. This monitoring and evaluation system should be institutionalised within an appropriate permanent body to facilitate cross-sectoral collaboration. Whenever possible, such a system should be based on existing monitoring and evaluation systems. The information generated by these systems must be accessible to all authorities that have an impact on land use.[48]

[43]cf. Tucker et al. (2013).

[44]de Groot et al. (2013).

[45]Nkonya, Mirzabaev, and von Braun (2016).

[46]UNCCD Global Mechanism (2016).

[47]Chasek et al. (2015).

[48]Global Mechanism of the UNCCD (2016a).

Moreover, land cover/use, land productivity dynamics and soil organic carbon databases and data processing methodologies must be further enhanced both at national and global levels (measurement accuracy, resolution, periodicity) to ensure effective monitoring of progress made towards the achievement of LDN targets. Although sustainable land management can be easily monitored using the land-cover/land-use change/land productivity indicators, further development of the soil organic carbon indicator is essential in connection with climate change policies.[49]

3.7 *Improve Enabling Environment*

Improving the enabling environment not only relates to the further development of instruments and capacity building of institutions but also includes awareness raising and communication to foster support of the LDN process.

To achieve LDN, possible (new) interventions (e.g., analysis of legal, economic, social and political enablers) and measures need to be considered.[50] It is also essential, to ensure an enabling environment and responsible governance of land resources, including land tenure, to establish mechanisms for integrated land-use planning and to have multi-stakeholder platforms and frameworks at local, national and regional levels to collaborate in planning, implementing, monitoring and evaluating LDN interventions.[51] Also, policies that incentivise sustainable land use need to be (put) in place. Moreover, the targets and proposed measures/ interventions all need to be transposed in the relevant spatial planning tools.

For the identification of measures to achieve the targets, the selection of 'bright spots' (success stories for further learning and communication on how to address land degradation), in addition to the conventional 'hot spots' (as areas for priority intervention), appeared to be successful in spreading the LDN concept in the 14 pilot countries.[52]

Engagement of all stakeholders in the process is essential to facilitate 'buy in' and ownership to the policies associated with the LDN targets set by governments. Such political support makes the upscaling of sustainable land management and restoration activities possible. For this to become a reality, LDN training and capacity building must be strategised.[53]

Having outlined what we consider are the necessary steps for implementing LDN at a national level, the following section reflects how the actual process in Germany to implement LDN started and where it currently (April 2017) stands. It did not (yet) follow the steps as outlined above. Rather, the stepwise approach was

[49]Global Mechanism of the UNCCD (2016a).

[50]Akhtar-Schuster et al. (2016b).

[51]Global Mechanism of the UNCCD (2016b).

[52]Global Mechanism of the UNCCD (2016b).

[53]Global Mechanism of the UNCCD (2016b).

developed within the mentioned German research project, which was in itself a step to start the implementation of LDN in Germany.

4 German Activities to Kick-Start the Implementation of LDN

4.1 SDG Implementation in National Policies

Following the adoption of the UN SDGs in September 2015, Germany has committed to implement the SDGs with high ambition. Germany was also among the first 22 countries that presented their progress at the UN High Level Political Forum (HLPF) in July 2016 in New York.

The German government has chosen the National Sustainable Development Strategy (*Nachhaltigkeitsstrategie*) as the key framework for achieving the SDGs in Germany. The first National Sustainable Development Strategy was adopted in 2002, setting out national sustainability goals and indicators. Since then, the government has reported on its implementation status every four years in the form of progress reports that also update the strategy's content. Every 2 years, the Federal Statistical Office publishes an independent indicator report with information about progress towards meeting the goals. The revised strategy, integrating Agenda 2030s ambition and goal structure, was published in January 2017.

The existing indicators of Germany's National Sustainable Development Strategy with relevance to land and soil include, for example, nitrogen surplus, area under organic farming and species diversity. Most importantly, the strategy includes an indicator on land take ('Built-up area and transport infrastructure expansion', in German: 'Flächeninanspruchnahme') with the objective to reduce expansion of built-up area and infrastructure to a maximum of 30 ha/day by 2020.

However, the German government also sees the need for a new indicator for land and soil, particularly to implement SDG target 15.3 on land degradation neutrality but also in order to support the French '4 per 1000' initiative presented at the UNFCCCs COP 21, which aims to enhance organic matter in soils.

The current draft of Germany's Sustainable Development Strategy announced that there is ongoing work to design an appropriate indicator. There is no official time frame, but a new indicator might become part of the revised version of the German Sustainability Strategy in 2018.

4.2 A Research Project Helps to Kick-Start the Process

To explore how the soil-related SDGs can be implemented in Germany, the German Environment Agency and the German Federal Environment Ministry commissioned the above-mentioned research project, 'Implementing the Sustainable Development Goals on Soils in Germany'.[54] It ran from October 2015 to July 2017 and was carried out by the Ecologic Institute. The project helped to initialise the national discussions on the options for implementing LDN in Germany. A key project objective was to discuss appropriate indicators that can complement the German Sustainability Strategy and help in monitoring the implementation of LDN in Germany.

The focus here is laid on *soil* indicators. Since the terms 'land' and 'soil' have many overlaps but also differences, this is important to note. The focus on soil is also due the fact that the English word 'land' does not translate clearly to German but can be translated to 'Boden' (soil), 'Fläche' (i.e., 'surface area') or 'Land' ('country' or 'countryside').

Within each country, the definition of indicators requires the LDN concept to be made more concrete, as well as the identification of the perceived main threats for land and soil (see Sect. 3). Therefore, a first step within the project was to conduct a literature review to identify the most important soil threats and soil functions in Germany. In addition, 40 expert interviews were carried out in order to collect expert opinions on the question of whether certain soil functions and soil threats can be prioritised over others. The interviews clearly showed that creating a general hierarchy of soil *functions* is neither possible nor desirable.

However, for soil *threats*, the majority of the experts who replied to this question argued that soil sealing and land take are of particular relevance for Germany. Other soil threats that were mentioned as particularly relevant were erosion, loss of soil organic matter, compaction, contamination and nutrient overload.

The report also compiled possible soil quality indicators, their strengths and weaknesses in the context of LDN, as well as an overview of existing databases and monitoring opportunities that could be applied.

The preliminary results of the report were then discussed at an expert workshop on 6 July 2016 at the German Federal Environmental Ministry. The debate showed that an agreement on one or two main indicators for LDN that might be used for the monitoring was not yet possible. However, the discussion confirmed the results of the expert interviews about the main soil threats. In addition, there was a strong interest to further investigate whether an indicator on land-use change (i.e., focusing on land management, compared to simply a biophysical indicator) can be used for monitoring LDN.

The principle idea of this approach is that information on land-use practices and land-use changes (such as conversion of grassland to arable land, sealing of land, extension of grazing periods, etc.) allow for certain conclusions about positive or negative impacts on soils, thereby saving a lot of effort compared to measuring the biophysical effects of every single case.

---

[54]http://ecologic.eu/12876.

Within the above-mentioned research project, we therefore developed an approach to use land-use change as a proxy indicator for LDN, taking Germany as an example.

5 Land Use Change as a Proxy Indicator for LDN

In this last section, we present an approach for how LDN can be assessed at national level, taking Germany as an example. To do so, we take land use and land use change as a proxy indicator and align it with the hemeroby concept. The general assumption of this approach is that changes in land use directly correspond with changes in the natural functions of soil and soil quality and that some land uses are more beneficial for protecting natural soil functions than others.[55]

Furthermore, we assume that the degree of human impact (which is 'hemeroby') strongly correlates with the ecological significance of natural soil functions. In other words, if soils are less disturbed by human activities (such as cultivation or sealing), soil functions can be better preserved or regenerated.

The approach also has the following benefits:

- It avoids biophysical soil indicators (such as erosion, compaction, etc.), which are difficult to measure and imply a high level of monitoring effort.
- It uses a simple approach, which is applicable to existing conditions in Germany, as well as to existing data recording.
- The approach avoids conflicting trade-offs between different soil functions, at least at the conceptual level. The focus lies on the preservation of natural soil functions and not on utility or other functions of soils. We assume that (at least in Germany) land degradation neutrality is only achieved if natural soil functions remain stable.

Below we describe the conceptual elements of the approach, which was already discussed with experts in interviews and in an expert workshop. However, what is presented here can only serve as a first conceptual approach of how LDN can be assessed through a land-use change indicator at national level. Further refinement of the indicator (value scales, consideration of restoration time within the values given, etc.) is needed before it can be applied in practice or be underpinned with political goals or instruments. While there is a positive response to the general concept by the responsible ministries for environment and agriculture in Germany, there is still no final agreement if the concept presented here will be used to develop a new soil quality indicator within the 2018 version of the German Sustainability Strategy.

[55]Cf. FAO and ITPS (2015), Azeez (2009), Mal et al. (2015), Paulsen et al. (2013).

5.1 Categories for Land Use and Land-Use Change

Assessing land use and land-use change in the context of land degradation neutrality requires a clear definition of land-use categories that allow for conclusions to be drawn about the potential impact on natural soil functions. The derived categories should reflect the land-use categories that are already used by statistical agencies and therefore build on existing monitoring activities. In Germany, the federal statistical agency (DESTATIS) differentiates between eight categories, which are further divided into several sub-categories. Data for these categories is regularly recorded by local cadastral land registers. The eight main land-use categories are as follows:

- building and adjacent open land (*Gebäude- und Freifläche*);
- commercial/industrial land (including mining land) (*Betriebsfläche darunter auch Abbauland*);
- recreational land (*Erholungsfläche*);
- traffic areas (*Verkehrsfläche*);
- agricultural land (including arable land, pastures, gardens, vineyards, peatlands, heaths, orchards, agricultural settlements and fallow land) (*Landwirtschaftsfläche*);
- forests (*Waldfläche*);
- water surfaces (*Wasserfläche*);
- other land uses (for example, cemeteries).

In the context of LDN, it is important to note that within the land categories for traffic and settlement, not all areas are sealed. This distinction is important as the sealing of land in most cases goes along with severe loss of soil functions, while, for example, the use of urban gardens, unpaved ways, etc. still belong to building areas but often have considerably less negative impacts than sealed soils and can even be more soil friendly than, e.g., some agricultural uses. In Germany at least, a great share of land (approx. 50%) classified as traffic and settlement consists of unsealed areas such as green areas or roadside vegetation.

Moreover, for ecological assessment of land use, it is necessary to further divide agricultural land in the various sub-categories listed above because each of them features significant differences in soil quality and many hold great shares in total land area.

Despite the need to consider existing national monitoring systems, it is also helpful to review other land use classifications, such as those used by the Intergovernmental Panel on Climate Change (IPCC). Most land and land-use change assessments conducted in academic and political contexts build on the IPCC categories of forestry, cropland, grassland, wetland, settlement and other land, which are further divided or adapted depending on the overall purpose of the assessment or the geographical context. In a recent German study, additional

Table 1 Comparing different land-use categories

IPCC categories	German system DESTATIS (basic categories, first level)	Sub-categories according to Untenecker u. a. (2017)	CORINE land cover (2. level)
Forest	Forest	Forest	Forests
Cropland	Agricultural land	Arable land Horticulture	Arable land Permanent crops Heterogeneous agricultural areas
Grassland		Grassland Heathland Shrub land	Pastures Scrub and/or herbaceous vegetation associations
Wetland	Water surface	Fen Peatlands Water body	Inland wetlands Maritime wetlands
Settlement	Settlements and open land Industrial and commercial land Land for recreation Traffic areas	Settlement	Urban fabric Industrial, commercial and transport units Mine, dump and construction sites Artificial, non-agricultural vegetated areas
Other land	Other uses	Abandoned land Fallow land Others	Open spaces with little or no vegetation

sub-categories have been defined in order to assess long-term changes in land use and management in several federal states.[56]

At EU level, the CORINE Land Cover System distinguishes between three levels of land-use categories. The first level divides land-use forms into one of the following groups: (1) artificial areas, (2) agricultural areas, (3) forests and semi-natural areas, (4) wetlands and (5). water bodies. The second level is highlighted in Table 1 (water bodies not included). The third level is an even further refinement of the second level consisting of 44 land-use categories in total.

All of the listed land-use classifications, however, do not distinguish between land-use intensity within each of the land-use categories. In other words, the division between categories might be very detailed, but it cannot be detected how, for example, intense forest management is within the category 'forest'. For LDN, however, this is very important as land-use intensity has a strong impact on soil functions and soil (quality) parameters.

Similarly to the question of the intensity of land management within one category, there might also be changes in land use that do not lead to a change within categories (e.g., changes within different agricultural land uses) but may still

[56]Untenecker et al. (2017).

be significant. The conversion of conventional farming into organic farming, for example, might happen without changing land-use categories (arable land remains arable land, the same for grassland), but the farm manager reduces inputs such as pesticides or artificial fertilisers and adapts soil management practices, which affect soil quality.

Therefore, a systemic approach of land-use change within the context of LDN must distinguish between two dimensions of land-use change:

1. changes between different land uses (for example, grassland conversion to arable land or sealing of agricultural land for settlements or roads);
2. changes with regard to soil management within a specific land-use category (for example, changing grassland, agricultural practices or forest management).

A system of land-use classifications will need to differentiate both dimensions due to aspects of severity of impacts and because of the time aspects of restoration and degradation. Specifically, land-use changes between categories of the first dimension often have a more dramatic impact on soil functions than changes in the second dimension. Deforestation, logging and conversion of arable land cause impacts right after the intervention (e.g., loss of soil organic matter, water content in the soil). Restoring activities, on the other hand, such as reforestation or rewetting of peatlands, which also belong to the first dimension of land-use change, normally take much longer. Similarly, effects of the second dimension usually become only visible over longer time spans (often years).

5.2 *Integration of the Hemeroby Concept into a Land-Use Classification*

The proposed approach significantly builds on Fehrenbach et al. (2015),[57] who applied the hemeroby concept to assess land use and land-use change in life-cycle assessments (LCA). Based on a hemeroby classification, they assign numeric values to certain land uses according to their 'naturalness'. The classification divides between seven hemeroby classes, from 'I natural' to 'VII non natural'. When changes between categories occur, the value descents (or ascents) by 0.5. Value 1 represents the maximum distance to a natural ecosystem; value 0 would mean that practically no human impact exists on the respective ecosystem (Table 2).

In order to make the hemeroby concept applicable for assessing LDN, we simplify the land-use categories used in Fehrenbach et al. (2015),[58] resulting in new LDN land-use categories that are more strongly aligned with the categories used by the Federal Statistics Agency (DESTATIS, see above) and which can be easily detected by modern remote sensing (RS) techniques. Also, we broaden the

[57]Fehrenbach et al. (2015).

[58]Fehrenbach et al. (2015).

Table 2 Classification of hemeroby for different land uses (adapted from Fehrenbach et al. 2015)[a]

Hemeroby class		Forestry	Agriculture	Other
I	Natural	–	–	Undisturbed ecosystem, pristine forest, no utilisation
II	Close to nature	Close-to-nature forest management	–	
III	Partially close to nature	Intermediate forest management	Highly diversified agroforestry systems	
IV	Semi-natural	Semi-natural forest management	Close-to-nature agricultural land use, extensive grassland, orchards, etc.	
V	Partially distant to nature	Mono-cultural forest	Intermediate agriculture, moderate intensity, short rotation coppices, fertilised grassland	
VI	Distant to nature	–	Large-area, highly intensified arable land in cleared landscape	Solar fields, windparks
VII	Non-natural	–	–	Long-term sealed, mining lands, landfills

[a]Adapted from Fehrenbach et al. (2015)

scale of values for the different land-use categories. When arranging the new land-use categories with the hemeroby classes, they obtain a numeric value for the natural soil functions (see Table 3).

Unlike Fehrenbach et al.,[59] we chose an iterative scale with 0.5 intervals from one category to another instead of dividing the categories in half. The broader scale allows for a more flexible placement of every category due to its value for the natural soil functions. According to the new scale, natural forests have a maximum value of 6 for natural soil functions, whereas areas for settlements and traffic have a value of 0. Contaminated areas and landfills represent the lower end of the scale with a value of −1.5. Wetlands and peatlands have the highest value, equal to natural forests, which means that they also have the highest value for soil preservation.

The intervals between the categories, however, still need to be further elaborated between soil scientists and need to be discussed with national policy makers. After all, any assignment of values will reflect societal values and priorities, which also puts requirements on the transparency of the process.

With assigning a soil value for all land-use categories, land-use change can be assessed in the context of LDN. With regard to a certain geographical region, land-use change can be calculated by the numeric interval between respective land-use

[59]Fehrenbach et al. (2015).

Table 3 Value for land uses for natural soil functions building on the hemeroby concept

Hemeroby class	Value for soil	LDN land-use category	Categories according to Fehrenbach et al. (2015), adapted
I	6	1. (a) Pristine forest (b) Wetlands and peatlands	Pristine forest
	5.5		
II	5		Close-to-nature forest management
	4.5	2. Forest	
III	4		Intermediate forest management
	3.5		
IV	3	3. Grassland	Monocultural forest
	2.5		
V	2		Intermediate agriculture, moderate intensity, short rotation coppices, fertilised grassland
	1.5	4. Arable land	
VI	1		Large-area, highly intensified arable land in cleared landscape, solar fields, windparks
	0.5	5. Land of no particular use,[a] solar fields	
VII	0	6. Areas for settlement and traffic infrastructure	Sealed areas, mining lands, landfills
	−0.5		
	−1		
	−1.5	7. Mining areas, landfills	

[a]Rocks, dunes, abandoned mining areas

categories per hectare of land area under consideration. For an illustration of this, see the following examples:

forest to arable land: −3.0/ha;
arable land to grassland: +1.5/ha;
arable land to areas of settlement: −1.5/ha;
areas of settlement to forest (reforestation): +4.5/ha;
areas of settlement to landfills: −1.5/ha.

In the LDN context, negative values can be defined as land degradation, while positive values enhance the quality of land through restoration or regeneration. If 'negative land use change' is unavoidable, for example if new settlements are built on former arable land (−1.5/ha), the effected hectares need to be compensated at another place, for example by unsealing soils and subsequent planting of trees (+1.5/ha). The German impact regulation (*Eingriffsregelung*) operates under similar conditions; however, methodologies differ significantly between the German federal states, and soils are only one subject for protection besides many others (such as protected species, habitats, water, air, etc.), which often leads to an underrepresentation of soils in overall impact assessments.

This simple calculation model can be further extended into an overall LDN balancing model for a certain region. However, spatial and temporal dimensions of degrading and compensating land-use measures still need to be elaborated, as well as more detailed graduations between land-use categories, which represent respective local conditions.

5.3 Consideration of Land-Use Intensity and Changes in Land Management

For the second dimension (management changes within one category, see above), changes in soil quality and soil functions need to be balanced as a result of changing land-use management within the same land-use category. If, for example, a farmer converts his practices from conventional to organic farming, this has positive effects on the soil quality (in the long run). Such changes have to find their way into the overall LDN balance, but the methodology should not object to the original calculation approach. Hence, additional indicators have to be found that qualify the changes in land management as to be transferred into numeric values that can be added to the original scale. In the following, we propose such indicators for the most important land-use categories.

5.3.1 Share of Sustainable Forest Management (Categories 1 and 2)

The management of forests (e.g., allowance/prohibition of clear cuts, composition of trees, share of deadwood left on the ground) is connected to the natural soil functions and hence can have an impact on soil quality. However, sustainable forest management is poorly defined, especially at supranational level. In Germany, there is therefore no monitoring of 'sustainably managed forests' and 'conventionally managed forests'.

Areas certified under FSC or PEFC could provide a basis, though, for identifying areas with more ambitious requirements for sustainable forest management. Both certification schemes claim to support sustainable forest practices. Moreover, over the last few years, the German federal states have supported the change to sustainable forest management, often funded by European programmes. So it is possible that the share of areas where these programmes have been taken place can be used for the identification of changes in forest management within the land-use category forest. Furthermore, we extended category 2 towards deciduous and mixed and coniferous forests because, with regard to soil chemistry as well as to hemeroby, coniferous forests can be assigned a lower value than deciduous and mixed forests.

5.3.2 Share of Organic Farming (Categories 3 and 4)

Under the greening regulation of the EU Common Agricultural Policy (CAP), all farms in the EU need to fulfil certain requirements to enhance the environment. Organic farms, however, are excluded from this regulation as they are assigned as 'green-by-definition', which means that their farming practices are per se considered beneficial for the environment. In fact, many studies show that organic farms usually have a positive humus content.[60] They often apply a more varied crop rotation than conventional farms. Other studies have also shown that nutrient deficits occur in soils under organic management, especially phosphorus, magnesium and potassium. However, these nutrients effect soil fertility and productivity but are less relevant for avoiding soil degradation with regard to LDN.[61] Therefore, an increasing share of agricultural land managed as organic farming would be a suitable indicator to further differentiate within categories 3 and 4 of the above-mentioned category. Some grassland is only suitable for extensive use because of limited nutrient supply, regular floods or other constraining factors. Some of them are not explicitly assigned or certified as organic; thus, a category of 'extensively managed grassland/pastures' can be useful in addition to 'organic grassland/pastures' (see Table 4).

5.3.3 Share of Unsealed Land in the Category of Areas for Settlement and Traffic (Category 6)

In Germany, areas for settlement and traffic also include unsealed areas (see above). On the one hand, these areas are sometimes officially declared as (potential) areas for settlement and traffic by regional agencies but have not yet been covered with buildings or road infrastructure. Such areas, which mostly occur in Eastern German communities where demographic changes lead to a decreasing local population, still appear as areas for settlement and traffic infrastructure in the statistics. On the other hand, these areas are rarely completely sealed since they also consist of green areas, parks, trees or wasteland. Hence, the category areas for settlement and traffic of the first land-use change dimension cannot be applied automatically to the unsealed components because, depending on whether they are sealed or not, their value for soil functions differs significantly. Therefore, the category should be further split into sealed and unsealed components of areas for settlement and traffic (with and without vegetation cover), representing different values for soil.

Moreover, such differentiation could potentially provide incentives for countries applying the system to leave areas unsealed. This would provide benefits not only for soil functions but also for the quality of life in urban areas.

---

[60] Hülsbergen and Rahmann (2015).

[61] Kolbe (2015).

Table 4 Extension of LDN land-use categories to second dimension

Value for soil	LDN land-use category, dimension 1	LDN land-use category, dimension 2
6	1. (a) Pristine forest (b) Wetlands and peatlands	
5.5		
5		Sustainable forest management
4.5	2. Forest (a) Deciduous and mixed forest	
4	(b) Coniferous forest	Intensive forest management
3.5		Organic and extensively managed grassland
3	3. Grassland/pastures	
2.5		Conventional (intensive) managed grassland Set-aside arable land
2		Organic farming (arable land)
1.5	4. Arable land	
1		Conventional farming (arable land) Unsealed areas of settlement and traffic infrastructure (covered by vegetation)
0.5	5. Land of no particular use,[a] solar fields	Unsealed areas of settlement and traffic infrastructure (no vegetation)
0	6. Areas for settlement and traffic infrastructure	
−0.5		
−1		
−1.5	7. Mining areas, landfills	Sealed areas of settlement and traffic infrastructure

[a]Rocks, dunes, abandoned mining areas

For category 5 (fallow land), land for photovoltaic plants, and category 7 (land for mining and landfills), no additional indicators are needed because changes within these categories do not have an essential effect on land and soil degradation parameters.

In correspondence to land-use change in dimension 1 (see above), land-use change within a category (dimension 2) would be calculated as follows:

sustainable forest management to intensive forest management:	−1/ha;
conventional farming to organic farming:	+1/ha;
extensive grassland to intensive grassland:	−1/ha;
(partial) unsealing within areas for settlement and traffic (without replanting):	+2/ha.

5.4 *Outlook on LDN Monitoring in Germany and Conclusion*

Making the proposed balancing approach operational requires a solid database for detecting land-use changes. Here, we drew on the land-use categories used by the federal statistical agency and the local cadastral land registers. In Germany, such categories are widely consistent, but in other countries a harmonisation of land-use categories might be needed as the first step towards a proper monitoring system. However, the system presented can also be used in other countries. The second step involves the question of data collection. Given the broad application and coverage as well as current developments towards further refinement of resolution and interpretation methods, it is recommended to streamline land-use (change) monitoring via remote sensing (RS) techniques and data.[62] Also, comparability between regions and countries is far more achievable with RS rather than using solely national cadastral data sets for balancing land-use changes and LDN. Establishing a rigorous and consistent monitoring system will only be possible through close cooperation between statistical and environmental agencies and research institutions that are familiar with using remote-sensing technologies and their applications.

However, it first needs to be seen if in 2017 further progress can be made to include an indicator about soil quality and land degradation neutrality within the German Sustainability Strategy, which will be revised until 2018 and serves as the main strategic document for the implementation of the SDGs. Furthermore, despite the very positive feedback of different stakeholders within the expert workshops and interviews, it is not yet decided by policy makers if the above-mentioned indicator and assessment concept that integrates land-use changes, the hemeroby concept and soil functions in one indicator will be used as an indicator to implement SDG 15.3.

References

Akhtar-Schuster M, Stringer L, Erlewein A, Metternicht G, Minelli S, Safriel U, Ommer S (2016a) Unpacking the Concept of Land Degradation Neutrality and Addressing Its Operation through the Rio Conventions. https://doi.org/10.1016/j.jenvman.2016.09.044

Akhtar-Schuster M, Stringer L, Sommer S, Metternicht G, Safriel U (2016b) Schriftliches Interview mit Dr. Mariam Akhtar-Schuster (DesertNet International, University of Hamburg, Germany); Prof. Dr. Lindsay Stringer (Sustainability Research Institute, School of Earth and Environment, University of Leeds); Dr. Stefan Sommer (JRC IES, Ispra, Italy); Prof. Dr. Graciela Metternicht (School of Biological, Earth and Environmental Sciences, UNSW Australia); Prof. Uriel Safriel (The Hebrew University of Jerusalem, Department of Ecology Evolution and Behavior, Edmond J. Safra Campus, Givat Ram, Jerusalem, Israel) am 29.04.2016

[62] Lausch et al. (2015) and Lausch et al. (2016).

Azeez G (2009) Soil Carbon and Organic Farming - A Review of the Evidence of the Relationship between Agriculture and Soil Carbon Sequestration, and How Organic Farming Can Contribute to Climate Change Mitigation and Adaptation. Soil Association. http://www.soilassociation.org/LinkClick.aspx?fileticket=SSnOCMoqrXs%3D&tabid=574

Caspari T, van Lynden G, Bai Z (2015) Land degradation neutrality: an evaluation of methods. Umweltbundesamt. https://www.umweltbundesamt.de/sites/default/files/medien/378/publikationen/texte_62_2015_land_degradation_neutrality_0.pdf

Chasek P, Safriel U, Shikongo S, Futran Fuhrman V (2015) Operationalizing zero net land degradation: the next stage in international efforts to combat desertification? J Arid Environ 112(Part A):5–13 https://doi.org/10.1016/j.jaridenv.2014.05.020

de Groot R, Blignaut J, van der Ploeg S, Aronson J, Elmqvist T, Farley J (2013) Benefits of investing in ecosystem restoration. Conserv Biol 27(6):1286–1293

Dernier L, Scherr S, Schames S, Chatterton P, Hovani L, Stam N (2015) The little sustainable landscapes book. Global Canopy Programme, Oxford, http://globalcanopy.org/sites/default/files/documents/resources/GCP_Little_Sustainable_LB_DEC15.pdf

FAO and ITPS (2015) Status of the world's soil resources: regional assessment of soil changes in Europe and Eurasia. Food and Agriculture Organization of the United Nations and Intergovernmental Technical Panel on Soils, Rome, Italy, http://www.fao.org/3/a-bc600e.pdf

Fehrenbach H, Grahl B, Griegrich J, Busch M (2015) Hemeroby as an impact category indicator for the integration of land use into life cycle (impact) assessment. Int J Life Cycle Assessment 20(11):1511–1527

Feldwisch N, Balla S, Friedrich C (2006) 'LABO-Projekt 3.05 Endbericht Zum Orientierungsrahmen Zur Zusammenfassenden Bewertung von Bodenfunktionen'. Länderfinanzierungsprogramm "Wasser, Boden und Abfall 2005". Bergisch Gladbach & Herne: im Auftrag der Bund-/Länderarbeitsgemeinschaft Bodenschutz (LABO). Global Mechanism of the UNCCD. 2016a. 'Achieving Land Degradation Neutrality at the Country Level. Building Blocks for LDN Target Setting'

Global Mechanism of the UNCCD (2016a) Achieving land degradation neutrality at the country level. Building Blocks for LDN Target Setting'

Global Mechanism of the UNCCD (2016b) Scaling up Land Degradation Neutrality Target Setting. From Lessons to Actions: 14 Pilot Countries' Experiences'

Gnacadja L (2012) Moving to Zero-Net Rate of Land Degradation. http://www.unccd.int/Lists/SiteDocumentLibrary/secretariat/2012/UNCCD%20ES%20Statement%20at%20PR%20in%20NY%20on%2026%20March%202012.pdf

Hülsbergen H.-J, Rahmann G (2015) 'Klimawirkungen Und Nachhaltigkeit ökologischer Und Konventioneller Betriebssysteme – Untersuchungen in Einem Netzwerk von Pilotbetrieben : Forschungsergebnisse 2013–2014. Thünen Report 29. Johann Heinrich von Thünen Institut, Braunschweig

IAEG-SDGs (2016) IAEG-SDGs — SDG Indicators. Sustainable Development Goals. http://unstats.un.org/sdgs/iaeg-sdgs/metadata-compilation/

IUCN (2015) Land Degradation Neutrality: Implications and Opportunities for Conservation. Nature Based Solutions to Desertification, Land Degradation and Drought. IUCN Draft Technical Brief 08/10/2015. http://cmsdata.iucn.org/downloads/tech_brief_land_degradation_neutrality.pdf

König W (2016) Telefonisches Gespräch mit Prof. Dr. Wilhelm König (Ministerium für Klimaschutz, Umwelt, Landwirtschaft, Natur- und Verbraucherschutz des Landes Nordrhein-Westfalen (MKULNV), Ref. IV-4 Bodenschutz, Altlasten, Deponien) am 24.03.2016

Kolbe H (2015) Wie ist es um die Bodenfruchtbarkeit im Ökolandbau bestellt: Nährstoffversorgung und Humusstatus?, Tagungsbeitrag, Gemeinsame Tagung des Verbands der Landwirtschaftskammern e. V. (VLK) und des Bundesarbeitskreises Düngung (BAD) am 21. und 22. April 2015 in Würzburg; http://orgprints.org/29539/1/Bodenfruchtbarkeit_%C3%96ko_BAD-VLK15.pdf

Lausch A, Blaschke T, Haase D, Herzog F, Syrbe R-U, Tischendorf L, Walz U (2015) Understanding and quantifying landscape structure–a review on relevant process characteristics, data models and landscape metrics. Ecol Model 295:31–41

Lausch A, Erasmi S, King DJ, Magdon P, Heurich M (2016) Understanding forest health with remote sensing-part I—a review of spectral traits, processes and remote-sensing characteristics. Remote Sens 8(12):1029

Mal P, Hesse JW, Schmitz M, Garvert H (2015) Konservierende Bodenbearbeitung in Deutschland Als Lösungsbeitrag Gegen Bodenerosion. Journal Für Kulturpflanzen 67(9):310–319. https://doi.org/10.5073/JFK.2015.09.02

NABU and DVL (2014) Stellungnahme Des Deutschen Verbandes Für Landschaftspflege (DVL) e.V. Und Des Naturschutzbund Deutschland (NABU) e.V. Zum Entwurf Des Gesetzes Zur Durchführung Der Direktzahlungen an Inhaber Landwirtschaftlicher Betriebe Im Rahmen von Stützungsregelungen Der Gemeinsamen Agrarpolitik – Direktzahlungen-Durchführungsgesetz (DirektZahlDurchfG)'. https://www.nabu.de/downloads/140211-nabu-stellungnahme-direktzahlungen.pdf

Ngo H (2015) IPBES Land Degradation and Restoration Assessment Deliverable 3b. Presented at the UNCCD COP12, Ankara. http://www.unccd.int/en/about-the-convention/the-bodies/The-CST/Documents/CST12-presentations/Item5a_IPBES-LDRA_PPT%20by%20Hien%20Ngo.pdf

Nkonya E, Mirzabaev A, von Braun J (eds) (2016) Economics of land degradation and improvement – a global assessment for sustainable development. http://www.springer.com/de/book/9783319191676

Orr BJ, Cowie AL, Castillo Sanchez VM, Chasek P, Crossman ND, Erlewein A, Louwagie G, Maron M, Metternicht GI, Minelli S, Tengberg AE, Walter S, Welton S (2017) Scientific conceptual framework for land degradation neutrality. A report of the science-policy interface. United Nations Convention to Combat Desertification (UNCCD), Bonn, Germany. ISBN 978-92-95110-42-7 (hard copy), 978-92-95110-41-0 (electronic copy)

Paulsen HM, Böhm H, Moos J, Fischer J, Schrader S, Fuß R (2013) Fruchtbarer Boden - Welchen Einfluss Die Landnutzung Auf Den Boden Hat. Berichte Aus Der Forschung - Fruchtbarer Boden FoRep 2/2013

Stavi I, Lal R (2015) Achieving zero net land degradation: challenges and opportunities. J Arid Environ 112(Part A):44–51. https://doi.org/10.1016/j.jaridenv.2014.01.016

Tucker G, Underwood E, Farmer A, Scalera R, Dickie I, McConville A, van Vliet W (2013) Estimation of the financing needs to implement target 2 of the EU biodiversity strategy. Report to the European Commission. Institute for European Environmental Policy, London

UNCCD (2012) Zero net land degradation. A sustainable development goal for Rio +20. UNCCD Secretariat policy brief. http://www.unccd.int/Lists/SiteDocumentLibrary/Rio+20/UNCCD_PolicyBrief_ZeroNetLandDegradation.pdf

UNCCD (2015) Land Degradation Neutrality. Resilience at Local, National and Regional Levels. http://www.unccd.int/Lists/SiteDocumentLibrary/Publications/Land_Degrad_Neutrality_E_Web.pdf

UNCCD SPI (2015) Pivotal Soil Carbon. UNCCD Science-Policy Interface (SPI). http://www.unccd.int/Lists/SiteDocumentLibrary/Publications/2015_PolicyBrief_SPI_ENG.pdf

UNCCD/Science Policy Interface (2016) Land in Balance. The Scientific Conceptual Framework for Land Degradation Neutrality (LDN). Science-Policy Brief 02. September 2016

UNCCD Global Mechanism (2016) Land Degradation Neutrality: The Target Setting Programme. Bonn. http://www.unccd.int/Lists/SiteDocumentLibrary/Publications/4_2016_LDN_TS%20_ENG.pdf

UNDESA (2016) Sustainable Development Goals. Sustainable Development Knowledge Platform. https://sustainabledevelopment.un.org/?menu=1300

UN General Assembly (2015) Transformation Unserer Welt: Die Agenda 2030 Für Nachhaltige Entwicklung. Siebzigste Tagung Der Generalversammlung; Tagesordnungspunkte 15 Und 116 (A/70/L.1). http://www.un.org/depts/german/gv-70/a70-11.pdf

Untenecker J, Tiemeyer B, Freibauer A, Laggner A, Luterbacher J (2017) Tracking changes in the land use, management and drainage status of organic soils as indicators of the effectiveness of mitigation strategies for climate change. Ecol Indic 72(January):459–472. https://doi.org/10.1016/j.ecolind.2016.08.004

Part IV
National and Regional Soil Legislation

Reform of the Icelandic Soil Conservation Law

Birkir Snær Fannarsson, Björn Helgi Barkarson, Jón Geir Pétursson, and Sigríður Svana Helgadóttir

1 Introduction

Soils provide the basics for multiple ecosystem services and land productivity. Soil resources are key natural resources, an indispensable component of terrestrial ecosystems and a foundation for livelihood and well-being in most societies, however, often partly overlooked due to its sub-surface physical character.[1] Governing soil resources, seeking to ensure their sustainable management and to restore degraded soils, is a demanding challenge that has been transforming from a more taken-for-granted approach to become acknowledged part of national environmental and natural resource policies. This brings focus on the policy and legislative frameworks that countries develop and adopt for their soil resource governance, seeking to facilitate soil reclamation, prevent degradation, and achieve sustainable use.[2]

There is, however, no blueprint available for effective soil resource policy and legislative framework. Countries have adopted a variety of approaches that are regarded as quite innovative, meeting their specific requirements and challenges.[3] Further, soil governance issues are intersected with land-use policies and practices that have evolved in different trajectories in different countries and regions. It is therefore the need to understand soil governance challenges relative to the multiple

[1]Barrios (2007).

[2]Hannam and Boer (2002).

[3]Hannam and Boer (2002).

B.S. Fannarsson
Icelandic Soil Conservation Service, Hella, Iceland

B.H. Barkarson (✉) • J.G. Pétursson • S.S. Helgadóttir
Ministry for the Environment and Natural Resources, Reykjavik, Iceland
e-mail: bjorn.helgi.barkarson@uar.is

H. Ginzky et al. (eds.), *International Yearbook of Soil Law and Policy 2017*,
International Yearbook of Soil Law and Policy,
https://doi.org/10.1007/978-3-319-68885-5_12

historical and institutional processes that has shaped and manifested the current policy and legal structures.

This chapter takes the case of Iceland, a volcanic island in the North Atlantic that constitutes an interesting case when examining soil governance from a policy and law perspective. The Icelandic soil policy and legislation is currently under review with the aim of making it fit better to contemporary challenges and facilitate more efficient soil restoration, prevent degradation, and promote sustainable use.

Our objective is to examine historical political and institutional factors shaping the contemporary soil governance challenges in Iceland and discuss how these have contributed to and shaped the proposed policy and legislative reforms. We start by developing a contextual framework for the understanding of the proposed reforms and then setting the focus on the case of Iceland, first with some basics of the geo–/biophysical soil issues, then give a brief historical institutional background of its soil governance development and outcomes before moving to outlining some of the key challenges that the proposed reforms are supposed to tackle.

2 Contextual Framework for Understanding Soil Resource Governance

We adopt a contextual framework based on natural resource governance models for our examination of the soil governance challenges in Iceland and the proposed reform.[4] Governance in this context concerns social priorities and decision making related to soils and is about forming institutional structures to guide and coordinate interactions between actors as with the natural resources at stake. It is hence about how we establish goals, how we define rules for reaching the defined goals, and finally how we control outcomes following from the use of these rules.[5]

We outline a framework model for understanding of a soil resource governance system[6] (Fig. 1). The framework as applied in this chapter has in essence three analytical domains, the historical legacies, the three interrelated social-ecological elements (physical attributes, institutions, and actors), and then the combined outcomes.

Firstly, soil governance needs to be understood against the historical legacies that have shaped the current structures. Being directly intersected with land use, there are always underlying historical processes that have shaped the current structures and approaches. Secondly, there are three main elements of the current soil resource governance systems, hence the soil geo–/biophysical attributes (which can be of multiple character within country), the actors involved with different powers for decision making (like governments, land owners, etc.), and

[4]Ostrom (2009).

[5]Vatn (2010).

[6]Vatn (2005) and Ostrom (2009).

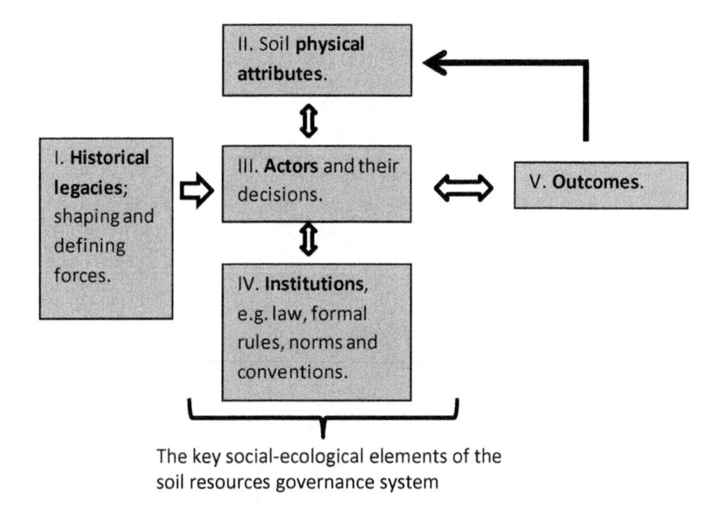

Fig. 1 A framework model for understanding of a soil resources governance system and its dynamics (adopted from Ostrom 1990; Vatn 2005; Ostrom 2009)

the institutions where we have both the formal (especially legal and by-laws) and the informal (different norms and conventions). These are the building blocks of any soil governance system. Thirdly, the interactions between and with the former two domains shape the outcomes of the soil resource governance system. This conceptual framework we use to guide our analysis in the chapter.

3 Soils of Iceland

Iceland is a 103,000 km^2 island in the North Atlantic Ocean. The climate is described as cold temperate in the lowlands and sub-arctic in the highlands and is strongly influenced by the warm Gulf Stream.[7] The country is mountainous with lowland areas along the coastline and river plains. Mean annual temperatures range from 2 to 6 °C, and annual total precipitation varies from 300 to 3500 mm yr^{-1}.

Since the settlement of Iceland in the ninth century, the status of Icelandic ecosystems, its soil, and vegetation has degraded dramatically. The famous words of Ari Þorgilsson, "the learned" in the *Book of the Icelanders*, that Iceland was "lushly vegetated from mountain to sea" when settlers arrived have often been quoted.[8] Sources from the first centuries, backed up by recent researches, estimate that the total vegetation cover has decreased from 60–70% to about 25%. The cover of natural woodlands decreased from estimated 25 to 30% at the time of the settlement to about 1%.[9] It seems clear that the interaction of volcanism and the

[7]Arnalds (2015).

[8]Þorgilsson (1887).

[9]Olafsdottir et al. (2001) and Thorsteinsson et al. (1971).

Table 1 Division of land in Iceland according to erosion classes (Arnalds et al. 2001)

Erosion class	km^2	% of whole
0 No erosion	4148	4.0
1 Little erosion	7466	7.3
2 Slight erosion	26,698	26.0
3 Considerable erosion	23,106	22.5
4 Severe erosion	11,322	11.0
5 Extremely severe erosion	6375	6.2
Mountains	9794	9.5
Glaciers	11,361	11.1
Rivers and lakes	1436	1.4
Unmapped	1010	1.0
Total	102,721	100

volcanic nature of Icelandic soils, harsh climatic conditions, and unsustainable land use led to catastrophic ecosystem degradation and erosion.[10]

Icelandic ecosystems evolved in the absence of large grazing animals. Fully vegetated ecosystems covered most of the country when humans arrived and initiated livestock grazing and wood harvesting. The vegetation varies from barren desert-like areas to lush birch woodlands. Plant production decreases rapidly with increased elevation. The volcanic geology of Iceland is a main factor influencing Icelandic soils, and therefore soils are mostly *andosols*.[11]

Soil conditions in Iceland are mainly influenced by volcanic activity, the cold maritime climate, soil erosion by wind, and water and cryogenic processes. Due to the long history of soil erosion, Icelandic soils are strongly affected by the steady flux of windblown materials from eroded areas.[12]

A national soil erosion assessment was initiated in 1990. The aim was to establish an overview of the soil erosion problem in Iceland and to provide guidance for further development of sustainable land use in Iceland. It was conducted by the Agricultural Research Institute and the Soil Conservation Service and was made at the scale of 1:100,000.[13] The assessment reveals that considerable, severe, and extremely severe soil erosion occurs on 22, 11, and 6.2% of the country, respectively, excluding mountains and glaciers. Desert areas are dominant surface types for most of the highland, representing >80% of the area within the interior. Soil erosion and related land degradation is therefore considered a major environmental problem in Iceland (Table 1).

For more details on the geo- and biophysical attributes of the soils of Iceland, we refer to Arnalds (2015), who gives a detailed and holistic overview of the various Iceland soil types, their properties, and status.

[10]Greipsson (2012).
[11]Arnalds (2015).
[12]Arnalds (2015).
[13]Arnalds et al. (2001).

4 Understanding Soil Governance in Iceland from a Historical Perspective

Since the settlement of Iceland and during the following 1000 years, it was a country of self-subsistence, to a large extent based on hay- and grazing-based livestock production in a harsh environment.[14] Winter survival of livestock was the main determinant of population size until the late nineteenth century.[15] Keeping the nation alive took its toll as the pressure of land use exceeded the ecological capacity, and catastrophic soil erosion and desertification devastated large parts of the country.[16] The scale of the degradation and soil erosion is unprecedented in Northern Europe.[17]

The capacity of the Icelandic society to respond to the degradation was limited, and no organized effort in stopping the massive land degradation occurred until late nineteenth century. Gradually and with intermittent extreme events, the degradation got worse and the majority of people generally believed that there was nothing that could be done to stem the tide of nature. Numbers of farms were abandoned, and often there remained no evidence of its former glory.[18]

In the nineteenth century, the situation is getting severe. Iceland's climate began to deteriorate after 1860, and some of the harshest years in the nation's history began around 1880. To give an indication of the hardship endured by the Icelandic nations, the year 1882 was especially bad and was often referred to as *The Sand Year*, or *The Winterkill Year*. It is ranked among the five coldest years over a 128-year period (1873–2001).[19]

As a result of these hardships, the cold years of 1890s and the volcano Askja eruption in 1874, the ecological and human consequences of volcanic hazards led to conditions where the population could no longer be accommodated or sustained within the Icelandic society, which was based on farming. Mass emigration to America became an outlet for the stressed Icelandic society in search of better living conditions, when it began in the 1870s and continued until about 1914.

Iceland was at the time ruled under the Danish administrative system. Iceland's first step on the road to independence was the foundation of the new parliament, Alþingi, in 1845. The old Alþingi, which had met each year at Þingvellir from 930 AD, had gradually declined under foreign rule and had finally been abolished in 1798. The new parliament had no legislative powers; it was simply an advisory body. In 1874, Iceland was granted its first constitution: Alþingi gained legislative

[14]Runólfsson and Ágústsdóttir (2011).

[15]Runólfsson and Ágústsdóttir (2011).

[16]Runólfsson and Ágústsdóttir (2011).

[17]Greipsson (2012).

[18]Arnalds (1988).

[19]Hanna et al. (2004).

powers, while executive powers remained in Denmark. The following years, the battle against land degradation began for real, one step at a time. Soil conservation and land reclamation (ecosystem restoration) finally got on the agenda.

4.1 Act on the Protection of Forests, Shrubs, Moss and Heather, 1891

In 1891, a bill was submitted to the Icelandic Parliament, Alþingi: *Bill for the protection of forests, shrubs, moss and heather*. The bill was intended to counter the destruction of the country's soil and vegetation. This bill never became law, but it was clear to the members of Parliament that the land's usage and situation was unacceptable, and one said: "... the worse the situation, the more the laws (lat. *pessimæ reipublicæ plurimæ leges*)."[20] In 1894, a statute was passed to preserve woodlands and lyme grass, granting county councils the authority to declare an area as protected. The county councils/committees representing local authorities (in Icelandic, sýslunefndir) were given the power to protect forest and lyme grass, but a 2/3 majority vote was required for the enactment of changes.

At this time, Iceland was one administrative unit under the Danish rule, called amt, with a district governor (amtman; in Icelandic, amtmaður) residing in Iceland. The amtman then had to accept this specific protection and publicize it unless he thought it would go against one's freedom of employment or if it would in other way violate the law and/or the fundamental principles of law.

4.2 Act on Protection against the Drifting Sand and Land Reclamation, No. 6/1895

In 1895, *an Act on protection against the drifting sand and land reclamation* was approved. This law was in many ways similar to the previously mentioned, no. 11/1894. Local authorities (in Icelandic, sýslunefndir) were given the power to establish agreements (for a region or a part of it) on protection against the moving sand and sand reclamation. If approved, the agreement was legally binding for everyone living in the area. These laws were not groundbreaking in any way.

[20]Iceland Parliamentary Debates (1891).

4.3 Act on Forestry and Protection against Soil Erosion, No. 54/1907

The *Act on forestry and protection against soil erosion* is generally recognized to mark the beginning of formal soil conservation in Iceland.[21] The statute was considered both short and weak. A director of forestation should be appointed, who should also be responsible for directing public measures undertaken to prevent blowing sand. Forestry officers should be appointed as required. A provision in Article 1 stated that "Forestry efforts shall be begun, with the aim of preserving and improving those forests and remains of forests still existing in Iceland, of growing new forest and of instructing Icelanders in care of forest and planting of trees. In connection therewith, protective measures shall be taken to prevent soil erosion and blowing sand wherever practicable." Another critical provision of the law stated that "Forest warden may be assigned the task of inspecting and recording drifting sand areas, and supervising implementation of drifting-sand measures that may be carried out." This is the only reference in the Act to land reclamation, and, in consequence, the legislation was considered a constraint on revegetation for years to come and delayed many important projects.[22]

At the beginning of the twentieth century, various changes took place in Icelandic society. In 1904, when Iceland was granted home rule by Denmark after more than 600 years under first Norwegian and then Danish rule, various movements made their mark on Icelandic society, often reflecting an era of innovation, new technology, and optimism: the trade-union movement was beginning to make its voice heard, youth associations nurtured a combination of nationalism and health culture, education became a priority with the Education Act of 1907 ensuring all children 4 years' free education. Various new developments that followed the Home Rule period, from 1904 to 1918, boosted the confidence and hopes of the Icelanders.

4.4 Act on Land Reclamation, No. 20/1914

The *Act on land reclamation* marked a significant turning point since it was the first legislation concerning only land reclamation. Since then, forestry had been a part of separate legislation. The law was passed because of the lack of direct provisions on these issues. Members of the Parliament pointed out that it was necessary for the upcoming work to have solid legal basis.[23] The State had by then a desire to have tenure over all lands being reclaimed, partly because farmers were not regarded as

[21]Landgræðsla ríkisins (1988b), Sigurjónsson (1958), and Olgeirsson (2007).

[22]Landgræðsla ríkisins (1988b), Olgeirsson (2007), and Crofts (2011).

[23]Iceland Parliamentary Debates (1914a) and Iceland Parliamentary Debates (1914c).

adequate stewards and often allowed degradation of soil and vegetation to occur. It is necessary to keep in mind that by that time there was a real danger of many farms and even bigger areas would in a few years be destroyed and deserted due to drifting sand.[24] Since it was the first legislation devoted to land reclamation, all of the provisions were new, but the most significant were the following:

- The Home Rule Government was allowed to have eroded areas, which might pose a threat of soil erosion to adjacent areas, fenced off with the aim of undertaking reclamation efforts both to fully revegetate the area and to prevent the erosion from spreading further. During the entire period that land reclamation efforts were underway, the Government had full control of such an area. The owner or person entitled to utilize the land could not demand compensation for any use he might otherwise have made of it.
- Areas where landowners, or other persons concerned, were prepared to make some financial contribution to the fencing cost were to be given priority in receiving aid. Three quarters of the cost of fencing, in excess of the contribution made by owners, would be paid by the Treasury and the remaining one quarter by the district's funds.
- The parties supervising the work were permitted to take without remuneration any material that could be used for reclamation purposes, such as stones, turf, and heather, without damaging the land itself.
- Then there were clauses on how landowners were to treat revegetated land after the return of it.

4.5 Act on Sand Reclamation, No. 45/1923

The *Sand Reclamation Act*, approved by the Parliament in 1923, was in many provisions similar to previous law, but in the light of experience some improvements were made. For the first time, the State was required to appoint a sand reclamation specialist to work on these matters.

Areas where there was ongoing land reclamation work were to be in the State's tenure, but the division of costs of fencing, revegetation, and maintenance was changed and should be divided evenly between the Treasury and the landowner. Local authorities were to ensure that the landowner paid his share of the cost.

One significant provision provided the State the resources to intervene when there was unacceptable land usage. The State was permitted to expropriate wind-eroded areas, which might pose a threat of soil erosion to adjacent areas if the landowner (or appropriate authority) either would not or could not finance its share of land reclamation. This was a big change attributed to bitter experience of farmers, who until that time were able to stop necessary land reclamation because

[24]Iceland Parliamentary Debates (1914b).

of limited finance or lack of interest.[25] The expropriate provision has twice been used, for the first time in the year 1933 and then in the year 1945.[26]

After the 1923 Act, increasingly more effort was put into the work of recovering soil and vegetation. Numbers of areas were fenced from sheep grazing, and the focus was on stopping the drifting sand in the neighbourhood of towns/villages with different reclamation measures, which, in some cases, was the prerequisite for continued residence. The town of Vík í Mýrdal was fenced off in 1933–1945, then the towns of Kirkjubæjarklaustur and Kópasker in 1934, Þorlákshöfn in 1935, and Sandgerði in 1939. Prior to that, the towns of Eyrarbakki in 1911–1922 and Bolungarvík in 1922 had been fenced off.[27]

In 1926, the State purchased the large farm Gunnarsholt in the south of Iceland as the base for sand reclamation, as it has been ever since. Formally, it became the headquarters of the Sand Reclamation Service in 1947. Gunnarsholt represents an excellent example of the story of land degradation and then land reclamation. Because of the drifting sand and land degradation, Gunnarsholt was abandoned in 1836 but resettled in 1839. Due to increasing sand drift, the farm was moved again in 1854. In 1925, 1 year before the State purchased the farm, Gunnarsholt was abandoned again. Since then, Gunnarsholt and adjacent areas have been revegetated successfully.[28]

4.6 Act Concerning Soil Reclamation and the Prevention of Drifting Sand, No. 18/1941

Yet again, indicating the political interest and relevance, the Parliament approved a new Act, the *Act Concerning Soil Reclamation and the Prevention of Drifting Sand*, in 1941. By that Act, the position of Director of the Sand Reclamation Service was established.

This Act had three key provisions. First, there were placed great emphasis on the State needs to own all the land reclamation areas. Expropriation was to be used to a greater extent. Deserted land was by then generally considered worthless, as well as the predictable cost of reclamation was generally significant. Farmers with the financial capacity to contribute to revegetation could, however, request that their land be revegetated without the need for expropriation. Second, land reclamation areas that were enclosed and being revegetated should in general be protected from grazing by sheep to improve the chances of success. Sheep in land reclamation areas should be impounded. The third significant change was also related to sheep grazing. There should not be more sheep, neither in the highlands nor in the lower

[25]Iceland Parliamentary Debates (1923).

[26]Fannarsson (2007).

[27]Landgræðsla ríkisins (1988b).

[28]Landgræðsla ríkisins (1988a).

areas, than were suitable for both the land and the sheep, and grazing in sandy areas should be limited to halt destruction of vegetation.[29]

In the years between 1940 and 1960, the State acquired many farms and some parts of farms based on the earlier mentioned government's policy to own all the land reclamation areas. During these years, there was extensive soil erosion and desertification occurring throughout the country. The biggest restoration projects were though in the southern and north-eastern parts of the country.

Numerous farms, in fact large areas, were under the sand attack, and to avoid the farms being abandoned and destroyed, something had to be done against the sand encroachment. As an example of the situation, farmers in Selvogur in 1943 stated in a letter to the Soil Conservation Service: "Last April, 1942, the weather was exceptionally bad and the sand drifted over fields and pastures ... great areas were black as coal because of the sand and it had to be shovelled in large quantities from the fields." Here is another example from the same year—where farmers in Meðalland area stated: "We wish the Sand Reclamation Service to fence and revegetate, as necessary, land where there is ongoing great soil erosion and desertification. This particular area is a part of our farmland and poses a great threat to our farms and our livelihood and without anything being done, it also threatens the future of others farms in the region." In this era, the ties between Denmark and Iceland were severed as Iceland became an independent republic on June 17, 1944.

4.7 Act on Land Reclamation, No. 17/1965

The *Act on Land Reclamation* from 1965 is still in force with only minor amendments. This Act marked a major turning point. The Soil Conservation Service of Iceland was formally established as the successor to the Sand Reclamation Service. The previous objective of sand reclamation was broadened to land in general.[30]

It was recognized that the State did not require ownership of all the land reclamation areas, and this led to greater emphasis on cooperation with landowners and grass-root collaboration with stakeholders. The purpose of the 1965 Act was to prevent degradation of vegetation and soil erosion and to promote revegetation of eroded and poorly vegetated land. The institute's work was divided into three chapters: (1) sand reclamation, (2) conservation of soil and vegetation, and (3) vegetation monitoring regarding land use, in order to prevent unsustainable use or damages.

The Act contains a few provisions about land reclamation and how it shall be obtained. The Soil Conservation Service shall either help farmers/landowners to revegetate the land themselves or by agreements temporarily take control of the land and return it to its owner after revegetation actions have transformed the

[29]Iceland Parliamentary Debates (1941) and Crofts (2011).

[30]Iceland Parliamentary Debates (1964).

condition of the land into acceptable state. Revegetation in Iceland may take years or even many decades to remedy in severely degraded areas.

The Soil Conservation Service of Iceland was to instruct landowners on how to use land in the best possible way. The institute was to make arrangements in case of natural disasters. The Act also permitted the establishment of Vegetation Conservation Committees in all administrative districts. A whole chapter in the Act is dedicated to instructions on how to legally establish *local land restoration organizations* all over the country and what their role was supposed to be. Finally, there were provisions about researches concerning grazing and what kind of grasses/ plants could be used for revegetation and legal liability in the form of fines.

5 Development in International Environmental Law

Since the implementation of the current Act on Land Reclamation in Iceland, international environmental law has grown and developed to the extent of being "one of the most remarkable exercise in international law making"[31] with conventions and agreements shaping environmental policy globally. Four conventions are referred to particularly in the new proposed legislation on soil law in Iceland. Those are the United Nations Convention to Combat Desertification in Countries Experiencing Serious Drought and/or Desertification, particularly in Africa (UN Convention to Combat Desertification); the Convention on Biological Diversity (CBD); the United Nations Framework Convention on Climate Change (UNFCCC); and the Convention on Wetlands of International Importance, especially as Waterfowl Habitat (Ramsar). In addition, the emergence of the principle of sustainable development has influenced policy and law makers around the world and is one of the primary goals of the proposed Icelandic legislation. This is of course not a complete list of international agreements, instruments, or policies that affect legislation relating to environmental issues in Iceland; they are, however, the most prominent ones that effect the soil legislation and policy.

The UN Convention to Combat Desertification was adopted in 1994, and Iceland became a party to it in 1997. The main objectives of the Convention, as stated in Article 2, are to "combat desertification and mitigate the effects of droughts." It is the only international Convention dealing directly with the protection of soil, reclamation of desertified land, rehabilitation of partly degraded land, and sustainable land management. The main tools are so-called National Action Programmes, which identify the major factors that contribute to desertification and measures that are necessary to combat it. The key part of the proposed legislation in Iceland is the making of such program, as well as subsequent programs, at the regional level.

The Convention on Biological Diversity (the CBD) main objective, as stated in Article 1 of the Convention, is the "conservation of biological diversity, the

[31] Birne et al. (2009).

sustainable use of its components and the fair and equitable sharing of the benefits arising out of the utilization of genetic resources." Iceland ratified the Convention in 1994. In its action program relating to the implementation of the Convention, Iceland stated its objective to protect and restore biological diversity in Iceland, prevent further degradation of biological diversity, ensure the sustainable use of biological resources, and restore the elements that have been degraded or lost because of human activity.

The United Nations Framework Conventions on Climate Change (UNFCCC) has been in force since 1994. The objective of the Convention, as stated in Article 2, is to achieve "stabilization of greenhouse gas concentrations in the atmosphere at a level that would prevent dangerous anthropogenic interference with the climate system." The Kyoto Protocol adopted under the Convention is a legally binding instrument that sets certain emission targets requiring parties to the Protocol to reduce their greenhouse gas emission within a certain period. Iceland adopted an action program on climate change in 2010, stating ten key actions to be taken to combat climate change. Among those actions are two that are relevant to the proposed legislation on soil law, to increase forestry and land reclamation and thereby sequester CO_2 in soil and vegetation. The latter action that relates directly to soil protection is restoration of wetlands and degraded ecosystems. Restoration of wetlands also serves to fulfill obligations under the CBD since it restores degraded biological diversity within degraded ecosystems. For a further discussion on restoration of wetlands, see Sect. 6.1 of the CBD.

The Ramsar Convention was adopted in 1971 and came into force in 1975. It is the first international instrument focused on protecting habitats.[32] The Ramsar Convention focuses on the conservation of wetlands for waterfowls; at the same time, it proposes the wise use of wetlands, recognizing the importance of wetlands for biological diversity and water production, making wetland ecosystems an important provider of services on which plants, animals, and humans survive. Restoring and protecting wetlands is therefore an important policy goal in the new proposed legislation.

Drawing conclusion from what has been stated in this chapter, the principle of sustainable development is the main thread that runs through international conventions and treaties relating to environmental affairs. The most quoted definition of the principle is probably the definition found in the famous report *Our Common Future* by the Brundtland Commission.[33] The Commission's conclusion was that "Sustainable development is development that meets the needs of the present without compromising the ability of future generations to meet their own needs." The principle has influenced legislation globally and is one of the primary objectives of the newly proposed soil legislation in Iceland. It sets forth the principle that land use shall be based upon sustainability in accordance with guidelines concerning different types of land uses, for example sheep grazing, tourism, and agriculture. The use of such guidelines is not unknown in Icelandic legislation; the

[32]Beyerlin and Marauhn (2011), p. 181.

[33]United Nations (1987).

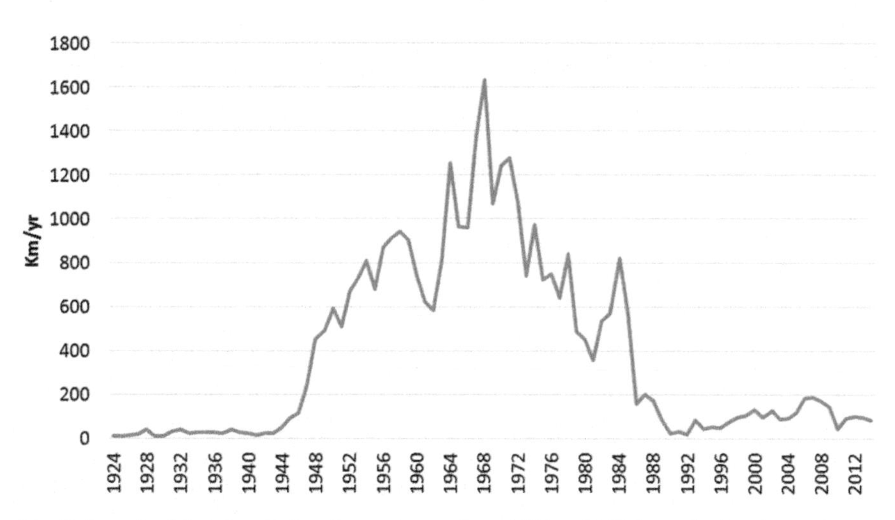

Fig. 2 Length of draining ditches dug each year from 1924 to 2013 for agricultural purposes

amount of cod quota each year is based on the principle of sustainable use, but it will be the first time that the principle of sustainable land use, irrespective of the type of land use, will be set forth in Icelandic law.

6 Contemporary Challenges

There are multiple challenges that drive the need to a revision and an update of the current soil conservation legislation in Iceland from 1965. Here we outline few of the most pertinent.

6.1 Climate Change

The challenge of climate change connects with soil conservation primarily in two ways. First, the erosion and degradation of soils rich in organic matter can cause the release of huge amounts of carbon dioxide (CO_2).[34] This applies to most ecosystems but especially to wetlands or peatlands, where organic matter is generally very high. For Iceland, from around 1950 until 1980, vast areas of wetlands were drained for agricultural purposes (Fig. 2). Some areas have been cultivated, but large areas remain idle, at most used for extensive grazing. These areas release large amounts of CO_2, a greenhouse gas, and there are great opportunities in restoring these wetlands.

[34]Oskarsson et al. (2004) and IPCC (2014).

Second, since degraded lands are abundant in Iceland, there is large potential of sequestrating carbon in soil and vegetation. This has been used as part of Iceland's action within the Kyoto Protocol.

Ecosystem restoration and soil conservation will remain as one of Iceland's key components in mitigating climate change. This also applies to afforestation and forest management.

6.2 Agriculture and Land Use

Land use and agriculture have been changing in Iceland in the last decades. Sheep farming has been a dominant land use for centuries, and after the mid-twentieth century, the number of sheep grew fast due to State support. In the 1970s, this led to overproduction of lamb and mutton and, in the context of this chapter, resulted in widespread overgrazing.[35] The legal structure to prevent land degradation and the State policy on agriculture were at that point and forward, aiming in opposite directions.

The early work of the Soil Conservation Service (SCS) involved erecting fences around desertified areas and seeding the area using species such as lyme grass to halt sand encroachment. These activities were largely undertaken by the SCS. In the late 1980s, the emphasis shifted to more participation approach, somewhat based on the "Landcare" concept.[36] This included launching a program focusing on farmer innovations in improving their private land.

In context of the results from the national soil erosion assessment and the fact that soil erosion occurs widely in Iceland, the SCS and the Agriculture Research Institute conclude that large part of the highlands of Iceland is not suitable for grazing, indicating the severity of the problem.[37] According to a research conducted in 1999 and 2000, 7% of the total number of sheep was being grazed on commons in these barren areas of the highland.[38] This only indicates that this type of land use in such a large part of the country is not of great national economic importance. Grazing of the highlands is, however, deeply rooted in Icelandic farming culture, and the grazing rights are affirmed in Icelandic law. Grazing management in the highland has proved to be a complicated issue. Also, farmers have argued that due to low grazing intensity compared to last decades, the impact on ecosystem succession and soil erosion is negligible.

Since the soil erosion assessment, land-use management has evolved. In the year 2000, a cross-compliance link between subsidies in Icelandic agriculture and land use was created.[39] To receive full payment from the State, sheep farmers shall graze

[35] Arnalds and Barkarson (2003).

[36] Barkarson and Jóhannsson (2009).

[37] Arnalds et al. (2001).

[38] Barkarson (2002).

[39] Arnalds and Barkarson (2003).

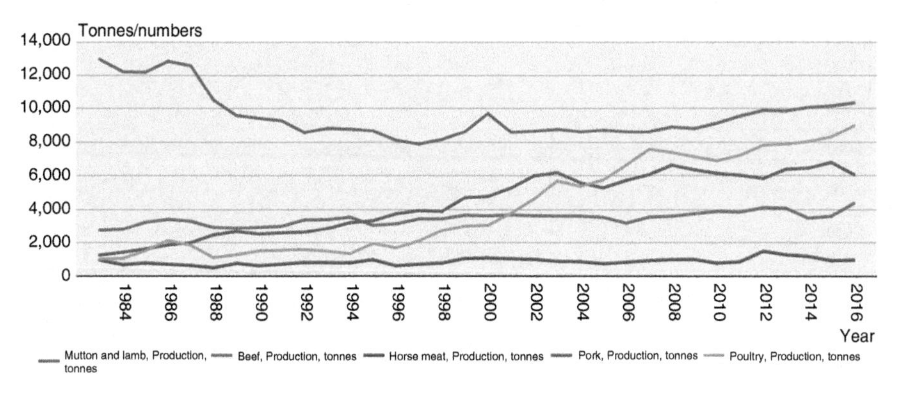

Fig. 3 Meat production in Iceland from 1983 (Statistics Iceland 2017)

land in a sustainable way. If the condition of their rangeland does not meet the criteria set in rules issued by the Minister of Agriculture, they need to develop and implement a land improvement plan. This program has been criticized by those who think the criteria are far too relaxed and also by those who oppose increased bureaucracy and State interference. The program has increased the involvement of farmers in soil conservation work, with focus on revegetation and growing emphasis on grazing management, although this is not voluntary but as a way to maintain income from sheep farming. However, this program has not completely prevented grazing of large areas of land where land condition is very bad.

The national meat production and consumption pattern has changed from mutton and lamb, being the prevailing meat type, with poultry and pork gaining a large share of the local consumption (Fig. 3). Since 1990, the proportion of the total production of lamb and mutton being exported has grown from 10–15% up to around 35%. Since the volume of production is not being regulated, this has made the producers of lamb quite vulnerable to fluctuations in international markets.

The production of cereals, primarily barley, has increased during the last two decades (Fig. 4). This is based on more land-based support from the State and also on progress in breeding and growing cultivating culture among farmers. Despite the production of cereals being very dependent on climate, there is a growing interest from farmers to be self-sustained in this regard. This means growing pressure on agricultural soils in some areas.

6.3 *Tourism*

Tourism in Iceland has increased dramatically during the last decade (Fig. 5). Surveys confirm that the majority of tourists visit Iceland for its natural values. As a consequence, this unrestrained increase in visitor number has caused pressure in many locations, with negative impact on vegetation and soils. Overall, this calls for a proactive response from conservation agencies, putting the conservation value

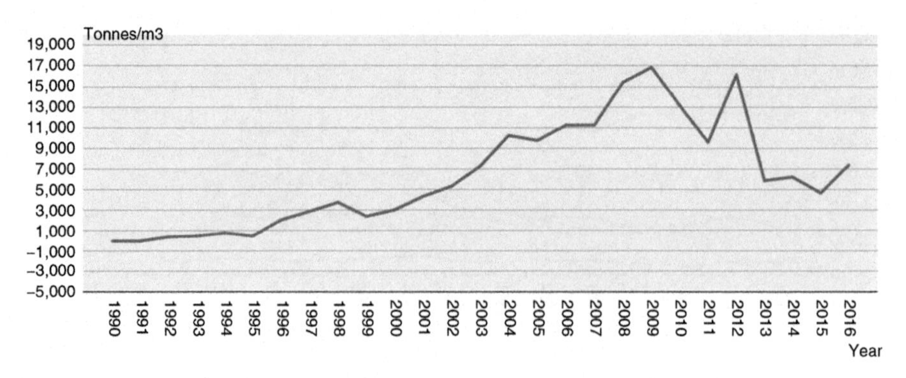

Fig. 4 Production of cereal grains from 1990 to 2016 (Statistics Iceland 2017)

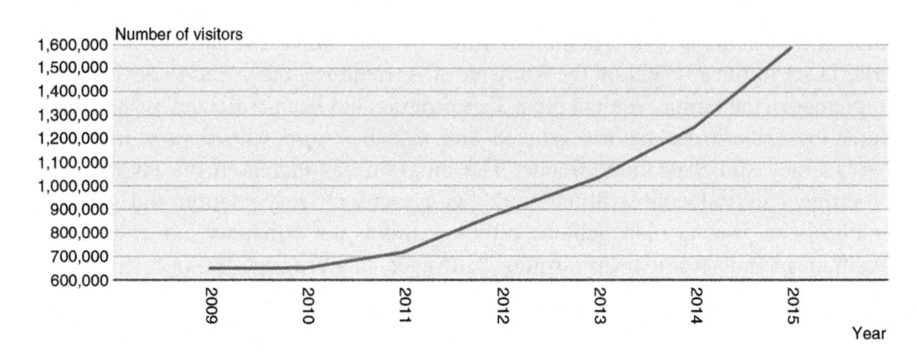

Fig. 5 International arrivals 2009–2015 (Statistics Iceland 2017)

as a primary target and building up expertise in the field of managing tourism in sustainable manner.

6.4 *Land Use, Planning, and Development*

Environmental law has evolved since the implementation of current soil conservation law of 1965. This applies to law covering nature conservation, planning, and environmental impact assessment, both on national and international scale. Soil conservation policy and application connects strongly to those subjects.

Since 1965, land use has changed from being primarily agricultural activities to more diverse operations. Ownership of land has also shifted from being primarily in the hands of conventional farmers to a state where individuals and enterprises see opportunities in owning land with a multiple agenda. The change in ownership of land in rural areas has raised different cultural and legal issues. One example of this is the provision in Icelandic law allowing general free roaming of livestock, making each land owner responsible for fencing off instead of making the owner of livestock responsible for impounding his livestock. This has as an example raised the cost of afforestation projects and has affected soil conservation work considerably.

6.5 *Coherence with Other Institutional Structures*

As pointed out by Barkarson and Jóhannsson (2009), complicated institutional structure makes a holistic approach to property planning in farming difficult. The number of small agencies with different roles in land-use-related activities forces farmers to approach different agencies for information and advice, including soil conservation, forestry, nature conservation, agri-environment support, and agricultural extension services. Barkarson and Jóhannsson (2009) suggest concerted efforts and coordination among the global information system laboratories and the research and extension service, run by different government agencies. It must be noted that this does not necessarily involve legislation reform but calls for clear political guidance.

7 Proposed Legal Reform

The work on reforming the Soil Conservation Act has been in the process for a few years in the Ministry for the Environment and Natural Resources, undergoing necessary public consultation and participation. In the year 2011, a specific committee developed a White Paper on the reform of the soil conservation law, which was presented in 2012 to the Minister of Environment and Natural Resources.[40] In 2014, the Minister initiated work on writing a bill that is in spring 2017 pending in the Icelandic Parliament.

The bill on Soil Conservation in process proposes fundamental changes to the Act on Land Reclamation, no. 17/1965.

Related to reforming the Soil Conservation Act, it must be noted that, in 2015, a new Act on nature conservation passed the Icelandic Parliament. The Nature Conservation Act gives the Environment Agency increased enforcement provisions to act if nature is under excess pressure. Also, parallel to the reform of the Soil Conservation Act, the Act on forestry has been reviewed as being originally from year 1955. The Forestry bill includes clauses on policy-making process for forestry and defines an afforestation program and sustainable use of forests. This links quite significantly with the Soil Conservation bill.

[40]Ministry for the Environment and Natural Resources (2012).

7.1 Principles and Objectives

The reform of the Soil Conservation Act is based on the principles of sustainable land use. The bill states the main objective to protect, restore, and improve the resources found in vegetation and soil and to ensure sustainable land management. This applies to biodiversity, ecosystem resilience, and function.

7.2 Institutions, Policy, and Consultation

The bill defines the Soil Conservation Service as the agency under the Minister, implementing and enforcing the law as is the current arrangement.

Policy is the fundamental tool for finding the path forward, and in this regard the bill states a ten-year Soil Conservation Plan (SCP) to be prepared, stating the vision and how soil conservation and ecosystem restoration can support society and improve ecosystem services. The development of the SCP shall include both dialogue and collaboration with stakeholders and relevant institutions and public participation. To facilitate further collaboration with stakeholders and local authorities, the bill also calls for the development of regional action plans, which define in more detail where and how the SCP is to be implemented in each region.

7.3 Sustainable Land Management

To reach the goal of sustainable land management, the SCS shall conduct monitoring on land condition, land use, and the progress of ecosystem restoration work being done. This shall be reported to the Minister and the public.

The SCS shall continue being responsible for reporting on unsustainable land use and advising land users on how to change their land use. However, the bill gives further enforcement to different government agencies, i.e. environment agency in the case of unsustainable tourism activities, and to traditional process of "ítala" in the case of livestock grazing. This narrows the scope of the authority given to the SCS from what it is today.

The bill directs developers and local authorities to avoid decisions leading to the degradation of vegetation and soils. This would apply, for example, to developments like road construction and energy-related activities such as reservoirs for hydrology power plants. If, however, construction or development leads to the degradation of vegetation and soils, measures should be taken to offset or compensate by restoration of ecosystems with similar structure and function.

7.4 Participation Approach in Soil Conservation

The Soil Conservation bill promotes a participation approach in soil conservation work but does not define specific programs in this direction. This is to be defined by the Minister, which gives the opportunity to review and adapt programs in line with the ten-year SCP. Also, the bill gives general direction to the SCS to work on halting soil erosion and to restore ecosystems, works that are to be further defined in the SCP.

Since the SCS does oversee degraded and desertified lands, either according to individual contracts with land owners or State-owned land (see Sect. 4.7), the bill defines the objectives of this caretaking. If these lands have been restored to good condition, they can be handed over or sold unless they have a particular ecological or social value.

Overall, the bill gives direction and objectives but allows for policy to be further defined and reviewed in light of monitoring in the SCP.

8 Conclusion

The Icelandic Parliament will review the bill and consult with stakeholders. The content might therefore change, and the bill can even be halted in the Parliamentary process, like what history tells us can happen due to differences in political directions. But soil conservation work in Iceland suffers due to outdated legal framework. The current legal framework was developed before the era of international environmental policy and conventions. New challenges have emerged through the shift in economic structure of the Icelandic society due to the growth in the tourist industry, which has increased the pressure on fragile ecosystems and susceptible soils. Also, pressing issues related to climate change call for action that may enhance the resilience of Icelandic land ecosystems. There are opportunities to strengthen the relationship between agriculture and soil conservation, keeping in mind that a large share of soil conservation work is now in the hands of farmers and other private land owners. This calls for further development in the participatory approach and of knowledge transfer. There is potential in restoring land ecosystems with various objectives and for different purposes. However, there is an urgent need for a broad discourse and consultation in the society, based on our best knowledge and experience, to develop a clear policy in soil conservation. Therefore, a reform of the Act is not only necessary but also pressing.

Assuming new law for soil conservation will be implemented, a new chapter should be written in a few years, reviewing the progress intended by the new law.

Acknowledgments The authors would like to express their appreciation for the opportunity to give an insight into how a 1000-year long history of settlement in the harsh nature of Iceland has affected Icelandic soils and ecosystems and its interplay with political winds and legislation. Also, and in particular, the authors want to thank Dr. Anna María Ágústsdóttir, specialist at the Icelandic Soil Conservation Service of Iceland, for her help and for providing suggestions for this article.

References

Arnalds A (1988) Brautin rudd. Saga landgræðslu á Íslandi fyrir 1907. In: Arnalds A (ed) Græðum Ísland. Landgræðslan 1907–1987. Landgræðsla ríkisins, pp 33–40

Arnalds Ó (2015) The soils of Iceland. Springer, Dordrecht. https://doi.org/10.1007/978-94-017-9621-7

Arnalds Ó, Barkarson BH (2003) Soil erosion and land use policy in Iceland in relation to sheep grazing and government subsidies. Environ Sci Policy 6(1):105–113

Arnalds Ó, Thorarinsdottir EF, Metusalemsson S, Jonsson A, Gretarsson E, Arnason A (2001) Soil erosion in Iceland. Soil Conservation Service and Agricultural Research Institute, Reykjavík

Barkarson BH (2002) Management of livestock grazing on central highland rangelands in Iceland. M.Sc. Thesis, University of Iceland. Reykjavík. English abstract

Barkarson BH, Jóhannsson M (2009) Arctic landcare. In: Catacutan D, Neely C, Johnson M, Poussard H, Youl R (eds) Landcare: local action – global progress. Nairobi, Kenya, pp 55–63

Beyerlin U, Marauhn T (2011) International environmental law. Hart Publishing, London

Barrios E (2007) Soil biota, ecosystem services and land productivity. Ecol Econ 64:269–285

Birne P, Boyle A, Redgwell C (2009) International law and the environment. Oxford University Press, Oxford

Crofts R (2011) Healing the land. Soil Conservation Service of Iceland, Gunnarsholt

Fannarsson BS (2007) Eignarréttarleg staða fasteigna sem verið hafa til uppgræðslu. University of Iceland, Reykjavík

Greipsson S (2012) Catastrophic soil erosion in Iceland: Impact of long-term climate change, compounded natural disturbances and human driven land-use changes. Catena 98:41–54. https://doi.org/10.1016/j.catena.2012.05.015

Gunnarsdottir S, Briem H, Gottfredsson M (2014) Extent and impact of the measles epidemics of 1846 and 1882 in Iceland]. [Article in Icelandic]. Laeknabladid 100:211–216

Hanna E, Jonsson T, Box JE (2004) An analysis of icelandic climate since the nineteenth century. Int J Climatol 24:1193–1210. https://doi.org/10.1002/joc.1051

Hannam ID, Boer BW (2002) Legal and institutional frameworks for sustainable soils. The World Conservation Union, Gland, Switzerland and Cambridge, UK

Iceland Parliamentary Debates (1891) Alþingistíðindi, 1891, A-deild, Reykjavík, bls 126

Iceland Parliamentary Debates (1914a) Alþingistíðindi 1914, A-deild, Reykjavík, bls 105

Iceland Parliamentary Debates (1914b) Alþingistíðindi 1914, B-deild, Reykjavík, bls 75

Iceland Parliamentary Debates (1914c) Alþingistíðindi 1914, B-deild, Reykjavík, bls 143

Iceland Parliamentary Debates (1923) Alþingistíðindi 1923, B-deild, Reykjavík, bls 1633–1634

Iceland Parliamentary Debates (1941) Alþingistíðindi 1941, B-deild, Reykjavík, bls 567–568

Iceland Parliamentary Debates (1964) Alþingistíðindi 1964, A-deild, Reykjavík, bls 565–566

IPCC (2014) 2013 supplement to the 2006 IPCC guidelines for national greenhouse gas inventories: wetlands. In: Hiraishi T, Krug T, Tanabe K, Srivastava N, Baasansuren J, Fukuda M, Troxler TG (eds) Switzerland

Landgræðsla ríkisins (1988a) Gunnarsholt á Rangárvöllum. In: Arnalds A (ed) Græðum Ísland. Landgræðslan 1907–1987. Landgræðsla ríkisins (Soil Conservation Service of Iceland), Reykjavík, pp 125–130

Landgræðsla ríkisins (1988b) Landgræðslan í 80 ár, 1907–1987. In: Arnalds A (ed) Græðum Ísland. Landgræðslan 1907–1987. Landgræðsla ríkisins, Reykjavík, pp 99–124

Ministry for the Environment and Natural Resources (2012) Tillögur að inntaki nýrra laga um landgræðslu. Greinargerð starfshóps til ráðherra (White paper on reform of the soil conservation law)

Olafsdottir R, Schlyter P, Haraldsson HV (2001) Simulating Icelandic vegetation cover during the Holocene - implications for long-term land degradation. Geogr Ann Ser A Phys Geogr 83A:203–215

Olgeirsson FG (2007) Sáðmenn sandanna: Saga landgræðslu á Íslandi 1907–2007. Landgræðsla ríkisins, Gunnarsholt

Oskarsson H, Arnalds O, Gudmundsson J, Gudbergsson G (2004) Organic carbon in Icelandic Andosols: geographical variation and impact of erosion. Catena 56:225–238. https://doi.org/10.1016/j.catena.2003.10.013

Ostrom E (1990) Governing the commons: the evolution of institutions for collective action. Cambridge University Press, Cambridge. ISBN 9780521405997

Ostrom E (2009) A general framework for analyzing sustainability of social-ecological systems. Science 325:419–422. https://doi.org/10.1126/science.1172133

Runólfsson S, Ágústsdóttir AM (2011) Restoration of degraded and desertified lands: experience from Iceland. In: Lal R, Sivakumar MVK, Faiz SMA, Mustafizur Rahman AHM, Islam KR (eds) Climate change and food security in South Asia. Springer, Dordrecht, pp 153–161. https://doi.org/10.1007/978-90-481-9516-9_11

Sigurjónsson A (1958) Ágrip af gróðursögu landsins til 1880. In: Sandgræðslan. Minnst 50 ára starfs Sandgræðslu Íslands. Búnaðarfélag Íslands og Sandgræðsla ríkisins, Reykjavík, pp 5–40

Statistics Iceland (2017) http://www.statice.is/statistics/business-sectors/agriculture/. Accessed 1 July 2017

Þorgilsson A (1887) The book of the Icelanders (Íslendingabók) by Ari Thorgilsson. Finnur Jónsson bjó til prentunar. Hið Íslenska bókmenntafélag, Copenhagen

Thorsteinsson I, Olafsson G, Van Dyne GM (1971) Range resources of Iceland. J Range Manag 24:86–93

United Nations (1987) Our common future – brundtland report. Oxford University Press, Oxford

Vatn A (2005) Institutions and the environment. Edward Elgar Publishing, Cheltenham

Vatn A (2010) An institutional analysis of payments for environmental services. Ecol Econ 69:1245–1252

Greek Soil Law

Natalia Charalampidou

1 Environmental Protection Law

Environmental Protection Law (hereinafter 'EnvPL') aims to bring into effect fundamental principles, along with criteria and mechanisms for environmental protection, so that human, both as a person and as a part of society, may live in a high-quality environment while protecting his health and fostering the development of his personality.[1] The basic principles of this law are pollution prevention, protection of human health, sustainable management of the environment and non-renewable resources, preservation of the ecological balance of natural ecosystems and fostering their reproduction ability and remediation of environmental pollution.[2] Soil protection, including its use according to its natural properties and reproductive ability, is stated as one of the main scopes, along with protection of water and air.[3]

Specific measures for soil protection are to be decided by the competent Ministries. On the basis of this statutory empowerment, the Ministries are free to enact a broad variety of measures of both preventive and remedying nature, which act against material and non-material contamination that is created by both man and nature. The statutory provision sets out that soil is to be protected against

[1]Section 1(1) EnvPL.

[2]Section 1(2) *ibid.*

[3]Section 1(3) *ibid.*

N. Charalampidou (✉)
Democritus University of Thrace, Komotini, Greece
e-mail: natalia.charalampidou@alumni.uni-heidelberg.de

H. Ginzky et al. (eds.), *International Yearbook of Soil Law and Policy 2017*,
International Yearbook of Soil Law and Policy,
https://doi.org/10.1007/978-3-319-68885-5_13

natural damage, corrosion, lack of soil aeration, reclamation, destruction of soil structure, salination, chemical depletion, over-fertilization, as well as wrong fertilisation and addition of toxic substances from fertiliser and pesticide use.[4]

Causing pollution or adverse effects to the environment may lead to penal sanctions, administrative fines and compensation. The administrative fines that may be imposed through administrative decision of the competent authorities are monetary fines of 500–2,000,000 euro. The precise amount depends on facts surrounding each case under consideration of seriousness and frequency of the violation in question.[5] Parallel to said fines, preliminary prohibition of the activity causing said damage may be ordered, until the appropriate preventive measures are in place. If this is impossible, permanent prohibition of same activity may be imposed. Permanent prohibition also applies in the event of non-compliance with ordered measures.[6]

Persons causing pollution or adverse effects to the environment are equally liable to compensation, unless it is proved that damage was caused by force majeure or deliberate (*δόλος* in the original, *Vorsatz* in German) action of a third party.[7]

Further, causing pollution or adverse effects to the environment in violation of this law may result to imprisonment of three months to two years and monetary fine.[8] Mental capacity of the perpetrator and the results of his actions effect punishment. Should pollution or adverse effects result to danger of death or serious physical injury, at least 1-year imprisonment and monetary fine are imposed,[9] whereas the maximum imprisonment sentence is ten years in the event of death or serious physical injury.[10] In the latter case, confidentiality of communications may be lifted while investigating the alleged crime.[11] Under the following circumstances, sentencing may be less strict. If pollution or adverse effects are caused by negligence (*αμέλεια* in the original, *Fahrlässigkeit* in German), only imprisonment of up to 1 year may be imposed.[12] Authorised activities that result to pollution and adverse effects do not face prosecution. Albeit, this does not apply to actions and omissions that exceed the authorised ones.[13] The perpetrator may receive a lighter sentence or even no sentence at all when on his own initiative and prior to authority's investigations essentially mitigates environment's pollution or adverse effects thereof. The same applies when for this purpose, he timely informs the authorities.[14]

[4]Section 11(1) *ibid.*

[5]Sections 31(9) and 21 Law No 4014/2011.

[6]Section 30(2) EnvPL.

[7]Section 29 *ibid.*

[8]Section 28(1)(a) *ibid.*

[9]Section 28(3)(a) 1st sentence *ibid.*

[10]Section 28(3)(a) 2nd sentence *ibid.*

[11]Section 4(1) Law No 2225/1994.

[12]Section 28(2) EnvPL.

[13]Section 28(1)(b) *ibid.*

[14]Section 28(6) *ibid.*

2 Legislation on Desertification and Drought

Combatting desertification and mitigating the effects of drought are the main objects of the United Nations Convention to Combat Desertification in Countries Experiencing Serious Drought and/or Desertification, particularly in Africa (hereinafter 'UNCCD'). UNCCD has been signed and ratified by Greece via Law No 2468/1997, and therefore it forms a part of the pattern of national laws on soil protection.

Desertification (*απερήμωση* in Greek, *désertification* in French, *Wüstenbildung* in German) is land degradation in arid, semi-arid and dry sub-humid areas resulting from various factors, including climatic variations and human activities,[15] whereas the phrase combatting desertification (*καταπολέμηση της απερήμωσης* in Greek, *lutte contre la désertification* in French, *Bekämpfung der Wüstenbildung* in German) includes activities that are part of the integrated development of land in arid, semi-arid and dry sub-humid areas for sustainable development aiming at the prevention and/or reduction of land degradation, rehabilitation of partly degraded land and reclamation of desertified land.[16] Drought (*ξηρασία* in Greek, *sécheresse* in French, *Dürre* in German) means a naturally occurring phenomenon that exists when precipitation has been significantly below normal recorded levels, causing serious hydrological imbalances that adversely affect land resource production systems,[17] while mitigating the effects of drought (*άμβλυνση των συνεπειών της ξηρασίας* in Greek, *atténuation des effets de la sécheresse* in French, *Milderung von Dürrefolgen* in German) stands for activities related to the prediction of drought and intended to reduce the vulnerability of society and natural systems to drought as it relates to combating desertification.[18]

Aiming to achieve said objects, parties to UNCCD ensure that decisions on the design and implementation of relevant programmes are taken with the participation of populations and local communities[19] while fostering international solidarity and partnership, improving cooperation and coordination at sub-regional, regional and international levels.[20] They further develop a better understanding of the nature and value of land and scarce water resources in affected areas and to work towards their sustainable use.[21] At the same time, they need to take into full consideration the special needs and circumstances of affected developing country parties, particularly the least developed among them.[22] Greece, as an affected country due to its biogeoclimatic characteristics and overexploitation of its natural resources,[23] stands

[15]Article 1(a) UNCCD.
[16]Article 1(b) *ibid.*
[17]Article 1(c) *ibid.*
[18]Article 1(d) *ibid.*
[19]Article 3(a) *ibid.*
[20]Article 3(b) *ibid.*
[21]Article 3(c) *ibid.*
[22]Article 3(d) *ibid.*
[23]Preamble to Ministerial Decree No 99605/3719/2001 (hereinafter '*DesertDe*').

under the obligation to give due priority to combating desertification and mitigating the effects of drought and allocate adequate resources in accordance with its circumstances and capabilities,[24] as well as to establish relevant strategies and priorities, within the framework of sustainable development plans and/or policies.[25] It further addresses the underlying causes of desertification and pays special attention to the socio-economic factors contributing to desertification processes[26] and promotes awareness and facilitates the participation of local populations, with the support of non-governmental organizations, in relevant efforts.[27] Lastly, it provides an enabling environment by strengthening laws and establishing long-term policies and action programmes.[28]

In response to said international obligations, DesertDe sets out the following. During drafting and applying land use and development plans under the applicable laws,[29] desertification is taken into account and sustainable land use is decided in order to prevent desertification and remediate its effects.[30] Specific preventive and remedial measures are taken, whereas in areas where the damage is beyond any repair, special management plans are applied.[31] The National Commission for Combating Desertification is entrusted with the task of preparing a campaign for informing productive sectors, competent authorities, non-governmental organisations and local authorities, as well as motivating citizens and affected social groups.[32]

In addition, governmental and non-governmental institutions having necessary know-how perform scientific research, needed inventories and mapping for the purpose of addressing desertification, along with developing relevant technology. Data are collected by monitoring stations in sensitive zones. This information is stored in a database under the responsibility of the National Commission and is available to any interested person.[33] Affected areas are specified in the best possible manner. General zoning of endangered areas aims to estimate the danger in a quantitative manner, specifying general measures at a national level, defining pilot areas and drafting general strategy against desertification.[34] On a local level, detailed programmes against desertification are prepared and integrated into the relevant development programmes of every area.[35]

---

[24] Article 5(a) UNCCD.

[25] Article 5(b) *ibid.*

[26] Article 5(c) *ibid.*

[27] Article 5(d) *ibid.*

[28] Article 5(e) *ibid.*

[29] These are Law No 2742/1999 and Law No 2508/1997.

[30] Annex section 1(6) DesertDe.

[31] Annex section 1(8) *ibid.*

[32] Annex section 1(9) *ibid.*

[33] Annex section 2(1) *ibid.*

[34] Annex section 2(2) *ibid.*

[35] Annex section 2(3) *ibid.*

3 Legislation on Agriculture

Statutory provisions on agriculture relate to soil pollution control as they regulate a human activity that involves direct interaction with the soil, and hence the same effects are expected. Agriculture law includes, in particular, legislation on fertilisers, that being Regulation (EC) No 2003/2003 and Law No 1565/1985 (hereinafter 'ReL'); livestock manure, that being Ministerial Decree No 1420/82031/2015, (hereinafter 'LiMaDe'); and sewage sludge, that being Ministerial Decree No 80568/4225/1991 (hereinafter 'SeSlDe').

3.1 Fertilisers

Fertilisers (*λίπασμα* in Greek, *engrais* in French, *Düngemittel* or *Dünger* in German), those being material, the main function of which is to provide nutrients for plants,[36] are regulated throughout the European Union by Regulation (EC) No 2003/2003. This regulation has been issued for the purpose of addressing the risk to the environment, as well as human and animal health, through contamination of fertilisers, such as the unintentional cadmium content in mineral fertilisers.[37] However, the main purposes that triggered its enactment were simplification and clarification of existing law,[38] securing free movement of such products throughout the European Union[39] and laying additional rules for ammonium nitrate fertilisers in view of public safety, health and protection of workers, as varieties of ammonium nitrate are used in the manufacture of products used as explosives.[40]

Although fertilisers are applied directly in the soil, the latter as such does not explicitly fall under the objective of this law. It is, though, affected in an indirect manner due to its inclusion in the general term environment. In particular, fertilisers that under normal conditions of use adversely affect human, animal or plant health or the environment do not enjoy free circulation in the market as they are not classified as 'EC fertiliser' (*λίπασμα EK* in Greek, *engrais CE* in French, *EG-Düngemittel* in German).[41] Further, EC fertilisers, which generally may not be prohibited, restricted or hindered from being placed on the market,[42] may be temporarily prohibited from said placing or subject to special conditions, when a

[36]Article 2(a) Regulation (EC) No 2003/2003.

[37]Recital no 15 to *ibid.*

[38]COM/2001/0508 final, para 1.1.

[39]CES0022/2002, para 3.2.

[40]*Ibid* para 1.4; Recitals no 11 and 12 to Regulation (EC) No 2003/2003.

[41]Articles 14(c), 3(2) and 5(2) *ibid*. Article 14 has been adapted from article 8 Council Directive 89/530/EEC. This had amended Council Directive 76/116/EEC, which has been repealed by Regulation (EC) No 2003/2003.

[42]Article 3(1) Regulation (EC) No 2003/2003.

Member State has justifiable grounds for believing that it constitutes a risk (*δυνητικό κίνδυνο* in Greek, *risque* in French, *Risiko* in German) to the environment or to the safety or health of humans, animals or plants.[43] Under Greek law, the Minister of Rural Development and Food has the power to make use of said safeguard clause and prohibit the circulation of fertilisers under same conditions.[44]

National law further prohibits the circulation of fertilisers containing substances not foreseen in the law but are toxic for the health of humans, plants, animals and generally the environment according to latest scientific and technical developments.[45] Circulation of fertilisers containing substances hazardous (*επικίνδυνα* in the original, *dangereux* in French, *gefährlich* in German)[46] for the health of humans, animals or plants and generally the environment is equally prohibited,[47] along with exposure of fertilisers to weather conditions that may produce changes to their natural or chemical characteristics, their package, and pollute the environment.[48] Lastly, farmers are not allowed to acquire and use fertilisers that are inappropriate, nocuous and hazardous for the health of humans, animals or plants and generally the environment, containing substances that pollute the environment or reduce fertility of the soil and the quality of agricultural products.[49] Violation of these obligations results in administrative fines of 3000 euro, 30,000 euro, 500 to 5000 euro and 3000 euro, respectively.[50]

---

[43]Article 15(1) *ibid.* Article 15 has been adapted from Article 9 Council Directive 80/876/EEC, which has been repealed by Regulation (EC) No. 2003/2003. Directive 80/876/EEC however makes mention of hazard ('*κίνδυνο*' in Greek, '*danger*' in French, '*Gefahr*' in German). Hence, the relevant threshold has been lowered in Regulation (EC).

[44]Section 11A FeL.

[45]Section 11(1)(h) *ibid.*

[46]Hazard has the meaning set out in Annex I Parts 2 (physical hazards, including explosives, flammable gases, aerosols, oxidising gases, gases under pressure, flammable liquids, flammable solids, self-reactive substances and mixtures, pyrophoric liquids, pyrophoric solids, self-heating substances and mixtures, substances and mixtures which in contact with water emit flammable gases, oxidising liquids, oxidising solids, organic peroxides and corrosive to metals), 3 (health hazards, including acute toxicity, skin corrosion/irritation, serious eye damage/eye irritation, respiratory or skin sensitisation, germ cell mutagenicity, carcinogenicity, reproductive toxicity, specific target organ toxicity—single exposure, specific target organ toxicity—repeated exposure and aspiration hazard), 4 (environmental hazards, including hazardous to the aquatic environment) and 5 (additional hazards, including hazardous to the ozone layer) Regulation (EC) No 1272/2008.

[47]Section 11(1)(i) FeL.

[48]Section 11(1)(p) *ibid.*

[49]Section 11(1)(r) *ibid.*

[50]Section 11(1)(h), (i), (p) and (r) *ibid.*

3.2 Livestock Manure

Livestock manure (*κτηνοτροφικά απόβλητα* in the original, *effluent d'élevage* in French, *Dung* in German) stands for waste products excreted by livestock or a mixture of water residues, feed, litter and waste products excreted by livestock[51] and is explicitly excluded from the notion of fertilisers.[52] The Code of Good Agricultural Practice, which provides proper guidance to everyone involved in agricultural and breeding activities for the purpose of environmental protection and especially prevention of pollution of groundwater and surface waters by nitrates,[53] encourages the use of livestock manure for increasing the soil's fertility and subsidizing chemical fertilisers (*χημικό λίπασμα* in the original, *engrais chimique* in French, *Mineraldünger* in German) that is differentiated from livestock manure through their manufacture by an industrial process.[54]

Certain mandatory obligations in relation to the use of livestock manure in the soil are set out in LiMaDe for specifying and materializing its objective. The first set of obligations relates to the area where this may be used, as well as to the relevant aim. Liquid livestock manure may be applied in cultivated areas for the aim of fulfilling cultivation needs mainly in nitrogen, as well as in other areas, as long as it has undergone proper mechanical treatment, for the purpose of developing native nitrogen-friendly vegetation according to the provided amount of nitrogen. The latter is defined as area of soil—plant filter.[55] Solid livestock manure may be applied in cultivated areas via direct injection with the aim of manuring and enhancing the soil with organic substances. It may further be disposed in facilities for drying, incineration, treatment and disposal of activated sludge or waste.[56] Liquid and solid livestock manure, when it has undergone no treatment, may be disposed in approved biogas or composting plants.[57] The second set of obligations relate to the manner of application. This must be controllable and in specific amounts so that the additional nitrogen in total does not exceed demands of cultivation per plant development stage and loading of waters with nitrogen is avoided. For this purpose, the amount of nitrogen in the soil at the time of cultivation, along with the following factors, are to be considered: the nitrate content of livestock manure; nitrogen losses during the application in the soil; the specific cultivation and its needs in nitrogen, depending on the plant development stage;

---

[51] Section 3(3)(c) LiMaDe. The national provision is narrower than the European one, which sets out that such waste products also in processed form fall under the notion of livestock manure. Article 2(g) Council Directive 91/676/EEC.

[52] Section 1(A) last sentence FeL.

[53] Section 1 LiMaDe and section 3 Ministerial Decree 16190/1335/1997.

[54] Sections 7(1) and 3(2)(c) LiMaDe. Section 3(2)(c) LiMaDe corresponds to Article 2(f) Directive 91/676/EEC.

[55] Section 7(2)(a)(i) and (ii) LiMaDe.

[56] Section 7(2)(b)(i) and (ii) *ibid.*

[57] Section 7(2)(b)(iii) *ibid.*

data of soil analysis related to the levels of nitrogen in the soil; the soil's qualities; the method and history of plot's manuring with nitrogenous fertilisers or livestock manure; the amount and quality of irrigation water; the irrigation method; the climate conditions and especially height and rain frequency.[58] It is further specified that the application of livestock manure takes place in a uniform manner through the existing irrigation system, other specialised mechanisms or practical manners,[59] whereas caution has to be demonstrated for avoiding losses or escapes.[60] Especially for plots with slope value of more than 8%, liquid livestock manure may only be applied through tickle irrigation or injection. It is rather encouraged to apply solid livestock manure in small doses that are injected into the soil during the application or immediately after thereof.[61] Last statutory obligation is that of keeping detailed records of manuring, including the exact amounts of applied livestock manure.[62]

The aforementioned statutory obligations are enhanced and clarified through a series of prohibitions equally set out in the law. Livestock manure that has not undergone any treatment, and thus has quality values not corresponding to the relevant national law, may not be applied in the soil or be disposed in surface or groundwater,[63] whereas application of excessive amounts of nitrogenous manure is prohibited.[64] More specifically, producers may not apply in the soil more than 250 kg of nitrogen from livestock manure per hectare during a 12-month period. This amount includes both treated livestock manure and manure from any animals that may be grassing in these areas.[65] Especially in vulnerable zones (*ευπρόσβλητη ζώνη* in the original, *zone vulnérable* in French, *gefährdete Gebiete* in German), this amount is limited to 170 kg per hectare.[66] Further restrictions relate to proximity to waters. Solid livestock manure may not be applied within 10 m from surface waters, whereas in the case of liquid livestock manure the area extends to 20 m. With reference to wells, springs and boreholes, the prohibition area further extends to 50 m for solid and liquid livestock manure.[67] In bare soil, sluice gates and neighbouring plots, any application or escape of livestock manure is prohibited.[68] Land application of livestock manure is further prohibited under specific weather conditions. This is the case when areas are frozen or covered with snow, as well as in soil saturated with water, which does not properly drain or is flooded.[69] Same prohibition applies when rain is expected within the next two days or when strong wind

[58]Section 7(4)(a) – (i) *ibid.*

[59]Section 7(9) *ibid.*

[60]Section 7(10) *ibid.*

[61]Section 7(14) *ibid.*

[62]Sections 7(4) last sentence and 5(1) *ibid.*

[63]Section 7(2) and (17) *ibid.*

[64]Section 7(4) 4th sentence *ibid.*

[65]Section 7(6) *ibid.*

[66]Section 7(7) *ibid.*

[67]Section 7(13) *ibid.*

[68]Section 7(15) *ibid.*

[69]Section 7(10)(a) *ibid.*

blows.[70] Further time-related prohibitions set out that livestock manure may not be applied within the time period of 1 November until 1 February with the exception of solid livestock manure for winter vegetables, flowers and cultivations under coating, treated manure from sheep and goats, solid manure from oxstalls and piggeries and treated manure from poultry houses for basic manuring for all necessary cultivations, including cereals, fodder plants and permanent crops. Especially for sand soils, the initiation of said time period is extended from 1 September.[71]

Non-compliance with the aforementioned statutory obligations and prohibitions may result to penal sanctions, administrative fines and compensation under sections 28, 29 and 30 EnvPL.[72]

3.3 *Sewage Sludge*

With the aim of making special arrangements to ensure that man, animals, plants and the environment are fully safeguarded from the harmful effects arising from uncontrolled use of sludge, Directive 86/278/EEC has been enacted and thereafter implemented in Greece through SeSlDe. Its scope is to regulate the use of sewage sludge in agriculture in such a way as to prevent harmful effects on soil, vegetation, animals and man, thereby encouraging the correct use of such sewage sludge.[73] In view of this, limit values for concentrations of heavy metals in soil[74] and in sludge,[75] along with the maximum annual quantities of the same, which may be introduced into the soil based on a ten-year average,[76] have been specified.

Measures and procedures ensuring adherence thereto have been equally set out in the law. These are differentiated depending on whether treated sludge is being produced or used. Production of such sludge requires a permit, which is issued upon

[70] Section 7(10)(b) and (c) *ibid.*

[71] Section 7(11) *ibid.*

[72] Section 7(1) Ministerial Decree 16190/1335/1997.

[73] Section 1 SeSlDe, which corresponds to Article 1 Directive 86/278/EEC.

[74] These are: Cadmium 1–3 mg/kg; Copper 50–140 mg/kg; Nickel 30–75 mg/kg; Lead 50–300 mg/kg; Zinc 150–300 mg/kg; and Mercury 1–1.5 mg/kg. Section 3 and Annex IIA SeSlDe, which corresponds to Article 4 and Annex 1A Directive 86/278/EEC.

[75] These are: Cadmium 20–40 mg/kg; Copper 1000–1750 mg/kg; Nickel 300–400 mg/kg; Lead 750–1200 mg/kg; Zinc 2500–4000 mg/kg; and Mercury 16–25 mg/kg. Section 3 and Annex IIB SeSlDe, which corresponds to Article 4 and Annex 1B Directive 86/278/EEC.

[76] These are: Cadmium 0.15 kg/ha/yr; Copper 12 kg/ha/yr; Nickel 3 kg/ha/yr; Lead 15 kg/ha/yr; Zinc 30 kg/ha/yr; and Mercury 0.1 kg/ha/yr. Section 3 and Annex IIC SeSlDe, which corresponds to Article 4 and Annex 1C Directive 86/278/EEC.

the condition that approval of environmental conditions[77] has been filed.[78] This permit contains, among other terms and conditions, provisions on the aforementioned limit values of heavy metals.[79] Permit holders are subject to specific statutory obligations. They are obliged to regularly provide users of livestock sludge with information on sludge analysis[80] that is performed by state laboratories or laboratories authorised by the state.[81] They further keep up-to-date records, which register the quantities of sludge produced and the quantities supplied for use in agriculture; the composition and properties of the sludge in relation to the parameters referred to in Annex II A; the type of treatment carried out, as defined in section 2(1)(b); and the names and addresses of the recipients of the sludge and the place where the sludge is to be used.[82] However, sludge from sewage treatment plants with a treatment capacity below 300 kg BOD5 per day, corresponding to 5000 person equivalents, which are designed primarily for the treatment of domestic waste water, may be excepted from the said obligations following a decision by competent authority.[83]

Use of livestock sludge also requires a permit.[84] The interested party submits an application along with a charter of the area where the use shall take place, information regarding the relevant cultivation species or the forest area, a certification of soil analysis and a certification of livestock sludge issued by the producer.[85] Competent authorities may, within their jurisdiction, inspect the area where the sludge shall be used with the aim of avoiding any soil degradation, as well as pollution or impairment of surface and underground waters.[86] The decision on granting or rejecting the application for a permit is issued within 20 days from the submission of the application. Should it be granted, it makes mention of terms and conditions on the limit values, the manner and methods for sludge transportation, the manner of application, any additional terms for the protection of public health and the environment, the area and cultivation species, the maximum values of treated sludge that may be used in the specific area and the general statutory prohibitions.[87] The general statutory prohibitions ban the use of sludge or supply for the use of the

[77]Approval of environmental conditions is regulated in Law No 1650/1986 and Ministerial Decree No 69269/5387/1990, which implemented, among others, Council Directive 85/337/EEC, as amended. This directive has been repealed by Directive 2011/92/EU, which has not yet been implemented in Greece.

[78]Section 4(1)(a) SeSlDe.

[79]Section 4(1)(b) *ibid.*

[80]The exact information is foreseen in Annex II A *ibid*, which corresponds to Annex II A Directive 86/278/EEC.

[81]Section 7(1) SeSlDe.

[82]Section 7(a)–(d) *ibid*, which corresponds to Article 10 Directive 86/278/EEC.

[83]Section 7(2) SeSlDe, which corresponds to Article 11 Directive 86/278/EEC.

[84]Section 4(2) SeSlDe.

[85]Section 4(2)(a.1) *ibid.*

[86]Section 4(2)(a.3) *ibid.*

[87]Section 4(2)(a.2) *ibid.*

same on grassland or forage crops if the grassland is to be grazed or the forage crops are to be harvested before a certain period has elapsed, which is specified taking particular account of their geographical and climatic situation but under no circumstances should be less than 3 weeks. Same prohibition applies on soil in which fruit and vegetable crops are growing, with the exception of fruit trees, as well as on ground intended for the cultivation of fruit and vegetable crops, which are normally in direct contact with the soil and normally eaten raw, for a period of 10 months preceding the harvest of the crops and during the harvest itself.[88] For the purpose of specifying the aforementioned terms and conditions in each permit, the competent authority takes account of the nutrient needs of the plants and that the quality of the soil and of the surface and ground water is not impaired. Where sludge is used on soils of which the pH is below 6, competent authorities further take into account the increased mobility and availability to the crop of heavy metals and, if necessary, reduce the limit values they have laid down in accordance with Annex I A.[89] For the purpose of ensuring that set limit values of concentration of one or more heavy metals in the soil are not exceeded as a result of the use of sludge,[90] the competent authority sets out in the permit the maximum quantities of sludge expressed in tonnes of dry matter that may be applied to the soil per unit of area per year while observing the limit values for heavy metal concentration in sludge, which they lay down in accordance with Annex I B. Alternatively, it ensures observance of the limit values for the quantities of metals introduced into the soil per unit of area and unit of time as set out in Annex I C.[91] Lastly, if a permit for using rough livestock sludge is granted, it specifies in its terms and conditions the manner of application, that being injecting or working into the soil. It further specifies the conditions for using said sludge.[92]

The competent authorities stand under the statutory obligation to undertake regular controls for ensuring that the aforementioned provisos are fulfilled.[93] Non-compliance with the aforementioned statutory obligations and prohibitions results to penal sanctions, administrative fines and compensation according to EnvPL.[94] Should, however, the material facts relate to the geographical area of the District of Athens or the one of Thessaloniki, the administrative sanctions set out in Law No 1515/1985 and Law No 1561/1986 are respectively applicable.[95]

---

[88] Section 5 *ibid*, which corresponds to Article 7 Directive 86/278/EEC.

[89] Section 4(2)(a.5)(ii)(a) and (b) SeSlDe, which corresponds to Article 8 Directive 86/278/EEC.

[90] As the Member States' obligation set out in Article 5(1) *ibid.*

[91] Section 4(2)(a.5)(i)(a) and (b) SeSlDe, which corresponds to Article 5(2)(a) and (b) Directive 86/278/EEC.

[92] Section 4(2)(a.4) SeSlDe, which corresponds to Article 6(a) Directive 86/278/EEC.

[93] Section 8(2) SeSlDe.

[94] Section 11(1) *ibid* and sections 28, 29 and 30 EnvPL.

[95] Section 11(2) SeSlDe.

4 Legislation on Environmental Liability

Environmental liability is regulated by Presidential Decree No 148/2009 (hereinafter 'EnvLL'), which has implemented the Environmental Liability Directive. On the basis of the polluter-pays principle, as a principle attributing solely financial liability deriving from remediation of environmental damage, this law aims to establish environmental liability.[96] In this respect, environmental damage (*περιβαλλοντική ζημιά* in the original, *dommage environnemental* in French, *Umweltschaden* in German) means damage to protected species, natural habitats and water damage, along with soil damage, which stands for any soil contamination that creates a significant risk of human health being adversely affected as a result of the direct or indirect introduction in, on or under land of substances, preparations, organisms or microorganisms.[97]

EnvLL applies to environmental damage caused by specific occupational activities,[98] including imminent threat of such, without prejudice to negligence (*υπαιτιότητα* in the original, *Verschulden* in German) of the operator.[99] It does not give private parties a right of compensation as a consequence of environmental damage or of an imminent threat of such damage.[100]

EnvLL foresees both preventive and remedial actions with specific obligations for the operators of said occupational activities. Imminent threat of damage (*άμεση απειλή ζημίας* in the original, *menace imminente de dommage* in French, *unmittelbare Gefahr eines Schadens* in German), which means a sufficient likelihood that such damage will occur in the near future,[101] triggers the need for preventive measures. In particular, when environmental damage has not yet occurred but said threat exists, the operator shall, without delay, take the necessary preventive measures. He is further obliged to inform the competent authority of all relevant aspects of the situation, as soon as possible.[102] The competent authority on its part may at any time require the operator to provide information, to take the necessary preventive measures or to follow precise instructions on same measures.[103] Equally, the competent authority may take itself the necessary preventive measures.[104]

---

[96]Section 2 EnvLL, which corresponds to Article 1 Directive 2004/35/EC. Siouti (2011), p. 295.

[97]Section 3(1)(a), (b) and (c) EnvLL, which corresponds to Article 2(1)(a), (b) and (c) *ibid.*

[98]These activities are: set out in Annex III EnvLL, which corresponds to Annex III Directive 2004/35/EC.

[99]Section 4(1)(a) EnvLL, which corresponds to Article 3(1)(a) Directive 2004/35/EC.

[100]Section 4(4) EnvLL, which corresponds to Article 3(3) Directive 2004/35/EC.

[101]Section 3(9) EnvLL, which corresponds to Article 2(9) Directive 2004/35/EC.

[102]Section 8(1) EnvLL, which corresponds to Article 5(1) and (2) Directive 2004/35/EC.

[103]Section 8(2)(a), (b) and (c) EnvLL, which corresponds to Article 5(3)(a), (b) and (c) Directive 2004/35/EC.

[104]Section 8(2)(d) EnvLL, which corresponds to Article 5(3)(d) Directive 2004/35/EC.

Where environmental damage has occurred, the operator is obliged to inform immediately the competent authority of all relevant aspects of the situation. Furthermore, he must take all practicable steps to immediately control, contain, remove or otherwise manage the relevant contaminants and any other damage factors in order to limit or to prevent further environmental damage and adverse effects on human health or further impairment of services.[105] The operator further proposes remediation measures to the competent authority for approval and performs his own evaluation for the seriousness and magnitude of the damage while providing all necessary information to the competent authority and economic and financial operators, providing the competent authority with financial security, if applicable.[106] The competent authority on its part may at any time require the operator to provide supplementary information on any damage that has occurred, as well as take itself, instruct the operator to take or give instructions to the operator concerning all practicable steps to immediately control, contain, remove or otherwise manage the relevant contaminants and any other damage factors in order to limit or to prevent further environmental damage and adverse effect on human health or further impairment of services.[107] The competent authority has further the power to give the operator instructions on the necessary remedial measures to be taken and itself take such measures.[108] The latter equally comes to effect when the operator may not be identified or when he is identified but does not comply.[109]

The operator bears the cost for preventive and remedial actions.[110] In the event that the competent authority has paid the expenses for such actions, it may recover the cost from the operator, who caused the damage, unless the expenditure required to do so would be greater than the recoverable sum or the operator cannot be identified.[111] However, the operator bears no cost when he can prove that the damage or threat was caused by a third party and occurred despite the fact that appropriate safety measures were in place. This equally applies when the damage or threat resulted from compliance with a compulsory order or instruction emanating from a public authority other than an order or instruction consequent upon an emission or incident caused by the operator's own activities.[112]

[105]Section 9(1)(a) and (b) EnvLL, which corresponds to Article 6(1)(a) Directive 2004/35/EC.

[106]Section 9(1)(c) and (d) EnvLL, which corresponds to Article 6(1)(b) Directive 2004/35/EC.

[107]Section 9(2)(a) and (b) EnvLL, which corresponds to Article 6(2)(a) and (b) Directive 2004/35/EC.

[108]Section 9(2)(c) and (d) EnvLL, which corresponds to Article 6(2)(c) and (d) Directive 2004/35/EC.

[109]Section 9(3) EnvLL, which corresponds to Article 6(3) Directive 2004/35/EC.

[110]Section 11(1) EnvLL, which corresponds to Article 8(1) Directive 2004/35/EC.

[111]Section 11(2) and (3) EnvLL, which corresponds to Article 8(2) Directive 2004/35/EC.

[112]Section 11(4)(a) and (b) EnvLL, which corresponds to Article 8(3) Directive 2004/35/EC.

5 Legislation on Waste

Waste is regulated by Law on Waste of 2012, which, along with implementing Directive 2008/98/EC, lays down measures to prevent or reduce the adverse impacts of the generation and management of waste and improve the efficiency of such use (Sect. 1 Law on Waste). Nevertheless, remediation of soil contaminated by waste is regulated by ministerial decrees issued prior to Law on Waste, which only partially upheld them.

5.1 *Legislation on Solid Waste*

Management of solid waste is regulated by Ministerial Decree No 50910/2727/2003 (hereinafter 'SolidWaDe'). In this context, solid waste means any substance or object that the holder discards or intends or is required to discard,[113] but for hazardous waste. Substances and objects that may constitute waste are divided into categories depending on them being residues or contaminated, having expired for appropriate use or their use being banned by law.[114] Regulating the management of solid waste aims to prevent or reduce negative effects on the environment and to ensure a high level of protection of the environment and public health while developing the principles of section 12 EnvPL into more specific rules and fully complying with Directive 75/442/EEC.[115] It further ensures that waste is recovered or disposed of without endangering human health and without using processes or methods that could harm the environment and, in particular, without risk to water, air, soil, and plants and animals.[116] The principles governing solid waste management are the precautionary principle, the principle of preventive action, the polluter-pays principle in relation to the liability of the waste producer, the principle of proximity and the principle of remediating environmental damage.[117]

Upon terminating waste management, the operator is obliged to remediate significant environmental damage caused therefrom and to restore the natural environment.[118] The competent authority reviews and approves the performed remediation and rehabilitation actions prior to authorising termination of operation. It also determines the time period during which aftercare measures are necessary,

---

[113]Section 2(a) SolidWaDe, which corresponds to Article 1(a) Directive 2006/12/EC.

[114]These categories are set out in Annex IA SolidWaDe, which corresponds to Annex I Directive 2006/12/EC.

[115]Section 1 SolidWaDe. 442/442/EEC has been repealed by Directive 2006/12/EC, which itself was later repealed by Directive 2008/98/EC.

[116]Section 4(1)(a) SolidWaDe, which corresponds to Article 4 Directive 2006/12/EC.

[117]Section 4(2) SolidWaDe.

[118]Section 9(1)(a) and (b) *ibid.*

taking into account the nature of the performed operation, the affected area along with risks for the environment and public health.[119]

5.2 *Legislation on Hazardous Waste*

Management of hazardous waste is regulated by Ministerial Decree 13588/725/06 (hereinafter 'HazWaDe'). Hazardous waste (in the original *επικίνδυνα απόβλητα*, in French *déchets dangereux*, in German *gefährlicher Abfall*) means waste that displays one or more of the hazardous properties listed in Annex III Law on Waste.[120, 121] Prior to 2012 and the enactment of Law on Waste, hazardous waste was waste resulting from specific treatments, processes and industries and displaying one or more of hazardous properties, which were classified in the European harmonised list.[122] It should be noted that common terminology throughout the Union serves the purpose of improving the efficiency of waste management activities.

HazWaDe aims to embody section 12 EnvPL and implement Directive 91/689/EEC in order to ensure a high level of protection of the environment and public health, especially through waste prevention, reduction and recovery and developing and using clean technologies that do not involve excessive cost.[123] The aforementioned scopes of managing solid waste equally apply on managing hazardous waste.[124] Measures taken encourage waste prevention and reduction, reuse and material recycling, energy recovery from waste and remediation of sites contaminated by hazardous waste.[125] The principles governing hazardous waste management are the same as the ones governing solid waste management.[126]

The operator bears the duty to remediate any soil pollution after termination of operation, along with the duty of taking all necessary measures, if the environment is effected during the course of operation.[127] The hazardous waste holder, along

---

[119]Section 9(1)(c) *ibid.*

[120]Properties of waste which render it hazardous are: H1 'explosive', H2 'oxidizing', H3-A 'highly flammable', H3-B 'flammable', H4 'irritant', H5 'harmful', H6 'toxic', H7 'carcinogenic', H8 'corrosive', H9 'infectious', H10 'toxic for reproduction', H11 'mutagenic', H12 Waste which releases toxic or very toxic gases in contact with water, air or an acid, H13 'sensitizing', H14 'ecotoxic', H15 Waste capable by any means, after disposal, of yielding another substance, e.g. a leachate, which possesses any of the characteristics listed above. Annex III Law No 4042/2012, which corresponds to Annex III Directive 2008/98/EC.

[121]Section 11(2) Law on Waste, which corresponds to Article 3(2) *ibid.*

[122]Annex to 2000/532/EC: Commission Decision.

[123]Section 1 HazWaDe.

[124]Section 4(1)(a) *ibid.*

[125]Section 4(2)(a), (b) and (c) *ibid.*

[126]Section 4(3) *ibid.*

[127]Sections 9 and 11(4)(e) *ibid.*

with the operator, may also be called to pay the bill for soil remediation in sites where hazardous waste management and disposal took place.[128] When these persons may not be identified or when pollution occurs in abandoned areas, the competent authorities proceed with soil remediation measures and may recover the costs, if identification succeeds.[129] If an accident causes such pollution, the competent authorities take all necessary measures for dealing with such accident.[130] In view of the *holder-pays-principle*, established in section 12(1) and (2)(a) HazWaDe, which does not originate from a directive of the European Union or the general environmental principles set out in EnvPL, it is crucial to specify the exact meaning of holder. Holding (in Latin: *detentio*), which is a notion distinct to possession (in Latin: *possessio*),[131] is established in Law of Rights in Rem[132] and includes both physical control over a thing (*corpus*—objective element) and the willingness of the holder to control it (*animus*—subjective element).[133] When the holder is not the owner of a thing, the holder's willingness to control it takes place in the name of another person (in Latin: *animus rem alii habendi*).[134] In the context of liability for soil remediation, these elaborations result into requiring willingness of the person to control hazardous waste. Hence, knowledge of hazardous waste being discarded is crucial for a holder to be held liable for the clean-up operation. If the person in physical control lacks such willingness, his holding has no legal effects and therefore may not be ordered to remediate contaminated soil or pay for such measures.

On the basis of HazWaDe, a study 'on investigation, assessment and remediation of uncontrolled contaminated sites and facilities with industrial and hazardous waste' has been concluded for the Ministry of Environment and Energy in 2008–2009 (hereinafter 'the Study'). The proposed methodology adopts a three-stage approach that begins from locating possibly contaminated sites and reaches to application of specific remediation measures and aftercare. The first stage involves applying certain criteria for the purpose of identifying possibly contaminated sites.[135] This shall lead to drafting catalogues of possibly contaminated sites.[136] The second stage involves detailed description of possibly contaminated sites, including taking samples and performing analysis, describing the effects on the environment, suggesting limit values according to international practice, referring to international standards for contaminated sites and presenting usual ways of humans exposing to contaminated sites and groundwater and possible

[128] Section 12(1) and (2)(a) *ibid.*

[129] Section 12(2)(b) *ibid.*

[130] Section 12(3) *ibid.*

[131] Mpalis (1961), p. 8.

[132] Section 974 Greek Civil Code.

[133] Vathrakokoilis (2007), p. 158.

[134] Mpalis (1961), p. 8; Georgiades (2010), p. 165.

[135] The Study p. 32.

[136] *Ibid* p. 33.

symptoms.[137] The third stage describes the decision-making procedure, including the procedure of evaluating hazard, and the basic factors for determining the appropriate remediation measures.[138]

The Study is a positive initiative for remediation of contaminated soil as it provides practical steps and technical guidelines for determination and remediation of soil contamination. However, certain issues raising therefrom must be addressed. The Study refers broadly and vaguely to the *European legislation on contaminated soil or waste management*,[139] although no legislation on contaminated soil has been enacted, as the proposal for a Directive establishing a framework for the protection of soil[140] was abandoned. Moreover, the European legislation on waste management has been implemented and thus became national law. Hence, the overall mention to said European legislation constitutes *contradictio in terminis*. Further, the legal basis of the proposed methodology is HazWaDe, and the aforementioned definition of hazardous waste applies. The suggestion that multiple definitions of the term hazardous waste exist, as purported in the Study,[141] is not accurate. Equally irrelevant is the reference to the methodology of the Ministry of Health of Australia and to Australian and New Zealand directives on quality of fresh and sea water for the purpose of evaluating risks,[142] along with the reference to the New Dutch List and the *Bundes-Bodenschutzverordnung* of Germany.[143] In addition, the term industrial waste does not correspond to a legal notion. Perhaps it was intended to address the issue of industrial emissions, which is, however, regulated in Ministerial Decree No 36060/1155/E.103/2013, which implemented Directive 2010/75/EU. Naturally, this legal instrument is not mentioned in the Study as it was issued after its conclusion. However, the ones in force at the time of composing the Study, those being Ministerial Decree Nos 15393/2332/2002, 11014/703/Φ104/2003 and 37111/2021/2003, implementing the then in force Directive 96/61/EC, are mentioned solely in the bibliography.[144] One more issue of paramount importance is that the Study does not seem to take into consideration the authority's restricted powers to order remediation of soil contamination only in sites where management and disposal of hazardous waste took place.[145] Neither does it consider the fact that HazWaDe may only unfold its effect from its enactment (2006) onwards. Rather, it

[137] *Ibid* p. 48 f.

[138] *Ibid* p. 127 f.

[139] *Ibid* p. 143.

[140] COM/2006/0232 final; Koutoupa – Regkakou (2008), p. 169 et seq.

[141] *The Study* p. 33.

[142] *Ibid* p. 96 et seq, p. 102 et seq.

[143] *Ibid* p. 110.

[144] *Ibid* p. 422.

[145] Section 12(1) HazWaDe.

analyses instruments for soil remediation generally throughout the country and for all adverse effects. The Ministry, acting on the Study, cites areas related to mining of metal ores as possibly contaminated sites,[146] despite the fact that these do not fall within the area of application of HazWaDe. Hence, the administration is acting beyond its statutory powers and therefore violates the principle of legality (original *αρχή της νομιμότητας*, in France *principe de légalité*, in Germany *Prinzip der Gesetzmäßigkeit der Verwaltung*).[147] In addition to this, local administration, which may be held liable for soil remediation, along with waste managers and holders,[148] when it bears the duty and cost of soil remediation, imposes further burden on the state budget, for actions that, to the extend explained above, lack legal base. In addition to that, the Study makes a brief and modest mention to the polluter-pays principle,[149] despite it being a fundamental principle of the HazWaDe and the Union policy on the environment.[150] Rather, the Study prefers to highlight, as suggested, financial instruments: programmes of the EU, the Global Environmental Fund, state aid—without any reference to guidelines on state aid for environmental protection,[151] taking loans, budget of local authorities and state budget.[152] In view of this, it could be said that the Study undervalues the importance of the 'polluter-pays-principle' and totally ignores the 'holder-pays-principle' and the principal incompatibility of state aid with the internal market.[153] Lastly, the Study mentions as legislation ensuring environmental compliance of the industry the *Environmental Liability Directive*, the *IPPC Directive* and the *SEVESO Directive*.[154] This is not accurate for a number of reasons. Only national legislation implementing Directives of the EU is binding and can create duties and rights between citizens and the administration or between citizens, but for the exceptional event of implementation failure and under strict conditions. Moreover, at the time that the Study was concluded, Directive 82/501/EEC (SEVESO) was already upheld by Directive 96/82/EC (SEVESO II). Most importantly, said Directives and the national laws implementing those in Greece do not regulate hazardous waste. They therefore fall outside the scope of the Study and may not be used for bringing into effect the proposed technical guidelines as that would be contrary to the

[146]See in this respect: http://www.ypeka.gr/Default.aspx?tabid=885&language=el-GR (last visited on 12.4.2017).

[147]Manesis (1967), p. 410 et seq; Manitakis (1994), p. 128; Venizelos (2008), pp. 414, 417; Spiliotopoulos (2011), p. 88; Maurias (2014), p. 69; Chrysogonos (2014), p. 378.

[148]Section 12 HazWaDe.

[149]The Study p. 416.

[150]Article 191(2) TFEU.

[151]See in this respect the now in force Guidelines on State aid for environmental protection and energy 2014–2020.

[152]The Study p. 417 et seq.

[153]Articles 107, 108 and 109 TFEU.

[154]The Study p. 416.

principle of legality (original *αρχή της νομιμότητας*, in France *principe de légalité*, in Germany *Prinzip der Gesetzmäßigkeit der Verwaltung*), as explained above.

6 The Practice, Especially with Regard to Soil Remediation

Following the above analysis of legal provisions regulating prevention and remediation of soil contamination, some randomly selected administrative acts relevant to that are portrayed in order to deliver an insight into the praxis.

6.1 The Trastic Enterprises Limited Case

In the Trastic Enterprises Limited case,[155] the company, which was an industry of *building materials (existing drilling)* according to the environmental conditions approved on 12.11.2008, was ordered to remediate the soil and remove *hazardous waste* according to section 9 EnvLL and section 12 HazWaDe. The existing condition, that of widespread *asbestos* material and waste, was considered to cause damage and adverse effects to human health and the environment. The legal bases cited in this act were EnvPL, EnvLL, HazWaDe, Ministerial Decree No 24944/1159/2006, Ministerial Decree No 8668/2007 (hereinafter 'ManPlHazWaDe'), Law No 4014/2011,[156] Law on Waste,[157] Law on Canalization,[158] Ministerial Decree No 8243/1113/1991[159] (hereinafter 'AsbDe'), Presidential Decree No 212/2006[160] and Ministerial Decree No 4229/395/2013.[161] Trastic had been in all probability operating after November 2008, when the environmental conditions had been approved, and had been rending facilities and equipment from ELLENIT S.A. Ellenit operated previously on the same grounds, manufacturing building materials with asbestos while having building permits (1967–1975), permit for liquid waste disposal (1996) and approved environmental conditions (1996). According to the last consolidated financial statements published in 1998, Ellenit suffered loss of 1 bln 708 mils drachmas. More recently (7.7.2016), it was

[155]Administrative act No ΒΙ09ΟΡ1Υ-ΣΨΨ (2014).

[156]This law partly implements Directive 2010/75/EU.

[157]This law implements Directive 2008/99/EC and Directive 2008/98/EC.

[158]This law has been harmonizing Directives 97/11/EC and 96/61/EC. However, Directive 97/11/EC has been repealed by Directive 2011/92/EU. Similarly, Directive 96/61/EC has been repealed by Directive 2008/1/EC, which itself has been repealed by Directive 2010/75/EU that has been implemented by Ministerial Decree No 36060/1155/E.103/2013, as explained above.

[159]This implemented Council Directive 87/217/EEC.

[160]This implemented Council Directive 83/477/EEC and partially Directive 2009/148/EC.

[161]This, along with Presidential Decree No 212/2006, has implemented Directive 2009/148/EC.

announced by the Greek Public Revenue Authority that Ellenit owes 18,860 mln euro to the state on taxes, social security and others.

Said pertinent facts and legal points raise the following issues. Citing a plethora of laws causes confusion, especially when no mention is made as to how the facts of the case correlate to specific statutory provisions. Most importantly, this identifies as a lack of explicit, specific and due reasoning, required for administrative acts under section 17(2) Rules of Administrative Procedure. This constitutes ground for annulment of the administrative act.[162] Furthermore, the cited legal provisions for soil remediation, those being sections 9 EnvLL and 12 HazWaDe, are not applicable in this case. EnvLL, which sets out obligation of soil remediation for the polluter, is not applicable for two reasons. Firstly, the environmental damage was in all probability caused by Ellenit, that being before November 2008. EnvLL is not applicable to damage caused prior to 1.5.2007.[163] Under the assumption that damage was continuing to be caused after 1.5.2007, allowing for the applicability of EnvLL, Trastic was not the polluter and hence may not be called to take measures or pay the bill under section 9(1)(b) and 9(2)(d) EnvLL. Secondly, EnvLL is not applicable as the described adverse effects to the environment were not caused by a professional activity listed in Annex III EnvLL, as required by section 4(1)(a) EnvLL. Thirdly, asbestos waste is regulated exclusively by AsbDe. Its application prevails due to the principle *lex specialis derogat legi generali.* Preventing asbestos emissions into the air, asbestos discharges into the aquatic environment and solid asbestos waste and reducing them at source was the sole duty of Ellenit, according to AsbDe. The competent authority had the duty to ensure that Ellenit was adhering to AsbDe but failed to do so. Lastly, in examining the application of HazWaDe, the competent authority has failed to give reasons for the substances in question to have been discarded and hence became waste, as well as to having properties that render them hazardous under the Law on Waste. Therefore, the reasoning for the application of HazWaDe raises concerns. Equally, the competent authority has failed to give reasons on Trastic Enterprises Limited being waste holder, which includes both physical control and willingness to control, as explained above.

At this point, the question arises whether any connection between Ellenit and Trastic would have any effect to the asbestos liability. It is quite possible that Trastic Enterprises Limited, the addressee of the administrative act, is the same person as Trastic Enterprises Limited, a company incorporated in Cyprus on 7.12.1998 (that being the last year that Ellenit filed consolidated financial statements) with company No HE 98923 and having its registered office at Kennedy 12, 1703 Nicosia, Cyprus. The Cyprus company has been stricken off the Registrar of Companies and Official Receiver of Cyprus on 11.1.2016. An indication of these

[162]See in this respect judgements of the Supreme Administrative Court Nos 40/1964, 2441/1983, 2027/1986, 2662/1999; Tachos (2008), p. 674; Spiliotopoulos (2015), p. 150 et seq.; Lazaratos (2014), p. 629.

[163]Section 19(a) EnvLL.

companies being closely related is the existence of *Employees' and Workers' Union of Ellenit-Trastic Plant in Thessaloniki*, as mentioned by MP Vasileios Geranides on 16.6.2009 and 8.4.2009, MP Evaggelia Amanatidou-Paschalidou on 3.4.2009, MP Sofia Kananidou and MP Ioannis Ziogas on 2.4.2009, before the respective planetary sessions of the Parliament.[164] Under the conditions that, firstly, Trastic had been incorporated for the purpose of avoiding the asbestos-related duties and, secondly, Trastic was continuing the business activities of Ellenit, according to relevant contract, this contract could be annulled due to abuse of right according to section 281 Greek Civil Code, or even the corporate veil could be pierced. Similar cases occurred in Germany prior to the enactment of *Bundes-Bodenschutzgesetz*, when contaminated soil was sold to foreign companies having no assets for the purpose of avoiding liability for soil remediation. The contract was annulled on the ground of being contrary to good faith according to section 138 German Civil Code.[165]

6.2 The F.A. Hellas S.A. Case

In the F.A. Hellas S.A. case,[166] the company was ordered to remediate contaminated soil situated within the grounds of the company's warehouse, but the precise legal provisions for this liability where not mentioned. The pollutant was estimated to be TPH. The legal bases cited in this act were EnvPL, SolidWaDe, HazWaDe, ManPlHazWaDe and the *Netherlands Ministry of spatial planning, Housing and the environment, 2000, Circular on Target Values and Intervention Values for Soil Remediation, 4th February 2000*. Further facts cannot be deduced from the face of said act.

The competent authority has failed to give reasons for TPH in question to have been discarded and hence became waste, as well as to having properties that render it hazardous under the then in force HazWaDe. F.A. Hellas could have been called as hazardous waste holder to remediate the pollution caused under section 12 (1) HazWaDe. Sadly, neither the addressee's capacity nor the specific statutory provision was mentioned in the said act, raising concerns as to due reasoning, as explained above. Equally concerning is citing the Dutch Intervention Values as only laws enacted by the Greek Parliament or Decrees issued by the President of the Republic or the competent Minister, following relevant statutory empowerment form binding national legislation. The authorities do not have the power to apply foreign legislation. In doing so, they act *ultra legem*.

[164] Minutes of planetary sessions of the Hellenic Parliament, available at: http://www.hellenicparliament.gr (last visited on 14.4.2017).

[165] Administrative Court of Second Instance in Mannheim (2008).

[166] Administrative act No 45ΟΨΟΡ1Κ-Η00 (2011).

6.3 The Abandoned Plant Case

In the abandoned plant case,[167] the competent authority ordered Apostolos Staurou, the land owner, to remediate a contaminated site according to section 9 EnvLL and the polluter-pays principle. The pollutant was oil, which had escaped from containers placed in a building and reached the soil. The legal bases cited in this act were EnvPL, EnvLL, Law on Waste, HazWaDe and ManPlHazWaDe. The area now in ownership of Mr. Stavrou was formerly owned by a state agency (*Κεντρική Υπηρεσία Διαχείρισης Εγχώριας Παραγωγής*), which had ceased operations in 1993, when it was placed under special compulsory winding up according to judgment of the Civil Court of Second Instance in Athens No 2695/1993.[168] Lastly, in the said act, the contaminated site is characterised as abandoned plant being accessible to anyone while being a danger to the environment and civil safety.

Apart from the repeated legal issue of citing a plethora of legal instruments and failing to specify the facts of the case that correspond to the applied legal provisions, the legal issue raised here is the capacity of Mr. Stavrou, a resident of Patras, being both the owner of the contaminated site in Chalkida and the polluter. No reasoning as to the manner in which he caused the pollution is provided in the same act. The fact that the area was characterised as an abandoned area contributes to the assumption that no operations were taking place and hence no operator was available. Moreover, his capacity as owner of the land does not suffice for ordering him to perform any remediation measures neither under the polluter-pays principle nor under the holder-pays principle according to section 12 HazWaDe, as no reasoning on controlling the site in question or him knowing about the pollution is offered.

6.4 The Bankrupted Plant Case

In the bankrupted plant case,[169] the competent authority ordered Demetrios Chantzis, Achilleas Chantzis and Christina Chantzis, co-owners of the contaminated site, to remediate the contaminated soil, but the precise legal provisions for this liability were not mentioned. The pollutants were organichlorine pesticide, such as DDD/DDE/DDT and heptaclor, TPH and asbestos. The legal bases cited in this act were EnvPL, Law on Waste, EnvLL, HazWaDe, ManPlHazWaDe, SolidWaDe, Law No 2939/2001,[170] Ministerial Decree No 52167/4683,[171] AsbDe, Presidential Decree No 212/2006 and Ministerial Decree No 4229/395/

[167] Administrative act No ΒΙΦΙΟΡ10-Π95 (2014).

[168] Section 27(3) Law No 3147/2003.

[169] Administrative act No 7ΦΞΦΟΡ1Υ-ΔΓΧ (2014).

[170] This law implements European Parliament and Council Directive 94/62/EC.

[171] This implements Directive 2008/68/EC.

2013. At the contaminated site, an undertaking was previously operating that was producing pesticides but had gone bankrupt.

In the present case, once more, a plethora of legal instruments was cited and the exact facts of the case that correspond to precise legal provisions were not mentioned. The exact legal provisions setting out the addressees' liability for soil contamination were equally not mentioned. Hence, it is questionable whether the statutory requirements on due reasoning are fulfilled. In addition, the addressees' capacity as co-owners of the land does not suffice for ordering them to perform any remediation measures under any of the cited legal instruments for the reasons explained above under the abandoned plant case.

6.5 The Dumping Waste Case

In the dumping waste case,[172] the competent authority ordered remediation of a site contaminated with dumped hazardous waste. The pollutants were a dumped tank with H2SO4 and dumped empty plastic containers. The legal bases cited in this act were EnvPL, EnvLL, HazWaDe, ManPlHazWaDe and Council Decision 2003/33/EC. The pollutants were dumped in the forest, and as no information on the polluter was available, the state proceeded with the remediation according to section 9 EnvLL.

The fact that no information on the polluter was available to the competent authorities led to the state undertaking the duty to remove the waste and bearing the relevant financial cost. Basing this duty on section 9 EnvLL, which sets out the remedial liability of the operator, makes little sense. Lastly, this act sets out that it shall be revoked or altered in the event that the polluter is later discovered. Under the assumption that the administrative act is promptly executed due to imminent threat to the environment, its revocation or alteration on said ground has no practical effect as the ordered measures would have been already performed. Hence, it would have been much wiser if this act had provided instead for the recovery of the remediation cost. Such a provision would also provide a fine practical implementation of the polluter-pays principle.

6.6 The Soil Pollution Case on Military Grounds

In the soil pollution case on military grounds,[173] the competent authority ordered soil remediation in and around a military base, which was contaminated with oil. The legal basis cited in this act was HazWaDe. The environmental study on

[172] Administrative act No B4Γ1OP10-Θ9Θ (2012).

[173] Administrative act No B496Ω96-990 (2012).

contaminated soil was already approved by the competent military authorities, and the local authorities were instructed with bringing into effect the remedial measures described in the said study. The cost was coved by European funding.

This is a classic case of escaping pollutants as it is safe to assume that the action or omission that caused the pollution took place in the military base and then it expanded to the neighbouring area. Such assumption is necessary as no further information on the exact pollutant, the reasons for it being considered to be hazardous or the manner of soil pollution is available from said administrative act and relevant documents publicly available. In addition, although it would be a great opportunity to examine the effect of pollution from escaped substances to the polluter-pays principle or even the holder-pays principle, these principles were neither applied, nor solid grounds for the deviation therefrom were offered. Lastly, it should be noted that HazWaDe does also apply to military facilities, which are typically excluded from the area of application of environmental legislation, as, for example, set out in section 5(1)(d) EnvLL.

6.7 *Identification of Potentially Contaminated Sites Case*

The Study has led the competent authority to proceed[174] with identifying, recording and providing an initial evaluation of potentially contaminated sites from 'industrial-hazardous waste' (in the original *βιομηχανικά – επικίνδυνα απόβλητα*) within specific areas of Greece having heavy industrial activity, as such activity results to 'industrial and hazardous waste' that cause pollution, when not properly managed.[175] This work was materialised through public procurement.[176]

In the tender offer, it is stated that basic principle of dealing with danger from all contaminated sites is to set out an integrated strategy prior to conducting investigation and taking technical measures. Such stage approach is considered to be 'international practice'.[177] The relevant methodology is already available to the competent authority,[178] in all probability, meaning the Study. The work is called to focus on sites having heavy industrial activity and especially sites that relate to industrial parks, sites with heavy industrial activity, ports, military sites, sites where temporary storage of hazardous waste takes place, mining of metal ores and others.[179] It is further stated that identifying and evaluating hazardousness and programming remediation of contaminated sites due to improper waste deposit or

[174] Administrative act No ΒΙΥ00-ΗΑΛ (2014), p. 3; Administrative act No 6ΝΚΙ465ΦΘΘ-3ΕΦ (2015), p. 3.

[175] Administrative act No 45Ο90-ΤΩΨ (2011), p. 40.

[176] Administrative act No ΒΛΛ40-ΑΨ2 (2013), p. 3.

[177] Administrative act No 45Ο90-ΤΩΨ (2011), p. 39.

[178] *Ibid* p. 39 et seq.

[179] *Ibid* p. 40.

contaminated sites are a target of national environmental policy set out in SolidWaDe and ManPlHazWaDe.[180] This work, an investment of 495,999.96 euro, that being gross fixed capital formation according to section 2 Protocol No 12 to TFEU and Annex I 3.102 Regulation (EC) No 2223/96 for the purpose of state budget, was materialised through public procurement and was completed in February 2017.[181] However, it is not available to the public.

The aforementioned reservations on the Study (see in this respect above under Sect. 5.2) equally apply here. In addition to that, the following deserve mention. The tender refers to SolidWaDe, which does not apply to hazardous waste, and fails to refer to HazWaDe, which sets out the fundamental provision on soil remediation, that being section 12 HazWaDe. Even if it is assumed that reference to HazWaDe in the Study suffices, its provisions on area of application are not taken into consideration as sites related to mining of metal ores are included in the work, despite of waste resulting from activities relating to mineral resources and quarries being explicitly excluded from the area of application (section 3(2)(c2) HazWaDe). The aforementioned stage approach is set out in ManPlHazWaDe in a slightly different manner. In particular, it is stated that contaminated and potentially contaminated sites are to be identified on a national level. Further investigation, remediation and restoration of sites are to take place according to the polluter-pays principle and section 12 HazWaDe (Annex 5.6 ManPlHazWaDe). Furthermore, the statement in the tender that an integrated strategy prior to conducting investigation and taking technical measures is necessary contradicts the practice of the administration of ordering soil remediation, as the above-examined cases prove. Lastly, the unavailability of the work does not seem to fall under any of the statutory exceptions to making environmental information public, such as protection of intellectual property, personal information, confidentiality or trade secrets (section 4(2) Ministerial Decree No 11764/653/2006).[182]

7 Conclusions

The legal provisions on preventive measures providing effective soil protection and remediation of soil pollution are in place. However, four points deserve attention. The first one is the manner that laws are applied by the competent authorities. As discussed above, the extensive citation of laws causes confusion and raises concerns as to due reasoning. In addition to that, laws are not properly applied, in the sense that owners are called to pay for soil remediation on the polluter-pays principle, no grounds on land holders being waste holders and hence triggering the application of HazWaDe are offered, no grounds on the identified pollutants

[180] *Ibid* p. 39.

[181] Administrative act No ΩP2X465X18-33K (2017).

[182] This implements Directive 2003/4/EC.

being hazardous and hence triggering the application of HazWaDe are offered. The second one relates to the equal ranking of the liability on the 'polluter-pays-principle' and the 'holder-pays-principle' in section 12 HazWaDe. In view of the environmental policy of the EU, as set out in Art. 191 TFEU, the general principles of the HazWaDe and the lack of any public law instrument enabling the holder to claim restitution from the polluter, equal ranking between the two seems to promote inequality. Hence, in practice, the competent authorities should first apply the polluter-pays principle and only when this is not possible will they proceed with the holder-pays principle while giving due reasoning. The third issue relates to the competent authorities failing to timely ensure legal compliance of waste holders and managers and hence fostering pollution prevention, as seen in the Trastic Enterprises Limited case, where asbestos waste has been laying around on the grounds of the undertaking for at least a decade. The fourth one is best illustrated through the above-explained cases and relates to matters not yet regulated in law. These include bankruptcy of the polluter, the effect of escaped substances on the polluter-pays principle, using legal instruments in order to avoid liability for soil remediation contrary to good faith, using the corporate veil in order to elude liability for soil remediation and historic pollution. The latter is problematic as EnvLL applies to soil pollution since 2007 and HazWaDe since 2006. Prior to that, in the event of soil pollution, the polluter and the state were obliged to remediation under statutory provisions of 1997,[183] whereas the oldest statutory prohibition of discarding and abandoning toxic and hazardous waste was issued in 1985.[184] Hence, the polluter who caused soil pollution in 1997 onwards could be ordered to remediate it, as he violated a statutory duty, as long as this is duly reasoned. That could also be purported for soil pollution caused in 1985 onwards. However, soil pollution caused by discarding and abandoning toxic and hazardous waste prior to that date may not be remediated by the polluter as at that time no statutory breach took place and therefore the act would breach the principle of legality. In addition to that, any law issued at present and ordering remediation on the basis of the polluter-pays principle or the holder-pays principle for soil pollution that occurred prior to 1985 would have retroactive effect. On this ground, such law would be contrary to the Constitution and hence void. The issues raised here deserve close scientific attention and statutory regulation that lawfully brings precaution and soil remediation into effect.

[183]Ministerial Decree No 19396/1546/1997.

[184]Ministerial Decree No 72751/3054/1985.

References

2000/532/EC: Commission Decision of 3 May 2000 replacing Decision 94/3/EC establishing a list of wastes pursuant to Article 1(a) of Council Directive 75/442/EEC on waste and Council Decision 94/904/EC establishing a list of hazardous waste pursuant to Article 1(4) of Council Directive 91/689/EEC on hazardous waste (notified under document number C(2000) 1147), available at: http://eur-lex.europa.eu (last visited on 12.4.2017), as amended

2003/33/EC: Council Decision of 19 December 2002 establishing criteria and procedures for the acceptance of waste at landfills pursuant to Article 16 of and Annex II to Directive 1999/31/EC, OJ L 11, 16.1.2003, p. 27

Administrative act No 45O90-ΤΩΨ (2011) 'Approval of the tender for international open adjudication for the work of 'Recording and Providing initial evaluation of potentially polluted sites from 'industrial-hazardous waste'' ('*Έγκριση α) του τεύχους της επαναπροκήρυξης δημοσίου ανοιχτού διεθνούς διαγωνισμού για την επιλογή αναδόχου του έργου «Καταγραφή και πρώτη αξιολόγηση επικινδυνότητας ρυπασμένων χώρων από βιομηχανικά-επικίνδυνα απόβλητα», β) της διάθεσης σχετικής πίστωσης*'), of 14.1.2011, available at: https://diavgeia.gov.gr (last visited 14.4.2017)

Administrative act No 45ΟΨΟΡ1Κ-Η00 (2011) 'Approval of on-site remediation of polluted area of the company F.A. HELLAS S.A. in Aspropyrgos, Attica' ('*Έγκριση εργασιών επιτόπιας εξυγίανσης/ αποκατάστασης του ρυπασμένου χώρου της εταιρείας Φ.Α. ΕΛΛΑΣ Α.Ε. που βρίσκεται στον Ασπρόπυργο Αττικής*') of 13.10.2011, available at: https://diavgeia.gov.gr (last visited 14.4.2017)

Administrative act No 7ΦΞΦΟΡ1Υ-ΔΓΧ (2014) 'Approval of in-situ remediation of area polluted with hazardous waste at the former plant of DIANA S.A. in Thessaloniki' ('*Έγκριση εργασιών επιτόπιας εξυγίανσης – αποκατάστασης χώρου ρυπασμένου από επικίνδυνα απόβλητα στο πρώην εργοστάσιο της ΔΙΑΝΑ Α.Β.Ε.Ε. στην Ευκαρπία του Δήμου Παύλου Μελά, Π.Ε. Θεσσαλονίκης, Π.Κ.Μ.*') of 23.10.2014, available at: https://diavgeia.gov.gr (last visited 14.4.2017)

Administrative act No Β4Γ1ΟΡ10 Θ9Θ (2012) 'Approval of remediation of area polluted from dumping of hazardous waste in Stavros' ('*Έγκριση εργασιών εξυγίανσης/ αποκατάστασης χώρου ρυπασμένου από ανεξέλεγκτη ρίψη πλαστικών περιεκτών με επικίνδυνα απόβλητα σε περιοχή της Δ.Κ. Σταυρού του Δήμου Διρφύων-Μεσσαπίων Ν. Εύβοιας*') of 9.8.2012, available at: https://diavgeia.gov.gr (last visited 14.4.2017)

Administrative act No Β496Ω96-990 (2012), 'Decision to materialize the work «Remediation of polluted soil of 9th Advanced Warning Squadron and restoration of the area»' ('*Λήψη απόφασης για την υλοποίηση του έργου «Εξυγίανση ρυπασμένων εδάφων 9ης ΜΣΕΠ και αποκατάστασης περιοχής»*'), of 30.4.2012, available at: https://diavgeia.gov.gr (last visited 14.4.2017)

Administrative act No ΒΙ09ΟΡ1Υ-ΣΨΨ (2014) 'Remediation of soil polluted with hazardous waste at the facilities of TRASTIC ENTERPRISES LTD in Thessaloniki' ('*Εξυγίανση – αποκατάσταση χώρου ρυπασμένου από επικίνδυνα απόβλητα στις εγκαταστάσεις της «TRASTIC ENTERPRISES LTD» στη Γέφυρα Θεσσαλονίκης*') of 28.4.2014, available at: https://diavgeia.gov.gr (last visited 14.4.2017)

Administrative act No ΒΙΦΙΟΡ10-Π95 'Approval of remediation of area polluted with oil at an abandoned plant of former KIPED (currently owned by Apostolos Staurou) in Chalkida' ('*'Έγκριση εργασιών εξυγίανσης/ αποκατάστασης ρυπασμένων χώρων με πετρελαιοειδή σε όροφο, υπόγειο και ακάλυπτο χώρο εγκαταλελειμμένου εργοστασίου της πρώην ΚΥΠΕΔ (νυν ιδιοκτησίας Αποστόλου Σταύρου), στη Χαλκίδα Π.Ε. Εύβοιας*') of 5.5.2014., available at: https://diavgeia.gov.gr (last visited 14.4.2017)

Administrative Court of Second Instance in Mannheim (2008), judgement of 20.1.1998 – 10 S 233/97, available at: https://openjur.de (last visited: 14.4.2017)

Chrysogonos K (2014) Constitutional law (in Greek). Sakkoulas Publications, Athens – Thessaloniki

Communication from the Commission — Guidelines on State aid for environmental protection and energy 2014-2020, OJ C 200, 28.6.2014, p. 1
Council Directive 75/442/EEC of 15 July 1975 on waste, OJ L 194, 25.7.1975, p. 47
Council Directive 76/116/EEC of 18 December 1975 on the approximation of the laws of the Member States relating to fertilizers, OJ L 24, 30.1.1976, p. 21
Council Directive 80/876/EEC of 15 July 1980 on the approximation of the laws of the Member States relating to straight ammonium nitrate fertilizers of high nitrogen content, OJ L 250, 23.9.1980, p. 7
Council Directive 82/501/EEC of 24 June 1982 on the major-accident hazards of certain industrial activities, OJ L 230, 5.8.1982, p. 1
Council Directive 83/477/EEC of 19 September 1983 on the protection of workers from the risks related to exposure to asbestos at work (second individual Directive within the meaning of Article 8 of Directive 80/1107/EEC), OJ L 263, 24.9.1983, p. 25
Council Directive 85/337/EEC of 27 June 1985 on the assessment of the effects of certain public and private projects on the environment OJ L 175, 5.7.1985, p. 40
Council Directive 86/278/EEC of 12 June 1986 on the protection of the environment, and in particular of the soil, when sewage sludge is used in agriculture, OJ L 181, 4.7.1986, p.6
Council Directive 87/217/EEC of 19 March 1987 on the prevention and reduction of environmental pollution by asbestos, OJ L 85, 28.3.1987, p. 40
Council Directive 89/530/EEC of 18 September 1989 supplementing and amending Directive 76/116/EEC in respect of the trace elements boron, cobalt, copper, iron, manganese, molybdenum and zinc contained in fertilizers, OJ L 281, 30.9.1989, p. 116
Council Directive 91/676/EEC of 12 December 1991 concerning the protection of waters against pollution caused by nitrates from agricultural sources, OJ L 375, 31.12.1991, p.1
Council Directive 96/61/EC of 24 September 1996 concerning integrated pollution prevention and control, OJ L 257, 10.10.1996, p. 26
Council Directive 96/82/EC of 9 December 1996 on the control of major-accident hazards involving dangerous substances, OJ L 10, 14.1.1997, p. 13
Council Regulation (EC) No 2223/96 of 25 June 1996 on the European system of national and regional accounts in the Community, OJ L 310, 30.11.1996, p. 1
Directive 2003/4/EC of the European Parliament and of the Council of 28 January 2003 on public access to environmental information and repealing Council Directive 90/313/EEC, OJ L 41, 14.2.2003, p. 26
Directive 2004/35/EC of the European Parliament and of the Council of 21 April 2004 on environmental liability with regard to the prevention and remedying of environmental damage, OJ L 143, 30.4.2004, p. 56
Directive 2006/12/EC of the European Parliament and of the Council of 5 April 2006 on waste, OJ L 114, 27.4.2006, p. 9
Directive 2008/1/EC of the European Parliament and of the Council of 15 January 2008 concerning integrated pollution prevention and control (Codified version), OJ L 24, 29.1.2008, p. 8
Directive 2008/68/EC of the European Parliament and of the Council of 24 September 2008 on the inland transport of dangerous goods, OJ L 260, 30.9.2008, p. 13
Directive 2008/98/EC of the European Parliament and of the Council of 19 November 2008 on waste and repealing certain Directives, OJ L 312, 22.11.2008, p. 3
Directive 2008/99/EC of the European Parliament and of the Council of 19 November 2008 on the protection of the environment through criminal law, OJ L 328, 6.12.2008, p. 28
Directive 2009/148/EC of the European Parliament and of the Council of 30 November 2009 on the protection of workers from the risks related to exposure to asbestos at work, OJ L 330, 16.12.2009, p. 28
Directive 2010/75/EU of the European Parliament and of the Council of 24 November 2010 on industrial emissions (integrated pollution prevention and control), OJ L 334, 17.10.2010, p. 17

Directive 2011/92/EU of the European Parliament and of the Council of 13 December 2011 on the assessment of the effects of certain public and private projects on the environment, OJ L 26, 28.1.2012, p. 1

European Parliament and Council Directive 94/62/EC of 20 December 1994 on packaging and packaging waste, OJ L 365, 31.12.1994, p. 10

Georgiades AS (2010) Law of Rights in Rem (in Greek). Sakkoulas Publications, Athens-Thessaloniki

Koutoupa–Regkakou E (2008) Environmental law (in Greek). Sakkoulas Publications, Athens-Thessaloniki

Law No 1515/1985 on Regulatory Plan and Programme for the Environmental Protection of the Greater Area of Athens ('*Ρυθμιστικό Σχέδιο και Πρόγραμμα Προστασίας Περιβάλλοντος της Ευρύτερης Περιοχής της Αθήνας*') FEK A 18, 18.2.1985, p. 361, as amended

Law No 1561/1985 on Regulatory Plan and Programme for the Environmental Protection of the Greater Area of Thessaloniki ('*Ρυθμιστικό Σχέδιο και Πρόγραμμα Προστασίας Περιβάλλοντος της Ευρύτερης Περιοχής της Θεσσαλονίκης*') FEK A 148, 8.9.1985, p. 2325

Law No 1565/1985 on Fertilizers ('*Νόμος υπ' αριθ. 1565 Λιπάσματα*'), FEK A 164, 26.9.1985, p. 2524, as amended ('*FeL*')

Law No 1650/1986 on Environmental Protection ('*Για την προστασία του περιβάλλοντος*'), FEK A 160 of 16.10.1986 p. 3257, as amended ('*EnvPL*')

Law No 2468/1997 on Ratification of the United Nations Convention to Combat Desertification in Countries Experiencing Serious Drought and/ or Desertification, particularly in Africa ('*Νόμος 2468 Κύρωση της Σύμβασης των Ηνωμένων Εθνών για την Καταπολέμηση της Ερημοποίησης στις Χώρες εκείνες που Αντιμετωπίζουν Σοβαρή Ξηρασία ή/ και Απερήμωση, ιδιαίτερα στην Αφρική*'), FEK A 32 of 6.3.1997, p. 463

Law No 2508/1997 on Sustainable Urban and Rural Housing Development and other Provisions ('*Βιώσιμη Οικιστική Ανάπτυξη των Πόλεων και Οικισμών της Χώρας και άλλες Διατάξεις*') FEK A 124, 13.6.1997, p. 4915

Law No 2742/1999 on Land Use Planning, Sustainable Development and other Provisions ('*Χωροταξικός Σχεδιασμός και Αειφόρος Ανάπτυξη και άλλες Διατάξεις*') FEK A 207, 7.10.1999, p. 4159

Law No 2939/2001 on Packages and Management of Packages and other products ('*Συσκευασίες και εναλλακτική διαχείριση των συσκευασιών και άλλων προϊόντων*'), FEK A 179, 6.8.2001, p. 2767

Law No 3010/2011 on Harmonization of Law No 1650/86 with Directives 97/11/EC and 96/61/EC, Marking Procedures and Regulation of Canalization and other provisions ('*Εναρμόνιση του ν. 1650/86 με τις οδηγίες 97/11/ΕΕ και 96/61/ΕΕ, διαδικασία οριοθέτησης και ρυθμίσεις θεμάτων για τα υδατορέματα και άλλες διατάξεις*'), FEK A 91, 25.4.2002, p. 1427 ('*Law on Canalization*')

Law No 4014/2011 on Environmental Authorizations of Plants and Activities ('*Περιβαλλοντική αδειοδότηση έργων και δραστηριοτήτων, ρύθμιση αυθαιρέτων σε συνάρτηση με δημιουργία περιβαλλοντικού ισοζυγίου και άλλες διατάξεις αρμοδιότητας Υπουργείου περιβάλλοντος*'), FEK A 209, 21.9.2011, p. 6215

Law No 4042/2012 on Protection of the Environment through Criminal Law – Harmonization with Directive 2008/98/EC – Frame of Production and Management of Waste ('*Ποινική προστασία του περιβάλλοντος – Εναρμόνιση με την οδηγία 2008/99/ΕΚ – Πλαίσιο παραγωγής και διαχείρισης αποβλήτων – Ρύθμιση θεμάτων Υπουργείου Περιβάλλοντος Ενέργειας και Κλιματικής Αλλαγής*'), FEK A 24, 13.2.2012, p. 231 ('*Law on Waste'*)

Lazaratos P (2014) Administrative procedural law (in Greek). Ant. N. Sakkoulas Publications, Athens

Manesis A (1967) Constitutional law, University Lectures (in Greek), Vol I. Sakkoulas Publications, Athens – Thessaloniki

Manitakis A (1994) Rule of law and judicial review of legislation (in Greek). Sakkoulas Publications, Athens – Thessaloniki

Maurias K (2014) Constitutional law (in Greek). P. N. Sakkoulas Publications, Athens

Ministerial Decree No 11014/703/Φ104 regarding the procedure for preliminary environmental assessment and evaluation and approval of environmental conditions (‘*Διαδικασία Προκαταρκτικής Περιβαλλοντικής Εκτίμησης και Αξιολόγησης (ΠΠΕΑ) και Έγκρισης Περιβαλλοντικών Όρων (ΕΠΟ) σύμφωνα με το άρθρο 4 του Ν. 1650/1986 (Α' 160) όπως αντικαταστάθηκε με το άρθρο 2 του Ν. 3010/2002 ‘Εναρμόνιση του Ν. 1650/1986 με τις Οδηγίες 97/11/ΕΕ και 96/61/ΕΕ ... και άλλες διατάξεις' (Α' 91)*’), FEK B 332, 20.3.2003, p. 4860

Ministerial Decree No 11764/653/2006 regarding access of the public to public authorities for environmental information in accordance with provisions of Directive 2003/4/EC of the European Parliament and of the Council of 28 January 2003 on public access to environmental information and repealing Council Directive 90/313/EEC (‘*Πρόσβαση του κοινού στις δημόσιες αρχές για παροχή πληροφοριών σχετικά με το περιβάλλον, σε συμμόρφωση με τις διατάξεις της οδηγίας 2003/4/ΕΚ «για την πρόσβαση του κοινού σε περιβαλλοντικές πληροφορίες και για την κατάργηση της οδηγίας 90/313/ΕΟΚ» του Συμβουλίου. Αντικατάσταση της υπ' αριθμ. 77921/1440/1995 κοινής υπουργικής απόφασης (Β' 795)*’), FEK B 327, 17.3.2006, p. 3981

Ministerial Decree No 13588/725/2006 Measures, Conditions and Restrictions for Managing Hazardous Waste in accordance with provisions of Council Directive 91/689/EEC of 12 December 1991 on hazardous waste. Replacement of Joint Ministerial Decision 19396// 1546/1997 Measures and Conditions for Managing Hazardous Waste (‘*Μέτρα όροι και περιορισμοί για την διαχείριση επικινδύνων αποβλήτων σε συμμόρφωση με τις διατάξεις της οδηγίας 91/689/ΕΟΚ «για τα επικίνδυνα απόβλητα» του Συμβουλίου της 12ης Δεκεμβρίου 1991. Αντικατάσταση της υπ' αριθμ. 19396/1546/1997 κοινή υπουργική απόφαση «Μέτρα και όροι για τη διαχείριση επικίνδυνων αποβλήτων»*’), FEK B 383, 28.3.2006, p. 4693 (‘*HazWaDe*’)

Ministerial Decree No 1420/82031/2015 Code of Good Agricultural Practice for the Protection of Waters against Pollution Caused by Nitrates from Agricultural Sources (‘*Κώδικας Ορθής Γεωργικής Πρακτικής για την Προστασία των Νερών από τη Νιτρορύπανση Γεωργικής Προέλευσης*’), FEK B 1709, 17.8.2015, p. 20495, as amended (‘*LiMaDe*’)

Ministerial Decree No 15393/2332 regarding Clasification of Public and Private Works (‘*Κατάταξη δημοσίων και ιδιωτικών έργων και δραστηριοτήτων σε κατηγορίες σύμφωνα με το άρθρο 3 του Ν. 1650/1986 όπως αντικαταστάθηκε με το άρθρο 1 Ν. 3010/2002 «Εναρμόνιση του Ν. 1650/86 με τις οδηγίες 97/11/ΕΕ και 96/61/ΕΕ κ.ά. (Α 91)»*’), FEK B 1022, 5.8.2002, p. 13765

Ministerial Decree No 16190/1335/1997 on Measures and Conditions for the Protection of Waters Against Pollution Caused by Nitrates from Agricultural Sources (‘*Μέτρα και Όροι για την Προστασία των Νερών από την Νιτρορρύπανση Γεωργικής Προέλευσης*’), FEK B 519, 25.6.1997, p. 5913

Ministerial Decree No 19396/1546/1997 on Measures and Conditions for Management of Hazardous Waste (‘*Μέτρα και Όροι για τη Διαχείριση Επικινδύνων Αποβλήτων*’) FEK B 604, p. 6821

Ministerial Decree No 24944/1159/2006 regarding Approval of Technical Standards for Management of Hazardous Waste (‘*Έγκριση Γενικών Τεχνικών Προδιαγραφών για τη Διαχείριση Επικίνδυνων Αποβλήτων σύμφωνα με το άρθρο 5 παρ. Β ΚΥΑ 13588/725/2006*’), FEK B 791, 30.6.2006, p. 11133

Ministerial Decree No 36060/1155/E.103 regarding Establishment of Rules, Measures and Procedures for Integrated Pollution Prevention and Control of Industrial Emissions (‘*Καθορισμός πλαισίου κανόνων, μέτρων και διαδικασιών για την ολοκληρωμένη πρόληψη και έλεγχο της ρύπανσης του περιβάλλοντος από βιομηχανικές δραστηριότητες, σε συμμόρφωση προς τις διατάξεις της οδηγίας 2010/75/ΕΕ «περί βιομηχανικών εκπομπών (ολοκληρωμένη πρόληψη και έλεγχος της ρύπανσης)» του Ευρωπαϊκού Κοινοβουλίου και του Συμβουλίου της 24ης Νοεμβρίου 2010*’), FEK B 1450, 14.6.2013, p. 21993

Ministerial Decree No 37111/2021 regarding Determination of the Manner of Public Information and Participation during the Approval of Environmental Conditions for Works and Activities (*'Καθορισμός τρόπου ενημέρωσης και συμμετοχής του κοινού κατά τη διαδικασία έγκρισης περιβαλλοντικών όρων των έργων και δραστηριοτήτων σύμφωνα με την παράγραφο 2 του άρθρου 5 του Ν. 1650/1986 όπως αντικαταστάθηκε με τις παραγράφους 2 και 3 του άρθρου 3 του Ν. 3010/2002'*), FEK B 1391, 29.9.2003, p. 19377

Ministerial Decree No 4229/395/2013 regarding Establishment and Operation of Undertakings performing Demolition Activities and Removing Asbestos from buildings, constructions, appliances, installations and ships, as well as Maintenance, Coverage and Confinement of Asbestos (*'Προϋποθέσεις ίδρυσης και λειτουργίας των επιχειρήσεων που δραστηριοποιούνται με την εκτέλεση κατεδαφιστικών έργων και εργασιών αφαίρεσης αμιάντου ή/και υλικών που περιέχουν αμίαντο'*), FEK B 318, 15.2.2013, p. 6561

Ministerial Decree No 50910/2727/2003 on Measures and Conditions for Managing Solid Waste - National and Regional Planning of Management (*'Μέτρα και Όροι για τη Διαχείριση Στερεών Αποβλήτων. Εθνικός και Περιφερειακός Σχεδιασμός Διαχείρισης'*), FEK B 1909, 22.12.2003, p. 26073

Ministerial Decree No 52167/4683 regarding implementation of Commission Directive 2010/61/EU of 2 September 2010 adapting for the first time the Annexes to Directive 2008/68/EC of the European Parliament and of the Council on the inland transport of dangerous goods (*'Προσαρμογή της Ελληνικής νομοθεσίας προς τις διατάξεις της Οδηγίας 61/2010/ΕΕ της Επιτροπής της 2ας Σεπτεμβρίου 2010 για την πρώτη προσαρμογή στην επιστημονική και τεχνική πρόοδο των παραρτημάτων της οδηγίας 2008/68/ΕΚ του Ευρωπαϊκού Κοινοβουλίου και του Συμβουλίου, σχετικά με τις εσωτερικές μεταφορές επικίνδυνων εμπορευμάτων'*), FEK B 1385, 20.1.2012, p. 359

Ministerial Decree No 69269/5387/1990 on Classification of Works and Activities, Content of Environmental Impact Assessment, Content of Special Environmental Studies and other provisions (*'Κατάταξη έργων και δραστηριοτήτων σε κατηγορίες, περιεχόμενο Μελέτης Περιβαλλοντικών Επιπτώσεων (ΜΠΕ), καθορισμός περιεχομένου ειδικών περιβαλλοντικών μελετών (ΕΠΜ) και λοιπές συναφείς διατάξεις, σύμφωνα με το Ν. 1650/1986'*), FEK B 678, 25.10.1990, p. 8141

Ministerial Decree No 72751/3054/1985 on Toxic and Hazardous Waste and Elimination of PCBs (*'Τοξικά και Επικίνδυνα Απόβλητα και Εξάλειψη Πολυχλωροδιφαινυλίων και Πολυχλωροτριφαινυλίων'*), FEK B 665, 1.11.1985, p. 6517

Ministerial Decree No 80568/4225/1991 regarding Methods, Conditions and Restrictions for Using Sewage Sludge from Processed Household and Urban Waste Water in Agriculture (*'Μέθοδοι, όροι και περιορισμοί για την χρησιμοποίηση στη γεωργία της ιλύος που προέρχεται από επεξεργασία οικιακών και αστικών λυμάτων'*), FEK B 641, 7.8.1991, p. 5564, as amended (*'SeSlDe'*)

Ministerial Decree No 8243/1113/1991 regarding Measures and Methods for Prevention and Reduction of Environmental Pollution by Asbestos (*'Καθορισμός μέτρων και μεθόδων για την πρόληψη και μείωση της ρύπανσης του περιβάλλοντος από εκπομπές αμιάντου'*), FEK B 138, 8.3.1991, p. 1109

Ministerial Decree No 8668/2007 regarding Approval of National Plan for Hazardous Waste Management (*'Έγκριση Εθνικού Σχεδιασμού Διαχείρισης Επικινδύνων Αποβλήτων'*), FEK B 287, 2.3.2007, p. 7089 (*'ManPlHazWaDe'*)

Ministerial Decree No 99605/3719/2001 on Acceptance of Greek National Action Plan Against Desertification (*'Αποδοχή του Ελληνικού Εθνικού Σχέδιου Δράσης κατά της Ερημοποίησης'*) FEK B 974, 27.7.2001, p. 13663 (*'DesertDe'*)

Mpalis G (1961) Law of Rights in Rem (in Greek). P. Sakkoulas Publications, Athens

Opinion of the Economic and Social Committee on the 'Proposal for a Regulation of the European Parliament and of the Council relating to Fertilisers', CES0022/2002, available at: eur-lex.europa.eu (visited on 13.4.2017)

Presidential Decree No 148/2009 on Environmental Liability with regard to Prevention and Remediation of Environmental Damage – Implementation of Directive 2004/35/CE of the European Parliament and of the Council of 21 April 2004, as amended (‘*Περιβαλλοντική Ευθύνη για την Πρόληψη και την Αποκατάσταση των ζημιών στο Περιβάλλον – Εναρμόνιση με την οδηγία 2004/35/ΕΚ του Ευρωπαϊκού Κοινοβουλίου και του Συμβουλίου της 21 ης Απριλίου 2004, όπως ισχύει*’), FEK A 190, 29.9.2009, p. 6527 (‘*EnvLL*’)

Presidential Decree No 212/2006 on the protection of workers from the risks related to exposure to asbestos at work (‘*Προστασία των εργαζομένων που εκτίθενται σε αμίαντο κατά την εργασία, σε συμμόρφωση με την οδηγία 83/477/ΕΟΚ του Συμβουλίου*’), FEK A 212, 9.10.2006, p. 2327

Proposal for a Directive of the European Parliament and of the Council establishing a framework for the protection of soil and amending Directive 2004/35/EC, COM/2006/0232 final - COD 2006/0086, of 22.9.2006, eur-lex.europa.eu (visited on 10.4.2017)

Proposal for a Regulation of the European Parliament and of the Council relating to fertilizers, COM/2001/0508 final – COD 2001/0212, para 1.1, available at: eur-lex.europa.eu (visited on 13.4.2017)

Regulation (EC) No 1272/2008 of the European Parliament and of the Council of 16 December 2008 on classification, labelling and packaging of substances and mixtures, amending and repealing Directives 67/548/EEC and 1999/45/EC, and amending Regulation (EC) No 1907/2006, OJ L 353, 31.12.2008, p. 1, as amended

Regulation (EC) No 2003/2003 of the European Parliament and of the Council of 13 October 2003 relating to fertilisers, OJ L 304, 21.11.2003, p. 1, as amended

Siouti GP (2011) Handbook of environmental law (in Greek). Sakkoulas Publications, Athens-Thessaloniki

Spiliotopoulos EP (2011) Text-book on administrative law (in Greek), vol I. Nomiki Vivliothiki Publications, Athens

Spiliotopoulos EP (2015) Text-book on administrative law (in Greek), vol II. Nomiki Vivliothiki Publications, Athens

Tachos A (2008) Greek administrative law (in Greek). Sakkoulas Publications, Athens – Thessaloniki

University of Crete, Department of Environmental Engineering, Laboratory of Management of Toxic and Dangerous Waste, Study on Investigation, Assessment and Remediation of Uncontrolled Polluted Areas and Facilities from Industrial and Dangerous Waste (‘*Μελέτη για τη διερεύνηση, αξιολόγηση και αποκατάσταση ανεξέλεγκτων ρυπασμένων χώρων / εγκαταστάσεων από βιομηχανικά και επικίνδυνα απόβλητα στην Ελλάδα*’), Chania 2008 – 2009, available at: http://www.epper.gr/el/Pages/eLibraryFS.aspx?item=1740 (last visited on 9.10.2016) (‘*The Study*’)

Vathrakokoilis BA (2007) Interpretation and Jurisprudence of Civil Code, Vol. IV Pt. A, Law of Rights in Rem. Nomiki Vivliothiki Publications, Athens

Venizelos E (2008) Lectures on constitutional law (in Greek). Ant. N. Sakkoulas Publications, Athens

The French Law on Biodiversity and the Protection of Soils

Maylis Desrousseaux

1 Introduction

As recalled by the UICN knowledge forum of 2010, the "Non-Regression" principle is an "International Law Principle known by Human Rights specialists requiring that norms which have already been adopted by States not be revised, if this implies going backwards on the subject of standards of protection of collective and individual rights." In France, this principle can be compared to what is referred to by the Constitutional Court as the "Ratchet effect," commonly defined as an instance of the restrained ability of human processes to be reversed once a specific event has occurred, analogous with the mechanical ratchet that holds the spring tight as a clock is wound up. Even though the French legal doctrine had observed a certain decline of the use of this effect,[1] environmental law has renewed it. Indeed, it has recently gained the field of environmental law, especially in France but also in Europe under a more political than legal aspect.

This rebirth is due to the fact that, for the past 40 years now, French environmental law has been growing and extending as it was a new legal area. Its purpose has genuinely been to improve the protection of natural elements and insure good quality of life, and in a way, we could only observe progress. For instance, environmental law, which started as the protection of nature, widened its scope in accordance with scientific progress in ecology, hydrology, biology, pedology, etc. Consequently, at national and European levels, from the protection of areas emerged the protection of species and their habitats, to finally integrate the multifunctionality of ecosystems and to preserve ecological corridors.

[1] Mathieu and Verpeaux (2004).

M. Desrousseaux (✉)
INRA, Paris, France
e-mail: maylis.desrousseaux@inra.fr

H. Ginzky et al. (eds.), *International Yearbook of Soil Law and Policy 2017*,
International Yearbook of Soil Law and Policy,
https://doi.org/10.1007/978-3-319-68885-5_14

Then came the idea that environmental law was too complex and inefficient and required to be simplified and also modernized. Lately, in France, denouncing the high complexity of environmental law in itself has become common for politicians who mention the subject. Yet the environment is complex, its degradations are multifactorial, and its interfaces, interrelations, and connections are constantly evolving. As a result, it seems logical for the law to be somehow complex, in a sense that it is facing multifactorial challenges. Also, within the past five years, we have observed a constant movement toward the reduction of administrative procedures and a limitation of the possibilities to seize the Court of Justice in the environmental, industrial, and land planning fields.

In the same time and at a superior level, the fragility of the progresses achieved by the international negotiations has been pointed out, and the "Non-Regression" principle comes as a solution. Strongly discussed during the French Congress of environmental lawyers in Limoges in 2011, while preparing recommendations for the Rio +20 Conference, this principle is inscribed under Recommendation n° 1 of the Declaration of Limoges. It demands that "To prevent any step back of the protection of the environment, States must, in the common interest of humanity, recognize the non-regression principle. To do so, States must take the necessary dispositions to guarantee that no measures can lower the level of protection reached so far." Despite all the efforts made by the community of environmental lawyers, it has not been included in the outcome document, "The future we want," of the Conference. It was, however, carried by the European Parliament resolution of 29 September 2011 on developing a common EU position ahead of the United Nations Conference on Sustainable Development (Rio+20).[2]

At the national level, it has recently gained a noticeable legal value and has been validated by the Constitutional Court. Indeed, the French law to reconquer biodiversity, nature, and landscapes, adopted in August 2016,[3] has inscribed the non-regression principle in Article L. 110-1, 9° of the Environmental Code. This article is important as it gives all the principles that should guide environmental protection measures and is applicable to the whole Code. More specifically, the non-regression principle applies to any laws or decrees and conveys the idea of a constant improvement of the law, in the sense of protection.

Before the publication of the law, opposition deputies and senators had seized the Constitutional Court, claiming that the principle violated various other principles protected by the Constitution of 1958 and the Declaration of Human Rights of 1789, such as the hierarchy of norms, the principle of intelligibility and accessibility of the law, and the precautionary principle protected by the French Environmental Chart of 2005.[4] However, following the observations of the government, the

[2]European Parliament resolution of 29 September 2011 on developing a common EU position ahead of the United Nations Conference on Sustainable Development (Rio+20).

[3]Loi n° 2016-1087 du 8 août 2016.

[4]Loi constitutionnelle n° 2005-205 du 1er mars 2005 relative à la Charte de l'environnement (JORF n° 0051 du 2 mars 2005 page 3697).

Court concluded the nonviolation of the abovementioned elements.[5] Based on the constitutional validity of the non-regression principle, we propose to analyze in this article the new perspectives that its implementation could offer to protect soils at the French level, and what limits it could encounter.

In the first part, the current state of soil protection in French law will be discussed to identify what sort of regression could occur (Sect. 2), while the second part will point out the progresses made in the very same law and susceptible to have a positive impact on the protection of soils: the protection of biological processes and the recognition of soils as an element of the French national common heritage (Sect. 3).

2 Can Soil Protection Regime Benefit from the Non-regression Principle?

The international context of soils reinforce the necessity to improve and not reduce the protection of soils (Sect. 2.1), while at the French level we will need to determine to what already-implemented protection framework the non-regression principle can refer (Sect. 2.2).

2.1 A Constructive International Context

Facing several rather political than legal debates, the future of soil protection regimes in France and at the European level is not engaged on the path of a forthcoming improvement. However, at the international level, nonlegally binding instruments have been designed, and as recalled in the foreword of the first volume of the international yearbook on soil law and policy: "It is no wonder that soils have been much more prominently featured in the outcome document of the Rio + 20 Conference 'The Future We Want' in 2012 and ultimately in Sustainable Development Goals (SDG) in 2015. In addition, the UN General Assembly declared the year 2015 as the first "International Year of Soils."[6]

For the moment, European Member States remain on the same position on the European Soil Framework Directive officially abandoned in 2014.[7] But it is true that a "shift" occurred in the public perception,[8] and in September 2016, a European citizen's initiative (ECI), carried by more than 400 environmental, agricultural, etc.

[5]Décision n° 2016-737 DC du 4 août 2016.

[6]Ginzky et al. (2016).

[7]*Official Journal of the European Union,* 2014/C 153/03, May 21st of 2014.

[8]Ginzky et al. (2016).

associations, launched a petition to "Save the soil." How could national legislation counter such recognition?

In this context, the non-regression principle appears to be a particularly supportive tool. It presents the advantage of being well studied in the field of human rights[9] and is also utilized in several legal orders, as mentioned above. Translated as the "Ratchet effect," it is equivalent to the method of the Standstill.[10] However, its definition and applicability is far from being simple, and some researchers who compared it to the "Babel Tower" have pointed out its "fuzziness" and highly scattered terminology.[11] The French government initially followed this position. The debates that preceded the adoption of the Law on Biodiversity of 2016 (*nb*: it took more than two years to be voted by the two parliamentary chambers) are very informative: the Green party unsuccessfully attempted to integrate the principle in the law. It was notably rejected after the declaration of Segolene Royal, the French Minister of the Environment at that time, explaining that "by including this principle in the law, we would risk highly complex litigations leading to the freeze of projects."[12] It was finally adopted, after being supported by the work of the Commission for a sustainable development.[13]

2.2 *The Difficulty to Identify a Framework of Reference to Measure a Regression*

Regression needs to be measured in order to estimate whether future laws and decrees will cause it. That is where the main difficulty lies, as when the regression of soil protection will have to be identified, the French protection framework will be too incomplete.[14] Currently in France, soils do not benefit from a specific legal framework of protection. Instead, they are scattered in numerous fields of the law, such as agricultural laws and industrial laws, or they are indirectly taken into account by other protection laws. It is the case, for instance, with water quality measures that have an impact on soil planning and soil activities.[15]

Consequently, to measure the regression of soil protection, we could imagine that the European soil charter of the Council of Europe of 1976, revised in 2003,[16]

---

[9]For one example, see the Decision of the French Constitutional Court 84-181 DC; comm. L. Favoreu, *RDP*, 1986, p. 395.

[10]Prieur (2012).

[11]Herve-Fournereau (2012).

[12]Official Journal of the French National Assembly, CR., n° 71, June 24th of 2014, amendment n° CD534.

[13]Gaillard (2014) and Bignon J (2015).

[14]Desrousseaux (2016).

[15]Farinetti (2013).

[16]Revised European Charter for the Protection and Sustainable Management of Soil, adopted by the Committee of Ministers of the Council of Europe at its 840th meeting on 28 May 2003.

could be the text of reference. This Chart applies to all kinds of soils and threats,[17] unlike, for instance, the soil protection protocol specific of the Alps (1991) or the UN Convention on Desertification (1994),[18] focused on specific threats or areas (i.e. the Alps). Its international status is not, in itself, a serious obstacle. According to Christophe Krolik, the non-regression principle can use international norms as a reference. *Vice versa*, it also applies to any "general and non-individual norm coming from a public authority, at a national level as well as a supranational level."[19] However, the fact that the European soil charter is not legally binding could be an issue for judges, even though they can use provisions of soft law to support decisions. In this context, we interrogate the feasibility to refer to the proposal of the European Commission for a Soil Framework Directive. As it was more ambitious than any French legal framework, regressions should then be easily qualified. Unfortunately, even if it was adopted by the European Parliament, the text never entered into force.

At this stage of our demonstration, we are facing the obstacle of the "life cycle" of a norm: when does the regression occur? This question asked by Nathalie Herve-Fournereau invites us to consider the process of conception of a norm followed by its various interpretations given by multiple categories of recipients. It also questions the legitimacy to elaborate a global framework and to make it the reference norm.

Indeed, the purpose of the Soil Framework Directive to build an efficient protection regime is not that obvious and needs to be demonstrated. Some authors discuss this idea by considering that the adoption of "Integrated regulations," or furthermore "Framework Directive", does not always guarantee the improvement of the existing measures, but also when those texts result from political will to simplify and improve the quality of the European legislation.[20] In the same order, global goals such as the good ecological and chemical status of water were not proven to be more efficient than discharge standards. As revealed during the adoption process of the Water Framework Directive, such improvements were not guaranteed regarding the existing regime.

Even if adopted in French law, the non-regression principle can be read in accordance with the European goal to reach a high protection level for the environment. It also guides its implementation at the national level by giving it a supranational strength: international and European initiatives always demand a high level of protection in terms of environment, which is not compatible with any regression that would lower environmental standards. Hence, the minimum level of protection should be translated as the maximum level of protection considering local circumstances.[21]

Consequently, protection frameworks should guarantee at least an equal level of protection in the hypothesis where they would be elaborated outside any previous

---

[17]Blum (1993).

[18]Brauch HG, Spring UO (2009).

[19]Krolik (2012).

[20]Kramer (2007).

[21]Herve-Fournereau (2012).

legal frame. They could also base their level on the qualitative requirements developed for other natural elements. Furthermore, the norm utilized as the norm of reference to estimate the regression should correspond, in terms of position in the hierarchy of norms, to the norm about to be modified. However, such control will not be applicable in the case where a public authority would consider to "get around" the principle by a succession of slight regressions that would lead in fine to a significant regression. Thus, the control of the reality of the regression mostly relies on the power given to judges and on a concrete and global appreciation of the situations.

Regarding these developed elements, we could then analyze the implementation of the non-regression principle, considering the recent evolutions of the French environmental law.

3 The Protection of Soils by Their Integration to the Protection of Biological Processes and to the National Common Heritage

The French law on biodiversity of 2016 allowed two significant changes in addition to the recognition of the non-regression principle: Firstly, it recognized the concept of biological process (Sect. 3.1), and secondly, it included soils among the elements constitutive of the common national heritage (Sect. 3.2). We will then focus on the evolution brought out by these two concepts and see them as a progression of the protection of soils.

3.1 The Protection of Biological Processes Is a Progress for the Protection of Soils

In the French Environmental Code, soils are not considered as a natural element. However, soils can be indirectly protected as they are the territorial basis of protected areas. For instance, the legislation of national parks[22] (Articles L. 331-1 and following of the Environmental Code) specifically describes "soils" as constitutive elements of the area. They are also protected for their function of habitats, for instance by the Natura 2000 regime of the Directive of 1992 (Articles L. 414-1 and following of the Environmental Code).[23]

This being said, the French system protecting biodiversity relies on lists, like most European countries. This means that to be protected, species need to be

[22]Law n° 60-708 of 22 July 1960 on national parks, revised in 2006.

[23]Directive 92/43/CEE.

previously identified. Yet soil biodiversity is largely composed of microorganisms, and most of the soil species are yet to be discovered according to soil specialists.[24] Thus, the current listing system appears to be inefficient in order to properly protect soil biodiversity. Instead, some researchers proposed to adapt the current mechanisms to protect biological processes.[25] This idea is far from being new as the listing system was already criticized in the 1980s. It accompanies the idea that environmental law and environmental sciences must interact increasingly.[26]

For these reasons, and considering that soil sciences still have a lot to discover and understand about soils, the recognition of biological processes by Article L. 110-1 of the French Environmental Code will certainly have an impact on soil protection. Such progress is not the first step of environmental law toward the protection of the functionality of ecosystems. The famous "Grenelle laws" of 2009 and 2010[27] had already confirmed the willingness of France to follow this path by creating the mechanisms of protection of ecological corridors and connections (Article L. 371-1 of the Environmental Code).[28] The next step will be to follow the implementation of this recognition and its binding value, but at the present time it is still too early to consider this.

Moreover, the same law brought out the new concept of "geodiversity," and it was then discussed that protecting soils was equal to protecting geodiversity.

As described by Tukiainen et al., "One coarse-filter strategy in conservation and protected-area management, conserving nature's stage, centers on the physical structures that underlie biotic processes and recognizes that geodiversity—the diversity of Earth surface forms, materials, and processes[29]—itself has conservation value and is related to biodiversity."[30,31] But soil scientists maintain the difference and refuse the first position of the French government which consisted in forgetting soils for the benefit of geodiversity.[32] This position was defended by a part of the Senate even though the definitions were slightly confusing.[33] To sum up, we can observe that soils are recognized twice: under the geodiversity concept and under the definition of the elements contributing to the national common heritage.

[24]Jeffery and Gardi (2010).

[25]Chaussod (1996) and Chenu (2014).

[26]Keiter (2004) and Naim-Gesbert (1999).

[27]Law n° 2009-967 and Law n° 2010-788.

[28]Bonnin (2011).

[29]Gray (2013).

[30]Anderson and Ferree (2010) and Lawler et al. (2015).

[31]Tukiainen et al. (2017).

[32]Official Journal of the French National Assembly, CR., n° 71, June 24th of 2014.

[33]Debates of January19th of 2016, 1st reading, Senate, discussion on the article L. 110-1 of the environmental Code.

3.2 *The Recognition of Soils as an Element of the National Common Heritage*

An amendment was initially voted to delete the mention of soils in the law on biodiversity, justified by the fact that soils were already understood under the definition of "geodiversity." Here is what we understand from the debates that took place in the two chambers of the French parliament: the progressive international recognition of soils in declarations and working documents has raised new perspectives, and the word "soil" or "soils" has always been utilized in French to identify several objects but the soil in itself. Nowadays, the word is progressively meeting its ecological definition and could have a new significance. To illustrate this hypothesis, an ordinance of September 2015[34] has deleted a sentence from the Land Planning Code, saying that the public authorities in charge of land planning were "responsible of the careful management of soil [*sol*]," even though it was well admitted that the soil targeted by the article was actually the surface and not the natural element.[35]

In France, private and public properties, administrative and national territories, rely on the soil, literally written "sol" in French. For instance, the Civil Code (Article 552) expresses that "the property of the soil [understood as a surface] implies the property of what is under and what is above." Property is a volume, and traditionally the owner is the master of its parcel within the limits of the law (Article 544), meaning that property right is sacred and constitutionally protected (Article XVII of the Declaration of Human Rights of 1789) and that this right is generally not compatible with the protection of the environment, and protecting soils would reduce the prerogative of the owners.[36]

This conception greatly differs from the perception that public authorities and the population have of water. Water is indeed recognized as an element of the national common heritage since the law on water of 1992.[37] Furthermore, some authors have observed that, thanks to this recognition, some terrestrial elements were implicitly included in this legal category. As Aude Farinetti explained, the concept of common heritage is "attractive" and the proclamation of water as an element of the common national heritage has created a protecting regime not only for water but also for the water environment, including the terrestrial elements such as wetlands, water beds, etc.[38]

Only a small step further was needed in order to include soils and to set the first stones toward a proper soil regime. It was achieved, thanks to the law on biodiversity, and the consequences are for now quite minimal. To pursue the parallel with

[34]Ordinance n° 2015-1174 of September 23 of 2015.

[35]Godfrin (2010).

[36]Remond-Gouilloud (1989).

[37]Law n° 92-3.

[38]Farinetti (2013).

the water regime, the category of the national common heritage has just been "juxtaposed" on the preexisting and various status of water (public and private regimes, common good, etc.). Thus, it does not directly or significantly impact private property rights and preserves the law from a possible sanction by the French Constitutional Court.[39]

4 Conclusion

We can conclude that the law on biodiversity of 2016 has significantly renewed the perception of the law on soils and is likely to bring future positive evolutions. This might eventually lead to the construction of adequate soil governance.[40] Furthermore, the recognition of soils by the French Environmental Code was accompanied with new concepts and principles such as the nonregression principle and the "Biological processes" concept, which will certainly impact the future of the whole environmental legal framework. This article has strived to express the fact that despite political obstacles coming from European and national levels, the protection of soils has gradually been gaining new material. This being said, another concept has already been designed and claims to go beyond the "non-regression principle": the "no backsliding" principle, defined as a progressive one, implying the obligation for the states to constantly and continuously improve their environmental laws, based on a predetermined time line.[41]

References

Anderson M, Ferree C (2010) Conserving the stage: climate change and the geophysical underpinnings of species diversity. PLoS One 5(7):e11554

Bignon J (2015) Report n° 607 of the Commission for a sustainable development and land use planning on the law on biodiversity, National Assembly, 602 p (French)

Blum W (1993) Soil protection concept of the Council of Europe and integrated soil research, chapter, integrated soil and sediment research: a basis for proper protection, vol. 1 of the series. Soil & Environment, pp 37–47

Bonnin M (2011) Protected areas and ecological networks: global environmental management or management of the conservation institutions? In: Aubertin C, Rodary E (eds) Protected areas, sustainable land? Ashgate, IRD editions, Surrey, 183 p

Brauch HG, Spring UO (2009) Securitizing the ground, grounding security, Desertification land degradation and drought. UNCCD issue paper n° 2, 52 p

Chaussod R (1996) The biological quality of soils, assessment and implications. Étude et Gestion des Sols 3(4) (French)

[39]Favoreu and Philip (2013).

[40]Montanarella (2015).

[41]Kerbrat et al. (2015).

Chenu C (2014) Soils and the protection of biodiversity, oral communication for the 2d French world soil day
Desrousseaux M (2016) The legal protection of soil quality. LGDJ, 502 p (French)
Farinetti A (2013) The legal protection of soil quality through water laws. Env. et DD., n° 6 (French)
Favoreu L, Philip L (dir) (2013) The great decisions of the Constitutional Court. Dalloz, 17e éd., p 157 (French)
Gaillard G (2014) Report of the Commission for a sustainable development and land-use planning on the project of law on biodiversity (n° 1847), 668 p (French)
Ginzky H et al (eds) (2016) International yearbook of soil law and policy, vol 1
Godfrin G (2010) The careful management of soil. Construction-Urbanisme, n° 10, étude 10 (French)
Gray M (2013) Geodiversity: valuing and conserving abiotic nature, 2nd edn. Wiley-Blackwell, Chichester
Herve-Fournereau N (2012) Le principe de non régression environnementale en droit de l'Union européenne: entre idéalité et réalité normative? In: Prieur M (dir) La non régression en droit de l'environnement. Bruylant, p 199
Jeffery S, Gardi C, et al (eds) (2010) European Atlas of soil biodiversity. European Commission, Office of Publications of the EU, Luxembourg
Keiter R (2004) Ecological concepts, legal standards, and public land law: an analysis and assessment. Nat Resour J 44:943
Kerbrat Y, Maljean-Dubois S, Wemäere M (2015) The international Climate Conference of Paris: how to build a time adaptable agreement? Int Law J, (Clunet) n° 4 (French)
Kramer L (2007) Mieux légiférer et déréglementation du droit de l'environnement. RDUE 4:801
Krolik C (2012) Contribution à une méthodologie du principe de non-régression. In: Prieur M, Sozzo G (eds). Bruylant, Bruxelles, pp 137–150
Lawler J, Ackerly D, Albano C, Anderson M, Dobrowski S, Gill J, Heller N, Pressey R, Sanderson E, Weiss S (2015) The theory behind, and the challenges of, conserving nature's stage in a time of rapid change. Conserv Biol 29:618–629
Mathieu B, Verpeaux M (2004) Chronicle on the jurisprudence of the Constitutional Court, La Semaine Juridique Edition Générale n° 51, doctr. 192 (French)
Montanarella L (2015) Govern our soils. Nature 528:30–33
Naim-Gesbert E (1999) Les dimensions scientifiques du droit de l'environnement. Bruylant, 810 p
Prieur M (dir) (2012) The non regression principle in environmental law. Bruylant (French)
Remond-Gouilloud M (1989) The right to destroy. PUF, 304 p (French)
Tukiainen H, Bailey J, Field R, Kangas K, Hjort J (2017) Combining geodiversity with climate and topography to account for threatened species richness. Conserv Biol 31(2):364–375

Implementing Land Degradation Neutrality at National Level: Legal Instruments in Germany

Ralph Bodle

1 Law and LDN

In order to assess how law can contribute to achieving land degradation neutrality (LDN), we need to define more precisely what the legal instruments would seek to achieve. The goal 'land degradation neutrality' on its own is not specific enough.

A basic intuitive understanding of LDN is that in a specific time and area, the total amount and quality of land resources should not become worse. This basically is what the UNCCD adopted as a definition of LDN in more scientific terms: LDN is a 'state whereby the amount and quality of land resources necessary to support ecosystem functions and services and enhance food security remain stable or increase within specified temporal and spatial scales and ecosystems'.[1] However, there are no universally used definitions of each of the three terms 'land', 'degradation' and 'neutrality' that provide clear guidance for implementing the concept as a whole. For instance, the SDG 15.3 does not provide a baseline against which 'neutrality' is supposed to be achieved. So far, much of the work at the international level has focused on clarifying these terms by developing definitions and indicators. This work is addressed in detail in a different chapter of this book.[2]

[1]CCD decision 3/COP.12, para 2. On the CCD's work see Boer et al. (2017), pp. 61–63; Minelli et al. (2017).

[2]Wunder et al. (2017), pp. XX-XX; see previously Ehlers (2017), pp. 79 et seq. The current indicator by the UN's Inter-agency Expert Group on SDG Indicators merely states 'Proportion of land that is degraded over total land area', see IAEG-SDGs (2016).

R. Bodle (✉)
Ecologic Institute, Berlin, Germany
e-mail: ralph.bodle@ecologic.eu

H. Ginzky et al. (eds.), *International Yearbook of Soil Law and Policy 2017*,
International Yearbook of Soil Law and Policy,
https://doi.org/10.1007/978-3-319-68885-5_15

At the international level, the UNCCD considers that the term 'land' comprises more than 'soil'.[3] In German and EU legal texts, there is no uniform translation or usage of the terms 'land' and 'land degradation'. Sometimes they are used synonymously with 'soil' and 'soil degradation'.[4] The German Federal Soil Protection Act and most laws generally use the term 'soil' and do not have or address a distinct 'land' category. The term 'soil' and the functions included by the law are broad enough to address the purpose of LDN.[5] In this chapter, we use the term 'soil' in this wider sense.

From a legal perspective, it is not necessary to know each LDN indicator in order to assess whether and how the existing legal instruments and techniques are suitable and adequate for achieving LDN. Instead, it is important to identify *conceptual* components of LDN to be addressed by law.

1.1 Offsetting as Part of LDN

The wording of SDG 15.3 does not say that all degradation should be avoided. Instead, SDG 15.3 accepts land degradation as long as the *total* amount and quality of land resources remains at least stable within a specific time frame and area. LDN involves a balancing approach by which degradation is set off against improvements.[6]

The UNCCD Secretariat's 'Zero Net Land Degradation' target had already included setting off degradation by restoration, as part of sustainable development.[7] Slowing down land degradation by sustainable land management went hand in hand with restoring ecosystems, and together they led to LDN.[8] On the other hand, The UNCCD argued that LDN was no 'license to degrade' and did not advocate for a 'grand compensation scheme'.[9] Market-based offset or compensation schemes had

[3]Wunder et al. (2017), pp. XX-XX, Sect. 1.

[4]See for instance the translation of Agenda 2030 by the German Translation Service for the UN: Deutscher Übersetzungsdienst der Vereinten Nationen (2015), p. 26.

[5]On the functions covered by the term "soil" in the Soil Protection Act see Erbguth and Schlacke (2016), pp. 373–375.

[6]See the "conceptual framework" elaborated by the UNCCD Science-Policy-Interface to assist the implementation of LDN: "The objective is that losses are balanced by gains", UNCCD/Science-Policy Interface (2016), p. 1; on offsetting in the "zero net rate of land degradation" concept see Desai and Sidhu (2017), p. 44.

[7]UNCCD/Science-Policy Interface (2016), pp. 9 and 12. The keyword "offset" is mentioned on page 24: "The achievement of land degradation neutrality, whereby land degradation is either avoided or offset by land restoration".

[8]UNCCD (2014), p. 12.

[9]UNCCD (2014), p. 12.

been proven to be complex, problematic and generally ineffective.[10] It is therefore not clear whether the balancing concept of LDN should allow *any* degradation to be offset in order to achieve LDN or whether there should be a threshold beyond which offsetting is undesirable and would not count towards LDN.

In order to offset degradation, it has to be determined *where and when* restoration may take place in order to qualify for offsetting and thus contributing to achieving LDN.

With regard to **where** neutrality is to be achieved, the SDGs address the UN Member States. Germany is to achieve LDN on its territory as a whole, and restoration would have to take place there in order to offset degradation. This makes sense as it corresponds with Germany's sovereign power to govern its territory. There are other governance models that are not based on legal or political boundaries. For instance, the EU's water framework directive requires EU Member States to jointly manage river basins as whole ecosystems, regardless of where they cross borders. In international law, there is practice and basic principles concerning the utilisation of shared resources such as rivers.[11] However, there is no indication that the UN intended to encourage States to engage in a global offsetting scheme beyond their own territory—even though SDG 15.3 refers to a land-degradation-neutral *world*.

If offsets require a close spatial link such as occurring in the same ecosystem, Germany will have to compile and aggregate the balance of all local and regional areas in which offsets occur, in order to know whether Germany as a whole achieves neutrality.

With regard to the **time frame** for LDN and offsetting, three questions need to be addressed: (1) what is the starting point against which LDN is to be achieved (baseline); (2) at which point in time is LDN to be achieved and (3) when does restoration have to occur in order to count as an offset against degradation?[12] SDG 15.3 does not answer the first question. It does not define the baseline, and the UNCCD's conceptual framework merely recognises that it must be established.[13] With regard to the second question, SDG 15.3 sets the year 2030 as the target date. Although strictly speaking LDN does not have to be *achieved* by 2030 because States should merely *strive to achieve* by then, we assume that Germany seeks to achieve the goal by that point in time. With regard to the third question, SDG 15.3 does not mention or require a temporal link between degradation and offsets. The UNCCD's definition merely mentions 'specified temporal...scales' but does not specify them.[14]

---

[10]UNCCD, Land Degradation Neutrality – Frequently Asked Questions (FAQs), http://www.unccd.int/en/programmes/RioConventions/RioPlus20/Pages/LDNFAQ.aspx, last accessed 11 June 2017.

[11]Cf. Durner (2001).

[12]Wunder et al. (2017), pp. XX-XX.

[13]UNCCD/Science-Policy Framework (2016), p. 4.

[14]The UNCCD's conceptual framework does not seem to address this issue.

In addition to where and when offsets may take place, there may be other requirements that have to be fulfilled in order for restoration to count for LDN. For instance, restoration may have to restore the same soil functions as the degradation. The UNCCD's conceptual framework suggests that 'counterbalancing is managed within the same land type', although it is not clear whether this is meant to be a normative statement.[15]

1.2 LDN Components to be Addressed by Legal Instruments

Based on previous conceptual work[16] and the analysis above, legal implementation of the LDN goal should distinguish the following *conceptual* components of LDN:

1. **Preventing** degradation: this involves legal rules and instruments preventing (further) degradation. The less land is further degraded, the less restoration has to occur as offsets in order to achieve neutrality. The simplest example is a prohibition of a particular activity that degrades land. The law could also require precautionary measures.
2. **Restoring** and rehabilitating degraded land: this pertains to legal rules and instruments requiring or enabling that land is restored or rehabilitated. For instance, the law might require that persons responsible for land degradation, e.g. through negligence, have to restore it. The underlying assumption that land functions can be restored to how it was is not necessarily tenable. However, such shortcomings have to be accepted to some extent, and practicable indicators have to be developed in order to use legal instruments to achieve LDN.
3. **Offsetting and land-use planning and management**: first, this includes legal rules and instruments that require or allow that degradation is offset by restoration. Such rules apply to specific cases of land degradation that has occurred or is about to occur. For instance, permission for a building project that would degrade a habitat may be granted only if the applicant restores or upgrades land to a functionally equivalent extent. Second, the law might require that land use is planned and managed in a forward-looking manner. It may require that land use and soil protection are considered as part of existing planning procedures, such as town and country planning. It could also require planning and management specifically for soil.

In addition, the concept of LDN requires **information**: for any offsetting, it is not only desirable but also inherent to know where soil has degraded or improved. Indicators are required that measure the status of land, degradation and restoration. The law might address the gathering of relevant information and the methodology to be used, in order to have reliable and coherent information about progress

---

[15]UNCCD/Science-Policy Framework (2016), p. 3.

[16]For instance, Altvater et al. (2015); Minelli et al. (2017), p. 88.

towards LDN, and a sound basis for political decision-making. Conceptually, this chapter addresses information systems as part of the category 'land use planning and management'.

1.3 Soil Protection and LDN in the German Legal System

Which legal instruments are available to protect soil and achieve LDN depends on the legal system they are part of.

For instance, Germany is a federal State with 16 States (*Länder*), in which the constitution distributes legal powers between the federal level, the State level and the municipal level. This is relevant, for instance, in order to determine whether and to what extent the federal level may regulate an issue at all or to what extent the States may deviate from existing federal laws. The division of powers is also relevant for considering which regulatory level would be competent (in addition to appropriate) to adopt potential new laws or measures regarding LDN. Last but not least, as a general rule, federal laws are implemented and enforced by the States, using their respective administrative structures and procedures. This can lead to different implementation rules, practices and effects in different parts of Germany.

Moreover, Germany has a highly detailed civil law system with courts deciding on minute details of interpretation of the numerous laws and administrative practice. Access to courts in order to challenge administrative decisions is possible in most cases and also affordable, although the process could be faster. Citizens and other actors affected by administrative decisions can take legal action, e.g. against refusal by the authority to grant a permit for an activity on the grounds that it would degrade soil.

Furthermore, Germany is also a Member State of the European Union (EU), which has the power to adopt legal acts that are binding on and sometimes directly legally applicable in Germany.[17] EU Member States have a legal obligation to transpose, implement and give legal effect to EU legislation, which overrides national law. Over the last decades, EU legislation has shaped many areas of environmental law in Germany and other Member States. However, while many EU environmental rules in other areas also address particular aspects of soil and land degradation, soil protection is one area where there is no specific overarching EU legislation. In 2006, the European Commission had made a proposal for a soil framework directive but withdrew it in 2014 because it saw no political chance for adoption.[18] In the absence of specific and comprehensive EU legislation, Germany is quite free to choose its approach to implementing SDG 15.3 and achieving LDN.

[17] Art. 288 Treaty on the Functioning of the European Union.

[18] European Commission (2014), p. 3.

2 Legal Instruments in Germany for Addressing LDN

This section further develops previous work that provided an initial overview of legal approaches to soil protection in Germany.[19] It analyses the legal mechanisms by which existing German law addresses each of the key components of LDN identified above[20]: (1) preventing degradation, (2) restoration and (3) offsetting, planning and management. It focuses on one specific soil threat: **erosion caused by agriculture**, and on the State of Mecklenburg-Vorpommern (MV).[21] In that State, almost half of the land is used for agriculture.[22] Soil erosion by wind and water is one of the main problems caused by it.[23] MV has adopted several laws that implement, supplement and sometimes deviate from federal laws.

2.1 Prevention

The starting point for preventing land degradation by erosion is the federal **Soil Protection Act (SPA)** of 1998.[24] German environmental law started addressing soil relatively late compared to other environmental issues. While many laws on other issues such as water, emissions and building also include provisions on soil, laws specifically addressing soil protection were first passed in the 1990s, with the federal Soil Protection Act (SPA) adopted in 1998.[25] The federal SPA bars the States from adopting deviating laws unless allowed by the SPA. Their remaining power is to fill the gaps that are not exclusively regulated by the SPA and to determine their respective administrative institutions and procedures.[26]

The SPA's objective is to protect soil against degradation and to pursue a precautionary approach. One significant shortcoming of the SPA is its scope of application: it is subsidiary to a whole range of other laws, including laws on waste, fertiliser and plant protection, infrastructure, town and country planning, building, mining and emission control. The SPA only applies to the extent that these laws do not regulate impacts on soil.[27]

[19] Altvater et al. (2015), pp. 39–52.

[20] See Sect. 1.

[21] Up-to-date versions of all federal laws mentioned in this chapter are available at https://www.gesetze-im-internet.de/. The State law of Mecklenburg-Vorpommern is available at http://www.landesrecht-mv.de/jportal/portal/page/bsmvprod.psml, last accessed 12.06.2017.

[22] Landesregierung Mecklenburg-Vorpommern, Antwort auf die Kleine Anfrage von Peter Ritter, Landtag Drs. 6/5329 of 13.05.2016, 2.

[23] UBA (2015), p. 46; Möckel et al. (2014), p. 113.

[24] Bundes-Bodenschutzgesetz.

[25] Schmidt et al. (2014), p. 389.

[26] Erbguth and Schlacke (2016), pp. 370–371.

[27] In this sense, the statement in Altvater et al. (2015), p. 43, that the SPA only applies if the respective sectoral law "does not contain any soil-related provisions", is not correct.

The SPA contains obligations to take precautionary measures, to prevent concrete risks of degradation and to restore degraded land.[28] Together with the Federal Soil Protection Regulation, a delegated statutory instrument, the SPA defines the instruments that are available to the authorities to monitor compliance and to enforce these obligations. It also authorises the federal States to determine soil protection areas.

Although there are no specific provisions in the SPA on erosion, it is included in the definition of degradation.[29] However, with regard to preventing erosion resulting from agriculture, § 17 SPA privileges agricultural land use in several ways: the standard that agriculture has to comply with is different from other land use with regard to precautionary measures, as well as with regard to averting concrete risks. In addition, the instruments and enforcement measures available to the authorities are significantly restricted.

Precautionary obligations apply when there is no indication of a concrete risk or hazard. Under the SPA, land owners, proprietors and operators are under a specific obligation to take precautionary measures against degradation. In contrast, agricultural land use merely has to comply with 'good agricultural practice'. The SAP does not define good agricultural practice but lists a few general principles. Those relevant for erosion provide that loss of soil matter should be avoided by land use that is appropriate to local circumstances and that structural elements, in particular hedges and shrubs, should be maintained. These principles are quite unspecific, even though the competent authority at State level has published several nonbinding guidance documents.[30] In addition, the enforcement measures for the default precautionary obligation are not available for agricultural land use. Moreover, not complying with good agricultural practice is not an administrative offence with potential penalties. The only instrument available to the authorities with regard to farmers who do not comply is to provide advice.[31]

In addition to precautionary obligations, the SPA also imposes obligations to *avert concrete risks* to soil. It also provides specific duties, powers and reference criteria for the authorities to investigate and determine whether and to what extent degradation has occurred.[32] However, the SPA provides legal privileges for agriculture that are similar to those for precautionary obligations. The authorities may take the measures available to it under the SPA only if neither other laws nor good agricultural practice provide for such measures. Insofar as measures under the SPA are possible, the statutory instrument provides indicators and thresholds for probing

[28]The key term is "adverse changes of soil" (schädliche Bodenveränderungen), which according to the SPA's definition requires (1) that soil function are adversely affected and (2) that these effects are capable of causing concrete risks, significant adverse effects or significant nuisance for individuals or the general public.

[29]The delegated regulation provides indicators for erosion by water.

[30]www.lms-beratung.de, last accessed 23.05.2017.

[31]Gröhn (2014), pp. 178–179.

[32]§ 9 BBodSchG.

and assessing erosion by water, but not for erosion by wind. And if the authorities take measures against a farmer that restrict agricultural land use, it may have to pay compensation.[33]

MV's State law that implements the federal Soil Protection Act does not provide additional relevant safeguards.[34]

The SPA is widely regarded as an ineffective instrument because it focuses on reactive control measures and contaminated soil, while it is weak on precautionary and planning aspects.[35] Specifically on agriculture, the SPA's regulatory technique is based on the self-interest of farmers to avoid degradation and erosion. The argument is that agriculture is too diverse to be regulated by standard rules. In contrast, critics argue that the farmers' self-interest might be guided more by productivity than by sustainability. It would be possible for authorities to use the different instruments at their disposal according to the individual case.[36]

The federal **Nature Protection Act** also prevents land degradation to some extent. A key rule provides that, in principle, interventions in nature have to be avoided and that unavoidable interventions are subject to a permit and have to be compensated.[37] Erosion caused by agriculture can qualify as an intervention in nature.[38] However, the Nature Protection Act also privileges agricultural land use: it states the legal assumption that activities in accordance with good agricultural practice do not constitute an intervention in nature for the purpose of this Act.[39] The Nature Protection Act defines principles of good agricultural practice in addition to those in the SPA. In the remaining cases where erosion qualifies as an intervention, the authorities can refuse the permit under certain conditions and impose fines for infringements.

At State level, MV's law on nature protection provides obligations to take precautionary measures against erosion on steep slopes and embankments, but these rules apply only to building activities and not to agriculture.

Other laws that protect erosion include, inter alia, **water law,** which protects by protecting riparian buffer zones. The federal Water Act[40] prohibits the turning over of all pastures within 5 m from the riverbank unless an exemption is granted.[41]

There are several **other sectoral laws** that can contribute to preventing land degradation by using different regulatory approaches.[42] For instance, the Federal

[33]§ 10 (2) BBodSchG.

[34]Landesbodenschutzgesetz—LBodSchG M-V.

[35]Erbguth and Schlacke (2016), pp. 366–367.

[36]Gröhn (2014), pp. 195–198.

[37]§ 13-17 Bundesnaturschutzgesetz—BNatSchG.

[38]Decision by the Federal Administrative Court, BVerwG, 11.07.2013 -7 A 20.11.

[39]§ 14 (2) BNatSchG.

[40]Wasserhaushaltsgesetz—WHG.

[41]§ 38 (4)-(5) WHG.

[42]Gröhn (2014) provides a comprehensive assessment; Altvater et al. (2015) provide an initial overview of selected instruments in English.

Forest Act provides that changing forest to a different land use requires a permit and lays down criteria for refusing such permits.[43] The Fertilising Act indirectly protects soils by regulating which types of fertilisers may be placed on the market and by prescribing criteria for their use.[44] However, its main purpose is to increase soil yield.[45] The Federal Immission Control Act defines detailed permit and monitoring procedures for industrial installations in order to prevent mainly air pollution and noise but also soil and water contamination. Based on EU legislation, its main feature is the integrative and cross-media regulatory approach, by which the permit authority has to consider all environmental impacts in an integrated manner.

At State level, MV's Act to Preserve Land Under Permanent Pasture[46] generally prohibits the turning up of land under permanent pasture, while exemptions require a permit, which may be granted under restrictive conditions. This law is based on the EU regulations on agricultural subsidies, under which maintaining areas of land under permanent pasture is one condition for receiving subsidies. The regulatory approach adopted by MV goes further because it applies not only to recipients of subsidies and changes the financial incentive into a regulatory prohibition.

Apart from obligations and prohibitions, the law also provides positive incentives to prevent degradation through **subsidies laws**. In the EU, the common agricultural policy provides an important source of income for farmers, including in Germany. They receive direct financial support that is partly conditional upon compliance with certain environmental requirements (cross-compliance) and carrying out 'greening' measures that benefit the environment and the climate. EU Member States have to set minimum standards, including protection of soil from erosion.[47] Accordingly, German federal law requires the States to classify and register agricultural areas according to erosion risk levels (see above) and also requires certain obligations according to the risk level. These include, for instance, a ploughing ban during certain periods, a ban on removing elements such as hedges and obligations regarding minimum soil cover. At State level, MV has financial support programmes for measures preventing soil erosion.

2.2 *Restoration and Rehabilitation*

Because the law will not be able to prevent further degradation completely, achieving LDN also requires restoration and rehabilitation of land. Legal

[43] § 9 Bundeswaldgesetz—BWaldG.

[44] Düngegesetz—DüngeG, together with technical provisions in delegated statutory instruments such as the Düngeverordnung—DüngV.

[45] Gröhn (2014), p. 231.

[46] Dauergrünlanderhaltungsgesetz—DGErhG M-V.

[47] Art. 93 and Annex II of EU Regulation 1306/2013: "good agricultural and environmental condition" no. 5.

instruments may, for instance, set **obligations to restore degradation**, e.g. by requiring that anyone who degrades land beyond a certain threshold, without being legally entitled to do so, has to restore it. Conceptually, this aspect of LDN is about restoring the actual land that is degraded, as opposed to offsetting degradation in one place by restoration somewhere else.

The German SPA was adopted at a time when *already contaminated* land was regarded as the key issue to be addressed. Restoration is at its core, with detailed provisions addressing investigation and assessment by the authorities, restoration plans[48] and enforcement powers, as well as technical criteria in a delegated statutory instrument.[49] When degradation has already occurred, the SPA requires to restore the land insofar as it is necessary to avert concrete risks or significant adverse effect or nuisance. The obligation is jointly on the legal entity or person causing the degradation, his legal successor, the land owner and the land holder.[50] In principle, this obligation applies to agricultural land use as well. However, as with precautionary obligations, the law privileges agriculture as the obligation to restore is subordinate to other rules regarding agriculture and to good agricultural practice. Although the SPA lists 'improvements to soil structure' as one of the elements of good agricultural practice, it does not specify details. This leaves very little room for the authorities to use their enforcement powers in cases of agricultural land use.[51]

When the obligation to restore applies, the law takes a tiered approach to the type of **restoration measures** required, depending on how the land is used and other factors. In the case of contamination, for instance, the priority is on de-contamination as opposed to leaving the contaminants in the ground and merely preventing harmful effects.[52] With regard to other degradation such as erosion, as a priority the person responsible has to reverse or mitigate the degradation, e.g. by supporting a sliding slope. Where such measures are not possible or unreasonable, the second-best option is to contain and mitigate the effects of the degradation without actually reversing it.[53] As with prevention orders, the authorities may have to pay compensation if the restoration measures restrict agricultural land use.[54]

The SPA also provides that the land owner may be ordered to take **de-sealing** measures if the land concerned is permanently out of use and the sealing contravenes determinations made in planning instruments.[55] However, besides legal difficulties in implementing this provision, it is subordinate to the de-sealing rules in the Building Code.[56] These provide, in particular, that municipal authorities may

[48]Ludwig (2011), pp. 176–209.

[49]Bundesbodenschutzverordnung—BBodSchV.

[50]§ 4 (3) BBodSchG.

[51]Möckel et al. (2014), pp. 152–153.

[52]§ 4 (5) BBodSchG.

[53]Erbguth and Schlacke (2016), pp. 386–387.

[54]§ 10 (2) BBodSchG.

[55]§ 5 BBodSchG.

[56]See Gröhn (2014), pp. 150–155.

require the land owner to acquiesce in dismantling and de-sealing measures under certain conditions. However, the land owner does not have to carry out this measure himself, and he is entitled to compensation.[57]

As with preventing erosion, the agricultural **subsidies rules** also provide incentives to restore erosion caused by agriculture. The conditions for receiving full support include requirements with regard to land cover in certain agricultural areas, in particular areas that have been set aside. Such areas have to be left to grow natural cover or be seeded for growing cover. This is a financial incentive to restore pastures, but it applies only to recipients of subsidies.

2.3 *Offsetting and Land-Use Planning and Management*

Because land will continue to be degraded, in addition to preventing degradation and restoring of land, achieving LDN requires that, overall, land degradation is at least offset by land restoration. At the *individual activity and project level*, legal instruments can contribute, for instance, by requiring that activities that degrade land are offset by restoring or otherwise improving other land somewhere else. At *area level*, legal planning instruments are a tool for managing land use, e.g. by determining which types of land use are permitted in which areas, or by setting qualitative or quantitative targets.

2.3.1 Legal Rules for Offsetting Land Degradation at Project and Activity Level

German law includes rules requiring that land degradation on one place is offset by compensatory measures somewhere else. Notably, the **'intervention rule' in the federal Nature Protection Act** requires that unavoidable interventions in nature are offset by balancing or substitution measures. The rule is designed for offsets regarding *individual projects*.[58]

In theory, if an intervention in nature is fully compensated, the balance stays neutral. However, this depends on how and by which method compensation takes place. The intervention rule applies a hierarchical **stepwise approach**: (1) in principle, interventions in nature that can be avoided have to be avoided. (2) Interventions that are unavoidable (in the legal sense) have to be compensated primarily either by balancing measures that achieve 'similar' nature functions or by substitution measures that achieve 'equivalent' nature functions. (3) If neither is possible, the authority decides whether, on balance, the interest of nature protection prevails over the applicant's interest to implement the project. (4) If the authority permits

[57] § 179 Baugesetzbuch—BauGB.

[58] Peters et al. (2015), p. 107.

the intervention, the person responsible for the intervention has to pay monetary compensation.

Conceptually, **the intervention rule contains all elements of LDN in one legal obligation**. It fully captures and translates into legal terms what the UNCCD's conceptual framework calls the 'LDN response hierarchy'.[59]

However, the intervention rule does not specifically address soil, although its scope might also apply to interventions caused by land degradation. However, it is worth a detailed look because it is an attempt at pursuing neutrality specifically through a legal obligation, because the rule specifically addresses offsets, and because its subject matter—nature protection—is close to land degradation. Which questions and problems are raised when the law regulates whether and how a particular degradation has to be offset? There is extensive practical experience in applying the intervention rule, including practical implementation and enforcement, detailed methodology and a large body of case law.

Not every modification of nature and landscape is an intervention in the sense of the law. The Nature Protection Act *defines* 'interventions' and the threshold from which the intervention rule applies. These are basically changes to land that alter its outward appearance, replace previous land use with a new one or effect certain changes to the ground water level and that are likely to adversely affect the objectives of the Nature Protection Act.[60] There is a legal assumption that agriculture in accordance with good agricultural practice is not contrary to these objectives. However, it may be an intervention if land use is changed in order to enable agriculture or make it more effective, e.g. when there is a change from one type of agricultural use to another.[61]

With regard to *timing*, when do the compensatory measures have to take place in order to fulfil the requirements of the intervention rule? Which time periods are included in calculating the two sides of the balance? The law requires that the measures are compensated 'within a reasonable time'. In principle, the compensatory measure has to be ensured at the time when permission for the intervention is granted and must not depend, e.g., on approval by third parties. Moreover, the compensatory measures should be carried out at the same time as the intervention and have to be maintained 'for a necessary time period'.

In terms of a *spatial proximity*, the law does not explicitly require a spatial link. However, according to the courts, there has to be a 'spatial-functional' link between the intervention and the compensatory measures. This may simply mean that there is a spatial link between the two.[62]

From a *functional* perspective, compensatory measures have to improve ecology at other sites and create a situation that is at least similar to the nature functions

[59]UNCCD/Science Policy Interface (2016), p. 3.

[60]§ 14 BNatSchG; Schmidt et al. (2014), pp. 431–432.

[61]Möckel et al. (2014), pp. 124–125.

[62]Schmidt et al. (2014), p. 434; BVerwG (Federal Administrative Court) judgment of 10.09.1998—4 A 35/97.

affected by the intervention. Merely maintaining and protecting existing land does not improve ecology and therefore does not qualify as compensation.[63] However, the functional link relates to nature protection as a whole. In other words, the law does not absolutely require that *land* degradation in one place is compensated by *land* improvement in a different place.

If compensatory measures are not possible and the authority decides, on balance, to allow the intervention, the person responsible for the intervention has to provide monetary compensation. The law provides that the payment has to be earmarked for nature protection measures.

The Nature Protection Act also authorises the States to maintain a system for 'banking', trading and pooling of compensatory measures. The State of MV, for instance, has done so in order to facilitate land-use-efficient offsetting. The law also provides for a register of compensatory and banking measures and the respective sites. This enables keeping track of and statistically analysing interventions and offsets.

The intervention rule is also relevant for municipal building planning instruments. The municipal planning authorities have to determine whether implementing a legal plan, such as a zoning by-law, would amount to an intervention under the nature protection law. If this is the case, the rules of the Building Code take over and determine the legal consequences of the intervention and the rules for compensation. In these cases, authorities have more flexibility than under nature protection law to decide whether, how and where interventions are compensated, and compensation areas and measures can also be determined in a planning instrument.[64]

Besides the intervention rule, **the State of MV** also has an offset provision with regard to the prohibition to turn over permanent pasture into farmland: an exemption may be granted if the applicant creates new permanent pasture on farmland somewhere else, which may belong to a different owner. The compensation areas have to be in the same county 'as a priority', and of the same size, but do not have to be functionally equivalent.[65]

2.3.2 Area-Based Legal Instruments for Planning and Land Management

At area level, area-based legal instruments include (1) designating specific protected areas in which particular legal restrictions and other requirements apply; (2) legal planning and management instruments that determine, e.g., targets, priorities, permitted land use; and (3) soil information systems:[66]

[63]BVerwG (Federal Administrative Court) judgment of 10.09.1998—4 A 35/97; OVG (Higher Administrative Court) Koblenz, judgment of 06.06.2000—8 C 11556/98.OVG.

[64]§ 18 BNatSchG, § § 1a (2) and (3), § 9 (1a), § 200a BauGB.

[65]§ 3 DGerhG M-V.

[66]Gröhn (2014), p. 191.

(1) Protected areas: the SPA authorises the States to provide in their legislation that their authorities may designate soil protection areas in cases where there is extensive degradation. The States may also provide for 'other area-based soil protection measures'.[67]

MV has used the delegated powers and enacted legislation allowing the designation of soil protection areas. The legislation defines, in a general way, which land-use restrictions, obligations and measures the authorities may prescribe when they designate a particular protected area. These may include that the soil has to be covered or planted.[68] However, for the most part, the list includes restrictions of land use rather than restoration. Therefore, under current legislation in MV, soil protection areas are mainly suitable for preventing (further) degradation but less so for restoration. In addition, precautionary area-based protection is difficult because one condition for designating areas is that extensive degradation already occurs. The powers delegated under the SPA provide more options to MV for area-based soil protection, but so far MV has not made full use of them.[69]

The federal Nature Protection Act also provides for several different types of area-based legal instruments. They can protect soil indirectly if the soil functions are necessary in order to serve the primary purpose to protect nature. To what extent these instruments actually manage land use therefore depends on the content of each individual designation.

Water law also provides for area-based instruments, notably water protection areas and flood areas. Similar to nature protection areas, they can indirectly protect soil. Water protection areas are mainly designated to protect the ground water from pollutants and contaminants and may involve, e.g., restrictions on fertiliser and pesticide use, on cultivation and on turning over pastures.[70] These legal obligations apply to agricultural use and may be stricter than the requirements of good agricultural practice. However, the extent to which water protection areas protect soil is limited because the water protection area and the designated legal consequences have to be necessary to protect *water*. In addition, if the designation restricts agricultural use, the authorities have to pay adequate compensation.[71] This limits the practical potential of using water protection areas for soil protection.

With regard to flood areas, the federal Water Management Act uses a regulatory technique that differs from the SPA: it not only enables the States to identify flood areas; it *requires* them to do so. The Water Management Act also defines the particular legal obligations for flood areas and does not delegate this to the States. Besides prohibitions on building and on designating building areas, specifically with regard to erosion, it is prohibited to change land use from pasture to arable land and from alluvial forest to any other use. However, exemptions are possible, and in

[67] § 21 (3) BBodSchG.

[68] § 9 (2) no. 5 LBodSchG M-V.

[69] Gröhn (2016), pp. 161–164.

[70] Gröhn (2014), p. 353.

[71] § 52 (5) WHG.

order to refuse an exemption, the authority would have to show that the particular degradation would specifically raise the flood risk—which is difficult.[72] In general, the potential of water law for land use and soil management is limited.

Protected areas can be important tools because they are often designated by a statutory instrument. In the hierarchy of rules, such instruments rank lower than formal statutes enacted by Parliament but higher than, e.g., local planning or by-laws.[73] For example, soil protection areas in MV are designated in the legal form of a statutory instrument. Therefore, the legal duties and restrictions determined in that soil protection area have to be complied with by any local planning order or by-law. The same goes, e.g., for nature protection areas and water protection areas.[74]

(2) General legal planning and management instruments: the federal government's Sustainability Strategy 2016 mentions LDN as its highest priority with regard to soil.[75] This is a political goal for which there is no formal legal land planning or management tool specifically on soil protection.[76]

Germany has a complex tiered system of planning laws and instruments. German planning law incorporates environmental concerns, including soil protection at many planning levels. There are legal planning instruments at federal, State, regional and local levels. Some planning instruments are overarching as they address and seek to reconcile all different competing interests, such as infrastructure, mobility and environment. Others address specific issues such as energy, building, nature protection or water. Their legal nature also differs and may, for instance, be binding on third parties or only on internal administrative decision-making. The cross-cutting federal Environmental Impact Assessment Act requires several planning instruments to include and take into account an environmental impact assessment, which includes impacts on soil.[77]

The federal Spatial Planning Act[78] is the starting point for *overarching regional planning* since it is mainly implemented by the States, which are authorised to enact deviating legislation.[79] The Spatial Planning Act aims at 'equivalent' living conditions in the respective spatial areas. Moreover, authorities have to aim at 'balanced' conditions overall, including in environmental terms. This does not mean 'equal' or 'homogenous' conditions. But it does exclude a radical offsetting approach by which only the overall balance of land degradation matters, regardless of whether the quality of soil and land is particularly degraded in some places and particularly

[72]Wasserhaushaushaltgesetz—WHG. See § 76 (2), § 78 WHG.

[73]But federal law of any type overrides State law.

[74]Durner (2005), p. 450; Gärditz (2016), p. 290.

[75]Bundesregierung (2016), p. 198.

[76]Möckel et al. (2014), p. 118.

[77]Gesetz über die Umweltverträglichkeitsprüfung—UVPG. Ludwig (2011), p. 123.

[78]Raumordnungsgesetz—ROG.

[79]The federal government has so far its own powers under the Spatial Planning Act only for planning instruments regarding the Exclusive Economic Zone in the sea off the German coast.

good in others. The Spatial Planning Act stipulates 'objectives' and 'principles' of spatial planning, which have different legal implications. The main legal difference is that the objectives defined in spatial plans have to be *complied with*, whereas principles have to be *taken into account* in subsequent administrative decisions such as granting of permits. Some of the principles in the Spatial Planning Act are relevant for land use and soil protection: for instance, nature, including soil, has to be used sparingly.[80]

In order to ensure a common approach to spatial planning, the relevant ministers of the different States jointly adopted non-binding guiding principles for spatial development. In MV, the State's planning law reiterates the aim of creating equivalent living conditions. Land that is particularly suitable for agriculture has to be preserved and farmed in an environmentally sound manner as much as possible. Damage to natural assets, including soil, has to be remediated as much as possible. These principles have to be taken into account in all spatial planning processes and balanced against other principles.[81] On this basis, MV has adopted a Spatial Development Programme for the whole State.[82]

Specifically with regard to planning instruments for soil, MV's Soil Protection Act states that the highest soil protection authority 'should' develop a 'soil protection program' containing targets and measures for MV. It has to take into account not only existing spatial planning but also 'national and international environmental programmes'.[83] This could provide an interesting legal hook for incorporating SDG 15.3 at the State level. The government of MV has been working on this programme for some time but so far has not published it. Once it is adopted, the State law provides that the soil protection programme will be part of the State's formal planning.

These instruments of overarching regional spatial planning law are a tool for directing and managing how areas are used and for prioritising or subordinating certain land uses vis-à-vis other uses. With regard to agricultural use and erosion, in particular, the steering effect of spatial planning is limited because for this use, there usually is no permit procedure in which the regional planning instruments would have to be taken into account.[84]

At *municipal level*, the available legal planning instruments are the central legal mechanism for steering local development and land use. They may prescribe the type and extent of permitted uses in the areas and individual parcels of land in the municipality. The federal Building Code provides several requirements regarding whether and how these planning instruments have to address environmental and soil protection, e.g. in the elaboration of planning instruments, balancing against other

---

[80] § 2 (2) no. 6 ROG.

[81] § 2 no. 3-4, § 3 Landesplanungsgesetz—LPlG MV.

[82] Landesverordnung über das Landesraumentwicklungsprogramm (LEP-LVO M-V) of 27.05.2016, GVOBl. M-V 2016, p. 322, as corrected in GVOBl. M-V p. 872.

[83] § 11 LBodSchG MV.

[84] Möckel et al. (2014), p. 138.

interests, the permitting process and de-sealing. At a general level, for instance, local planning instruments have to include and take into account an environmental impact assessment. More specifically, several provisions address in particular the growing concern about increasing land usage and sealing. There are rules prioritising inner city development for further building over using forests or agricultural land. Another key provision restricts building in areas that are outside existing plans and already built-up areas.[85] The 'intervention rule' is also relevant to municipal planning, although the authorities have more flexibility regarding compensatory measures. Under certain conditions, the Building Code excludes compensation in developed areas in order to provide an incentive to build there rather than in previously undeveloped areas.[86]

There is also a special 'soil protection clause' that contains particular requirements for municipal planning instruments.[87] It is intended to give soil protection more weight in the planning and balancing process. However, the soil protection clause does not impose absolute restrictions and appears to have not made much difference in practice.[88] A recent amendment to this provision has added an obligation to provide a reasonable justification if a plan intends to change agricultural or forest land use. The justification should be based on an analysis of the potential for using land in already developed areas instead. It remains to be seen whether this regulatory approach of increased transparency could strengthen land management at the municipal level.

Besides the overarching planning instruments, there are *planning instruments for specific subject matters*. In nature protection law, the federal Nature Protection Act provides that landscape planning is mandatory for States. The plans are to determine and justify the objectives and measures for nature protection in the areas to which they apply. Measures determined in landscape planning instruments may apply not only to buildings and construction but also to agriculture. They may include improving soil quality and its rehabilitation, and municipalities may, for instance, prescribe measures to prevent erosion.[89] The State law in MV provides that all levels, from State level down to the municipal level, have to do landscape planning. However, the plans are binding only on the authorities, which have to comply with them or take them into account.[90]

(3) Soil information systems are needed in order to provide the information necessary for pursuing LDN.[91] At national level, in 2015 the Federal Environment

---

[85] § 35 (3) BauGB.

[86] See above Sect. 2.3.1.

[87] § 1a (2) BauGB.

[88] Gröhn (2014), pp. 286–288.

[89] Möckel et al. (2014), p. 140.

[90] § 11 (1) BNatSchG, § 11 (3) NatSchAG M-V.

[91] Wunder et al., "Implementing land degradation neutrality (SDG 15.3) at national level: general approach, indicator selection and experiences from Germany".

Protection Agency prepared a comprehensive report on the state of soil in Germany, and the federal government reports once every electoral term on progress in soil protection.[92] The federal government's revised Sustainability Strategy of 2016 specifically mentions developing better indicators in order to support implementing measures towards LDN.[93]

MV has a mandatory land information system that contains not only contaminated land but also degraded land and potentially degraded land.[94] Federal law also requires MV to classify and register agricultural areas according to erosion risk levels.[95] The register[96] is mandatory, but its background is governance by financial incentives: subsidies under the EU's Common Agricultural Policy scheme are conditioned on, inter alia, applying farming practices that limit erosion. Accordingly, the competent ministry in MV has to inform annually farmers who apply for subsidies about the areas that are vulnerable to erosion.

In addition, MV has established a register of actual erosion *occurrences* on agricultural land and has issued detailed guidance on how to assess erosion by water and define response measures.[97] Citizens can submit occurrences via a website.[98]

3 Assessment and Conclusions

Despite a range of legal provisions and instruments in German law that protect soil, the absence of an overarching holistic concept is a fundamental shortcoming also with regard to LDN. It is apparent, e.g., in the Soil Protection Act's subordination to several other laws and its lack of teeth in respect of implementation and enforcement.[99]

With regard to **preventing** land degradation, there are many legal rules and mechanisms, but they are scattered, and the piecemeal approach does not facilitate implementation. Regarding agriculture in particular, it is legally privileged in several respects. Agricultural land use usually does not require a permit, which means that a number of laws do not apply. Moreover, the standards of conduct to be applied to agriculture are often limited to those of 'good agricultural practice',

[92] UBA (2015); Bundeskabinett (2013). On the current status of and further need for monitoring and indicators specifically for LDN see.

[93] Bundesregierung (2016), pp. 197–198.

[94] https://www.lung.mv-regierung.de/insite/cms/umwelt/boden/geologie_fis_boden.htm, last accessed 17.02.2017.

[95] § 6 AgrarZahlVerpflV, implemented by § 6 ff. AgrarreformUmsetzLVO M-V.

[96] http://www.gaia-mv.de/gaia/feldblockkataster, last accessed 17.02.2017.

[97] Ministerium für Landwirtschaft, Umwelt und Verbraucherschutz Mecklenburg-Vorpommern (Hrsg.) (2016).

[98] https://www.gaia-mv.de/dBAK/Meldung/, last accessed 10.06.2017.

[99] Erbguth and Schlacke (2016), p. 369.

which are vague and difficult to monitor and enforce. The legal situation is better in the case of permanent pasture and subsidy requirements.

The existing rules on **restoration** mainly focus on contaminated land and increasingly on de-sealing. The SPA takes an interesting regulatory approach by providing a range of measures and indicators to assess risks and determine the response and procedure, including in complex cases. However, restoration under the SPA is linked to the land's particular function and does not go beyond averting risks from the degradation.[100]

With regard to **offsetting at project and activity level**, there are few legal rules on offsetting soil degradation in one place with soil improvements in another. The 'intervention rule' in the Nature Protection Act provides a model and long experience in how offsetting can be required and implemented. The register of compensation measures could be a model for managing offsetting more comprehensively at a later stage. Similar to the SPA, however, the intervention rule is not well suited for addressing agriculture because its scope of application is limited.

Although Germany has a complex system of legal planning tools at different spatial areas and sectors, there is currently no specific **legal land planning and management** instrument that ensures a precautionary approach.[101] The SPA is barely suited to provide sustainable and precautionary soil management.[102] At the State level of MV, the government has for years not fulfilled its obligation to adopt a soil protection programme that would be part of the State's legal planning instruments. But it would be possible to enshrine an objective such as LDN in a law.[103]

German law provides for different area-based instruments, all of which require from the outset that the area in question needs to be protected in certain respects—e.g., to protect nature or water quality.[104] Amongst these instruments, protected areas can be important tools because they are often designated by a statutory instrument. However, the effectiveness of protected areas both under the SPA and under the nature protection laws depends on the terms of the respective individual designation.

The law provides only few possibilities to address erosion caused by agriculture because the planning instruments do not apply to agriculture or they have little steering effect.

In general, environmental issues such as soil and land use play a mainly defensive part in legal planning processes: environmental law and considerations *de facto* mainly come into play as restrictions to other uses such as building roads, industrial activities etc.[105] Some argue that in order to go beyond this approach and move towards a holistic eco-spatial management, the overarching regional spatial

[100]Erbguth and Schlacke (2016), pp. 384–385.

[101]Reese (2015), p. 21.

[102]Möckel et al. (2014), p. 117.

[103]Gröhn (2016), p. 159.

[104]Möckel et al. (2014), p. 38.

[105]Gärditz (2016), pp. 290–291.

planning instruments were crucial. Their function was to balance environmental interests with competing other interests—a function that isolated issue-specific planning instruments such as for nature or water protection are unable to fulfil.[106] On the other hand, currently, soil protection and land-use management are just some of the many environmental aspects to be considered in overarching planning processes. **Soil and land often lose in the balancing processes** and are deferred, in particular when there is no formal land planning instrument specifically on soil protection that could give them more legal weight. This would not necessarily change under a holistic eco-spatial management approach. Therefore, as a first step towards achieving LDN, it might be more effective to enable the authorities to use the existing planning law instruments of German law more effectively for the purpose of soil protection.

References

Altvater S et al (2015) Legal instruments to implement the objective 'land degradation neutral world' in international law. German Federal Environment Agency (Umweltbundesamt), Dessau-Roßlau, Germany. Available at: http://www.umweltbundesamt.de/publikationen/legal-instruments-to-implement-the-objective-land

Boer B et al (2017) International soil protection law: history, concepts and latest developments. In: Ginzky H et al (eds) International Yearbook of Soil Law and Policy 2016. International Yearbook of Soil Law and Policy. Springer International Publishing, Cham, pp 49–72. doi: https://doi.org/10.1007/978-3-319-42508-5

Bundeskabinett (2013) Dritter Bodenschutzbericht der Bundesregierung - Beschluss des Bundeskabinetts vom 12. Juni 2013. Bundeskabinett. Available at: http://www.bmub.bund.de/fileadmin/Daten_BMU/Download_PDF/Bodenschutz/dritter_bodenschutzbericht_bf.pdf. Accessed 23 May 2017

Bundesregierung (2016) Deutsche Nachhaltigkeitsstrategie. Neuauflage 2016. Available at: https://www.bundesregierung.de/Content/DE/Infodienst/2017/01/2017-01-11-Nachhaltigkeitsstrategie/2017-01-10-Nachhaltigkeitsstrategie_2016.html. Accessed 23 May 2017

Desai BH, Sidhu BK (2017) Striving for land-soil sustainability: some 1 legal reflections. In: Ginzky H et al (eds) International Yearbook of Soil Law and Policy 2016. International Yearbook of Soil Law and Policy. Springer International Publishing, Cham, pp 37–45. doi: https://doi.org/10.1007/978-3-319-42508-5

Deutscher Übersetzungsdienst der Vereinten Nationen (2015) UN Doc A/RES/70/1 vom 21.10.2016: Transformation unserer Welt: die Agenda 2030 für nachhaltige Entwicklung (deutsche Übersetzung). Available at: http://www.un.org/Depts/german/gv-70/band1/ar70001.pdf. Accessed 11 July 2016

Durner W (2001) Common goods. Statusprinzipien von Umweltgütern im Völkerrecht. 1. Aufl. Nomos, Baden-Baden

Durner W (2005) Konflikte räumlicher Planungen. Mohr Siebeck, Tübingen

Ehlers K (2017) Land degradation neutrality and the UNCCD: from political vision to measurable targets. In: Ginzky H et al (eds) International Yearbook of Soil Law and Policy 2016. International Yearbook of Soil Law and Policy. Springer International Publishing, Cham, pp 73–84. https://doi.org/10.1007/978-3-319-42508-5

[106] Gärditz (2016), p. 299.

Erbguth W, Schlacke S (2016) Umweltrecht, 6th edn. Nomos, Baden-Baden
European Commission (2014) Withdrawal of obsolete Commission proposals, OJ C 153, 21.5.2014, pp 3–7
Gärditz KF (2016) Umweltschutz durch rechtliche Raumstrukturierung. Zeitschrift für Europäisches Umwelt- und Planungsrecht 4:290–299
Gröhn K (2014) Bodenschutzrecht - auf dem Weg zur Nachhaltigkeit: Konkretisierung der Schutzziele und Harmonisierung der Regelungsfülle, 1st edn. Nomos, Baden-Baden
Gröhn K (2016) Flächenhafter Bodenschutz -Steuerungsmöglichkeiten zur Erreichung neuer Nachhaltigkeit. In: Gesellschaft für Umweltrecht ed. Dokumentation zur 39. wissenschaftlichen Fachtagung der Gesellschaft für Umweltrecht e.V. Berlin 2015. Erich Schmidt Verlag, pp 155–165
IAEG-SDGs (2016) IAEG-SDGs—SDG indicators. Available at: http://unstats.un.org/sdgs/iaeg-sdgs/metadata-compilation/. Accessed 2 June 2016
Ludwig R (2011) Planungsinstrumente zum Schutz des Bodens. Duncker & Humblot, Berlin
Minelli S et al (2017) Land degradation neutrality and the UNCCD: from political vision to measurable targets. In: Ginzky H et al (eds) International Yearbook of Soil Law and Policy 2016. International Yearbook of Soil Law and Policy. Springer International Publishing, Cham, pp 85–104. https://doi.org/10.1007/978-3-319-42508-5
Ministerium für Landwirtschaft, Umwelt und Verbraucherschutz Mecklenburg-Vorpommern (Hrsg.) (2016) Erosionsereigniskataster Mecklenburg-Vorpommern. Bodenerosion durch Wasser. Available at: http://www.lms-beratung.de. Accessed 14 Feb 2017
Möckel S et al (2014) Rechtliche und andere Instrumente für vermehrten Umweltschutz in der Landwirtschaft. Umweltbundesamt, Dessau-Roßlau. Available at: http://www.umweltbundesamt.de/publikationen
Peters H-J et al (2015) Umweltrecht, 5th edn. Kohlhammer, Stuttgart
Reese M (2015) Klimaanpassung und Raumplanungsrecht. Zeitschrift für Umweltrecht 15–27
Schmidt R et al (2014) Umweltrecht, 9th edn. C.H.Beck, München
Umweltbundesamt (UBA) (Hrsg.) (2015) Bodenzustand in Deutschland zum "Internationalen Jahr des Bodens". Dessau-Roßlau. Available at: https://www.umweltbundesamt.de/sites/default/files/medien/378/publikationen/bodenzustand_in_deutschland_0.pdf
UNCCD (2014) Land degradation neutrality – resilience at local, national and regional levels. Available at: http://www.unccd.int/en/resources/publication/Pages/default.aspx. Accessed 23 May 2017
UNCCD (n.d.) Land degradation neutrality - frequently asked questions (FAQs). Available at: http://www.unccd.int/en/programmes/RioConventions/RioPlus20/Pages/LDNFAQ.aspx. Accessed 13 July 2016
UNCCD/Science-Policy Interface (2016) Land in balance: the scientific conceptual framework for land degradation neutrality (LDN). UNCCD Science-Policy Interface policy brief. Available at: http://www.unccd.int/Lists/SiteDocumentLibrary/Rio+20/UNCCD_PolicyBrief_ZeroNetLandDegradation.pdf. Accessed 10 Feb 2016
Wunder S et al (2017) Implementing Land Degradation Neutrality (SDG 15.3) at national level: general approach, indicator selection and experiences from Germany. In: Ginzky H et al (eds) International Yearbook of Soil Law and Policy 2017. Springer, Heidelberg, pp XX–XX

The Community Consultation in the Scope of the Land Management in Mozambique

Eduardo Chiziane

Abbreviations

CC	Community consultation involving the local people and authorities and investors
DNT	National Direction of Land and Forest
DUAT	The right to use and explore land acquired by occupation in itself, according to the customary practices or permission of the Public Administration
FCT	Forum for Land Consultation
LC	Local community
LL	Land Law, Law Nr. 19/97, from October
LLR	Land Law Regulation from Decree of the Council of Ministers Nr. 66/98, from December 8

1 Introduction

Consultation with local communities is a mandatory legal procedure in Mozambique that is carried out within the land use right (DUAT) in order to confirm if the land is free and has no occupants.

The need to harmonize the land laws and local State bodies has led to the authorities making changes to the Land Law Regulation regarding the consultation process, by Decree Nr.43/2010 of 20 October, amending Article 27 (2) and the Ministerial Diploma Nr. 158/2011, of June 15, which lays down specific procedures for consultation at community level.

The Land Consultation Forum (FCT), which is the Government's consultation organ for consolidating the policy and regulatory framework for access to the use of land, established by Decree Nr. 42/2010, from October 20, at its First Ordinary Session, held on March 4, 2011, in Maputo, recognized the issue of consultation

E. Chiziane (✉)
Eduardo Mondlane University, Faculty of Law, Maputo, Mozambique
e-mail: eduardo.chiziane@uem.mz

H. Ginzky et al. (eds.), *International Yearbook of Soil Law and Policy 2017*,
International Yearbook of Soil Law and Policy,
https://doi.org/10.1007/978-3-319-68885-5_16

with the local communities as being one of the subjects of land legislation that deserved an in-depth debate on its clarification.

This chapter is the result of this motivation and aims at facilitating the debate on the community consultation model and provides the basic information needed for discussion by the members of the FCT. Added to the analysis performed in the political and legal framework, the present study aims also at summarizing the key issues in impact on secure access to land and the investment promotion.

This chapter addresses the legal framework of the community consultation (Sect. 2); the problems relating to community consultation (Sect. 3); problem analysis and proposed solutions (Sect. 4); the innovations of Ministerial Diploma Nr. 158/2011, from June 15 (Sect. 5); and conclusions (Sect. 6).

2 The Legal Framework of the Community Consultation

2.1 The Legal Texts

Community consultation is regulated by different legal texts. However, we have selected the normative instruments that more directly establish the legal regime of the community consultation in the context of the attribution of land rights, especially the following ones:

- Constitution of the Republic of Mozambique—2004;
- Resolution of the Council of Ministers Nr.10/95, from October 17, which approves the National Land Policy;
- Law Nr. 19/97, from October 1, Land Law;
- Decree of the Council of Ministers Nr. 66/98, from December 8, which approves the Regulation of the Land Law;
- Ministerial Diploma (Ministry of Agriculture and Fisheries) Nr. 29-A/2000, from March 17, which approves the Technical Annex to the Regulation of the Land Law;
- Decree of the Council of Ministers Nr. 50/2007, from October 16, which alters Article 35 of the Regulation of the Land Law;
- The approval of Resolution Nr. 70/2008, from December 30, on Procedures for Presentation and Appraisal of Investment Proposals Involving Land Extensions of More than 10,000 ha;
- Decree of the Council of Ministers Nr. 43/2010, from October 20, which amends paragraph 2 of Article 27 of the Regulation of the Land Law;
- Ministerial Diploma Nr. 158/2011, from June 15, approving the procedures for consulting local communities in scope of DUAT titling.

2.2 *Brief Presentation of the Legal Regime of the Community Consultation Procedure, Its Meaning and Limits*

Article 13.3 of the Land Law prescribes: "The process of titling the right to use the land includes the opinion of the local administrative authorities, preceded by consultation with the respective communities for confirmation that the area is free and has not occupants." Article 27 of the Land Law Regulation 1998 reaffirms that the titling process matters require an opinion of the District Administration and consultation with local communities. But neither the Land Law nor the regulation clarifies anything about the procedures for consultation with the community. Article 27 (2) of the RLT only lists the entities that will do the "joint work"; we believe that such "joint work" is that of the community consultation. In this context, the legislator indicates a set of entities that must carry out the said "joint work," namely, the Geography and Cadastre Services, the District Administrator or its representative, and the local community. This joint work should produce a result that should be written down (community consultation minutes) and signed by a minimum of three and a maximum of nine local community representatives.

The regulation of the consultation processes had to be improved. Therefore, very lately, the Government approved the Ministerial Diploma Nr. 158/2011, of June 15, which approves the procedures for consultation of the local communities in relation to DUAT qualification.

Despite the approval of this Ministerial Diploma, the legislation remains ineffective in achieving its objectives: the land conflicts show in some cases the perverse effect of applying the Land Law and its respective Regulation.[1]

Ministerial Diploma Nr. 158/2011, from June 15, has failed to effectively fill the regulatory gap. This will be explained in the subsequent sections.

3 The Problems Related to Community Consultation

The following main problems can be identified with regard to the community consultation model:

1. clarification of the concept and meanings of local community;
2. the nature of the powers legally attributed to the local community;
3. the administrative or regulatory measures necessary to promote investment in community land;
4. the implications and corrective measures required by Decree Nr. 50/2007, from October 16;

[1]Baleira et al. (2008), pp. 48–49.

5. the eligibility criteria, profile, methods of appointment, and control of LC representatives;
6. the innovations of Ministerial Diploma Nr. 158/2011, from June 15, approving procedures for consultation of local communities under DUAT diploma.

4 Problem Analysis and Proposed Solution

This section addresses problem nos. 1–4 as listed above in Sect. 3:

4.1 *Clarification of the Concept and Meanings of Local Community*

The definition of local community is contained in Article 1 of the Land Law (LL): "Local Community: grouping families and individuals, living in a territorial district of locality or lower, aiming at safeguarding common interests through the protection of housing areas, agricultural areas, whether cultivated or fallow, forests, sites of cultural importance, pastures, water sources and expansion areas." According to the definition, "community" has three elements: the population element, the territory, and the shared socioeconomic interests.[2] This definition applies to land law enforcement.

However, there is another notion of local community in Decree Nr.11/2005, from June 10, which applies to the organization and operation of the local State bodies. Thus, we believe that some conflicts are caused by having several legal definitions of local communities and that the most important notion is the scope of application.

4.2 *The Nature of the Powers Legally Attributed to the Local Community*

Article 24, paragraph 1, of the Land Law—1997, establishes that LCs participate (a) in the management of natural resources, (b) in conflict resolution, (c) in the titling process, and (d) in the identification and definition of the limits of the land occupied by them. In the exercise of the powers referred to in (a) and (b) above, the LCs apply, among others, customary norms and practices (n. 2, of Article 24 of the Land Law—1997).

[2]Quadros (2004), p. 58.

This provision therefore confers on the LCs powers for the management of natural resources, where it can include the power of attribution, modify and extinguish rights to land under customary law.[3]

Section 4.1 showed that an LC is a multifaceted, political, economic, social, or cultural entity. The LCs are consulted in the field of titling, participate in the resolution of land conflicts, etc. However, reality shows that LCs do not play a proactive role in the use of their lands.

The central question at this point is whether the LCs under Article 24 of the Land Law from 1997 can exercise powers in the planning of community land and act, in the allocation of the right to use communal land, in order to guide investment and settlements "to their lands." The power of "natural resource management" attributed to LCs includes the power to establish the land-use plan and to establish rules for the occupation, use, and transformation of community soil. Thus, there is no doubt that LCs pursuant to Article 24 of the Land Law—1997 can allocate, modify, and extinguish rights of third parties to land in the community areas according to a land-use plan previously approved by them, according to customary law, which gives people the right to land use from the occupation process itself, without the need of an official written document of the title and registration. This means that the absence of the title or registration does not mean that there is no right to use the land.

This interpretation could stimulate the effective use of communal lands, combining several options—partnership, assignment for exploration, and allocation of land rights. In the latter case, it is possible to stimulate investment without the need to establish partnerships between third parties and the community.

4.3 The Administrative or Regulatory Measures Necessary to Promote Investment in Community Land

The National Land Policy establishes as one of its fundamental principles "the promotion of foreign national private investment without harming the resident population and ensuring benefits for this and for the national public treasury." One of the most effective mechanisms for making this equation operational is the establishment of economic "partnerships" between private investors and other actors through the implementation of the investment projects. For this purpose, Article 13.3 of the Land Law Nr.19/97, from October 1, establishes a mechanism for consultation with the local communities that, without neglecting the main objective of "confirmation if the (intended) area is free and has no occupants," constitutes a space for dialogue between the holders of the rights acquired by occupation and the new applicants. This may actually result in the establishment

[3]Today is the customary and constitutional right protected by Article 4 of the 2004 Constitution.

of partnerships or in some form of "promotion of domestic and foreign private investment without prejudice to the local population."

For the purposes of establishing partnerships, the Land Law, together with Article 27 (3) of Land Law Regulation of 1998, also provides that "other rights or (the District Administrator's) opinion shall include the terms by which the partnership shall govern between the owners of the right to use land acquired by occupation and the applicant."

The partnership regime is also addressed by Resolution Nr. 70/2008, from December 30, on "the Procedures for the Presentation and Evaluation of the Proposals for investment involving the Land Extension of more than 10,000 ha, which obliges in addition to the Community Consultation Act, the partnerships between the holders of Land Use Rights (DUATs) for occupancy on the intended land and the investor."

Finally, it is still important to point out that this entire legal framework should be used in an interesting and creative way in the specific context of the Law of Public-Private Partnerships (PPP) for an effective and nondiscriminatory promotion of economic development and transformation of rural areas in Mozambique. The Public-Private Partnerships Act aims particularly at large-scale projects, i.e. projects whose value exceed the sum of 12,500,000,000.00 MT (twelve thousand and five hundred million meticais). Its principles include the "adaptation to existing legal frameworks," "fairness in the sharing of benefits accruing to each undertaking between the parties contractors, stakeholders and interested parties or affected parties," the "establishment and the transfer of know-how to workers and Mozambican managers," and the "continuation of programmes, projects or actions of responsibility, sustainability and social development among the communities places."

This means that the law provides real opportunities for involving local communities, including by establishing partnerships for development projects based on land use and other natural resource investments.

The issue of promoting investment and partnerships between the local communities and private investors is also linked to the concepts of spatial planning and land-use plans. The Law for Ordination of the Territory, Law Nr. 19/2007, from July 18, includes principles such as "... legal certainty as a guarantee that in the elaboration, modification and execution of territorial planning and management instruments, the fundamental rights of citizens are legally protected, promoting the stability and observance of the legal regimes instituted."

The legal basis for the elaboration of the Land Use Plans in the community areas by LC is Article 24, paragraph 1, of the Land Law from 1997, which states that the LC participate in the management of natural resources. A proper management of natural resources by LC implies the possibility for them to plan the land use. The Land Use Plan allows to harmoniously define areas for economic, social, private or public, individual or collective initiatives. It also allows defining and establishing the principles and rules for the occupation, use, and transformation of soil.[4]

[4]Calengo (2005), p. 223.

This whole process is fundamental for the establishment of secure partnerships between private investors and local communities as it minimizes potential concerns with regard to the legal limitations of DUAT.

4.4 The Implications and Corrective Measures Required Under Decree Nr. 50/2007 of October 16

Decree Nr. 50/2007 changes the requirements regarding the titling process of the right for use of land that the local communities acquired by occupation.

One of the modalities for acquiring the right to use and explore land is occupation by local communities, according to customary norms and practices that do not contravene the Constitution. Initially, the procedure for titling the community land right included the following steps: community name, demarcation, administrator's opinion, dispatch of the Governor of the Province, and payment of a fee for the procedure.[5]

This legal regime reflected the idea that communities are owners of DUAT, and therefore the titling process only "*seeks to recognize the existing rights in favour of communities*."[6] Land legislation encourages the registration of these rights, especially in communities where there are investment projects and conflicts over land or in cases of applications submitted by the communities themselves.

The previous system permitted the local communities to acquire DUAT by occupation with relative bureaucratic ease, according to customary norms and practices, rights guaranteed by Article 12, paragraph a), of Law Nr. 19/97, from October 1, and Article 35 of the Regulation of the Land Law, approved by Council of Ministers Decree Nr. 66/98, from December 8; the process was fully carried out at local level, including an administrator's opinion and a final order of recognition by the Provincial Governor.

However, the lack of a title (Article 13 of Land Law Nr. 19/97) and/or registration (Article 14 of Land Law Nr. 19/97) is without prejudice to DUAT acquired by occupation, under the terms of subparagraph a), Article 12, of the Land Law Nr. 19/97. If necessary, or at the request of the local communities, the areas where land is acquired by occupation according to customary practices may be delimited and entered in the National Cadastre of Lands, Article 9, Nr. 3, of the Land Law Regulation, Decree Nr. 66/98, from December 8.

[5]Article 35 of the Decree of the Council of Ministers Nr. 68/98, from December 8. Published in the Bulletin of the Republic, Nr. 48, Series I, third Supplement of December 08, 2008.

[6]Chiziane et al. (2008), p. 41 and stgs.

This regime was radically changed with the approval of the Council of Ministers Decree Nr. 50/2007, from October 16.[7] Its single article changes Article 35 d) of the Regulation of the Land Law from 1998 and provides that the process related to DUAT titling will contain a "*decision of the competent authority for the area, as defined in a) of paragraph 1, a) of paragraph 2 and paragraph a) of paragraph 3 of article 22 of Law nr. 19/97, from October 1*."

According to the previous rule, in point d) of Decree of the Council of Ministers Nr. 66/98 of December 8, only the Governor's decision could recognize that the land belonged to the community, regardless of the size of the area.[8] However, under the current rule introduced by Decree Nr. 50/2007, from October 16, one decision is taken by the competent authority of the area, and a different authority has to approve the titling (Governor, Minister of Land, and Council of Ministers), which raises concern about whether or not this new measure is in accordance with the Land Law from 1997.[9]

Given that DUAT confirmation of the communities merely recognizes a right that already exists, the decision-making power should be maintained at the level of the Governor because this person is close to the local communities. DUAT confirmation should not include the intervention of the Council of Ministers because this entity is far from the communities, which impedes communication. Basically, the new regime involving different authorities in granting DUAT, depending on the size of the required area, could make sense for such DUAT applications. However, this case is about the mere recognition of an existing right.

The new rule introduced by Decree Nr. 50/2007, from October 16, appears to be contrary to Article 250 (1) of the Constitution, which establishes the principle of deconcentration. It also appears to violate the principles of reducing bureaucracy and deconcentration established by Law Nr. 8/2003.[10] The Council of Ministers is a distant administrative and political organ in relation to the local communities.

An additional argument relates to the form of the recognition of DUAT of communities, established by the single article of Decree Nr. 50/2007, from October 16, which is by "decision." The authorities indicated in Article 22 of Law Nr. 19/97, from October 1, may grant or refuse the claim of the community based on criteria

[7]Approves the amendment of article 35 of the Regulation of the Land Law, approved by Decree Nr. 68/98 from December 08. Published in the Bulletin of the Republic, Series I, Nr. 41, eight Supplement, from October 16, 2007. The legal validity of this Decree was widely questioned during the 10 year Commemorative Conference of the Land Law, co-organized by Centre for Judicial Training of the Ministry of Justice, UEM Faculty of law, National Union of Peasants (UNAC), October 17–19, 2007, in Maputo. See Final synthesis of the Conference (not published).

[8]For acquisition of DUAT on request Land Law No. 19/97, article 22 distributes the powers according to the size of the area required, as follows: paragraph 1, point a), the Provincial Governor authorizes requests for DUAT of areas Up to 1000 ha, ... (2), point (a), the Minister of Agriculture authorizes DUAT applications for areas between 1000 and 10,000 ha ... en 3, line paragraph a) the Council of Ministers authorizes applications in areas beyond the maximum limit of the Minister of Agriculture ....

[9]Chiziane et al. (2008) op.cit., p. 42.

[10]Chiziane (2009), p. 31, in www.ul.pt.

such as expediency, which provides them with broad decision-making powers. This seems contrary to the existing Law of Land, which recognizes rights to access and use of land for communities without setting such restrictive conditions for the approval of the enjoyment of such right. This calls into question the legal validity of Decree Nr. 50/2007, from October 16.

Since Decree Nr. 50/2007, from October 16, was approved, the titling of the community areas, which was already difficult, stagnated. In fact, there is no indication that a community's land has been recognized since.

In the interest of good land management by the communities, we believe that Decree Nr. 50/2007, from October 16, should be revoked. We recommend the restoration of the previous regime, where the Provincial Governor would be the competent authority and has the exclusive power for the issuance of the final order of DUAT titling in favor of the communities.

5 The Innovations of Ministerial Diploma Nr. 158/2011, from June 15, Approving the Procedures for Consultation of Local Communities Under DUAT Titling

The Community Consultation under the old regime of 1988 involved Cadastre Services, their representative, and the owners or occupants of adjacent lands. In the regime of 2011, the Community Consultation involved the representative of the Cadastre Services, members of the village and locality, the Advisory Councils, owners or occupants of adjacent lands. In terms of process, the old regime was not clear and lacked a clear indication of the legal consequence of noncompliance with the consultation procedures. The new regime has two phases. The first consists of a public meeting to provide information to LC on the request to acquire DUAT and the identification of limits of the plot. The second is to be held 30 days after the first meeting, with the objective of LC's pronouncement on the availability of the area and the realization of the undertaking or exploration plan.

The 2011 Diploma introduced changes to the consultation procedure, which are summarized in the following table

Relevant aspects	Old regime—Procedures, consultation of LCs/RLT-1998	New regime—Procedures, consultations with LCs/Decree Nr. 43/2010 and DM Nr. 158/2011
1. Participants during the consultation	(1) Cadastre Services, (2) District Administrator or its representative, (3) local communities, (4) owners or occupants of adjacent lands (Article 27, paragraph 2, RLT)[a]	(1) Administrator of the District or its representative, (2) representative of the Cadastre Services, (3) members of the Village and Locality Advisory Councils,

(continued)

Relevant aspects	Old regime—Procedures, consultation of LCs/RLT-1998	New regime—Procedures, consultations with LCs/Decree Nr. 43/2010 and DM Nr. 158/2011
		(4) members of the local community, (5) owners or occupants of adjacent lands, and (6) the applicant or his representative
2. Composition of local community representation	A minimum of three (3) and a maximum of nine (9) representatives of local community (Article 27, Nr. 2, of the RLT)	By three (3) to nine (9) representatives of LC, Decree Nr. 43/2010
3. Phases of the consultation	The LC Consultation did not have a clear indication of the phases. It merely provided that it was a joint work, which could be translated into holding meetings, and there was no clear indication of the objectives of the joint work. According to Article 27, paragraph 2, of the RLT-1998, "joint work will be carried out involving Cadastre Services ... and LCs" ... Article 13.3 of the LT-1997 indicates the purpose of the community consultation: "*for effects of confirmation if the area is free and has no occupants.*"	The Local Community Consultation comprises two phases: the first consists of a public meeting to provide information to LC on the request to acquire DUAT and the identification of limits of the plot. The second is to be held 30 days after the first meeting, with the objective of LC's pronouncement on the availability of the area and the realization of the undertaking or exploration plan (Article 1 of the Ministerial Diploma Nr. 158/2011)
4. Power for signature of the minutes of consultation	It is attributed to the representatives of LC and to the owners or occupants of the bordering lands, according to Article 27 of the RLT-1998	It is attributed to the members of the Consultative Councils of Population and of Locality, Article 2, Nr. 2, of Ministerial Diploma Nr.158/2011
5. Validity of consultations	It lacked a clear indication of the legal consequence of noncompliance with the consultation procedures by RLT-1998. However, by Article 361 of RAU, the breach of procedures had the effect of annulling the act of consultation	Consultations that do not respect the procedures are not valid, Article 5 of the Ministerial Diploma Nr. 158/2011
6. Financing of the consultation process	In RLT-1998, there was no indication of who was responsible for financing the consultation process	The Consultation process is financed by the applicant according to Article 4, paragraph 1, of Ministerial Diploma Nr. 158/2001. "At the beginning of the consultation process, the applicant deposits a payment amount from the costs relating to the consultation procedure are deducted."

[a]Land Law Regulation

Further meetings may be held where additional information is provided to the local community (Article 1, paragraph 2, of Ministerial Diploma Nr. 158/2011). A copy of the Minutes of Consultation, after the opinion by the District Administrator, is delivered to the local community, according to Article 2, paragraph 3, of Ministerial Diploma Nr.158/2011, from June 15.

The Decree of the Council of Ministers Nr. 11/2005, from June 10, on the composition and principles of constitution of the Advisory Councils of Population and Locality establishes the following procedure.

Community participation and consultation is implemented through the local councils at the levels of district, administrative post, town, and village (Article 117). The local councils, the community authorities, the representatives of economic, social, and cultural interest groups chosen by the local councils or forum of lower echelon in proportion to the population of each territorial level are members. The head of each local body can invite influential persons from civil society to join the local council in order to ensure the representation of the various actors and sectors. The representation of women should never be less than 30% (Article118).

The local council of locality is composed of a minimum of ten (10) and a maximum of twenty (20) people (Article 119). Decree Nr.11/2005 does not establish the composition of the local town council.

The new procedures for consultation with of local communities in the DUAT process established by Ministerial Diploma Nr.158/2011, from June 15, fill important gaps in the previous procedure. However, the new rules do not establish rules for designating community representatives or for the accountability of the community representatives to the members of the local community.

6 Conclusions

1. LCs under Article 24 of the Land Law—1997 may assign, modify, and extinguish rights to land in the community areas to third parties according to a land-use plan previously approved by them under customary law.
2. Decree Nr. 50/2007, from October 16, should be revoked, and it is recommended that the previous regime be restored, where the Provincial Governor would be the exclusive authority in the issuance of the final communities.
3. The new procedures for consultation of local communities within the framework of DUAT qualification established by Ministerial Diploma Nr.158/2011, from June 15, 2011, do not lay down rules on the methods of designating the community representatives or the mechanisms for monitoring representatives of the community by members of the local community.

References

Baleira S, Tanner C et al (2008) Pequisa sobre parcerias entre comunidades locais e investidores do sector privado, Matola

Calengo A (2005) Lei de Terras anotada e comentada. CFJJ, Maputo

Chiziane E (2009) The trends of re-concentration and re-centralization of the administrative power in Mozambique, p 31
Chiziane E et al (2008) Relatório de Assistência Jurídica e Organizacional do Programa Chipanje Chetu – Distrito de Sanga, Provincia de Niassa. WWF, Maputo
Quadros M (2004) Manual de Direito de Terra. CFJJ, Maputo

Reports

MINAG (DNTF) e MCA –Moçambique «Avaliação das necessidades da Administração de terras», Outubro, 2010.
MINAG (DNTF) e MCA –Moçambique «Analise ao Quadro de Politicas e Legislação sobre terras de Moçambique: Proposta de Roteiro para o Processo de Consulta», Outubro, 2010.
MINAG (DNTF) e MCA –Moçambique «Analise Institucional e Planos de Trabalho para Capacitação», Março, 2011.
MINAG (DNTF) e MCA –Moçambique «Contribuição para a Estratégia Nacional de Administração de Terras», Março, 2011.

Legislation

Constituição da República de Moçambique – 2004.
Resolução do Conselho de Ministros nº10/95, de 17 de Outubro, aprova a Politica Nacional de Terras.
Lei nº 19/97, de 1 de Outubro, aprova a Lei de Terras.
Diploma Ministerial (Ministério da Agricultura e Pescas) nº 29-A/2000, de 17 de Março, aprova o Anexo Técnico ao Regulamento da Lei de Terras.
Decreto do Conselho de Ministros nº 66/98, de 8 de Dezembro, aprova o Regulamento da Lei de Terras.
Decreto do Conselho de Ministros nº 50/2007, de 16 de Outubro, Altera o artigo 35 do Regulamento da Lei de Terras.
A aprovação da Resolução n° 70/2008, de 30 de Dezembro, sobre os Procedimentos para Apresentação e Apreciação de Propostas de Investimento Envolvendo Extensão de Terra Superiores a 10.000 hectares.
Decreto do Conselho de Ministros nº 43/2010, de 20 de Outubro, Altera o nº 2 do artigo 27 do Regulamento da Lei de Terras.
Diploma Ministerial n° 158/2011, 15 de Junho, que aprova os procedimentos relativos a consulta às comunidades locais no âmbito da titulação do DUAT.

Who Owns Soil Carbon in Communal Lands? An Assessment of a Unique Property Right in Kenya

Nelly Kamunde-Aquino

1 Introduction

One of the main and most impactful ways of intervening against climate change is to mitigate its effects by reducing greenhouse gas (GHG) emissions,[1] which cause global warming. The most notorious of these GHGs is carbon dioxide (mostly shortened as 'carbon'), whose effect of global warming is exceedingly high in comparison to the other GHGs. Importantly, carbon is a useful building block of all life on earth, which exists in various forms in the carbon cycle. The problem occurs when carbon is located in places where it is not needed, a situation that is caused by excessive human activities, which in turn cause the carbon to be emitted into the atmosphere spawning unpleasant unpredictability in weather patterns, drought, drops in water body levels, unexpected climatic disasters, among others.

There have been many international and national efforts to combat climate change. The key convention, under which climate change mitigation activities, among others, are carried out, is the United Nations Framework Convention for Climate Change (UNFCCC). One of its key objectives is the stabilisation of greenhouse gas concentrations in the atmosphere so as to avoid climate interference.[2]

[1]Green House Gas (GHG) gases are those which absorb and emit radiations. They include: Water vapor (H. 2O); Carbon dioxide (CO) Methane (CH), Nitrous oxide (N. 2O), Ozone (O), Chlorofluorocarbons (CFCs), Hydro fluorocarbons (incl. HCFCs and HFCs).

[2]UNFCCC Art.2.

N. Kamunde-Aquino (✉)
Department of Private Law, Kenyatta University School of Law, Nairobi, Kenya
e-mail: nelly.kamunde@gmail.com

H. Ginzky et al. (eds.), *International Yearbook of Soil Law and Policy 2017*,
International Yearbook of Soil Law and Policy,
https://doi.org/10.1007/978-3-319-68885-5_17

This is supplemented by the Kyoto Protocol.[3] Kenya is party to the UNFCCC and the Kyoto Protocol, which it ratified in 1994 and 2005, respectively. The Kyoto protocol does not have specific provisions on soil carbon emission reduction, but it mentions generally on the role of agricultural land in emission reductions.[4] However, it was proposed at the time of drafting its text that progressive efforts are going to be made in order to set up specific guidelines and rules as far as agricultural soils and land-use change is concerned.

Kenya is also a party to the Paris Agreement, 2015, and has taken steps towards its implementation in a number of areas, including agriculture.

2 The Link Between Climate Change, Carbon Rights and Community Land Tenure

Climate change activities that target soil carbon focus on limiting the release of soil carbon into the atmosphere through various activities such as sustainable agriculture through initiatives such as climate smart agriculture, alongside other possible emission reduction and sequestration[5] efforts such as those of reducing emissions from deforestation and forest degradation (REDD), where the focus is to maintain carbon in the tree and prevent it from being emitted into the atmosphere, and others such as Clean Development Mechanisms (CDM).[6]

There are a variety of ways through which the economisation of soil can be achieved. For instance, in order to reap the benefits of monetising climate change interventions through soil carbon, there has to be certain tools that measure the 'intervention' so as to justify the basis of economic gains. The interventions count when we speak of climate change reservoirs that either sequester carbon from the atmosphere or retain it in the soil. These interventions may include a broad spectrum of interventions such as sustainable land management, the enhancement of soil organic matter, among others, which may be specifically coined into programmes and structures such as land use and land-use change (LULUCF), sustainable agriculture and land-use management (SALM), among others.

Soil carbon (besides other forms of carbon labelled as ocean carbon, forest carbon, atmospheric carbon, etc.) is interesting in this discussion as it contains the second-highest volume of carbon sinks.[7] Scientists and policy makers have

---

[3]See generally the Kyoto Protocol (1995).

[4]Article 3(4) Kyoto Protocol.

[5]This relates to soil carbon capture from the atmosphere and its long term storage.

[6]CDM mechanisms are one of the market based climate change approaches that were introduced under the Kyoto Protocol. The mechanism functions like a form of Foreign Direct Investment (FDI) for emission reduction projects that are situated in Africa and whose payment and establishment comes from developed countries.

[7]The term carbon sink relates to a situation where the amount of carbon that is emitted is lesser than the one that is absorbed. It also relates to a situation of acting as a carbon reservoir.

observed that maintaining soil carbon sinks could tremendously reduce carbon emissions. Carbon in the soil is an important component of the soil without which what we know as soil matter would not be in existence. The problem comes in when that carbon is released into the atmosphere from various activities such as draining of peats, cutting down trees, negative unsustainable agricultural activities such as burning of vegetation, among others.

In order to curb the problems that are specifically caused by soil carbon emissions, various countries have engaged in a number of steps, including agroforestry, effective grazing, land management and land restoration, among others. Other activities include those that increase carbon sinks, primarily through sequestering carbon in the soil.[8]

It is important to mention the centrality of agricultural practices to climate change and the opportunity that good agricultural practices avail towards reducing carbon emissions. It has been observed that from all sources of soil carbon, agriculture accounts for approximately 14%[9] of the global anthropogenic greenhouse gas emissions.[10] Importantly, also, almost all climate change interventions that stream from agriculture target soil carbon. It is also important to note that agriculture lies fairly low when it comes to financing climate-relevant projects in the agricultural sector.[11]

Soil carbon would also be attractive to Africa and to Kenya specifically since it seems to be a double-edged sword. Farmers would benefit by better farming methods, therefore a higher yield, and they would also get support because of engaging in a climate change mitigation initiative. This is contrary, for instance, to mere forest conservation.[12]

Those who want to invest in carbon markets by undertaking agricultural-based activities are strongly affected by the tenure regime that obtains in a jurisdiction. For instance, investors will shy off a system where ownership rights cannot be ascertained and whose benefit distribution system is not guaranteed, as happens often in the case of communal land tenure. Take, for instance, an investor who wants to initiate a soil carbon project over a hundred hectares whose land titles cannot be accessed.

The complexity of the concerted efforts of reducing soil carbon emissions (and other carbon emissions generally) comes in when we speak of the right and duty holders. For starters, there is undoubtedly a global benefit to all people for averting climate change. In other words, mitigating against the effects of climate change is for public good. However, the interventions that need to be made, say for instance by the State, must be made all over: in private and public spaces and by public and

[8] *ibid.*

[9] Foucherot and Bellassen (2011), p. 2.

[10] *Ibid.*

[11] GOK (2014).

[12] Ringius (1999), p. 5.

private persons. There is no challenge presented when the State makes interventions for climate change in public places like national parks, public forests or public bare lands.

The challenge comes when there is need to ensure that private rights over land[13] are tampered for the sake of climate change, for instance by limiting one's rights to cut trees, burn vegetation or drain out mangrove peats. When it comes to forest carbon, for instance, for certain jurisdictions, the policy makers create room for State intervention, which has the consequence of limiting the private holder's rights.

In Kenya, the landowners would fall in either of three categories[14]: public landowners (including State agencies, parastatals, statutory bodies, among others, which hold the land in trust for the public), private landowners (who hold the land as individuals, and the land is registered in a person's name, company's name, etc), communal lands (where land is held collectively by a group of people). The most obvious way to conclude this discussion would be to say that soil is inextricably linked to soil carbon and to land, and therefore, automatically, he who has ownership over the land has the ownership of the carbon. However, as will be discussed later, this would be oversimplifying the problem to a point of rendering the status of carbon inconsequential. This is because of the inherent nature of carbon and also the fact that the consideration is not for carbon *per se* but on the carbon emission reduction.

Kenya faces a further peculiar problem. Even if a simple link would be made where communal lands yield communal rights over communally owned lands, and the same applies to the other forms of tenure (public and private), there would still be difficulty of ascertaining whether to consider carbon as real or personal property, and consequently the appertaining rights of use, dominion or control. As far as the assignment of an owner is concerned, Kenyan Law assigns land rights in their absolute form to the holder of the land (subject of course to statutory limitations). In this way, community landowners should be the automatic owners of related carbon rights. It is also important to note that there are other inherent challenges of investing in communal lands in Kenya, which have led to people shying away from investing in community lands. This includes management issues over community lands, gender-related inequality, non-formalisation of communal lands and, most recently, loose legal framework and poor transitional law design between the Trust Lands Act and the Land (Group Representatives) Act on the one hand and the Community Land Act (hereafter CLA) on the other hand. These challenges if not well addressed and prepared for, will only worsen the peculiar nature of assignment of carbon rights when it comes to carbon sequestration initiatives. It is important to

---

[13]With regard to this discussion, one can consider private rights for Kenya to include all rights that are not public, including communal land rights.

[14]Article 61, Classification of land: (1) All land in Kenya belongs to the people of Kenya collectively as a nation, as communities and as individuals. (2) Land in Kenya is classified as public, community or private.

state that this situation is reflective of many African countries, where just like in Kenya 60% of the land mass falls under communal landownership.[15]

Following the recent developments of enacting the CLA in 2016, the Climate Change Act, the ongoing discussions on carbon finance policy, the progressive efforts towards enacting the REDD+ strategy for Kenya, the drafting of the regulations under the CLA, the development of the Medium Term Plan (2017–2020),[16] it is true to say that the time is ripe to discuss where Kenya is going in terms of strategic policies and laws. The discussions herein also call for approaches that may be used to chat a way forward as far as carbon rights over community land are concerned. As at the time of writing this paper, the relevant rules under the Community Land Act were being discussed under the Ministry of Lands. As such, it is hoped that the discussions herein will influence the discussion of defining the property rights that touch on carbon from a juridical point of view.

3 Justification for Assessing Soil Carbon Property Rights

The issue of rights arises with regard to soil carbon in a number of ways, one of which is the assignment of benefits. Under the Kyoto Protocol, member States have emission reduction commitments[17] for developed countries, which call upon them at times to engage in carbon credit trading schemes with developing countries. For Kenya, most of the schemes are based on clean energy such as wind energy, hydroelectric production, reforestation and afforestation, among others. While the Kyoto Protocol indicates that there can be GHG reduction through land use, land-use change and forestry (LULUCF), which may contribute to the exploitation of soil carbon, there are presently no assigned targets for developed States to use these methods for achieving their targets.

In the Kenyan project on soil agricultural land-use management (SALM),[18] the farmers were facilitated by the BioCarbon Fund under the World Bank, enabling them to participate in the carbon market through engaging in sustainable agricultural activities. This does not prevent the farmers to sell the credits to another purchaser. The proprietary interests here do not cause any obvious challenges since most of the land parcels were registered as private land, and therefore the benefits were distributed in accordance with pre-agreed sharing agreements between the registered owners. The situation would be different if the land was owned communally as in the case of community land.

[15]Kameri Mbote (2005), p. 4.

[16]Medium Term Plans (MTP), refers to the periodical progressive time plans aimed at achieving Kenya's vision 2030(a policy based blue print for economic realisation).

[17]This is a task force that was launched pursuant to Section 4(e) of the Community Land to prepare specific laws under the Act that will among others, guide the registration process.

[18]World Bank (2013).

4 The Legal Status of Carbon

The discourse of soil carbon forms part of the general discussion of property rights over carbon that can be said to be a fundamental aspect of the climate change debate.[19] Carbon has been considered as new property, which does not necessarily follow the traditions of real and personal property as we know it.[20]

The need to clarify carbon rights goes back to the general agenda for States to reduce their GHG emissions[21] through a variety of mitigation alternatives. Climate change mitigation efforts take a variety of forms, some of which incorporate the marketisation of emission reduction (for instance through emissions trading schemes). Introducing a market component into mitigation efforts was debated and considered through intense inquiry into the environmental economic theory. This paper acknowledges that this form of climate change mitigation is not without its fair share of critics.

The use of trading schemes for carbon most definitely introduces the question of carbon property rights.[22] Let us, for instance, take the carbon credits earned from a sustainable soil management project. Who has the duty to make the intervention, and who will benefit from it? The preliminary response to these questions may seem astonishingly simple. However, there are many subsequent questions that need clearer answers. Some of those questions include: does the State have priority rights and duties of making mitigation interventions since carbon is given, in some instances, the status of a national natural resource?[23] Can the landowner decline to participate in such an intervention? Does the government have any right to claim from the sequestration that is carried out over privately owned lands such as communal lands or private lands?

Part of the questions that are under consideration, and which have been thought of by other scholars,[24] is if a person engages in a sequestration exercise that touches on soil carbon and obtains some carbon credits, whether that should be considered as a property right, who the proprietor of that right will be and against whom the right is enforceable.

A specific assessment of soil carbon is important for a number of reasons. First and as stated earlier, there are huge amounts of carbon in the soil. Second is the fact that the release of soil carbon out of negative agricultural practices poses a further

[19]Barnes and Quail (2009).

[20]Charles (1990), p. 225.

[21]Kyoto Protocol.

[22]Allan and Baylis (2006), p. 104.

[23]Carbon is listed as a gaseous Mineral under the Mining Act, Laws of Kenya.

[24]See for instance Allan and Baylis (2006).

real threat such as food security and also the cumulative consequences of climate change. This is particularly the case since agriculture is the main economic activities for many African countries, whereby up to a third of its GDP comes from agriculture and about 40% of its export is agriculture based.[25]

4.1 The Legal Status of Soil Carbon Under Kenyan Law

A discussion of soil carbon tenure rights and its conclusion in Kenya can only be deduced from statutory interpretation and analysis as there is still no law that amply describes soil carbon. Kenya was the first to obtain carbon credits issued under sustainable agricultural land-use management (SALM) carbon accounting methodology, under the Kenya Agricultural Carbon Project. This provides a practical way, to the extent possible, to discuss the relevant issues on soil carbon.

In discussing soil carbon tenure in Kenya, Chapman, Kamunde *et al.*[26] identify that under Kenyan law, soil carbon can either be considered to be part of the soil, as a collectivity that forms the land (belonging to its rightful owners), or be considered as a special resource that belongs to the State and that is subject only to the control and ownership of the State.[27]

In the ordinary course, the former is mostly the situation that obtains in Kenya, whereby the rights of the owner over the land are absolute, with exception only to certain specifics in law such as minerals under the Mining Act and Aviation Laws, among others. The unprecedented nature of clarifying soil carbon cannot, however, presume this simple analogy. One of the reasons for a further inquiry as far as carbon is concerned is the fact that what is valued and eventually monetised is not the carbon itself but the carbon emission reduction. This calls for a consideration as to who has the right to engage in soil carbon emission reduction and who benefits from its reduction, where benefits are available. In order to consider the issue of soil carbon tenure, the Kenyan legal situation avails three options: constitutional interpretation, common law and statute law analysis. Each of these will be considered below in turn.

4.2 Carbon Under the Constitution and Statute Law

Like many countries, the Constitution does not mention the word 'carbon' in its text. In defining land, the Constitution states certain content that is intrinsically found in the land and that is part of the soil, which would constitute land, including

[25]Ringius (1999), FAO (1996), FAO (1996).

[26]Sophie et al. (2014), p. 110.

[27]Section 6, Mining Act, (2016).

rocks, minerals, fossil fuels and other sources of energy.[28] This would be perhaps the most likely place to mention whether soil carbon can be part of the 'land ingredients'. However, despite the specific mention of 'soil carbon', the Constitution still have some guidance that will shed light to how soil carbon may be considered.

Proprietary right arises out of what the Constitution defines as property, and that is 'any vested or contingent right to, or interest in or arising from- ... land ...'.[29] Land is then defined through a listing of what it contains, and this includes natural resources completely contained on or under the surface,[30] and natural resources is defined to include rocks, minerals, fossil fuels and other sources of energy. The question to this chain of definition leads us to say, at least as far as the text 'carbon' is concerned, that it can be considered in two possible ways in the Constitution. Firstly, it is a natural resource that is found in land (under the surface), and therefore the owner of the carbon is automatically the owner of the communal land, private land or public land. Secondly, it can be considered that carbon, as far as definitions go, is a mineral, which is a State resource, in light of the chain of text and phrase definitions below.

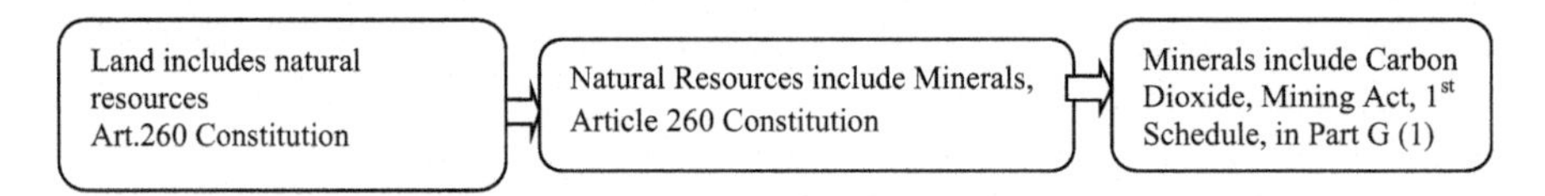

In this scenario, the undisputable situation would be that carbon is an exclusive property of the State, held in trust for the people, as stated in the Constitution.[31]

Another way of looking at the State's proprietary interest over soil carbon is through the limitations that statutes like Petroleum (Exploration and Production) Act[32] provide, which give priority rights to the State. The Mining Act[33] further states that every mineral is the exclusive property of the State, vested in the national government and held in trust for the people of Kenya.[34] This position has also been

---

[28]Article 260, Constitution of Kenya.

[29]See Constitution definition of Property under Art. 260.

[30]Section 260 of the Constitution of Kenya, the interpretations section, states the following: "land" includes:

(a) the surface of the earth and the subsurface rock;
(b) any body of water on or under the surface;
(c) marine waters in the territorial sea and exclusive economic zone;
(d) natural resources completely contained on or under the surface; and the air space above the surface.

[31]Article 62(2), Constitution of Kenya.

[32]Petroleum (Exploration and Production) Act, Cap 308, Laws of Kenya, which states that all petroleum that lies in the earth's strata belongs to the government.

[33]Mining Act; Trading in Unwrought Precious Metals Act, Cap.309; Diamond Industry Protection Act, Cap.310.

[34]Article 6(c) Mining Act.

affirmed by judicial interpretation, where, for instance, the court in the case of Oil Turkana Limited was sharp in holding that the company had no absolute drilling rights.[35]

The State may also claim a right to make interventions and limit individual property rights on the basis that such limitation is useful for the public good in reducing climate change. For instance, the State may introduce rules on soil governance, which would be put in place in order to maintain the quality of soil and to ensure productivity and food sustainability. In order to introduce soil governance, there is a need to appreciate tenure as the key issue relating to how property rights are considered. In the Kenyan context, there is a need to consider the tenure options as far as soil is concerned.

4.3 Carbon Under Common Law

Another way of considering carbon tenure in Kenya is under the principles of common law, to which Kenya subscribes by virtue of the Judicature Act.[36] The understanding of common law would mean that 'whoever's is the land, it is theirs all the way to Heaven and all the way to hell', as the Latin maxim translates.[37] The Constitutional definition of land and natural resources, discussed above, gives further affirmation that soil carbon is an intricately joint part of the land.

Common law jurisdictions embrace the maxim that 'whatever is attached to the land becomes part of the land'.[38]

There is no doubt that soil is inextricably linked to the land and that soil carbon is an imperative component of the soil. Australia, like Kenya, considers that whoever owns land, owns all the resources thereon, including the carbon sequestration potential of the land. The crown in Australia (like the State in Kenya), however, reserves certain controls and ownership over minerals below the surface.[39]

With regard to the legal status of carbon internationally, and in comparison with other countries, different jurisdictions have different ways of perceiving carbon credits. These are the options that give rise to the considerations in this article, in that considerations must be made of the kind of legal context that the pre-existing laws avail so that one does not offend them in charting out a future discourse for climate change interventions.

The upshot of the common law understanding is that of the 'bundled rights system' where the amalgam of interests are considered collectively. The special

---

[35] *Oil Turkana Limited (Previously Known As Turkana Drilling Consortium Limited) & 3 Others V National Oil Corporation Of Kenya & 4 Others* [2013] eKLR.

[36] S.3 Judicature Act, Laws of Kenya.

[37] '*cuius est solum, eius est usque ad coelum et ad inferos'.*

[38] This reference is normally identified in Latin as' *quicquid plantatur solo, solo cedit.*

[39] Cite the law and property rights forest/in Ongoing publications....

nature and purpose of carbon emission reductions, however, presents likelihood to overhaul this position, which overhaul can only be justified by the exigencies prevalent in the Kenyan jurisdiction.

5 Communal Land Tenure in Kenya

> Unless the reality of land tenure arrangements are taken into account and adequate tenure security is provided for all stakeholders, the introduction of climate change mitigation measures may fail to meet their objectives. Indeed, they may cause harm.

While there are three forms of land tenure in Kenya (including public and private), this paper focuses on communal land tenure in affirmation of the fact that communal landholdings consist of over 60% of all landholdings in Kenya.[40] These lands constitute arid and semi-arid lands (ASALs), where agricultural activities are minimal, and this to a certain extent reduces the agriculturally related risks of soil carbon depletion as a result of negative agricultural practices. This situation could be the basis of seeking to make interventions to preserve soil carbon in communal lands. It is also observed that the nature of communally held lands limits the risks of subdividing the lands into minimum acreage holdings, which is a determent for soil carbon investments for climate change. To exemplify this scenario, the Kasigau REDD+ project in Kenya[41] is able to derive benefits, partially because of the fact that the concerned land is held communally under group ranches, which provide for opportunities to exploit large tracks of land.[42]

This therefore means that in order to exploit the market-based alternative for climate change mitigation with regard to soil carbon, the potential for impactful interventions that is presented by the community lands in Kenya must be considered.

Exploiting the property rights over community land is also important in light of the risk of rights disenfranchisement that the community land dwellers are prone to. Most occupiers of community land have maintained traditional lifestyles and at times suffer economic and educational marginalisation. The landholding of

---

[40] GOK (2009).

[41] http://www.wildlifeworks.com/saveforests/forests_kasigau.php (last accessed on 19th June 2017).

[42] The use of the term 'group ranch' pertains to the formalisation of community land in the now repealed Land (Group) representatives Act. The lands that are registered here, are required by law to transition into the Community Land Act of 2016.

community land is also founded on traditional and customary basis[43] or one's belonging to a group[44] that sometimes places little or no value to formalisation.

The history of community land rights in Kenya is erroneously perceived to have a lower degree of tenure in comparison to privately held land. This is because communal landownership is oftentimes based on informal landholding, which is seen as a determent to investors as it becomes difficult to engage in land-related commercial transactions.

Prior to the formalisation of communally held land in Kenya, the system that people used to attach rights and claims to land was based on customary landownership. There is vast indisputable literature that indicates that prior to the colonial period, land was held under customary law in Kenya.[45] This system was set on a path of slow disintegration by the introduction of various agrarian policies by the colonial government based on the logic that individualisation of parcels will provide economic empowerment through growing of cash crops.[46] These were the first steps towards a journey of haphazard patchwork of laws that only responded to the need to individualise community land to a certain extent. The inevitable consequence of the efforts then, and the efforts now, is that there is a high display of statutory disharmony, among other problems.

5.1 The Law on Community Land Tenure in Kenya

Community land tenure was introduced under the 2010 Constitution as a new form of land tenure.[47] While this tenure was not entirely new, the Constitution introduced into its text that community landholding is a third category of land tenure (besides private and public tenure),[48] which must be seen to be of equal strength with other forms of tenure.

Community land under the Constitution of Kenya consists generally of land that the community has legitimate rights over whether those rights have been formalised or not. For the formal acquisition, community land is that which is acquired legitimately through transfer or lawful registration or otherwise declared so by law. Importantly, the Constitution also recognises land that is acquired by informal but legitimate means through management or usage, as in the case of community

[43]S.2 of the Land Act defines customary land rights as rights conferred by or derived from Kenyan customary law whether formally recognised by legislation or not; s.2 of the CLA defines customary land rights as those rights that are conferred by or derived from African Customary law, customs or practices provided that they are consistent with the constitution.

[44]Okoth-Ogendo (1979).

[45]Kenyatta (1938), p. 25.

[46]GOK (1954).

[47]GOK, Ministry of Lands (2011), 4.

[48]Article 61(2) Constitution of Kenya.

forests, grazing areas or shrines by hunter gatherer communities, and also land that is acquired ancestrally.[49]

Customary land tenure is a phrase that is closely related to community land tenure, in that the members of the community may share a common custom and further that the attachment that is used to legitimise their claim over the land is based on a construct of customs.

It is important to note that the Constitution commendably expounds on the various legitimate ways of landownership, which is in line with the Food and Agricultural Organization's (FAO's) Voluntary Guidelines on the Responsible Governance of Tenure of Land, Fisheries and Forests (VGGT),[50] which, among others, recommend the broadest possible forms of tenure. In other words, legitimacy of community land rights is not diminished by the absence of registration (as per the Constitution). However, the owner who fails to formalise his/her land may suffer investor disinterest and inability to explore his/her land economically. In addition, the CLA makes it compulsory for one to register community land within the reasonable transitional time stipulated thereunder.[51]

Prior to the recently repealed laws[52] that majorly regulated communal land tenure, it is important to note that there have been many laws that were dedicated, through various periods of time, to the governance of communal lands. More recently, and prior to the establishment of the CLA, 2016, land that is held communally was recognised under the Trust Lands Act and the Land (Group Representatives) Act.

The Trust Lands Act was enacted for the purposes of setting apart certain lands to be held by the counties. The repealed Act provided for compensation if such land was taken away and also allowed a good measure of user rights for the community.

The L (GR) Act, on the other hand, was enacted to provide for the incorporation of those who were elected to represent a collectivity of people who owned land jointly. The Act provided, to a certain extent, the form of electing the representatives and the process of decision-making for the sake of the group. This law was predicated on the simple logic that sometimes a collectivity of land could include over 200 people, which would cause a logistical disaster in terms of administration.

These laws, however, brought forth challenges such as the uncertainties of which kind of land ought to be registered in which regime (Trust Lands Act or L (GR) Act). As such, the possibility of having duplicity, and opting to register on the law that best favoured the situation, did not only create tenure insecurity but also lowered investor confidence as far as community land investments were concerned.

---

[49]Section 63 Community Land Act.

[50]Tenure and REDD+ Developing enabling tenure conditions for REDD+ Authors: Ann-Kristin Rothe & Paul Munro-Faur, available at http://theredddesk.org/sites/default/files/resources/pdf/UN-REDD%20Programme%20Policy%20Brief%20Tenure%20and%20REDD%2B.pdf, (last accessed on 22nd July 2016); "Voluntary Guidelines" on tenure (Committee on World Food Security, May 2012).

[51]Section 11, Community Land Act.

[52]The Land (Group Representatives) Act and the Trust Land Act.

Under the L (GR) Act, there were challenges where those who were selected to be group representatives did not consult or failed to sufficiently exercise their fiduciary duty vis-à-vis the members of the community. The Trust Land Act, among other challenges, was the percolation and abuse of power by the county council (this structure has now been revoked and replaced by the county governance structures under the 2010 Constitution), where they would not exercise the 'trust' arrangement of holding the land for the benefit of the community members but would even subdivide and sell it corruptly.

In order to curb these and other problems and to create secure tenure over community land, the Constitution of 2010 under Article 63(5) required that a law be passed to legislate on community land rights,[53] and this was the journey that led to the enactment of the Community Land Act, 2016, and that repealed the Land (GR) Act and the Trust Lands Act.

The CLA was enacted to enhance, ascertain and formalise community land rights in Kenya. Among other things, this Act formalises customary land rights. Section 14 of the CLA states that a customary right of occupancy in community land shall in every respect be equal in status and effect to a right of occupancy granted in any category of land.[54]

It is important to note with patience that the enactment of CLA was not a silver bullet to all community land tenure issues in Kenya but is a fair beginning. The process of applying the Act, just like the revision of other land laws, must suffer and hopefully survive the pangs of difficult transitional hurdles (between the old laws and the new laws),[55] the political realities surrounding land and general statutory disharmony.

6 Conclusion: The Future of Soil Carbon Regulation in Kenya.

Although indirectly, Kenya's policies, laws and regulatory frameworks seem to be fairly receptive to climate change mitigation through soil carbon.[56] The black letter of the law, in assessing carbon soil tenure and its attendant component, soil carbon, has unfortunately not been given judicial analysis in Kenya, and therefore any

[53]Schedule 5, Constitution of Kenya, specifying that among other legislation to be drafted by the Kenyan Parliament was the Community Land Act.

[54]Section 14, Community Land Act.

[55]Transition from the old communal land tenure regime is provided for under Sections 47, CLA and the Transitional Provisions.

[56]National Climate Change Strategy, National Climate Change Action Plan, National Climate Change Response Strategy, National Climate Change policy.

attempt to give a conclusive status here would be an exercise in prophesy, no matter how academically sound.

In order to make decisions as to the property value of carbon, for the sake of soil carbon and all carbon emission sequestration efforts, a number of important considerations must be made. Firstly, and most importantly, homage must be paid to the fact that Kenya is a common law country where property has always been viewed by statute and the court institutions in a certain way. In this regard, however, soil law regulatory frameworks must balance between common law validation, statutory clarification and the inherent special character of soil carbon.

In the end, there is a need to consider whether Kenya needs a specific provision on soil carbon, over and above the statutory hints and common law interpretations above. For instance, Germany[57] and the US[58] have a general law on soil protection and governance, from which a mandate to engage in soil carbon sequestration activities may be derived. It is, however, considered that soil carbon emission reductions must consider aspects that require more detail than mere protection and governance. This means that what is available in Kenya as statutory and treaty[59] provisions should receive support from more detailed laws since what is available does not devote the requisite attention to the special issue of soil carbon and its role in emission reduction and climate change in general.

There are three possible ways that are discussed for considering the property rights:[60]

> ....
>
> a. bundled rights system: that the carbon in the tree is held automatically by the person who owns the land, either publicly, privately or communally, and that the transfer of sequestration rights goes hand in hand with the land;
> b. unbundled rights system: that the right to benefit from carbon sequestration is separable and exploitable by itself and not attached to the land; and
> c. that the State is the overall owner of the carbon as a State resource.
>
> ....

While the discussions above were in reference to tree carbon for purposes of assessing carbon tenure for REDD+, this paper concurs with the three possible suggestions with the necessary modifications in light of the fact that soil carbon may seem to be more intricately attached to the land in comparison to tree carbon. Soil carbon seems to be more amenable to the approach of the bundled rights system as far as the Kenyan situation is concerned. This is also agreeable to the socio-political realities surrounding land tenure, which would make it densely difficult to separate a right over land and award it to the State or some other entity.

---

[57] See Generally the German Soil Protection Act.

[58] see US soil Conservation Act of 1935.

[59] Convention on Biological Diversity, UNFCCC, etc....

[60] See Kamunde, Unpublished, Tenure in the Context of REDD+ in Kenya (unpublished) draft of 30th Aug 2016. Paper in file with the Author.

The position it therefore appears in is that the soil and all its components ought to be owned by the person who has a claim to the land. However, the State in various ways may introduce a number of limitations to provide a fair and carbon-specific balance between the rights of landowners and the interests of the State in meeting its climate change obligations for the sake of the public. This should, however, not underestimate the uniqueness of carbon property rights in general and particularly soil carbon.[61]

The social context in Kenya also indicates that people will not consider sequestration efforts viable unless they see a direct economic potential. The extent to which forests are cleared in Kenya (see, for instance, the Mau Forest) to make room for agricultural practices is a strong indication that people prefer farming alternatives that seem to favour economic survival more clearly. While deforestation should be discouraged at all costs, it is critical to seek the compromise of agricultural activities, where possible, for the sake of soil carbon sequestration. This will ensure climate change mitigation and also create room for sufficient local income.

Based on the Kenyan experience with regard to insecure land titles, this paper observes that delineation carbon property rights ensuring registrable interests is important, and securing such rights through law to guarantee that the gains towards climate change mitigation are not taken away is imperative. While this may not be enacting specific special laws, as is the case for South Australia,[62] it is important to grant carbon rights some formal recognition, without diminishing the rights of the landowner. In addition, registrable carbon sequestration rights must be explored so as to ensure that such rights are not perforated with the hurdles that flood the Kenyan land sector. As such, a system of non-registration of rights such as the one that obtains in New South Wales cannot work for Kenya. A formalisation of rights (whether formal or informal) should ask whether a carbon cadastre is possible, considering various issues: the nature of these rights, who holds them, when and the duration within which they are acquired, how they were acquired, where they are located and what their dimensions are.

Further, there is a need to ensure preservation for the land structures and philosophies under Kenyan law, which generally indicate conservativeness in land property rights. Any suggested systems therefore, particularly in rural farmlands, should not be too unfamiliar to the locals. This is in addition to the fact that there are certain alternatives under common law that may avail channels for the formalisation of soil carbon property rights. Australia has explored its legal devices to come up with a viable option for carbon tenure,[63] such as the alternative that closely resembles *profits a prendre*. However, clarification through guidelines and policy formulations may assist to ensure that in the event of a dispute, the courts will not be oblivious to the uniqueness of carbon property rights.

Ultimately, a proposition can be made with reference to community land rights, that the community is the sole proprietor of the bundle of rights that obtain over

[61] Hepburn (2008), p. 245.

[62] *Ibid.* 246.

[63] Kamunde (2016), p. 29.

community land but for which the State can make special restrictions and carbon-related requirements to ensure optimum emission reduction benefits.

Finally, as is the case for all carbon emission reduction initiatives, there is a need to offer clarity to the 'asset regime' of carbon property rights. In doing so, focus should be on considering a reasonable degree of statutory and policy harmony. This paper observes that there is a need to embrace all-encompassing sustainable land use and soil governance structures that will subsequently usher soil carbon sequestration. At the same time, it will be crucial to delineate principles that will incentivise behaviour towards the optimum soil carbon sequestration and the enhancement of soil carbon sinks.

References

Articles/Books

Allan T, Baylis K (2006) Who owns Carbon? Property Rights Issues in a Market for Green House Gases

Barnes G, Quail S (2009) Property rights to carbon in the context of climate change. University of Florida, March, 2009

Charles R (1990) The new property after 25 years. Univ San Francisco Law Rev 24:223

Foucherot C, Bellassen V (2011) Carbon Offset Projects in the Agricultural Sector. CDC Climate Research, Climate Report, Research on the economics of climate Change

Hepburn S (2008) This article is based on a seminar given by the author. Carbon rights as property interests: categorization, creation and emissions trading. Australian National University College of Law Seminar, Canberra

Kameri-Mbote P. Land tenure, land use and sustainability in Kenya: Towards innovative use of property rights in wildlife management IELRC Working Paper 2005

Kamunde N, An Assessment Of The PLR Framework Governing Tenure In The Context of REDD+ In Kenya, (23rd August 2016) (Unpublished, In File With the Author)

Kenyatta J (1938) Facing Mount Kenya: the traditional life of the Gikuyu. Kenya Publications, Nairobi

Okoth-Ogendo HW (1979) Land tenure and its implication for the development of Kenya Semi-Arid Areas. Institute of Development Studies, University of Nairobi, Nairobi

Ringius L, UNEP Collaborating Centre on Energy and Environment (UCCEE) (1999) Soil carbon sequestration and the CDM Opportunities and challenges for Africa

Sophie MC, Kamunde-Aquino N, Kago C, Kiguatha L, Idun Y, Creating an Enabling Legal Framework for REDD+ investments in Kenya, June 2014

Tenure and REDD+ Developing enabling tenure conditions for REDD+ Authors: Ann-Kristin Rothe & Paul Munro-Faur

Cases

Oil Turkana Limited (Previously Known As Turkana Drilling Consortium Limited) & 3 Others V National Oil Corporation Of Kenya & 4 Others [2013] eKLR.
LLC v U.S. Fish & Wildlife Serv., 2006 U.S. Dist. LEXIS 29334, 9-10 (W.D. La. 2006).

Government Documents

Annex C: Government of Kenya Climate Change Activities, Adam Smith International Community land rights recognition (CLRR) model: for the recognition, protection and registration of community rights to land and land based resources, Ministry of Lands, September 2011, Land Reform Transformation Unit.
GOK, Agricultural Land Act.
GOK, National Climate Change Action Plan
GOK, National Climate Change Policy
GOK, National Climate Change Response Strategy
Kenya Vision 2030, (2007).
Mining Act, Cap 306, Laws of Kenya, Published by the National Council for Law Reporting (NCLR) and Government Printers.

Reports

Food and Agriculture Organization of the United Nations (1996) The State of Food and Agriculture, 1996. FAO, Rome
FAO (2012) Voluntary Guidelines" on tenure (Committee on World Food Security)
Africa, Vivid economics, KIPPRA (Kenya Institute for Public Policy Research and Analysis).
Swynnerton RJM (1955). The Swynnerton Report: A plan to intensify the development of African agriculture in Kenya. Government Printer, Nairobi

Statutes

Community Land Act, Laws of Kenya (2016) Published by the National Council for Law Reporting (NCLR) and Government Printers.
Constitution of Kenya, Revised Edition 2010, Published by the National Council for Law Reporting (NCLR) and Government Printers.
German Soil Protection Act (1998). Long title: Act on Protection against Harmful Changes to Soil and on Rehabilitation of Contaminated Sites, Federal Soil Protection Act *(Bodenschutzgesetz, BBodSchG).*
Judicature Act, Cap 8, Act No. 16 of 1967, Published by the National Council for Law Reporting (NCLR) and Government Printers.
Land (Group Representatives Act) *Now Repealed,* by the Community Land Act, 2016.
Petroleum (Exploration and Production) Act, Cap 308, Published by the National Council for Law Reporting (NCLR) and Government Printers, Laws of Kenya.

Trust Land Act, *Now* Repealed by the Community Land Act, 2016, Published by the National Council for Law Reporting (NCLR) and Government Printers.
US soil Conservation Act of 1936

Treaties

Convention on Biological Diversity (1992), [1993] ATS 32/1760 UNTS 79/31 ILM 818 (1992).
United Nations Framework Convention for Climate Change (U.N.FCCC) 1771 UNTS 107/ [1994] ATS 2/31 ILM 849 (1992).
UNFCCC (1997) Kyoto Protocol to the United Nations Framework Convention on Climate Change adopted at COP3 in Kyoto, Japan, on 11 December 1997

The Use of Property Law Tools for Soil Protection

Jessica Owley

> Literally speaking, the health and the productivity of the ground that we stand on will largely determine the future prosperity and security of humankind.[1]

1 Introduction

While there is no doubt among conservationists that protection of the soil is important for the health and prosperity of the planet,[2] it is a resource that fails to garner the same attention as other aspects of our natural world. Even in the realm of land conservation, an area that one might think would be dominated by questions of soil health and conservation, the discussion is more likely to center on scenic values, water, biodiversity, and other ecosystem services. Such an approach may be particularly shortsighted in a world marked by climate change.

Soil conservation practices can do more to mitigate climate change than other approaches, and the role of healthy soil in adaptation efforts is unquestionable.[3] As protecting the soil can be more forward thinking than making land-conservation decisions based on other environmental characteristics, soil protection can be viewed as environmental protection writ large.

[1] UNCCD Executive Secretary Monique Barbut in September of 2015 in response to the 2030 Agenda for Sustainable Development; http://sdg.iisd.org/news/unccd-unep-release-infographic-and-video-to-welcome-adoption-of-sdg-target-on-ldn/.

[2] Davidson and Janssens (2006), Lal (2004).

[3] Pacala and Socolow (2004).

J. Owley (✉)
University at Buffalo, State University of New York (SUNY), Buffalo, NY, USA

Universidad Pontificia - Comillas (ICADE), Madrid, Spain
e-mail: jol@buffalo.edu

H. Ginzky et al. (eds.), *International Yearbook of Soil Law and Policy 2017*, International Yearbook of Soil Law and Policy,
https://doi.org/10.1007/978-3-319-68885-5_18

As conservationists have considered how to incorporate climate change into their planning, some have advocated for shifting from a focus on species or even on the historical or present ecosystems to thinking about "conserving nature's stage."[4] That is, some conservation scientists recommend looking at the geophysical characteristics of a landscape, advocating that we shift our conservation focus from current ecosystems and refugia to something that considers the potential ecosystem makeup, and these researchers assert that the best way to make that assessment is by considering soils, geology, elevation, and similar characteristics.[5] This suggests that looking at the geological composition of the landscape and the components of the soil will provide better indicators of which lands are worthy of protection. Indeed, they argue that examination of the conditions beneath the surface may be the most useful in determining which ecosystems (and their associated services) will be able to thrive as the climatic conditions change.[6] This approach to conservation (which is gaining broader acceptance)[7] highlights the importance of soil integrity as an element of ecosystem protection.

The international community has recognized the importance of soil protection. We can see this most prominently in the Sustainable Development Goals and the United Nations Convention to Combat Desertification. The Sustainable Development Goals (a project led by the United Nations with the support of over 190 countries) set 17 global goals with 169 total targets within those goals.[8] Goal 15 is to "[p]rotect, restore and promote sustainable use of terrestrial ecosystems, sustainably manage forests, combat desertification, and halt and reverse land degradation and halt biodiversity loss."[9] This goal clearly ties to the conservation of soil, and the goal's 12 targets with their indicators show an even stronger link.[10]

Target 15.3 gives a clear mandate from the international community as it seeks to "combat desertification, restore degraded land and soil, including land affected by desertification, drought and floods, and strive to achieve a land degradation-neutral world."[11] The Sustainable Development Goals set a date of 2030 for meeting this target and lists "proportion of land that is degraded over total land area" as the indicator for determining whether the goal has been met (but does not dictate what this proportion must be).[12] Referred to as Land Degradation Neutrality or LDN, Target 15.3 has been a focus of the UN Convention to Combat Deserti-

[4]Anderson and Ferree (2010), Lawler et al. (2015a).

[5]Anderson and Ferree (2010).

[6]Id.

[7]Lawler et al. (2015b).

[8]United Nations (2017b).

[9]United Nations (2017a).

[10]Id.

[11]Id.

[12]Id.

fication (UNCCD) and the United Nations Environment Programme (UNEP).[13] Thus identified as a major concern of our era, the negative trends on land and soil has harmed food and water security and reduced the ability of communities to be resilient in the face of climatic changes.[14] Moreover, soil health is indicative of agricultural productivity.[15]

Achieving land degradation neutrality is no easy task, and the established work plan begins with a target-setting process where countries determine national baselines, set land degradation targets, and explore measures to reach those targets.[16] Currently, over 100 countries are involved in this effort.[17] The hardest step in this process is undoubtedly trying to determine (and then implement) the strategies that actually conserve the land. Once we acknowledge that soil conservation is important, what can we do to prevent degradation and promote and protect healthy soils? How can we actually reach the land neutrality targets, and equally importantly, how can we ensure that we do not backslide once those targets have been achieved? How do we maintain the vigilance needed to ensure that successful land protection projects remain in place? Additionally, once we agree that soils should be protected, we need to identify which soils. Which areas should we choose? How should we choose them? Acknowledging that some development is not only inevitable but desirable, we do not want to protect all soils to the detriment of other societally beneficial land uses.

This chapter seeks to explore these questions through the lens of property law, with a focus on the development of property law in common law countries but with an acknowledgment that these arrangements occur all over the world, with different terms but similar concepts. This chapter begins in Sect. 2 with a brief glimpse into how countries are addressing soil conservation before investigating, in Sect. 3, public and private approaches to land degradation. From there, the chapter touches briefly on the contract law approach to conservation in Sect. 4 before describing in detail the property law approaches in Sect. 5—the heart of the chapter. These property law tools vary from complete control over a parcel to limited ability to control certain aspects of land use for a limited time. There is a variety of property law tools available, and this area continues to develop. Land conservationists, however, must consider their choice of property tool carefully as the restrictions can be hard to change and may not always be consistent with changing societal needs. Thus, the chapter ends in Sect. 6 with a note of caution.

[13]United Nations Convention to Combat Desertification (2017).

[14]FAO (2015), Allouche (2010).

[15]Doran and Safley (1997), Kibblewhite et al. (2008).

[16]See Chasek et al. (2015).

[17]UNCCD (2016), IUCN (2015), United Nations (2015).

2 Soil Protection and Land Conservation

To understand the role that law can play in soil conservation, we can begin by thinking about the measures that countries might use to meet those land conservation antidegradation targets. To consider a few examples (but far from an exhaustive list), we see that some countries are focusing on protection of forestland and planning to implement projects to restore degraded forests.[18] Other countries are working to promote sustainable land management. For example, in Kenya, the government is promoting sustainable land management as a way to achieve land degradation neutrality.[19] Colombia, which identifies its major soil concern as erosion, has created a national policy for sustainable soil management.[20] Some projects focus on simply ceasing restrictive activity, while others call for more active land management behavior. How do we protect such land though, and once you have implemented your on-the-ground projects, how do you ensure that they remain in place?

For most countries, we do not yet have information about land protection goals or implementation. We are quite a way off from having implementation plans or rules for conservation. Most countries are at the stage of formulating their plans. They are identifying their priorities and setting goals, but they have not drilled down into all the details of how the targets will be reached. Countries set land degradation neutrality targets and then work with the United Nations and other agencies to help meet those targets. This chapter can help with that phase. As we seek to protect land and prevent further degradation, the mechanisms available through property law may facilitate building rules for ecological management.

3 Approaches to Land Conservation

There are a number of legal tools at our disposal for soil protection. We can largely think of the approach as following either a public route or a private one. In terms of public land protection, we see governments controlling the land uses and practices on government-held land. We also see the government acting as a regulator. In legislating for the public health, safety, and welfare, governments at all levels constrain individual behavior with the hope of improving environmental conditions. Thus, the public land conservation route involves the government either constraining its own behavior or constraining the behavior of people within its jurisdiction. Unfortunately, our public regulations have focused on pollution control and land use without addressing soil health and protection directly.

[18]E.g., República de Costa Rica (2015), Republic of Indonesia (2015).

[19]UNCCD (2017a).

[20]UNCCD (2017b).

On the private actor side, we look to nonprofit organizations and individuals who do not have the power to legislate as governments do. Nor do they hold the power to acquire land for public use through eminent domain or similar mechanisms. This leaves private actors with the more traditional private law tools of torts, contracts, and property law. Tort law has not been particularly helpful in soil conservation, offering only limited assistance in the realm of nuisance law. Contract law and property law have been more promising, with the use of property law flourishing in the context of land conservation.

This section considers both public and private land conservation strategies from the legal perspective. It begins with an overview of the public conservation efforts and then addresses the private methods. The following section goes into more detail about the workings and potentials of property law as a land conservation avenue.

3.1 Public Land Conservation

A primary tool of land conservation worldwide is government ownership of at-risk areas. The government then places these special areas under its protection and limits the activities that can be done on the land. In this way, we have something that looks like state action, but, in reality, the state is behaving more like a run-of-the-mill private landowner who can decide what she wants to do with her land and chooses to be as environmentally protective as she wants to be. A difference occurs between the private landowner and the state as landowner because many jurisdictions view the state as having an obligation to protect the land and other natural resources on behalf of its citizens. Sometimes called the public trust doctrine, this theory places differing level constraints on government behavior in different countries.[21]

Beyond acting as a conservation-minded proprietor or landowner, the government can also protect soils by regulation of harmful activities, something common and uncontroversial in most countries. For example, government agencies can mandate specific agricultural techniques or limit the amount of developable land area on a parcel. The effectiveness of such a technique depends on the strength of the governmental institutions, on other legal structures and restrictions, and on the capacity of the governmental entities involved. It may be particularly hard, for example, for government officials to monitor agricultural or forestry practices. Such an action clearly requires a lot of funding and staff—not something always in ample supply. In fact, constraints on public entities and their lack of capacity to monitor the land have led to (1) government entities building partnerships with NGOs and seeking their assistance in implementing or enforcing public goals and (2) NGOs setting out on their own to protect soils based on their belief that the government is

[21]Owley and Takacs (2016).

not doing an adequate job on this task. In the first scenario, we see that even what we label as public land conservation can be dominated by private action. In the second, we see a frustration with public action leading to an increase in private action. Either way, the pattern seems to be an increasingly important role for private conservationists.

3.2 *Private Land Conservation*

Contrasted with public efforts are the private law avenues for land conservation. As noted above, this often occurs where conservationists are dissatisfied with the public methods or extent of conservation. Where a private organization or individual seeks to protect soils, however, the legal tools available for such protection vary. NGOs do not have the same ability to pass laws and promulgate regulations to shape land use as government entities do. Nor do they have the power of acquiring land through eminent domain. Instead, NGOs and the government agencies that they work with are turning to other legal tools that are better fits, in what we think of as the realm of private law. Most markedly, this occurs in the realm of contract law and property law.

When we discuss private land conservation efforts, we need to identify the private parties we are talking about. While one might envision wealthy individuals with an interest in environmental protection taking action in this realm,[22] we are mostly referring to nongovernmental organizations: NGOs of different sizes and styles that come together to use property and contract law to protect the land. Notice that these NGOs differ from other environmental NGOs because of the tools they use. Less likely to spend energy on lobbying politicians or using public pressure or litigation to achieve their goals, these NGOs are less likely to be seen organizing a protest or circulating a petition (although of course some of them engage in such activities). Instead, we see a class of NGOs that focus their time and energy on securing property rights to land (or entering into contracts with landowners). We are particularly interested here in the NGOs that use property tools. In the United States, these organizations are labeled land trusts, perhaps because of their link to older trustee organizations but also because of their link to older ideas of trust lands in both the United States and the United Kingdom (school trust lands, the National Trust, etc.). In Latin America or Europe, these groups are often labeled land custodians, land stewards, or custodial entities. Phrases that also seem to indicate a standard of care regarding the land that is something more than landownership.

While we are chiefly interested in this chapter with exploring the property law tools available for soil conservation, a brief foray into contract law can help illustrate the options available and why property law has become the tool of choice.

[22]And we do see this with wealthy families like the Rockefellers and individuals like Ted Turner. Brechin (2015) Turner Foundation (2017), Turner Enterprises (2017).

4 Contracts for Land Conservation

Contract law can serve as an avenue for land conservation. Landowners can enter into contracts with NGOs that take the form of payments for ecosystem services.[23] The landowner agrees to engage in certain land-use practices that enhance or preserve the soil (or perhaps agrees to refrain from engaging in activities that would damage the soil).[24] In return (as consideration), the NGOs provide the landowners with payments or some other valuable item (facilitating a permit application, for example).[25] The contract agreement protects the individual parcel to which it applies and can be tailored to fit the circumstances involved. Thus, the contract can be a more fine-tuned tool than a government regulation, which is likely to apply more broadly. Contracts are voluntary, though, and will only be useful where the landowner is willing to be bound by one. The parties involved negotiate the terms, and it may be hard to create coherent rules across parcels as the landowners involved might have different goals for the land or ecology.

There is, however, an even greater concern than coherency with contract law in that a contract only binds the parties that enter into the contract. In the realm of soil conservation, this means that only the landowner who signed the contract will be required to comply with the terms of the contract. If the landowner sells the property, the restrictions will not be enforceable against the subsequent landowner. Therefore, whenever a landowner wants to change land uses, she need only sell the land to wipe the restrictions away. Thus, a more desirable restriction is one that binds even the subsequent landowners. But we do not do that with contract law—we do not like to require things of nonparties. This is where property law comes in handy for NGOs that believe that the government is not doing enough through public law.

5 Property Restrictions for Land Conservation

Unlike contract law, property rights in the land can be associated with a parcel and not solely with an individual. Property rights come in many shapes and sizes, and this section describes different types of property rights before delving more deeply into the partial nonpossessory property rights that have become the favored tool of land conservationists in the United States that we now see spreading across the globe.

For purposes of our discussion here, we can group property rights into full or partial and possessory or nonpossessory. The sections below describe each type and

[23]Mercer et al. (2011), Boyd and Banzhaf (2006).

[24]Owley and Takacs (2016) at 79.

[25]Owley (2006).

illustrate how the nonpossessory partial interest in land has become a particularly useful tool for conservation.

5.1 Possessory Interests

5.1.1 Full Fee Simple Absolute

One of the most straightforward ways to use property rights to protect land is through purchase of the land. In the common law system, we use the phrase "fee simple" to describe present possessory interest in the land. Where a landowner has the present possessory rights in a parcel, she has the ability to make decisions regarding the use of that land subject to general rules and regulations of the governments that have jurisdiction over the land and also in accordance with general laws of landownership, like nuisance, that prohibit a landowner from using her property in ways that unreasonably harm neighboring property. Holding title to the land bestows the landowner with freedom of action regarding the land and can enable land preservation alongside active conservation and management. Where an NGO wants extensive control over the activities on the land, this might be the best option. Moreover, if an NGO identifies that a parcel could potentially benefit from active soil conservation measures, the NGO would be most secure in holding the title to the land. With title, there will be little objection to whichever land conservation measures the NGO determines will be most beneficial.

Yet the ability to conserve soil through ownership by NGOs or conservation-minded individuals is limited. An obvious obstacle is the expense. Not only can land be expensive to purchase, but added costs come from management of the land. The NGOs not only need money for land purchases but also need capacity to staff, manage, and monitor the soil conservation efforts. In some places, this may mean being full-time occupiers of the land. Lack of vigilance could lead to interlopers. Additionally, depending on the interventions needed, the staff may need training or equipment that can add to the expenses.

Conservation efforts through fee simple ownership are also limited to where there are willing sellers. Without the government power to condemn land, NGOs can only gain title where landowners are willing to sell. This may not be beneficial for strategic soil conservation. Even where conservationists can identify the parcels that are most important for soil conservation, it does them little good if landowners are unwilling to sell those parcels. This means that strategic or important areas could go unpreserved with energy (and funding) being applied to marginal soils. This limitation makes it difficult to take a holistic or strategic approach to soil conservation.[26]

[26]A 2017 report from the Brookings Institute suggest that current patterns of conservation easement use are not likely to preserve lands strategically to meet environmental goals and instead are more likely to maximize private goals like tax savings. Looney (2017). Recent work from

Where NGOs hold fee simple title, land may be taken completely out of production and even result in a removal of people from the land. Such a development can change the patterns and composition of communities. Where we think of conservation as originating in a park-like concept that separates people from the land, NGOs can take on the aspect of a neocolonial power dictating land uses and community makeup. This may be particularly pronounced when international or foreign NGOs (or individuals) purchase land. Decision making may be coming from people who have no experience working the land or working with the people. While this concern has lessened over the years as NGOs have improved their working relationships within the communities they operate, we still see objections in the United States and across the world when private nonprofit organizations shape the landscape.[27] In some jurisdictions, this can have an added financial dimension as, in many cases, NGOs pay lower taxes and, in some cases, no property tax. The lower tax base can result from the status of the landowner as an NGO or the limitations on land use that restrict the development of the highest and best use of the land. To alleviate such concerns, NGOs sometimes make voluntary tax-like payments in the regions where they own land to avoid impacts on schools on other social services that might occur with reduced public funding.

5.1.2 Co-ownership (in Fee Simple Absolute)

Because of the inherent limits of soil conservation by fee simple ownership, conservationists began to explore partial property rights. Is there a way for us to get some of the same soil conservation benefits without needing to be the owner of the land?

In some cases, this might occur with a possessory interest like joint ownership. Where a property owner holds a portion of the property rights, she has the ability to control or at least influence conduct on the land. In the common law tradition, this ability is clear. Co-owners of property that all hold possessory interests have the right to use and occupy the whole.[28] Furthermore, co-owners have an obligation to each other to prevent abuse of the land and will be liable to one another for damage done to the land.[29] These background principles, however, may do little to prevent negative impacts on the soil. In the eyes of most courts, there is nothing unreasonable about traditional exploitation of the land (through forestry, agriculture, or grazing) and little that would limit development of natural resources. At some

economists at the University of Wisconsin and North Carolina State, however, argue that there is no evidence that conservation easements are concentrated on lower quality lands because conservation easement holders serve a gatekeeping function that prevents such a pattern from developing. (Parker and Thurman 2017).

[27] 7-49 Powell on Real Property § 49.01 (2017).

[28] 7-50 Powell on Real Property § 50.03 (2017).

[29] Id., 7-50 Powell on Real Property § 50.06 (2017) l; Watts v. Krebs, 131 Idaho 616, 962 P.2d 387 (1998).

level of exploitation, some methods would become unsupportable, particularly if they started to look like nuisance or waste, but court interpretations of such behavior are uncertain, and seeking to constrain land by simply holding a small percentage of the present possessory interest would be ill-advised.

5.1.3 Defeasible Fees

One of the easiest and earliest examples one learns about when studying property law is the defeasible fee. A defeasible fee looks similar to fee simple ownership because the property rights holder has the ability to presently occupy and use the land. However, the word defeasible indicates that there is a way the landowner could lose her property rights. That is, upon the occurrence of a certain action (or the failure of a required occurrence), the holder of the defeasible fee loses her property right, and it is then transferred to another.[30] This happens when a landowner places constraints on her land at the time of conveyance. This could happen in a deed transfer at any time, but we see it most commonly in wills. Here are some examples of how one might write a defeasible fee restriction for soil conservation:

- I leave my land to my son Jaime so long as he continues to employ soil conservation techniques.
- All of my property to my sister Victoria and her heirs, but if the soil quality reduces appreciably during their tenure, then to The Nature Conservancy.

Through language such as this, present-day landowners can make long-term decisions regarding their land. In these examples, both Jaime and Victoria have a right to hold and possess the land. Indeed, their rights look similar to the rights of a fee simple landowner. Yet the rights are not as complete because of the possibility that they will lose their property rights. In this way, defeasible fees constrain the activity of the landowner. The person creating the defeasible fee (often a person writing a will, but again the transfer need not be upon the death of a former owner) plays a powerful role, setting the agenda for the future of the land. The length of their control is limited by the law of the jurisdiction. Some countries may not allow such dead hand control, while others may be quite willing for a person living today to decide what may or may not happen on her land far into the future. A limitation of conservation by defeasible fee is the rigidness of the constraint. Careful drafting is needed to enable desirable changes to the land. For example, a restriction that

[30]Defeasible fees and future interests (discussed below) are much more complicated than these simple examples indicate. What is important to understand for the purposes of this chapter is that one can place constraints on land uses when conveying land. Additionally, in some jurisdictions a landowner can voluntarily constrain her rights by converting her fee simple absolute into a defeasible fee. 1-13 Powell on Real Property § 13.02 (2017); 1-13 Powell on Real Property § 13.05 (2017). The contours of such conveyances and how long the constraints might last differ by jurisdiction and should always be confirmed with legal counsel.

requires a tobacco farm to continue to be a tobacco farm might hamper development in a world that no longer wants as much tobacco or hamper the ecology of a region that no longer has the right conditions for tobacco farming. Such rigid constraints can be particularly troublesome as climate change, and the uncertainty of the exact scope and location of its effects, shapes our land.

5.2 *Nonpossessory Property Rights*

The limitations on possessory ownership lead us to examine nonpossessory interests. These can take the form of either present or future interests. The essential feature of a nonpossessory interest is that the holder of that interest does not have the ability to presently occupy and use the land. This does not make it less of a property right, though. The holder still has a valuable right that she can buy and sell. Moreover, she also has the ability to constrain the actions of possessory holders to protect her rights.

5.2.1 Future Interests

A concept that often confuses students and laypeople is the idea of the future interest. Although not recognized in all countries, the future interest is a nonpossessory interest that can be invoked to achieve some soil protection goals. The holder of a future interest holds a valuable property right today—one that can be bought, sold, and passed on to heirs like other property rights. The holder does not yet have the right to use and possess the land, however. This means that associated with every future interest is another property holder who has the present possessory interest. The future interest is waiting in the wings (sometimes patiently and sometimes no) for her nonpossessory property right to become possessory.

Depending on the jurisdiction in which you are operating, these property rights can take many forms and be rather complicated. For the purposes of this chapter, however, we need only understand that a known future landowner can have a say over what the current landowner can do on the property. Generally, the limitations are embodied in the doctrine of waste. This doctrine limits the ability of the current landowner to use the property in such a way as to hamper the future landowner. This can prevent destruction of a house, depletion of natural resources, and similar behavior. Unfortunately, the doctrine is coarse and cannot work to protect soils unless it can be shown that the depletion of the soil by the present landowner is an unreasonable and destructive use of the property. Reasonable and customary land uses (perhaps the ones that have led to environmental problems to begin with) will not meet the threshold of harm needed for the future interest holder to be able to constrain the activity.[31]

[31]Pappas (2014) at 745.

5.2.2 Servitudes

More useful than the future interest is a present interest even if that interest is nonpossessory. Indeed, the present nonpossessory interest is an attractive tool for many, meeting many land conservation goals.

The most classic example (and in the most widespread use) is the easement. An easement allows someone to have a right in someone else's land without actually becoming the owner or permanent occupier of the entire parcel land.[32] The most common example is an access easement. Your neighbors have the right to drive across your land to access their home. The electric company has a right to place and maintain electrical poles and wires on your property. Easements can take several forms and could include things like a right to hunt or gather wood. The easement holder might be an individual, a family, a business, or a group.

Traditionally, easements gave someone other than the fee simple landowner the right to do something on the land that the landowner would have otherwise been able to prevent. That is, they gave the easement holder an affirmative right. In some narrow circumstances, some jurisdictions also recognized negative easements. A negative easement prohibits a landowner from engaging in an otherwise lawful activity on her land. The most common examples of negative easements are prohibitions on disrupting free flowing water or sunlight. Where enforceable, negative easements are few, and the options are specifically enumerated by statute.

Traditional easements also were constrained as to who was recognized as a legitimate easement holder. As a property rights arrangement, easements were viewed as agreements binding the land. Such an arrangement has the benefit of creating agreements that remain tied to the land regardless of who owns the land, but they also limit the number of people or entities that can enter into an easement. To begin with, one has to be a landowner, but some jurisdictions go even further and limit the acceptable parties to adjacent landowners. A common exception to this rule is the utility easement. All states in the United States recognize the ability of utilities to hold easements for pipelines, power lines, cables, and similar equipment. Yet they do not recognize this based on the status of the utility as a landowner. In property law terms, we label such easements as "in gross" as opposed to "appurtenant." In gross easements do not have a benefited parcel of land, only a benefited person or entity. Such easements were historically disfavored and maybe limited by statute or common law.[33]

The limitations of the easement for protection of the soil thus become clear. In many jurisdictions, there is a limitation on using easements to control the *landowner's* behavior; they are more focused on the *easement holder's* behavior. One could envision an affirmative easement that allowed the easement holder to engage in activities to protect the soil, but without accompanying restrictions on the landowner that may not be that fruitful. Additionally, operating by affirmative

[32] 4-34 Powell on Real Property Chapter 34.syn (2017).

[33] Id.

action may be more cumbersome (and more expensive in terms of the staffing needed) than enforcing a negative restriction on behavior. The limits on who can enter into such an agreement also presents a quandary as it would only enable landowners to do so and might require purchases of small anchor parcels beside any land where one seeks an easement. The impracticality of such a rule for NGOs seeking to protect the soil is obvious.

The limitations of easements gave birth to two additional types of servitudes: real covenants (also labeled restrictive covenants) and equitable servitudes. These restrictions look a bit more like contracts than easements do. Indeed, they are sometimes called promises regarding the land. While the terms may look like contracts, the essential difference is that real covenant and equitable servitudes can bind future landowners and have life beyond the original parties to the agreements. However, various limitations on the use of these tools also limit their utility for soil conservation. Without delving too deeply into the potential variations in every jurisdiction, we can highlight a few concerns. Many restrictive covenants can only be enforced with damages. That is, when a landowner breaks a promise, the court just requires the landowner to make a payment for what it calculates to be the value of the promise. It does not require the landowner to actually change behavior and implement the soil conservation measures.

Another hindrance occurs in jurisdictions that put rigid limitations on who can enforce the agreement over time. In particular, we see restraints on who assumes the burden or the benefit of the promise with strict rules on transferability (or in property law terms, whether the agreements will run with the land). The conclusion then is that such servitudes can provide an avenue for soil conservation, but one has to look very carefully at the laws of the jurisdiction to ensure that the tool does not come with limitations that impede soil conservation efforts.

5.2.3 Conservation Easements

Discontent with the options above, conservationists began to seek out ways to use partial property rights to achieve land conservation goals but without the complications described above. In the United States, this led to the birth of the conservation easement. The use of the word "easement" places the tool in the context of servitudes, but it would be a mistake to think of it as a traditional easement because it has different rules and lifts many of the restriction associated with easements.

A conservation easement then is a nonpossessory property right that limits landowner behavior with the goal of producing a conservation benefit.[34] The agreements must have the purpose of producing one of the conservation benefits enumerated in the statute that governs that jurisdiction. There is no requirement that the agreements actually yield a benefit, just that they seek to do so.[35] While the list

[34]Cheever and McLaughlin (2015).

[35]Owley and Doane (2017).

of acceptable conservation purposes varies slightly by jurisdiction, they generally follow the pattern of the Uniform Conservation Easement Act (UCEA). Acceptable purposes for conservation easements under the UCEA "include retaining or protecting natural, scenic, or open-space values of real property, assuring its availability for agricultural, forest, recreational, or open-space use, protecting natural resources, maintaining or enhancing air or water quality, or preserving the historical, architectural, archaeological, or cultural aspects of real property."[36]

Conservation easements have the benefit of being enforceable not just with damages like a real covenant but also with injunctive relief—meaning that one can actually force the landowner to comply with the restrictions. Conservation easement enabling acts set forth acceptable holders as governmental agencies and nonprofit organizations that have conservation as part of their central purposes.[37] This obviates the requirement that holders have to own an anchor parcel or indeed need to be landowners at all.

Finally, the statutes confirm the ability of both the benefit and the burden of conservation easements to run with the land, that is, changes to the identity of the landowner or the conservation easement holder to not hinder enforcement of the agreement. A hallmark of conservation easements and what has made them especially attractive to conservationists is the fact that they are usually perpetual. Indeed, in the United States, three states (California, Hawaii, and Florida) require the agreements to be perpetual.[38] Most states make it the default duration, and only one state prohibits it (North Dakota limits conservation easements to 99 years).[39]

While conservation easements are most popular in the United States (where we see the first laws enabling them), they are growing in popularity across the globe, although they sometimes have a different name. They are now well developed in

[36]UCEA § 1(1).

[37]Again, the exact contours of eligible holders vary by jurisdiction. For example, some places require certain tax status for the NGOs, others specifically identify Native American tribes as eligible holders. Owley (2012c). The Uniform Conservation Act (a model act that nearly have of the U.S. States have adopted) lists the following acceptable holders:

> (i) a governmental body empowered to hold an interest in real property under the laws of this State or the United States; or (ii) a charitable corporation, charitable association, or charitable trust, the purposes or powers of which include retaining or protecting the natural, scenic, or open-space values of real property, assuring the availability of real property for agricultural, forest, recreational, or open-space use, protecting natural resources, maintaining or enhancing air or water quality, or preserving the historical, architectural, archaeological, or cultural aspects of real property

[38]California, Hawaii, and Florida require conservation easements to be perpetual (California Civil Code § 815.2(b); Hawaii Revised Statute § 198-2(b), Florida Statutes Ch. § 704.06(2)) as does the Internal Revenue Code for those hoping to associate their conservation easement with a tax deduction (Internal Revenue Code § 170(h)(2)(c)). *See* Korngold (2007).

[39]N.D. Cent. Code sec. 20.1-02-18.2; *see* Wachter v. Commissioner of Internal Revenue, 142 Tax Court No. 7 (March 11, 2014).

Canada and Australia. They are growing in New Zealand, Spain, and Scotland. There is a proposed law enabling them in England and Wales. Kenya, Chile, Columbia, and Mexico are developing similar ideas. And these are but a few examples. We also see the idea being exported by companies and intergovernmental organizations that are investing in climate change adaptation and other environmental projects abroad. They want guarantees that their projects will have longevity and are requiring conservation-easement-like arrangements to achieve that goal.[40]

The benefits of conservation easements for soil conservation are clear. With this tool, conservationists (government agencies or NGOs) can tailor restrictions to individual parcels, implementing conservation programs across the landscape. Landowners get to remain on their land, and community composition does not change. Instead, communities receive payments or other amenities to make smaller changes to behavior. The conservationists can pay for exactly what they want to implement. On some parcels, land uses might be curtailed severely where in others it is a smaller restriction on certain farming techniques. Scientists can work with the lawyers to craft agreements that best achieve soil conservation goals. The agreements last forever and are often enforced by private organizations, limiting the strain on the public coffers as government agencies can remain on the sidelines if they so desire.

5.2.4 Other Ideas

Beyond the options discussed here, conservationists have been exploring other ways that concepts from property law might protect land. These ideas are still theoretical and experimental, so we do not yet have a full understanding of how they might work. Some have argued that real estate options could be a way to achieve environmental goals. An option gives someone a right to acquire land for a certain time period but does not obligate that person to acquire the land. When real estate is burdened by an option, the landowner cannot materially change or degrade the land without violating the terms of the option. As the penalty is usually paying back the option price and the conservation is a passive one, it may not suit the needs of many who are working in land conservation. Yet the existence of this idea shows the efforts underway to explore new ways to achieve land conservation goals through property rights trends. We also see proposals for options to purchase conservation easements, annuity easements, and moveable easements. There are likely many other creative arrangements connected to ideas of private law developing around the world.

[40]Owley (work in progress).

6 Conclusion

While the development of private law tools to protect soil offers encouraging news for those seeking more ways to protect the land, we also need to be cautious about the use of these tools.

First of all, turning from public law to private law (even if the line between them is a blurry one) can raise concerns about democracy. These property law arrangements are available to government agencies but mostly reside in the hands of NGOs. This tool actually enables those NGOs to circumvent public plans for the landscape. For example, elected officials may create a development plan that protects some areas but allows development in others. A nonprofit organization may be unhappy about the planned development and use conservation easements to prevent development even in the area chosen for that purpose. Where we agree with the NGO, we may like the tool, but we need to recognize that it is not a democratic tool. Indeed, patterns of usage in the United States suggest that the individuals most likely to benefit from conservation efforts of this type are wealthy landowners who had little intention of engaging in destructive practices to begin with.[41]

Even where government agencies are using the tool, they may be doing so to prevent future changes by other government officials. Elected officials could often achieve the same goals by regulation. They may choose property tools over regulation because it appears more politically palatable or because they can draw upon the power of NGOs for assistance, but some local governments have also stated that they use conservation easements because they are more permanent than legislation that can be changed by the next legislature and want to prevent future politicians from making decisions that the current politicians do not like.[42] To a supporter of conservation, this may seem like a desired outcome. To a supporter of democracy, this is unquestionably problematic. If we think that partial property rights hinder government action, then we may have a problem. If we think that partial property rights can disrupt community efforts to make decisions, we might have an issue with that too.

Permanence may be both part of the solution and part of the problem. One of the most attractive aspects of partial property rights is the ability for the agreements to be perpetual. Generally, we view a contract as only binding the parties that enter into the contract. But property rights are something different. Property rights are agreements regarding the land and that stay with the land. This is attractive from a land conservation standpoint because they enable long-term protection of the land. Transferring the ownership of the land (or the ownership of the partial nonpossessory property right) does not remove the protection. This gives us some peace of mind for soil conservation: protections put in place will not easily disappear.

[41]Owley (2012a), Looney (2017).

[42]Owley (2012b).

However, it is not just that they do not easily disappear, they also do not change that easily either. The details of the agreements are written today, with today's goals and today's knowledge. They are often written in static terms, seeking to preserve the status quo or protect specific landscapes and practices. They offer little room for changing societal goals, but perhaps even more troublesome they offer little room to adjust to changing environmental circumstances or changing information. New studies that provide better guidance on soil management, for example, may not be able to influence conduct on land encumbered by a conservation easement as that agreement already sets the rules. Jurisdictions differ on the degree to which they are willing to allow changes or adjustment to such agreements, but the trend is toward only allowing agreements that remain in line with stated purposes. This means that a measure that is more protective of the soil on a conservation easement that seeks to protect the soil will probably be allowed (but not necessarily so), but a change from protecting species to protecting soil would not be permitted even if studies reveal that such efforts would be a better use of the land.[43]

In the end, this chapter presents a complicated story. The development of private law tools to protect the soil is exciting. It demonstrates a new energy to achieve conservation goals with engagement of new (and more players). People are thinking creatively about what can be done to improve the world we live in. Yet the excitement of using property law tools sometimes leads organizations to quickly tie up the land with agreements that have complicated and uncertain implications. As with all legal strategies, we must think carefully before assessing which tool is right for the project we want. In our current world, we must always assume change. As things get worse (or better) for soils, will we be able to achieve our goals with the tools we have chosen?

References

Allouche J (2010) The sustainability and resilience of global water and food systems: political analysis of the interplay between security, resource scarcity, political systems and global trade. Food Policy 36(Suppl 1):S3–S8

Anderson MG, Ferree CE (2010) Conserving the stage: climate change and the geophysical underpinings of species diversity. PLoS One 5(7):e11554

Boyd J, Banzhaf S (2006) Resources for the Future, What Are Ecosystem Services?: The Need for Standardized Environmental Accounting Units. http://www.rff.org/rff/Documents/RFF-DP-06-02.pdf

Brechin E (2015) Rockefellers a force in conservation. Ellsworth American, 8 June 2015. http://www.ellsworthamerican.com/rockefeller/rockefellers-a-force-in-conservation/

[43] A related concern may be present in the United States where conservation easements tend to list multiple purposes. We don't know what would happen where efforts to meet the various purposes conflict. There is no clear way to determine which purposes should take precedence. Owley and Rissman (2016).

Chasek P, Safriel U, Shikongo S, Fuhrman VF (2015) Operationalizing zero net land degradation: the next stage in international efforts to combat desertification? J Arid Environ 112:5–13

Cheever F, McLaughlin NA (2015) An introduction to conservation easements in the United States: a simple concept and a complicated mosaic of law. J Law Prop Soc 1:107

Davidson EA, Janssens IA (2006) Temperature sensitivity of soil carbon decomposition and feedbacks to climate change. Nature 440:165–173

Doran JW, Safley M (1997) Defining and assessing soil health and sustainable productivity. Biol Indic Soil Health

Food and Agriculture Organization of the United Nations (FAO) (2015) Soils store and filter water. http://www.fao.org/3/a-bc272e.pdf

IUCN (2015) Land degradation neutrality: implications and opportunities for conservation, Technical Brief 2nd edn, November 2015. IUCN. Nairobi, 19p

Kibblewhite MG, Ritz K, Swift MJ (2008) Soil health in agricultural systems. Philos Trans R Soc B Biol Sci 363(1492):685–701

Korngold G (2007) Solving the contentious issues of private conservation easements: promoting flexibility for the future and engaging the public land use process. Utah Law Rev 2007:1039

Lal R (2004) Soil carbon sequestration to mitigate climate change. Geoderma 123:1–22

Lawler J, Watson J et al (2015a) Conservation in the face of climate change: recent developments, F1000 Research 2015 4 (F1000 Faculty Rev): 1158

Lawler JJ, Ackerly DD et al (2015b) The theory behind, and the challenges of, conserving nature's stage in a time of rapid change. Conserv Biol 9(3):618–629

Looney A (2017) Charitable contributions of conservation easements. Brookings Institute, https://www.brookings.edu/wp-content/uploads/2017/05/looney_conservationeasements.pdf

Mercer E et al (2011) Taking stock: payments for forest ecosystem services in the United States, forrest trends: ecosystem marketplace 1. Available at www.forest-trends.org/documents/files/doc_2673.pdf

North Dakota Central Code § 20.1-02-18.2

Owley J (2006) The emergence of exacted conservation easements. Nebraska Law Rev 84:1043–1112

Owley J (2012a) Tribes as conservation easement holders: is a partial property interest better than none? In: Rosser E, Krakoff S(eds) Tribes, land and the environment. Ashgate Press, pp 171–191

Owley J (2012b) Neoliberal land conservation and social justice. IUCN Acad Environ Law E-J 3:6–17

Owley J (2012c) Use of conservation easements by local governments. In: Salkin P, Hirokawa K (eds) Greening local government. A.B.A. Publishing, pp 237–255

Owley J (work in progress on file with author) Exporting American Property Law

Owley J, Doane C (2017) Exploiting conservation lands: can hydrofracking be consistent with conservation easements? Kansas Law Rev 66 (forthcoming)

Owley J, Rissman AR (2016) Trends in private land conservation: increasing complexity, shifting conservation purposes and allowable private land uses. Land Use Policy 51:76–84

Owley J, Takacs D (2016) Flexible conservation in uncertain times. In: Kundis Craig R, Miller SR (eds) Contemporary issues in climate change law and policy: essays inspired by the IPCC, pp 65–104

Pacala S, Socolow R (2004) Stabilization wedges: solving the climate problems for the next 50 years with current technologies. Science 305:968

Pappas M (2014) Anti-waste. Ariz Law Rev 56:741

Parker DP, Thurman WN (2017) Tax Incentives and the Price of Conservation (unpublished manuscript on file with author) CHECK FOR UPDATE BEFORE PUBLICATION

Powell on Real Property (2017)

Republic of Indonesia (2015) Indonesia—land degradation neutrality national report. http://www.unccd.int/en/programmes/RioConventions/RioPlus20/Documents/LDN%20Project%20Country%20Reports/indonesia_ldn_country_report.pdf

República de Costa Rica (2015) Costa Rica Degradación neutral de la tierra informe nacional. http://www.unccd.int/en/programmes/RioConventions/RioPlus20/Documents/LDN%20Project%20Country%20Reports/costa-rica-ldn-country-report.pdf
Turner Enterprises Inc. Turner Ranches. http://www.tedturner.com/turner-ranches/ (last visited Feb. 16, 2017)
Turner Foundation. http://www.turnerfoundation.org/ (last visited Feb. 16, 2017)
UNCCD (2016) Land Degradation Neutrality: The Target Setting Programme. http://www.unccd.int/Lists/SiteDocumentLibrary/Publications/4_2016_LDN_TS%20_ENG.pdf
UNCCD (2017a) Kenya launches roadmap to set land degradation neutrality targets. http://www2.unccd.int/news-events/kenya-launches-roadmap-set-land-degradation-neutrality-targets
UNCCD (2017b) Colombia advances on the SDG 15 agenda on "life on land" through the implementation of the National Policy for Sustainable Soil Management. http://www2.unccd.int/news-events/colombia-advances-sdg-15-agenda-life-land-through-implementation-national-policy-0
Uniform Conservation Easement Act (UCEA). http://www.uniformlaws.org/shared/docs/conservation_easement/ucea_final_81%20with%2007amends.pdf
United Nations (2015) Resolution adopted by the General Assembly on 25 September 2015, 70/1. http://www.un.org/ga/search/view_doc.asp?symbol=A/RES/70/1&Lang=E
United Nations (2017a) Sustainable Development Goals, 17 Goals to Transform our World. http://www.un.org/sustainabledevelopment/sustainable-development-goals/
United Nations (2017b) Sustainable development goal 15. https://sustainabledevelopment.un.org/sdg15
United Nations Convention to Combat Desertification ((UNCCD) (2017) Land Degradation Neutrality. http://www.unccd.int/en/programmes/RioConventions/RioPlus20/Pages/Land-DegradationNeutralWorld.aspx
Wachter v. Commissioner of Internal Revenue, 142 Tax Court No. 7 (March 11, 2014)
Watts v. Krebs, 131 Idaho 616, 962 P.2d 387 (1998)

Thrifty Land Use by Spatial Planning Law: Considering the Swiss Concept

Yvonne Franßen

1 Introduction

Due to the country's topography, only about 30% of Swiss territory is suitable for human settlement. Therefore, Switzerland is one of the most densely populated countries in Europe. In the Alpine country, the continuing demands of land use have become a target of public criticism[1] and triggered lively debates on measures to promote sustainable development.[2] Population growth, improved living standards with ever-increasing individual mobility, as well as municipal competition to attract new inhabitants and companies, are responsible for the extent of new land sealing.[3] The tax system and subsidies are viewed as further possible causes.[4] Strong economic driving factors hinder the implementation of Swiss planning laws and support urban sprawl.[5] By reforming land-use law, Switzerland is aiming to limit land consumption. During the last few years, the Swiss parliament has passed more

Dipl.-Ing. Yvonne Franßen (urban planner) Employee in an urban development authority in Berlin, Germany; responsible for land use planning.

[1]Schwick et al. (2010), p. 13.

[2]Waltert and Seidl (2013), p. 178.

[3]Schwick et al. (2010), pp. 11, 14.

[4]Waltert and Seidl (2013), pp. 178 et seq. Taxation of housing properties at lower market value, weak enforcement of the partial commuter flat-rate and polluter pays principle are fiscal incentives influencing land usage. Two spatial reasons are oversized building areas and weak absorption of added values of property owners. Since 2014 amending spatial law respects these planning aspects.

[5]Auer et al. (2014), p. 11.

Y. Franßen (✉)
Urban Development Authority, Berlin, Germany
e-mail: yfran@gmx.de

H. Ginzky et al. (eds.), *International Yearbook of Soil Law and Policy 2017*,
International Yearbook of Soil Law and Policy,
https://doi.org/10.1007/978-3-319-68885-5_19

and more legislative amendments to stop urban sprawl and to strengthen closer building in settlement areas.

In the following pages, the Swiss planning concept is described as far as it relates to thrifty land-use management, and innovative aspects are highlighted and annotated in the conclusions. The central idea of Swiss land-use law is that one legally binding zoning plan covers the whole municipal territory. The zoning plan differentiates between building and nonbuilding areas. Oversized building areas must be reduced. Furthermore, Swiss municipalities are entitled and under obligation to claim at least 20% of the added value that the property owners can realize when their land values increase as a result of spatial planning. Previously, most municipalities did not claim the added value not to scare investors.

2 Swiss Federal System

Switzerland's federal system entails the division of responsibilities between the Confederation, cantons, and municipalities. The subsidiarity principle applies: all state functions that are not regulated in the Swiss Constitution are the responsibility of the cantons. The cantons enjoy the highest degree of sovereignty within the Swiss Confederation. The municipalities are the smallest political units. Canton laws determine municipal tasks.

3 Swiss Planning Levels and a Brief Planning History

Land-use planning takes place at all levels. Each administrative level has its own planning laws and zoning regulations. In order to establish an open and dynamic planning process, too strict hierarchical structures of decision taking between the planning levels needs to be avoided.[6] In Switzerland, planning takes place top down and from the bottom to the top at the same time. This cross-level cooperation is referred to as "principle of counter-acting streams of power." This principle requires harmonized and interconnected collaboration between the planning levels. This principle does, however, not oblige that contradictions between the different planning levels are completely avoided.

In Switzerland, the first calls for spatial planning systems were expressed in the 1930s. Social interaction in the regions and the necessary coordination of transregional infrastructure projects (e.g., railway network) required national and local plans. In the 1950/1960s, population growth made reliable planning for new settlements necessary.[7]

---

[6]Entire text paragraph: Gilgen (2012), p. 107.

[7]Entire text paragraph: Devecchi (2016), pp. 26–27.

Significant land-use problems such as the separation between building areas and nonbuilding areas were unsolved until the 1970s.[8] The Federal Council defined provisional nature protected areas in 1972.[9] In comparison to other state legislation, the Swiss Federal Law on Spatial Planning was passed rather late in 1980. But from the very start, the Swiss concept has based on land use as thriftily as possible.

3.1 Swiss Confederation

3.1.1 Federal Planning Laws

In 1969, the Federal Constitution was complemented by an article about spatial planning. The federal government was thus authorized to adopt the basic land-use legislation. This framework law had to implement legal obligations for land use to be as thrifty as possible and for sustainable settlement in regions. Key actions of spatial planning were conferred on the cantons. This corresponds to their guaranteed sovereignty in the Swiss Federal Constitution (Article 3). In 1969, Article 26, which confers property guarantees, was added to the Federal Constitution as well.[10]

The Swiss Federal Law on Spatial Planning ("Bundesgesetz über die Raumplanung"—RPG) was established in 1980 and primarily addresses the federal government and canton levels. Only a few regulations are relevant for the municipalities. Land-use law has been reformed often and at increasingly shorter intervals during the last 35 years. In 2014, a fundamental amendment was made.[11] Regulations on land protection (Article 1) were tightened[12]: the separation of building areas and nonbuilding areas complemented the legal obligation for thrifty land use. "Thrifty land use" means sustainable management of land but also optimum spatial allocation of land usages.[13] Article 1 RPG emphasizes that land-use planning should be environmentally compatible, first and foremost, but must also serve human development.[14] This order of priorities is remarkable and seems to be the heart of the Swiss planning concept.

Since 2014, the goals of land-use planning (Article 1 RPG) have been complemented by promoting inner urban development and by further

[8]Danielli et al. (2014), p. 43.

[9]Devecchi (2016) p. 27.

[10]Gilgen (2012), p. 85.

[11]See a brief overview of legislative reform in Switzerland in: Hengstmann and Gerber (2015), p. 241.

[12]Swiss Acts see: https://www.admin.ch/gov/de/start/bundesrecht.html. This homepage includes old Act versions. RPG see: https://www.admin.ch/opc/de/classified-compilation/19790171/index.html.

[13]Waldmann and Hänni (2006), Article 1 N 12.

[14]Waldmann and Hänni (2006), Article 1 N 22.

intensification of compact settlement structures. Cantons are also obliged to reform their laws: where future planning decisions increase the value of certain land spots, for example, by qualifying nonbuilding areas as building, Swiss municipalities are required by the new legislation to demand at least 20% of the added value. Municipalities must calculate the increased value at the time when the new land-use right comes into force. The owner, however, only has to pay at the time when he sells his property or builds on his own.[15] The legislation clarifies that municipalities are obliged to use the obtained added value for compensation (land falls in value as a consequence of losing building rights), for greening and renaturation measures.

Furthermore, spatial planning must be based on extensive analysis of regional and local situations and must identify planning results.[16] Politicians need statistical data and predictions for their decisions.

The Swiss Federal Law on Spatial Planning (Article 2 RPG) equally binds the Confederation, cantons and municipalities to synchronizing their planning. Spatial activities and planning should not take place in isolation or randomly between the different planning levels.[17] The RPG defines the planning instruments:

- cantonal structure plan (Richtplan) binding for public authorities;
- state conceptional plan and sectoral plan ("Konzeptplan" and "Sachplan") binding for public authorities;
- legally binding zoning plan (Nutzungsplan), especially for the municipalities.

"These three main instruments of spatial planning are interdependent and interrelated ('counter-current principle')."[18]

The Swiss Federal Law on Spatial Planning (RPG) was complemented by the "Raumplanungsverordnung" (RPV) in 2000. This executive order governs in particular the balancing of interests in cases where the competent authorities have choices (Article 3). Cantons must apprise federal authorities of how building in the nonbuilding areas is affecting spatial development and the landscape (Article 45). Legal standards (Articles 34 et seq.) govern the construction and management of buildings in nonbuilding areas. (These detailed regulations in enabling law are exceptional.) Generally, building in nonbuilding areas is forbidden. Permission can be issued if the application complies with the legal standards. The cantons are the building permit authorities. Individual municipal interests are avoided. The RPV, implementation regulations, legal and technical guidelines complete the sustainable aims of Swiss spatial law.

[15]Lezzi (2014), p. 136. Since 1980 the Swiss Federal Law on Spatial Planning recognizes the delivery of the added value. But since 2014 the municipalities have been obliged to require at least 20% or more according to cantonal law. A Federal Court decision has allowed up to 60% of the added value. Since 2014 the Swiss Confederation can impose sanctions against cantons which do not reform their laws. See Lezzi (2014), p. 138.

[16]Waldmann and Hänni (2006), Article 1 N 53.

[17]Waldmann and Hänni (2006), Article 2 N 14.

[18]Danielli et al. (2014), p. 51.

3.1.2 Federal Planning Instruments

In accordance with Article 13 RPG, the federal government has two planning instruments; both are binding for the public authorities: the conceptional plan and the sectoral plan. These instruments serve the relevant authorities for their spatial planning projects; they do not create new planning rights.[19] Conceptional and sectoral plans must be observed by cantons and municipalities during planning activities.[20] Spatial plans by federal authorities concern technical infrastructure, the military, landscape, etc. The competent authorities work out their conceptional or sectoral plans in cooperation with other federal and cantonal authorities. Sectoral and conceptional plans set priorities, times, and financial resources for the realization of measures, whereby sectoral plans also point out specific goals of space, times, and actions (Article 14 RPV). In 2012, a new instrument, the spatial development concept, "Raumkonzept Schweiz," was created. This spatial plan is a general guidance and a strategy document covering whole Switzerland and is worked out in cooperation with all spatial actors; it is not legally binding.[21]

The federal government is authorized to pass "Sondernutzungspläne"—special zoning plans[22]—under the law of the appropriate federal authorities, such as the Air Traffic Act and water protection law.[23]

3.2 *Cantons*

3.2.1 Laws on Cantonal Spatial Planning

According to the Swiss Constitution (Article 75), the cantons are the main spatial planning authorities. Both federal laws, the "Raumplanungsgesetz" and the "Raumplanungsverordnung," give specific planning tasks to the cantons. Instructions for the thrifty management of land use are addressed to the cantons particularly.

Cantons pass their own land-use laws, mostly in combination with building regulations. As a consequence, 26 different cantonal land-use laws exist in Switzerland. These cantonal spatial laws give detailed rulings on regional structure plans and zoning plans. Most of these laws also authorize municipalities to pass their own zoning and building laws.[24]

---

[19]Waldmann and Hänni (2006), Article 13 N 1.

[20]Danielli et al. (2014), p. 53.

[21]See: http://www.are.admin.ch/themen/raumplanung/00228/00274/index.html?lang=de.

[22]See Sect. 3.3.2.1.

[23]Gilgen (2012), p. 432.

[24]Gilgen (2012), p. 111.

The Swiss Federal Law on Spatial Planning, RPG, which is an enabling act, has transferred rights or rather legal obligations to demand the added value from property owners if the value increases due to new land-use rights (Articles 5, 15 a ff. RPG).

3.2.2 Cantonal Planning Instruments

The regional structure plan ("Richtplan"; Article 6 et seq. RPG) is the central instrument for land-use development in Switzerland. These cantonal plans are only binding for public authorities. The "Richtplan" must respect and consider planning by the state and the municipalities. Essentially, regional structure plans define the medium- to long-term development goals concerning nature, agriculture, settlement, recreation areas, etc.[25] Planning must be based on comprehensive analysis and prognosis as keynote of the Swiss concept. The determination of building areas must be established on population growth and employment development forecasts based on data from the Federal Statistical Office. Cantons must review their planning goals at least every 10 years. If necessary, the regional structure plan must be adapted. On the one hand, spatial planning is process planning, and on the other hand, it should offer legal certainty to those concerned. Consequently, regional structure plans may only be changed if basic conditions have changed substantially.[26]

Cantonal structure plans involve local and regional concerns and essential guidelines for the municipalities and prepare the municipal development.[27]

The objects and contents of the cantonal structure plans must at least cover spatial development, and coordination of all spatial actions, but also the implementation, timing, and financing of measures. Since 2014 (Article 8 a RPG), the "Richtplan" has to explain how settlement areas will be restricted or reduced and how inner urban settlement will be specified (Article 15 RPG). There must be coordination between settlement and traffic development. The Swiss Federal Council approves the regional structure plans and coordinates between the cantons.

In addition, cantons can establish special zoning plans (*Sondernutzungsplan*) per cantonal laws. These special zoning plans concern structured actions.[28] For example, Spatial Law of the Canton Graubünden enumerates special zoning plans for traffic facilities, service companies of cantonal importance, etc.[29]

---

[25]Danielli et al. (2014), p. 65.

[26]Gilgen (2012), p. 87.

[27]Danielli et al. (2014), p. 66.

[28]See Gilgen (2012), p. 432.

[29]See for example Spatial Law of Canton Graubünden (KRG), Art. KRG Artikel 15.

3.2.3 Building Permission

The Swiss Federal Law on Spatial Planning rules the legitimacy of projects in nonbuilding areas: agricultural buildings are permitted in agriculture zones (Article 16 et seq. RPG), and special constructions or land uses are only permitted in nonbuilding areas as an exception (Article 24 et seq. RPG), for example landfill sites or mountain restaurants. In accordance with the Swiss Federal Law on Spatial Planning, these strict regulations[30] meet the goals of thrifty land use. As part of the Swiss concept, cantons are the permit authorities for building applications in nonbuilding zones. In the case of permission in nonbuilding areas, the cantons enter notes in the land registers (Article 44 RVP): for example, temporary construction or usage permit and responsibility for building demolition.

3.3 Municipalities

3.3.1 Laws on Municipal Spatial Planning

Some regulations of the national spatial laws (RPG and RVP) are addressed to the municipalities: especially the obligation to take responsibility for land should be emphasized. Cantonal spatial laws and other regulations lay down how the municipalities have to undertake planning. For example, types of land-use zones are established and defined. Most cantons have transferred the right to pass local spatial laws to the municipalities.[31] The municipal zoning laws concretize the cantonal guidelines concerning the type of land use, building density and level of disturbance, etc.

3.3.2 Municipal Planning Instruments

Legally binding zoning plans ("Nutzungsplan"; Articles 14, 21 RPG) are the main municipal planning instruments. Compared to special zoning plans, the zoning plan is a framework plan.[32]

Some cantons empower their municipalities to design local structure plans that are binding for public authorities.

[30]Muggli (2014) sees a contradiction in Swiss policy: on one hand the government strengthens the duty for economical use of land and on the other hand it allows more exceptions for building in nonbuilding areas, pp. 68, 69, 75.

[31]Danielli et al. (2014), p. 82.

[32]Gilgen (2012), p. 477.

The Legally Binding Zoning Plan

The zoning plan defines land-use rights legally binding for the whole municipal territory on parcel level (Article 14 RPG).[33] For that, zoning plans are the key instruments for the thrifty management of land. They must observe national and cantonal objectives and principles.

The main land-use types are building areas, agriculture areas, and nonbuilding areas. But each canton and municipality can establish as many subtypes as practically required. Thus, every region or local authority can adapt the planning instrument to the specific area perfectly.

Main land-use types are as follows:

- Building areas

According to the expected needs of settlements, building areas should take into account the needs for the next 15 years[34]; oversized sites must be reduced. The new classification of building land is subject to certain conditions: the land has to be developed and ready for building within 15 years (Article 15). This regulation restricts space consumption and public debt[35] for infrastructure projects like road building. The Federal Council Noise Protection Regulations ("Lärmschutz-Verordnung"—LSV) enumerates four sensitivity levels assigned to different building zones (Article 43 LSV). The regulations differentiate between noise emission limits and alarm noise emission limits. Planned noise emission level must be 5 dB below the noise emission limit. This accounts for the fact that various noise sources lie outside the maximum limits in the long run.[36] If the noise limit values are exceeded, house construction or changing the building usage can only be permitted under extremely stringent requirements. In building areas, the building density and area of buildings are specified.

- Agricultural areas

To ensure continuing adequate food supplies and ecological compensation, agriculture areas must be protected. The agriculture areas also have to preserve the landscape and recreation areas. As consequence, building in agriculture areas is generally forbidden (Article 16 RPG).

[33] Gilgen (2012), pp. 432, 87.

[34] For calculating real needs, the government has stipulated requirements (Art. 30 a RPV) and directives. The extent of building areas depends on analysis and projections of the development of population and employees, infrastructural needs and financial budgets. See: Gilgen (2012), p. 493.

[35] Gilgen (2012), p. 492.

[36] Danielli et al. (2014), p. 111.

– Nature protection zones

Nature protection zones include water bodies, banks, near-natural landscapes, places of historical interest and monuments, as well as habitats worthy of protection for fauna and flora (Article 17 RPG).

Procedural Regulations
Zoning plans should be checked if conditions have changed significantly and must be adapted if necessary. This guarantees dynamic developments on the one hand and a degree of certainty on the other hand. Local land-use plans consist of drawings and supplementary regulations.[37] The cantons approve the municipal zoning plans (Article 26 RPG).

Special Zoning Plan
Zoning plans are framework plans and cover the whole municipality. The special zoning plan is an extra type of land-use plan. It defines land use for a part of the municipality or for one plot only. Special zoning plans can diverge from the zoning plan because this framework plan cannot consider all the possibilities of spatial planning goals. Special zoning plans might govern building design, road construction projects, or areas concerning future developments.

Federal authorities and cantons can enact special zoning plans in addition to the communities as well.

Land Exchange

Planning might require the reorganization of plots. The competent authorities are empowered to enforce land exchanges (Article 20 RPG). The formation of new plots pursues the target of the best land use or the exchanging of "ripe for construction" areas for nonbuilding land.[38] This measure severs the interest of thrifty land use.

3.3.3 Local Building Permission

Within building areas, municipalities are responsible for building permission and land-use permission. Permits are the central instrument for realizing spatial plans and goals.[39] The cantons are the building permit authorities in nonbuilding areas.

Nobody can be compelled to perform construction in Switzerland. Nevertheless, since the law changed in 2014, cantonal planning laws must regulate construction obligations relating to public interests (Article 15 a RPG). Realization of land-use rights before enacting new building areas corresponds with the idea of thrifty land use.

---

[37] Danielli et al. (2014), p. 84.

[38] Danielli et al. (2014), p. 128.

[39] Gilgen (2012), p. 111.

4 Conclusions

Land is a limited resource. Spatial planning has to deal with different and sometimes conflicting private and public interests. Especially, in the process of adopting legally binding plans, stakeholders are determined to push their financial, common, economic, or local interests. Soil/land protecting concerns are often underrepresented for commercial or political reasons. Sprawl and increasing land use continue.

As nature or agriculture is rarely the major source of revenue for the local governments, thrifty land use does usually not prevail in clash of interests. Spatial planning has to balance asymmetric conflicts of interests.

Legally binding spatial plans are the central instrument for thrifty land use because they define which types of land use is approvable: building or agricultural use or any human use in behalf of nature and land protection.

National spatial legislation determines the framework for land use and building law. The municipalities can only work out their plans in accordance to national scope for spatial planning. If national law allows land consumption and sprawl, local authorities will admit building and land consumption according to spatial acts and local interests. Sometimes they even cannot stop sprawl due to a lack of legal rights. First of all, state governments must restrict the planning and permission law system.

Increasing land consumption forces state governments evermore to tighten their planning laws or to enact other measures. During the last years, Switzerland was forced to reform the Federal Law of Spatial Planning more and more often, although thrifty land-use aims were firmly established from the very beginning. But still commercial interests clash with land-protecting aims.

Nevertheless, the Swiss spatial concept is exemplary:

- *Land use planning should be environmentally compatible first and foremost, but must also serve human development*: this objective is the principle of the Swiss planning law. Implementing this measure in practice is confronted with the well-known clashing interests. But as laid down as principle, it influences spatial planning laws and legally binding plans. In Germany, for example, the protecting items are in the inverse order, and as a consequence, building is the main issue of German land-use plans.
- *Legally binding plans covering the whole municipal territory and must serve thrifty land use*: plans covering the whole territory must respect all concerns and interests regarding space and time before they become legally binding. States, like Germany, who only enact legally binding plans for a part of the municipal area or few plots will never really shape the local development. They only react on single building concerns. Detailed surveys and balanced development planning and, in consequence, limiting the increasing land consumption will hardly be realizable if legally binding plans have only few plots in sight.
- *Reducing oversized building areas and requiring the added land increase*: legally binding differentiation between building areas, nonbuilding areas, and

protecting areas combined with the duty to reduce oversized building areas are the most effectively rules for thrifty land use. In Switzerland, restricting these rules and the commitment to require the added value from the owners, whose land value increased as a consequence of new land-use rights, seems to be very successful.

- *Spatial planning must be based on extensive analysis of regional and local situations and must identify planning results*: sensible and forward-looking planning must base on regional and structural survey and must respect substantial modifications. Otherwise, it cannot be efficacious but only can react to powerful influences. The Swiss Federal Law on Spatial Planning and other rules are stringent concerning comprehensive and current data collections and analysis. Without detailed surveys, land consumption cannot be determined, evaluated, and controlled. Switzerland seems to be successful with its rules and implementation.
- *Spatial activities and planning should not take place in isolation or randomly between the different planning levels*: interaction and coordination between the planning levels are fundamental conditions and needs if all spatial concerns should be regarded. If lower planning levels must respect the planning targets of the higher level, local tasks and objects cannot be featured as required. Or the regional aims are so vague and unverified that they cannot lead the local development effectively. Especially, a close collaboration between the planning levels can guarantee interconnection of spatial tasks like reducing land consumption and delimits rivalries between local authorities.
- *Permits are the central instrument for realizing spatial plans and goals*: law admission standards or exceptions from legally binding plans can contradict thrifty land-use aims. Therefore, assignment of building permits in nonbuilding areas to the Swiss cantons considers that local authorities have local and not overall agendas. But so far as the cantonal or regional/national aims go together with the local purpose, for example for financial and economic reasons, the control function gets lost. Wide admission policy and laws challenge thrifty land uses. But also the Swiss government relaxes rules. Exceptions from plans are still a weak point in the Swiss planning system, although rules were restricted:
- Transfer of 20% of the added value due to modifications of planning rights to the respective municipalities is an important tool because it enables the municipalities to conduct necessary compensation measures.

5 Prospects

In future, individual claims will go on demanding additional land consumptions. Insofar as thrifty land use is accepted by the society as essentially for survival, government concerns can be implemented. Public debates might raise the citizen's awareness for thrifty land use and tightening spatial planning laws. That's the way it occurred in Switzerland basically successful.

References

Auer A et al (2014) Fünf Thesen zur Raumplanung. Verlag Neue Züricher Zeitung, Zürich

Danielli G, Sonderegger R, Gabathuler C (2014) Kompaktwissen Raumplanung in der Schweiz. Rüegger Verlag, Zürich

Devecchi LU (2016) Zwischenstadtland, Zur politischen Steuerung der suburbanen Entwicklung in Schweizer Gemeinden. Transcript Verlag, Bielefeld

Gilgen K (ed) (2012) Kommunale Raumplanung in der Schweiz. vdf Hochschulverlag, Zürich

Hengstmann A, Gerber JD (2015) Aktive Bodenpolitik – Eine Auseinandersetzung vor dem Hintergrund der Revision des eidgenössischen Raumplanungsgesetzes, FuB 6/2015, pp 241–250

Lezzi M (2014) Mehrwertabgabe gemäss Artikel 5 Raumplanungsgesetz, Blätter für Agrarrecht, Heft 3, pp 133–144

Muggli R (2014) Ist der Föderalismus an der Zersiedlung schuld? Verlag Neue Züricher Zeitung, Zürich

Schwick C, Jaeger J, Bertiller R, Kienast F (2010) Zersiedlung der Schweiz – unaufhaltsam? Quantitative Analyse 1935 bis 2002 und Folgerungen für die Raumplanung. Bristol-Stiftung, Zürich

Waldmann B, Hänni P (2006) Handkommentar RPG 2006. Stämpfli Verlag, Bern

Waltert F, Seidl I (2013) Wie das fiskalische System die Zersiedlung fördert: Eine Untersuchung der flächenrelevanten Steuern und Subventionen in der Schweiz ZfU 2/2013, pp 178–196

The Soil Conservation Protocol of the Alpine Convention: Why Was the Adoption Possible?

Sebastian Schmid

1 Introduction

In May 2014, the European Commission decided to formally withdraw the Proposal for a Directive Establishing a Framework for the Protection of Soil.[1] It had already been put off in the years before when a group of Member States—the United Kingdom, France, Germany, the Netherlands and Austria—were firmly opposed to the framework. Nonetheless, soil protection continues to be the subject of discussion and controversy since the Union and its Member States have agreed to 'reflect as soon as possible on how soil quality issues could be addressed using a targeted and proportionate risk-based approach within a binding legal framework' in the 7th General Union Environment Action Programme.[2]

In context of the above, the question arises whether the Soil Conservation Protocol of the Alpine Convention can serve as a model for European legislation on soil protection.[3] By now, this international treaty is the only binding and comprehensive international framework on the conservation and restoration of soils. What are the reasons that can explain the successful adoption of the Soil

This paper is based on a legal opinion drawn up at the request of the German Federal Environment Agency in 2016.

[1]OJ 2014, C-153/3 and OJ 2014, C-163/15 (corrigendum).

[2]OJ 2013, L-354/171. See also OJ 2014, C-163/15, fn. 1 'The Commission remains committed to the objective of the protection of soil and will examine options on how to best achieve this.'

[3]See already Markus (2017), p. 149.

S. Schmid (✉)
Constitutional Court of Austria, Vienna, Austria
e-mail: Sebastian.Schmid@uibk.ac.at

H. Ginzky et al. (eds.), *International Yearbook of Soil Law and Policy 2017*,
International Yearbook of Soil Law and Policy,
https://doi.org/10.1007/978-3-319-68885-5_20

Conservation Protocol? Why did the contracting parties of the Alpine Convention find a compromise on this controversial issue, whereas agreement still seems far off on European level?

2 The Difficulty of Issuing Rules on Soil Protection

The European Union is not alone in its struggle to achieve comprehensive laws on soil protection. For the United States of America, there is nothing new about the insight that the law of soil conservation exists as an 'unwieldy number of disconnected federal and state statutes. For the most part of administrative character, they have been dictated by regional requirements, departmental exigencies, and the desire to avoid constitutional issues.'[4]

In Austria, efforts to develop a national soil strategy date back to the '70s of the last century. Up until now, this unsatisfying legal situation has persisted since neither the federal legislature nor state legislatures have the exclusive competence in soil matters.

In Germany, the Federal Soil Protection Act came into force in 1998,[5] after it had already been on the agenda of former federal governments, for instance, as a declared aim stated in the Environmental Programme 1971.[6] In parallel, several German states enacted their own soil protection laws, which aimed at the establishment of instruments of control and prevention but were criticised for being ineffective due to cautious regulations.[7] A main reason for the length of this legislation process was the fragmented separation of powers between the federal state and the 'Länder'.[8]

The following overview summarises the main causes for the difficulties that often come along with legislation on soil protection.[9]

2.1 *Complexity of the Subject Matter Concerned*

Unlike air, soil is not a homogeneous environmental medium. It is a mixture of minerals, organic matter, water, gases, liquids and organisms. Thus, it has different functions: it is a habitat for soil organisms, transforms organic materials into humus and is, thus, a necessary precondition for plant growth. Due to its regulatory function, soil is a recycling medium for water; it absorbs harmful substances

[4]Milde (1951), p. 45.
[5]German Federal Law Gazette I 1998, p. 502.
[6]See Fokuhl (1994), p. 49.
[7]Stollmann (1996), p. 367.
[8]Peine (1992), p. 353.
[9]See e.g. Odendahl (2001), p. 99 ff.

from the air and effectively neutralises them. Besides these ecological functions, soil—literally—is the basis for our buildings and other physical structures. As a deposit for minerals and energy resources, it has always been exploited by humans. Besides, soil is an archive for the history of nature and civilisation, and we also spend our leisure time running and hiking on natural or semi-natural ground.

2.2 *Variety of Issues Regulated by Soil Protection Laws*

As a consequence of the complexity of soils, legal provisions concerning soil protection have to deal with a wide variety of issues, and thus they are usually complex and diverse. Soil protection law can usually be found in many different legal acts.[10] The most common way of categorisation is to distinguish between quantitative and qualitative soil conservation. In detail, five regulatory approaches can be identified.[11]

According to the *spatial* approach, natural and near-natural soils are regarded as worth protecting by size and quantity. Such provisions can often be found in acts on spatial planning, as well as in soil conservation acts.

In other legal provisions, soil is treated as an *environmental medium* such as air and water. It shall be protected against any type of pollution or other adverse effects. Acts on environmental and soil protection often include regulations of that kind.

When a *pollutant-oriented* approach is chosen, certain harmful substances are in focus of a regulation; their adverse effects on the environment, including soils, shall be restricted. Plant protection product regulations are an example of which.

Other regulations that can be regarded as part of the soil protection law enact provisions on *industrial plants* because of the exhaust emission that goes along with their operation. Plant approval procedures and environmental impact assessments are examples of that kind of soil conservation law.

Finally, provision on air and water protection can be regarded as *mediated* soil protection. Because of the interaction and interdependence between these environmental media, minimising the pollution of air and water also means protection of soils. Forestry laws, acts on nature protection or specific procedural laws, such as environmental impact assessments, can be put in this category.

[10]For an overview see Mayrhofer (2015), p. 10.

[11]See Schmid (2015), p. 8; Duschanek (1989), p. 15 f.

2.3 Soil Protection: A Horizontal Issue

The outline above makes clear that soil protection is a so-called horizontal issue. Traditionally, it is not a subject matter codified in one book as is the case, for example, for legislation on water or nature protection. Instead, immediate and mediate soil protection law can be found in many different acts. Yet unification is regarded as advantageous because it increases the effectiveness of soil conservation, and it serves to create a regulatory system that benefits the long-term efforts of conservation. Nevertheless, the existing fragmentation causes difficulties for the process of combining the various provisions on soil conservation into one code of law.

2.4 Division of Powers

One reason for the fragmentation of soil protection law and for the difficulties of legislation on soil conservation is division of powers, in federal states between the different regional authorities, within the European Union between the Union itself and the Member States. The Commission has sought to put the Proposal for a Directive Establishing a Framework for the Protection of Soil on a sound legal basis. It has been argued that the provisions of this Directive relate to environmental protection, and consequently, the legal basis chosen was article 192 para. 1 TFEU.[12] With regard to the principle of subsidiarity, the Commission was clearly trying to provide detailed arguments why the objectives of the proposal cannot be sufficiently achieved by the Member States.[13]

2.5 Ownership of Land

Soil conservation relates to the diverging relationship between land property and environmental protection.[14] The private interests of landowners conflict with the—alleged—public interest of soil protection.[15] In contrast to air and water, soil is more or less static by nature—a fact that might have supported the traditional and somehow emotional view on land as reflection of privacy and self-determination. So in contrast to other environmental media, land can be subject to possession and ownership, which is not the case with air or water. It would hardly enter someone's mind to say, 'This is my air', whereas it is only natural to say 'This is my land'.

[12] Art. 175 para. 1 TEC.

[13] See COM(2006) 232 final, p. 6.

[14] Blasberg (2008), p. 10 f.

[15] Roellecke (2005), p. 82.

Further, in other fields of environmental protection, namely nature protection, the conflict between the interests of landowners and nature conservation is less pronounced as compared to soil conservation. This is notably due to the methods that appear to fit the purposes of the respective field of law: according to the traditional concept of nature protection, the designation of specific protection areas is regarded as an appropriate means to achieve the objectives. This is generally of low impact for the group of landowners since only a certain number of them are concerned by this measure. Soil conservation, in contrast, has to take place in a broader spatial scope in order to reach its goals. Urban sprawl can only be contained at regional level; banning certain fertilisers does not make sense when this prohibition only applies to a limited area.

In accordance with that, the Commission has pointed out that '[s]oil is a resource of common interest to the Community, although mainly private owned, and failure to protect it will undermine sustainability and long term competitiveness in Europe'.[16] For all of these reasons, legislators had great difficulties in the past in regulating soil conservation issues.

2.6 *The Territory as Part of National Sovereignty*

At first sight, soil is an immobile environmental medium, whereas air and water constantly cross borders. Pollution and degradation of soils, thus, are limited to the country concerned and do not interfere with the interests of others. However, there are many reasons why the opposite is true: 'Soil degradation in one Member State or region can have transboundary consequences. Indeed, dams are blocked and infrastructure is damaged downstream by sediments massively eroded in another country farther upstream. Equally, groundwater bodies flowing through bordering nations can be polluted by contaminated sites on one side of the border. [...] This would imply that the costs to restore environmental quality are borne by a Member State different from that where the soil degrading practice occurred.'[17]

2.7 *Lack of Soil Awareness*

So far, it has often been argued that lack of soil awareness is one reason for the inadequate protection. Pollution and degradation of soils are sometimes not immediately visible, and they can be covered up by certain measures like fertilisation. The hesitancy in the conservation of soil and in creating awareness about soil as

[16]See COM(2006) 232 final, p. 2.

[17]See COM(2006) 232 final, p. 6.

medium worth protecting is also highlighted by the fact that regulation on other fields of environmental law has a long tradition, whereas legislation on soil issues is a relatively new area of law.

3 The Soil Conservation Protocol: Negotiating Process and Contents

The Environment Ministers of the Alpine states agreed on common goals and priorities in the field of soil protection at the International Alpine Conference in Berchtesgaden in 1989. Considering the risks for soils that could arise from ongoing sealing and erosion, the results can be summarised as follows: economical use of soil and land (n. 46), use of soil-conserving production technologies in agriculture and forestry (n. 46), promotion of a form of agriculture in the Alpine region that suits local conditions (n. 47), prevention of natural risks (n. 47), conservation of the vegetation cover (n. 47), setup of national soil inventories based on common criteria in order to determine which actions have to be taken (n. 48 and n. 49), definition of common guidelines and standards of soil protection (n. 50).

Following this wide-ranging outcome of the first Alpine Conference of the Environment Ministers, the Alpine states agreed in 1991 to enter into a new international treaty, the Alpine Convention. Its article 2 para. 2 lit. d addresses soil conservation and states that 'the objective is to reduce quantitative and qualitative soil damage, in particular by applying agricultural and forestry methods which do not harm the soil, through minimum interference with soil and land, control of erosion and the restriction of soil sealing.'

The preliminary fixings of the Alpine Conference 1989 and the Alpine Convention created a sound basis to conduct negotiations on a Soil Conservation Protocol. The consultations started in 1991 under the leadership of the German delegation. Good fortune or strategy—the fact that at the same time the adoption of a German 'Bundes-Bodenschutzgesetz' was intensively discussed definitely supported the progress of the negotiations. Negotiations were led with great skill, and the outcome noted a difference in detail as compared to other protocols of the Alpine convention, e.g. the Protocol 'Spatial Planning and Sustainable Development'.

A first draft was prepared in October 1993; others followed in August 1994, in September 1994, in December 1995, in July 1996 and in July 1997.[18]

The negotiations on article 14 of the Soil Conservation Protocol (Effects of tourism infrastructure) are an example of how the contracting parties tried to come up with a wording that would meet all of the expectations involved.

The last draft adopted at the level of officials included two versions of article 14. All contracting parties except the French delegation were in favour of an alternative

[18]The preparatory work on the Soil Conservation Protocol has not been published. I wish to express my gratitude to Peter Haßlacher, one of the negotiating parties of the Protocol, who provided the necessary information for this analysis.

(so-called *Montafoner Fassung*), which stated, for instance, that permits for the construction and levelling of ski runs are not granted in forests with a protective function and for fragile areas and that included restrictions for the use of vehicles for the grooming of ski runs. Obviously, this proposal was rather far-reaching with regard to its soil protection contents. The other alternative was rather general in scope, suggesting to the contracting parties (only) that the effects of tourist infrastructures shall be observed and that they shall act with a view to stabilising soils that had been affected by intensive touristic use. The final version is a compromise text composed of both alternatives. The other contracting parties managed to convince France on political level to agree at least in part to the 'Montafoner Fassung'. So a strong effort was needed to formulate article 14 in a reasonably meaningful way and to avoid the mere bringing together of hollow phrases.

It is no coincidence that article 14 of the Soil Conservation Protocol—one of the few provisions of relatively clear content—played an important role in the application of the Alpine Convention by the courts in Austria. In the leading case concerning the direct applicability of the Convention Protocols in Austria in 2003, the authorities did not grant the permission to merge two ski resorts by way of construction measures on the legal basis of this provision.[19]

When comparison is made between the first and the final draft of the Soil Conservation Protocol, two developments can be identified.

First, the contracting parties made an effort to consolidate and clarify the articles of the Protocol from the first to the final draft. Accordingly, some provisions of the First Draft have been reworded in a more precise manner, and others have been restructured in terms of their content. One example for a specification is article 9 para. 3 of the Protocol. In the First Draft, this provision reads as follows: 'Utilised agricultural moor soils shall be managed in a way that the loss of soil organic matter is prevented and sustainable use of grassland is guaranteed.' At that time, the focus was put on the usage of moor soil and not so much on conservation and restoration of a near-natural state. In accordance with para. 1 of the same provision, which clearly spells out that the 'Contracting Parties undertake to preserve high moors and lowland moors', para. 3 was reworded in 1997: 'On principle, moor soils shall not be utilised or, when used for agricultural purposes, shall be managed so that their characteristic features remain intact.' This change in the formulation establishes a hierarchy of goals: first and foremost, moors shall not be utilised at all.[20] However, the conservation of moors depends in some cases on agricultural maintenance, for instance, to prevent the formation of woodland and scrub. Furthermore, the treatment of wetlands with low environmental impact methods is often a traditional way of farming and, thus, shall not be prevented generally. In these cases, when moor

[19]Austrian Constitutional Court (Verfassungsgerichtshof) 22.9.2003, B 1049/03-4; Independent Environmental Tribunal (Unabhängiger Umweltsenat) 22.3.2004, US 6B/2003/8-57 (*Mutterer Alm*); Austrian Administrative Court (Verwaltungsgerichtshof) 8.6.2005, 2004/03/0116.

[20]For a comprehensive interpretation of this provision, see Schmid (2007), p. 158.

soils are used agriculturally, the technics and methods of usage shall guarantee the functionality of moor soils in order to preserve them in their special characteristic.

An example for the restructuring of the First Draft is article 11, 'Designation and management of Alpine areas threatened by erosion'. In this provision, contents of the former articles 10, 13 and 14 have been united in order to create a 'focal point' for the particular issue of erosion.

On the other hand—this second development seems to be of particular importance for the purposes of this paper—certain contents of the First Draft have disappeared from the version that finally came into force. Unsurprisingly, rules adopting regulations in controversial areas, i.e. tourism and agriculture, were affected by this revision. The wording has been toned down, or the provisions have even been deleted from the First Draft. Initially, the contracting parties agreed to limit the uncontrolled expansion of tourist infrastructure by defining geographical limits (article 11 para. 1). They also decided that, basically, no tourist facilities will be developed in glacial zones (article 11 para. 2). The use of vehicles for the grooming of ski runs was limited to periods when the vegetation cover is sufficiently protected by snow (article 12 para. 2). None of these provisions 'survived' the negotiation process, and hence they did not come into force.

The same applies to certain articles in the initial draft on agricultural issues, which were deleted by and by. Neither the requirement that Alpine pasture areas have to be managed in a form that suits local conditions (article 13 para. 3) nor the provision that certain plant production products shall be generally forbidden (article 17 para. 4) was enacted.

Following a similar trend, the wording of some articles was changed in a manner that undoubtedly made clear that the obligations resulting from the Protocol should become less restrictive. Clear commitments by the contracting parties were replaced by declarations of intent. According to the clear wording of the First Draft, the contracting parties 'undertake' to designate, provide and make sure that certain measures are taken.[21] At the end, these provisions only stated that the Alpine states 'shall' see to it, provide or make sure that, for instance, engineering techniques are used in endangered areas that are as compatible with nature as possible.[22]

These examples disclose a process of ongoing softening of the initial drafts. Therefore, provisions have become less binding, and all contracting parties were able to accept the outcome in the end. However, this is certainly quite a common development when international contracts are being negotiated.

[21]See art. 5, 7 para. 1 and 15 First Draft.

[22]See art. 6, 7 para. 2 and 10 para. 2 Final Draft.

4 'Recipe of Success' of the Soil Conservation Protocol

Legislation on soil protection often faces difficulties; some of the reasons have been mentioned. So what is the 'recipe of success' of the Soil Conservation Protocol of the Alpine Convention? Why did the contracting parties of the Alpine Convention find a compromise on this controversial issue, whereas agreement still seems far off on European level? The following outline briefly sets out why the negotiators were able to arrive at a compromise on this international treaty.

4.1 Homogeneity of the Application Area

The uniform natural conditions within its area of application are undoubtedly a fact that contributed to the conclusion of the Soil Conservation Protocol.[23] The Alps extend over the territories of Austria, France, Germany, Italy, Liechtenstein, Monaco, Slovenia and Switzerland. However, geographically and with regard to environmental circumstances, soils within this vast area are exposed to similar risks, for instance:

- erosion due to mass slides, mudslides, landslides, avalanches and floods;
- instability of soils as a consequence of global warming and the thawing of permafrost;
- soil formation, as well as the regeneration of impaired soils, happens very slowly;
- continuing soil sealing and soil consumption that, for instance, increase the risk of flood since retention areas are lacking,
- land shortage because almost only the valley floors are suitable for living and economic purposes; or
- soil compaction as a consequence of touristic usage of soils, for instance, the levelling of ski runs.

In contrast, legislation on soil protection within the European Union has to deal with a variety of soils and, consequently, with all sorts of different risks from desertification to restriction of sewage sludge. It has already been pointed out that 'soil' is a weak basis as the common denominator for this multitude of different rules.

[23] Markus (2017), p. 162.

4.2 *Awareness of the Need to Take Measure on Soil Protection*

In the late ‘80s and early ‘90s of the twentieth century, soil protection became a fashion topic of environmental politics. So when the negotiating process on the Soil Conservation Protocol started, all contracting parties—at least implicitly—shared the opinion that soils have to be legally protected. An undertaking, such as this international treaty, was regarded as a useful supplement to national efforts on soil protection. The leader of the Swiss delegation held in 1995, for instance, that ‘the conservation of soils as a livelihood resource and a living environment for humans had only been recognised as an important public task ten years ago. The Soil Conservation Protocol, thus, has to be considered as a thoroughly positive means to increase the awareness and to coordinate actions between the Alpine states.’[24]

It seems that the favourable climate for soil policy has changed since then.[25] Today, in times of economic crises, environmental protection and legislation are regarded as a brake on economic initiatives. Common welfare can only be reached via economic growth. Pursuing this argument, the positive effects of ever larger ski resorts prevail over the environmental interests affected. As a consequence, the political commitment to introduce sophisticated regulation on soil protection is limited because the advantages and usefulness of such legislation is doubted.[26]

4.3 *Expectation with Regard to the Soil Conservation Protocol*

An additional reason why the adoption was politically agreeable was that legal authorities and political actors of the contracting parties widely regarded the Protocol as an instrument of information policy, as a tool to coordinate actions on soil protection and to cooperate between the Alpine states.

It is significant that most of the legislators of the contracting parties have stressed during the ratification process that changes of national law do not go along with the approval of the treaty: ‘The key objectives have already been taken into account by national legislators and authorities.’[27] ‘There is no need to implement the rules of the Protocols by federal or provincial law since corresponding provisions of national law are already in force. The protocols are mainly being implemented by

---

[24]Explanatory note relating to the Soil Conservation Protocol of the Alpine Convention, Zi/En/ 26.04.95 (unpublished).

[25]Schmid (2016), p. 47 f.

[26]Schrader (2009), p. 134 f.

[27]Explanatory note of the Austrian Parliament relating to the Soil Conservation Protocol, 1096 BlgNR 21. GP 32.

political programmes.'[28] 'None of the provisions of the Protocols is directly applicable and, thus, they do not directly influence the national laws.'[29] These statements illustrate that the Soil Conservation Protocol and the other Protocols of the Alpine Convention have not been acknowledged as so-called black letter law but have rather been considered as a policy document in order to show good will in the field of soil protection.

However, signing a treaty under these circumstances and based on such a reception was a Pyrrhic victory. Since their coming into force as legal documents, the Alpine Convention and its Protocols fight for recognition by the courts. Apart from Austria, national authorities have not effectively based their decisions on the treaties, and it is not foreseeable that things are changing soon.

4.4 *Widely Accepted Content*

Provisions in international treaties are usually divided into directly and indirectly applicable regulations.[30] The effects of the former on the national legal systems are usually regarded as more severe since these articles and paragraphs have to be applied by administrative authorities and courts in the same way as Acts of the national parliaments. In contrast, indirectly applicable law in international treaties is either addressed to the legislators who have to implement it before it is applied by authorities or, when such provisions are couched in too general terms, they are (only) taken into account as aid to construction of national law.

In Austria, only some provisions of the Soil Conservation Protocol are generally regarded as directly applicable. These are clear prohibition clauses: 'The Contracting Parties undertake to preserve high moors and lowland moors' (article 9 para. 1). Mountain forests that offer a high degree of protection to their own location or, above all, to human settlements, transport infrastructures, croplands and similar areas 'shall be preserved in their original locations' (article 13 para. 1). Permits for the construction and levelling of ski runs 'are not granted for fragile areas' (article 14 para. 1).

Besides, the Soil Conservation Protocol includes many provisions that formulate objectives and that are, thus, target oriented. The ecological functions of soil 'shall be safeguarded and preserved both qualitatively and quantitatively on a long-term basis' (article 1 para. 2). 'The measures to be taken are aimed specifically at soil utilisation which suits its location, at the economical use of land resources, at the avoidance of erosion and detrimental changes to the soil structure, and at

[28] Explanatory note of the German Bundestag relating to the Soil Conservation Protocol, BR-Drs 224/02, p. 6.

[29] Report of the French National Assembly, 2 June 2004, N° 1634, p. 11.

[30] Müller (2016), p. 8 ff.

minimising the input of substances harmful to the soil' (article 1 para. 3). The use of peat 'shall be discontinued completely in the medium term' (article 9 para. 1).

Such objectives in legal provision have to be placed in relation to the issue that is regulated. Due to the multifunctional role of soil, many policies are involved and conflicts of interest are unavoidable. In response to such a situation, legislators help themselves by introducing 'balancing-clauses' according to which authorities have to set contrary interests in relation in order to decide which particular interest is prevailing in the specific case. Objectives laid down in legal acts have to be considered in such decisions requiring the weighing of interests.[31] As a consequence, such balancing clauses leave administrative authorities and courts a wide margin of appreciation since the process of balancing almost always includes a subjective element. Controversial objectives, thus, can easily be bypassed when this way of applying legal provisions is prescribed.

That pertains, in particular, to the Soil Conservation Protocol, that, in effect, the contracting parties had to exceed a low threshold when signing a contract that includes to large extent target-oriented provisions. The risk of considerable impact on the national legal systems was low.

4.5 No Need to be Afraid of the Compliance Mechanism

A typical handicap of international law is the lack or at least the weakness of a legal enforcement regime. Although the breach of international law is illegal, non-compliance is usually (only) subject to political sanctions, for instance, in the form of 'naming and shaming'. As a consequence, modern international treaties enforce compliance regimes in order to increase their effectiveness and the culture of compliance.[32]

The legal regime of the Alpine Convention has also established a compliance mechanism. The Compliance Committee—a body constituted of a maximum of two representatives per contracting party of the Alpine Convention—was established to monitor the compliance of the national authorities with the legal regime of the Alpine Convention. However, although this institution is committed to elaborating the Convention and its Protocols,[33] its visibility seems to have suffered from the profile it was given and its closeness to politics. Decisions are

---

[31]See Austrian Federal Administrative Court (Bundesverwaltungsgericht) 28.8.2014, W104 2000178-1 (*Kronhofgraben*); Baumgartner (2016), p. 53; Schmid (2014), p. 695; Müller (2016), p. 20 ff.

[32]An example is the Aarhus-Convention and its Compliance Committee (ACCC). Its rulings—although not legally binding—have turned out to be valuable from the content point of view and have, consequently, already been cited by national courts in order to support their decisions with further legal arguments (see e.g. the judgement of the German Administrative Court [BVerwG] 5.9.2013, 7 C 21.12 [ECLI:DE:BVerwG:2013:050913U7C21.12.0]; also see Alge [2012], p. 109).

[33]See the record of decisions and recommendations at www.alpconv.org (20.6.2016).

reached on very rare occasions, and they are of little significance for the understanding of the Protocols.[34] Thus, the effectiveness of the compliance mechanism of the Alpine Convention is very low, which also might have been a reason why approval of the Soil Conservation Protocol by the contracting parties was not too difficult. In contrast to that, soil legislation on European Union level is backed up by the jurisprudence of the European Court of Justice, which is known to be 'dynamic' and sometimes unpredictable.

5 Concluding Remarks

When the Soil Conservation Protocol was signed in 1998, it represented 'cutting-edge' legislation on soil protection. However, agreement was achieved, on the one hand, because controversial issues had been omitted and, on the other hand, because of the specific structure of the Protocol, which, for instance, contains many target-related provisions and which is not subject to an effective compliance mechanism. Both of which have been important precondition for the signing of the treaty.

When reflecting on the Soil Conservation Protocol and on the question what lessons generally can be drawn from it for legislation on soil protection, a somehow sobering conclusion can be reached. In analysing the 'success' of the Soil Conservation Protocol, both sides of the coin have to be taken into account: a successful compromise at the price of reduced legal force and of contents that are watered down.

References

Alge T (2012) Aarhus-Entscheidung: Österreich unter Handlungsdruck. Recht der Umwelt 109–110

Baumgartner C (2016) Naturschutzrechtliche Interessenabwägung bei Stromleitungsanlagen unter besonderer Berücksichtigung des Energieprotokolls. In: Essl J, Schmid S (eds) Das Protokoll "Energie" der Alpenkonvention. Verlag Österreich, Wien, pp 53–63

Blasberg D (2008) Inhalts- und Schrankenbestimmungen des Grundeigentums zum Schutz der natürlichen Lebensgrundlagen. Springer-Verlag, Berlin, Heidelberg

Duschanek A (1989) Verfassungsfragen des Bodenschutzes. In: Duschanek A (ed) Beiträge zum Bodenschutz. Österreichischer Wirtschaftsverlag, Wien, pp 15–31

Fokuhl C (1994) Rechtliche Situation und Entwicklung des Bodenschutzes in Deutschland. Naturschutz und Landschaftsplanung 26(2):49–52

Markus T (2017) The Alpine Convention's soil conservation protocol: a model regime? In: Ginzky H et al (eds) International yearbook of soil law and policy 2016. Springer, pp 149–164

Mayrhofer W (2015) Bodenschutz – Die Alpenstaaten im Vergleich. Die Alpenkonvention 79(2):10–11

[34]See the decision of the Compliance Committee on art. 6 para. 3 Tourism Protocol, ImplAlp/2014/20/6a/3, published at www.alpconv.org (20.6.2016).

Milde KF (1951) Legal principles and policies of soil conservation. Fordham Law Rev 20(1):45–78
Müller A (2016) Unmittelbare und mittelbare Anwendung und Wirkung des Energieprotokolls der Alpenkonvention. In: Essl J, Schmid S (eds) Das Protokoll "Energie" der Alpenkonvention. Verlag Österreich, Wien, pp 7–26
Odendahl K (2001) Bodenschutz nach Völkerrecht: Bestandsaufnahme und Entwicklungsperspektiven. Archiv des Völkerrechts 39:82–109
Peine FJ (1992) Die Gesetzgebungskompetenzen des Bundes für den Bodenschutz. Natur und Recht 14(8):353–360
Roellecke G (2005) Natur- und Denkmalschutz durch privates Eigentum? In: Deppenheuer O (ed) Eigentum. Ordnungsidee, Zustand, Entwicklung, Springer-Verlag, Berlin, Heidelberg, pp 81–91
Schmid S (2007) Alpenkonvention und Moorschutz. Recht der Umwelt 158–166
Schmid S (2014) Anmerkung zu BVwG 28.8.2014, W104 2000178-1/63E. Zeitschrift der Verwaltungsgerichtsbarkeit 692–700
Schmid S (2015) Bodenschutzrecht in Österreich. Die Alpenkonvention 79(2):7–10
Schmid S (2016) Auf der Suche nach dem effektiven Alpenkonventionsrecht. In: Haßlacher P (ed) 25 Jahre Alpenkonvention. Innsbruck-Igls, pp 40–48
Schrader C (2009) Neue Instrumente des Bodenschutzes. In: IUR, ÖWAV (eds) Jahrbuch des österreichischen und europäischen Umweltrechts 2009. Manz, Wien, pp 133–150
Stollmann F (1996) Die Bodenschutzgesetze der Länder. Natur und Landschaft 71(9):367–370

Part V
Cross-Cutting Topics

Soil Health, Sustainable Land Management and Land Degradation in Africa: Legal Options on the Need for a Specific African Soil Convention or Protocol

Robert Kibugi

1 Introduction

This chapter explores the situation on soil health management, and the responses to land degradation in Africa, and examines whether the current multiplicity of international law instruments is sufficient or whether an Africa-specific legal instrument is required to address soil health in the context of sustainable land management and in response to land degradation. The chapter has drawn its inspiration from the first volume of the International Yearbook of Soil Law and Policy (2016), in which the welcome note from Africa written by Mr. Shem Shikongo from Namibia[1] focused on the continuing prevalence of land degradation across Africa, despite multiple international legal options and in spite of the opportunity for Africa to frame a bespoke legal and policy approach to address land degradation, food security, and related challenges, such as climate change. In that welcome note, the author summed up what, in his opinion, was the challenge of development and land degradation that continues to confound the African continent.[2]

> One of Africa's most urgent challenges today is food security; this is amid all the other urgent developmental challenges that are just as pressing and urgent. To make the matter worse, a hungry people are an angry people, and this worsens governance and Africa's ability to remain focused and to deal with the real issues without following the dreams and topics of others that do not serve our interests. Despite many efforts to assist Africa on her

[1]Directorate of Tourism and Gambling, at the Ministry of Environment and Tourism, of the Government of Namibia.

[2]Shikongo (2016), pp. 6–7.

R. Kibugi (✉)
University of Nairobi, Nairobi, Kenya
e-mail: rmkibugi@gmail.com

H. Ginzky et al. (eds.), *International Yearbook of Soil Law and Policy 2017*,
International Yearbook of Soil Law and Policy,
https://doi.org/10.1007/978-3-319-68885-5_21

> road toward sustainable development with a focus on the international sustainable development agenda, the battle has not been won. The problems and challenges continue on the continent driven by, among others, loss of culture, loss of language, loss of traditional knowledge, globalization, conflict and war, unclear land and resource tenure, livelihood and poverty, population growth, and use of foreign legal systems that fail to take into account our customary laws and culture. All of these challenges have an impact on Africa's ability to deal with her soils.

The author further attempted to establish a causal link as to why the challenge appears to persist:[3]

> Is it perhaps because of the fact that Africa is on a road for which the script she is using is not African—thereby addressing issues from a position of inherent disadvantage given the fact that Africa has given up its competitive and comparative advantage? How can African indigenous knowledge systems and true culture come to our aid in this struggle against the onslaught of soil degradation given its deleterious effect on Africans and the rest of the world? Is there a way out? The picture looks bleak but what is the road ahead and what can we conclude?

Against this background, Mr. Shikongo contends that "The reality is that Africa is still the least 'messed-up' continent in the world with regards to soils, ecosystems and biodiversity," and he further argues that Africa can be developed to become the "green lung of the World" and serve as a global carbon sink.[4] In this context, and among other recommendations made to enhance how Africa acts to address land degradation, and soil health, Mr. Shikongo argues that there is a "...need for the establishment of an Africa wide protocol or convention on soils ..."[5] It is this recommendation and background that form the basis of the assessment, evaluation, and propositions made in this chapter.

Section 1 is the introduction; Sect. 2 evaluates the situation on soil health and land degradation in Africa. Section 3 reviews the applicable global international law, and Sect. 4 of the chapter explores African level treaty law and policy approaches on soil and land governance. Section 5 is an inquiry into the need for an Africa-specific treaty instrument on land degradation, including examination of options. Section 6 provides a conclusion to the chapter.

2 Evaluating the Soil Health and Land Degradation Situation in Africa

According to the FAO, in a 2015 report,[6] although Africa has a diverse range of soils and land-use systems, very large areas, particularly in West Africa, experience unsustainable systems of land use and erosion, together with widespread low

[3]Shikongo (2016), pp. 6–7.

[4]Shikongo (2016), p. 7.

[5]Shikongo (2016), p. 7.

[6]FAO (2015), p. 231.

fertility that is caused by nutrient depletion, which is considered to be the chief biophysical factor limiting small-scale farm production.[7] The report further highlights the situation in North Africa, noting that the region is considered as the most water scarce place in the world and is particularly vulnerable to the impacts of climate change, increasing drought, declining soil fertility, and declining agricultural production, given that scarcity of land and water resources.[8]

2.1 *Overview of the Drivers of Land Degradation in Africa*

From a broader continental perspective, the 2010 African Union (AU) Framework and Guidelines on Land Policy in Africa put the situation in a geographical and ecological context,[9] noting that a key aspect of Africa's land question is that in spite of extensive dependence on farming, not much of the continent's land is arable or potentially arable. Thus, according to this AU land policy guidelines, large parts of the continent are deserts or semi-arid and/or facing ecological damage, and in many instances, unequal distribution of land has relegated a growing population of small-scale farmers onto marginal areas, leading to increased physiological pressure and land resource degradation, including deforestation.[10] In the narrow context of addressing desertification, the United Nations Convention to Combat Desertification (UNCCD), in the Regional Implementation Annex for Africa, has identified the conditions that have made the African continent particularly vulnerable to, and impacted by, land degradation and desertification:

(a) the high proportion of arid, semi-arid, and dry subhumid areas;
(b) the substantial number of countries and populations adversely affected by desertification and by the frequent recurrence of severe drought;
(c) the large number of affected countries that are landlocked;
(d) the widespread poverty prevalent in most affected countries, the large number of least developed countries among them, and their need for significant amounts of external assistance, in the form of grants and loans on concessional terms, to pursue their development objectives;
(e) the difficult socioeconomic conditions, exacerbated by deteriorating and fluctuating terms of trade, external indebtedness, and political instability, which induce internal, regional, and international migrations;
(f) the heavy reliance of populations on natural resources for subsistence, which, compounded by the effects of demographic trends and factors, a weak technological base, and unsustainable production practices, contributes to serious resource degradation;

[7] FAO (2015), p. 231.
[8] FAO (2015), p. 233.
[9] African Union (2010), p. 5.
[10] African Union (2010), p. 5.

(g) the insufficient institutional and legal frameworks, the weak infrastructural base, and the insufficient scientific, technical, and educational capacity, leading to substantial capacity-building requirements; and
(h) the central role of actions to combat desertification and/or mitigate the effects of drought in the national development priorities of affected African countries.

These conditions, according to a 2015 study by UNEP and Economics for Land Degradation (ELD), are part of the underlying drivers of land degradation in Africa.[11] This study on the economics of land degradation in Africa classifies the drivers of degradation as biophysical, socioeconomic, or human pressures.

Biophysical drivers of land degradation include the following[12]:

1. The high prevalence of drylands is one, where the majority of African countries have their agricultural lands in semi-arid regions and which are more susceptible to land degradation and desertification, given the fragility of dryland soils.
2. Inherent land quality – this refers to the ability of the land to perform its function of sustainable agriculture and enable it to respond to sustainable land management, with the ELD/UNEP study showing extensive coverage of land with low soil resilience, which is soil that is permanently damaged from degradation.
3. Type of soils – the 2013 Soil Atlas of Africa[13] adopted the approach of the World Reference Base for Soil Resources through its 32 WRB Reference Soil Groups (RSG), allocated on the basis of dominant identifiers that are usually the factors or processes that most clearly condition the formation of the soil.[14] According to the Soil Atlas for Africa, over 60% of the soil types represent hot, arid, or immature soil assemblages,[15] and agricultural production in much of Africa is hampered by the predominance of inherently low soil fertility, fragile ecosystems that do not support intensive agriculture.

   The three drivers above, all focused on fragility of soil, soil resilience, and soil fertility, demonstrate that, *a priori*, the soils in Africa are inherently vulnerable to degradation, and most have low levels of resilience.
4. Climate Variability, as most African countries experience large variations in rainfall, both throughout the year and between years, and they are subject to frequent extremes of flooding or drought, both of which contribute to soil erosion and land degradation.

Socioeconomic drivers are summarized as (a) poverty, with Africa having a higher share of low-income countries relative to other world regions and poverty acting as a driver of land degradation when farmers, herders, and others who depend directly on land resources cannot wait for soils and vegetation to recover and resort to inappropriate land management, and a vicious cycle in which rising land

[11] ELD and UNEP (2015), p. 24.
[12] ELD and UNEP (2015), pp. 22–29.
[13] Soil Atlas of Africa (2013), pp. 49–50.
[14] Soil Atlas of Africa (2013), p. 50.
[15] Soil Atlas of Africa (2013), p. 50.

degradation and lost livelihood drives people to put increasing pressure on fragile resources[16] and (b) population growth and density, where over the past 50 years, Africa has experienced continuous and rapid population growth increasing by nearly 300% since the early 1960s and doubling between 1982 and 2009, while in the same (1982–2009) period, the area of agricultural land (arable land plus land under permanent crops) has not grown in parallel over this period.[17] According to the Soil Atlas of Africa, population growth has also exacerbated intensity of productive pressure on soil such that while in 1962 each hectare of cultivated land in Africa had to support 1.91 persons, by 2009, this had risen to 4.55 persons for each hectare.[18] Another socioeconomic pressure identified by the ELD/UNEP study is deforestation, to make way for agriculture, biomass energy, infrastructure development, and timber production. Deforestation remains a key challenge, such as seen in Kenya, where activities in the agriculture sector have been identified as the greatest single cause of forest conversion, together with charcoal production.[19] A 2013 study on the drivers of deforestation reported that "while increased incomes of farmers could have resulted in increased investment in farm inputs," it was possible "that the remarkable performance of the sector could actually be due to more land being put under agriculture."[20]

2.2 The Impact of Climate Change on Soil Health and Land Management

Climate change represents a major global challenge to sustainable development and poses a specific obstacle to Africa due to the vulnerability of production systems, including land. According to the Fifth Assessment Report (AR5),[21] changes are occurring in the distribution and dynamics of all types of terrestrial ecosystems in Africa, including deserts, grasslands and shrublands, savannas and woodlands, and forests, and there is a small overall expansion of desert and contraction of the total vegetated area and a large increase in the extent of human influence within the vegetated area, accompanied by a decrease in the extent of natural vegetation.[22] AR5 further records that continuing changes in precipitation, temperature, and carbon dioxide associated with climate change are very likely to drive important future changes in terrestrial ecosystems throughout Africa.[23]

---

[16]ELD and UNEP (2015), p. 30.

[17]ELD and UNEP (2015), p. 31.

[18]Soil Atlas of Africa (2013), p. 152.

[19]Kenya (2013), p. 73.

[20]Kenya (2013), p. 36.

[21]IPCC (2014).

[22]IPCC (2014), p. 1213.

[23]IPCC (2014), p. 1214.

The Soil Atlas of Africa addresses the nexus between soil, as part of the terrestrial ecosystem, and climate change, noting that climate is a major soil-forming factor, and therefore climate change is likely to alter the key natural processes that underpin the capability of soils to perform their key functions.[24] In this context, temperature, rainfall, and changes in atmospheric carbon dioxide also affect soil ecology and organic matter levels, which in turn determine soil characteristics such as structure, water regimes, pH, and nutrient levels. For instance, higher temperatures also promote a more rapid breakdown of organic matter in the soil due to a thermal boost to microbial activity, which accelerates the release of carbon dioxide and methane into the atmosphere through increased soil respiration. Intense rainstorms or prolonged periods with lower levels of precipitation could lead to increased vulnerability to erosion, which can result in the loss of topsoil nutrients and stored carbon either as greenhouse gas emissions to the atmosphere or as particulate organic carbon. Additionally, the rate of organic matter decomposition also depends on soil moisture levels, which could affect soil structure and acidity levels, which, together with organic matter content, define the ability of a soil to store water and sustain many of the organisms that live within it.[25]

African countries are cognizant of the challenge faced from climate change, evident from the fact that when the Paris Agreement opened for signature on April 22, 2016, 47 out of 54 African countries appended signature.[26] A review of a random sample of Nationally Determined Contributions (NDCs) submitted by individual African countries (sample drawn from South, East, West, and North Africa) to the Secretariat of the United Nations Framework Convention on Climate Change (UNFCCC), as required under article 3 of the Paris Agreement, is as follows:

1. Algeria NDC—in the NDC submitted in September 2015, Algeria noted its national circumstances as being affected by desertification and land degradation, with most of the country being arid or semi-arid, and the areas receiving more than 400 mm of rain per year are located in a narrow strip along the coast, not-exceeding 150 km large. The country reported that it would, among other actions, implement adaptation measures to fight against erosion and rehabilitate its degraded lands as part of the efforts to combat desertification and integrate the impacts of climate change into sectorial strategies, in particular for agriculture, water management, public health, and transport.[27]
2. Kenya NDC—in the NDC submitted by Kenya in July 2015, the country in reporting on national circumstances, noted that more than 80% of the country's landmass is arid and semi-arid land (ASAL) with poor infrastructure, and other developmental challenges, and that the economy is highly dependent on climate-

[24]Soil Atlas of Africa (2013), p. 151.

[25]Soil Atlas of Africa (2013), p. 151.

[26]Online: http://www.un.org/en/africa/osaa/advocacy/climate.shtml (accessed on 5 June 2017).

[27]Algeria (September 2015), 8.

sensitive sectors such as agriculture that is mainly rainfed, energy, tourism, water, and health. The NDC notes that for Kenya, climate hazards, including droughts and floods, have caused considerable losses estimated at 3% of the country's gross domestic product (GDP).[28] Therefore, key actions will include mainstreaming climate change adaptation in land reforms[29] and making progress toward achieving a tree cover of at least 10% of the land area of Kenya.[30]

3. Ghana NDC[31]—in the 2015 NDC, Ghana identifies sustainable land use, including food security, as a strategic area of action, and this includes actions traversing agriculture and sustainable forest management,[32] extending to addressing land degradation through REDD+ measures that will enhance reforestation and afforestation.[33]
4. Botswana NDC—in the NDC submitted in 2015, Botswana identifies climate smart agriculture as a priority area of adaptation action, which includes interventions pertinent to soil and land management, such as low to zero tillage, and multicropping to increase mulching, which reduces evapotranspiration and soil erosion.[34] Other adaptation priorities include sustainable land management, with Botswana pointing to existing national initiatives such as the Sustainable Land Management in Ngamiland and Central Districts, which is aimed at enhancing resilience and reducing the vulnerability of communities to climate change.[35]

The Soil Atlas of Africa sums up the challenge of climate change, and the impact on land degradation and soil health, noting that much of the population of sub-Saharan Africa is at high risk to the impacts of climate change as many people live in rural areas where income and employment depend almost entirely on rainfed agriculture, especially since the continent already has a highly variable and unpredictable climate that is acutely vulnerable to floods and droughts.[36] According to the Soil Atlas, a third of the people in the region live in drought-prone areas, and floods are a recurrent threat in several countries, and with climate change, large parts of the region will become drier, increasing the number of people at risk of hunger and poverty by tens of millions.[37]

It is important to highlight that, as noted in AR5, the level of pressures and climate change vulnerability facing Africa makes it necessary to highlight possible risks of maladaptation, which in the context of land degradation is a necessary

[28]Kenya (July 2015), p. 1.
[29]Kenya (July 2015), p. 5.
[30]Kenya (July 2015), p. 2.
[31]Ghana (September 2015).
[32]Ghana (September 2015), pp. 2 & 7.
[33]Ghana (September 2015), p. 13.
[34]Botswana (2015), p. 2.
[35]Botswana (2015), p. 4.
[36]Soil Atlas of Africa (2013), p. 151.
[37]Soil Atlas of Africa (2013), p. 151.

caveat to ensure that African efforts are structured to avoid outcomes that may serve short-term goals but come with future costs to society, such as through policy or legal interventions that may favor economic growth over resilience, and themselves act as stressors.[38] AR5 gives an illustration from Simiyu wetlands in Tanzania, where adaptation strategies resulted in farmers moving to intensify farming in wetter parts of the floodplain, as a coping strategy to take advantage of the higher soil water content and as a result bringing about maladaptation through negative impacts on natural vegetation.[39] The foregoing analysis of the situation of soil health and land degradation in Africa, although succinct and high level, demonstrates how complex this is and that the challenges traverse a number of scientific and socioeconomic complexities. In addition, the challenges and situation discussed here traverse national boundaries and, at the global, continental, and regional levels, are governed by various legal instruments, which are reviewed in the next section—in order to further appreciate whether, as contended by Shem Shikongo, it is necessary to have an Africa-specific soil legal instrument.

3 Reviewing International Legal and Policy Instruments on Soil and Land Degradation Applicable to Africa

In this section, the chapter reviews several international legal and policy instruments governing soil and land degradation and evaluates whether their application is sufficient for African states, with respect to developing an endogenous primary focus on degradation, in light of challenges of low soil resilience, biodiversity loss, and climate change. In addition, African countries are part of the global compact to implement the Sustainable Development Goals (SDGs), including action on climate change, halting biodiversity loss, and action toward land degradation neutrality.

3.1 World Soils Charter and World Soil Policy

FAO adopted the World Soil Charter in 1982,[40] and in the same year, UNEP published the World Soils Policy.[41] FAO reported in 2015 that whereas the principles set out in these two soft law instruments have provided valuable guidance for national government pursuing actions on sustainable land management, it has been difficult to assess their practical impact.[42] The World Soil Charter (WSC) was, in

[38]IPCC (2014), p. 1235.

[39]IPCC (2014), p. 1235.

[40]FAO (1982).

[41]UNEP (1982).

[42]FAO (2015), p. 225.

2015, revised by the Global Soil Partnership (GSP), which restated the principles to recognize the critical role of soil, as highlighted indicatively below:

(a) Soils are a key enabling resource, central to the creation of a host of goods and services integral to ecosystems and human well-being (Principle 1).
(b) Soil management is sustainable if the supporting, provisioning, regulating, and cultural services provided by soil are maintained or enhanced without significantly impairing the soil functions that enable those services (Principle 3).
(c) The implementation of soil management decisions is typically made locally and occurs within widely differing socioeconomic contexts. The translation of good global soil governance into specific measures appropriate for adoption by local decision-makers requires multilevel, interdisciplinary initiatives by many stakeholders (Principles 4).

The 2015 WSC revisits the roles and actions that can be taken by governments and international organizations. The WSC specifically curves a role for international organizations, urging them to assist governments to establish appropriate legislation, institutions, and processes to enable them to implement appropriate sustainable soil management practices, a call that is likely suitable for the African Union, as well as the various regional economic communities in Africa. The revised WSC also calls on governments to take into account the role of soil management practices in planning for the adaptation and mitigation of climate change, and maintaining biodiversity.

From a policy and legislative perspective, the WSC's call on governments to mainstream and integrate principles and practices of sustainable soil management into policy guidance and legislation at all levels of government, ideally leading to the development of a national soil policy, is valuable. This is because the WSC further proposes that governments should put in place and pursue measures intended to overcome obstacles to the adoption of sustainable soil management associated with land tenure, and the rights of users, and supporting research that will provide scientific backing to develop and implement sustainable soil management programs relevant to end users. For African governments, this is an instrumental approach, especially the call to link land use (through sustainable soil and land management) with tenure rights, but nonetheless, the focus on scientific research, without taking into account the role of African indigenous and traditional knowledge, could impact the uptake of research solutions by farmers, for instance.

3.2 The United Nations Framework Convention on Climate Change and Related Instruments

The global climate change governance framework is administered through the 1992 United Nations Framework Convention (UNFCCC), together with the 1997 Kyoto Protocol, and the 2015 Paris Agreement. The Paris Agreement, in article 4, introduces a modification to the principle of common but differentiated responsibilities, with developing countries called upon to voluntarily take up mitigation targets and move over time toward economy-wide emission reductions, in light of different national circumstances. This is to be achieved through all Parties communicating to the UNFCCC Secretariat their Nationally Determined Contributions (NDCs), as required by article 4 of the Paris Agreement.

From the brief review of a sample of African NDCs, above, it is clear that sustainable land management is considered an integral element of climate change adaptation and mitigation actions. Further, this could be enhanced if African countries were to actively implement the letter and spirit of article 6 of the Paris Agreement, which urges Parties to pursue voluntary cooperation in implementing NDCs in order to promote sustainable development and environmental integrity. It is worth noting that from the NDCs reviewed for this chapter, cooperation among African countries is not evident as a priority. This is very closely related to global actions on the reduction of emissions from deforestation and degradation (REDD+), as evident from Decision 1/CP.16 (UNFCCC), urging developing country Parties, when developing and implementing their national REDD+ strategies, to address, among others, the drivers of deforestation and forest degradation and land tenure issues,[43] which may often traverse national borders.

As part of the Warsaw Framework for REDD+, countries are required to prepare and submit a National Forest Reference Emission Level/National Forest Reference Level (FREL), which is relevant to addressing degradation and deforestation.[44] In their FREL submissions to the UNFCCC, several African countries have demonstrated the linkage between climate change actions and soil management by addressing how soil carbon pools factor into their FRELs. In the January 2017 submission to the UNFCCC, Ghana identified changes in soil carbon as being relevant to post-deforestation land use and based its estimation on the use of soil factors that account for how the soil is tilled, method of management, and soil inputs in the post-deforestation land use.[45]

Zambia, in its FREL submission, reported that the national forest inventory undertaken had conducted a comprehensive sample of soil characteristics, including soil organic carbon, but very little was so far known regarding soil carbon dynamics in Zambia, and as such it remained unclear how soil organic carbon behaves within the deforestation activity chosen by Zambia. Consequently, Zambia excluded soil carbon pools from its FREL,[46] an approach similarly taken by Ethiopia, which reported difficulty in obtaining accurate data on a national scale.[47] Why is this important in the context of soil management and dealing with degradation in Africa? First, it is because REDD+ addresses challenges of

[43]UNFCCC Decision 1/CP.16, para 72.

[44]UNFCC Decision 13/CP.19.

[45]Ghana (2017), p. 78.

[46]Zambia (2016), pp. 4–5.

[47]Ethiopia (2016), p. 5.

degradation and deforestation that arise directly from land use and land management choices and further because the 2006 IPCC guidelines for National Greenhouse Gas Inventories established the link between changes in soil carbon with human activities and alterations in carbon dynamics of forest soils, for instance through forest management activities such as rotation length, choice of tree species, drainage, harvest practices (whole tree or sawlog, regeneration, partial cut, or thinning), site preparation activities (prescribed fires, soil scarification), and fertilization.[48]

3.3 The Convention on Biological Diversity

The 1992 Convention on Biological Diversity (CBD) has set out its objectives, in article 1, as the conservation of biological diversity, the sustainable use of its components, and the fair and equitable sharing of the benefits arising out of the utilization of genetic resources. The CBD, through article 6(b), calls on States Parties to "integrate, as far as possible and as appropriate, the conservation and sustainable use of biological diversity into relevant sectoral or cross-sectoral plans, programs and policies." Further, article 10(a) of the CBD calls on the Parties to "integrate consideration of the conservation and sustainable use of biological resources into national decision-making."

The CBD's Strategic Plan for Biodiversity 2011–2020 and the Aichi Biodiversity Targets, adopted by Parties in 2010, identified as Strategic Goal B the reduction of direct pressures on biodiversity and promoting sustainable use as priorities. Aichi Target 5 set the objective that by 2020, the rate of loss of all natural habitats, including forests, should be at least halved and, where feasible, brought close to zero and that degradation and fragmentation are significantly reduced.[49] During the 16th Conference of Parties (CoP16) in December 2016, the CBD adopted Decision 13/3, on strategic actions to enhance the implementation of the Strategic Plan for Biodiversity 2011–2020 and the achievement of the Aichi Biodiversity Targets.[50] A key plank of this decision was the adoption by the CBD of mainstreaming as a legal tool for enhancing the fulfillment of the objectives of the Convention, and the Parties recognized that, among other sectors, agriculture, forestry, fisheries and aquaculture, and tourism depend heavily on biodiversity, and these sectors impact biodiversity through various direct and indirect drivers, such that loss of biodiversity can impact these sectors negatively.[51] This decision further recognized the role

[48]IPCC (2006), 4.2.2.

[49]CBD (2010), para 13.

[50]CBD (2016).

[51]CBD (2016), preamble (c).

of land tenure and land use and urged Parties[52] to apply the FAO Voluntary Guidelines on the Responsible Governance of Tenure (VGGT).[53]

The VGGT establish a nexus between land use outcomes such as sustainability or degradation and land tenure systems and argue that tenure systems determine who can use which resources, for how long, and under what conditions, such that governance of tenure is a crucial element in determining if and how people and communities are able to acquire rights, and associated duties, to use and control land, and forests, and weak governance adversely affects sustainable use of the environment.[54] The role of the CBD in soil governance is reinforced by the mainstreaming approach, which is part of the Strategic Goals and Aichi Targets as Strategic Goal A, to address the underlying causes of biodiversity loss by mainstreaming biodiversity across government and society. This is to be achieved through Aichi Target 17, which requires Parties to develop, adopt, and implement, as a policy instrument, a participatory and updated National Biodiversity Strategy and Action Plan (NBSAP). Taking into account that soil and its organisms are part of biodiversity and its components, the NBSAP as a national policy tool for mainstreaming biodiversity, as well as the Aichi Targets, demonstrates that the CBD internalizes soil governance but that, as pointed out in CBD Decision 13/3 referred to above, it will be necessary to enhance mainstreaming in key sectoral activities that impact soil health and sustainable land management, such as agriculture. For Africa, the CBD is a valuable legal tool, especially through the implementation of the NBSAPs. However, it is also necessary to have in place legal mechanisms that focus on land degradation and sustainable land management, as the point of entry, from which CBD mainstreaming can draw complementing support. Such a possibility is evident through the provisions of the 1968 and 2003 African Conventions on the Conservation of Nature and Natural Resources, which are examined further in Sect. 4 of this chapter.

3.4 The United Nations Convention to Combat Desertification

The United Nations Convention to Combat Desertification (UNCCD) has the objective, in article 2, to combat desertification and mitigate the effects of droughts, particularly in Africa, through effective action at all levels, in the framework of an integrated approach consistent with Agenda 21,[55] with a view to contributing to the achievement of sustainable development. Chapter 12 of Agenda 21 is entitled "Managing Fragile Ecosystems: Combating Desertification and Drought," and

[52] CBD (2016), para 7.

[53] FAO (2012).

[54] FAO (2012), v.

[55] Agenda 21 (1992).

Programme Area B of Chapter 12 was dedicated to "Combating land degradation through, inter alia, intensified soil conservation, afforestation and reforestation activities." As a basis for action in Agenda 21's Programme Area B, states were called upon to launch preventative measure areas that are not yet affected or are only slightly affected by desertification; implement corrective measures to sustain the productivity of moderately desertified land; and take rehabilitative measures to recover severely or very severely desertified drylands.[56] This approach by Agenda 21 is consistent with article 2(2) of the UNCCD, which notes that achieving the objective of the UNCCD will involve long-term integrated strategies that focus simultaneously, in affected areas, on improved productivity of land and the rehabilitation, conservation, and sustainable management of land and water resources, leading to improved living conditions, in particular at the community level. The focus on the UNCCD has been enhanced from just paying attention to desertification, in recent years, to approaching the global challenge through the concept of land degradation neutrality (LDN), which has then been linked to the Sustainable Development Goals (SDG 15.3).

The International Working Group, appointed under the UNCCD, has provided a science-based definition of land degradation neutrality (LDN) as a state whereby the amount and quality of land resources necessary to support ecosystem functions and services and enhance food security remain stable or increase within specified temporal and spatial scales and ecosystems.[57] The scientific conceptual framework for land degradation neutrality[58] creates a target for land degradation management, promoting a dual-pronged approach of measures to avoid or reduce degradation of land, combined with measures to reverse past degradation. The objective is that losses are balanced by gains in order to achieve a position of no net loss of healthy and productive land. Thus, the concept of neutrality involves counterbalancing anticipated losses with measures to achieve equivalent gains.[59] The scale of implementation of LDN, at which neutrality is to be achieved, is the individual land type, within the landscape—for example, a catchment. This means that direct correlation of LDN measures and targets with ecosystem types and the specific biophysical and human challenges facing such an ecosystem type are critical considerations in structuring LDN implementation approaches, and targets. Thus, according to the LDN conceptual framework, it is important to couple management of land degradation with existing land-use planning, such that land-use planning takes into account the likely outcomes of land use and land management decisions, so that anticipated degradation can be counterbalanced by interventions to reverse the impacts of land degradation elsewhere in order to achieve land degradation neutrality.[60]

[56] Agenda 21 (1992), para 12.15.

[57] Decision 3/COP.12., para 2.

[58] UNCCD (2016).

[59] UNCCD (2016), p. 1.

[60] UNCCD (2016), p. 3.

The broadening of the legal scope of the UNCCD to include land degradation is supported by Boer et al., who contend that through UNCCD Decision 8/COP.12, the geographical scope of the Convention was extended *de facto* as the decision noted that a significant proportion of land degradation occurs beyond arid, semi-arid, and dry subhumid areas and further recognized that Parties may use the UNCCD to guide their policies relating to desertification/land degradation and drought (DLDD) and voluntary targets when striving to achieve land degradation neutrality (LDN) at national and subnational levels.[61]

This approach provides more legal tools to African nations as they get underway to implement the SDGs and an identification of indicators to guide the implementation of SDG 15.3. A global indicator list is now contained in the Report of the Inter-Agency and Expert Group on Sustainable Development Goal Indicators (E/CN.3/2017/2), Annex III, which was adopted at the 48th session of the United Nations Statistical Commission held in March 2017. The indicator that is identified for SDG 15.3 is the proportion of land that is degraded over total land area, but Knut Ehlers argues that this indicator is not sufficiently precise as there is currently no globally accepted definition of the term land degradation but a multitude of views colored by local perspectives and subjective interests.[62] Ehlers further raises doubt on whether it will be possible to successfully establish priorities and to interpret and fully technically refine the SDG targets at the global level.[63] In seeking consistency with the 2030 Agenda for Sustainable Development, which states that different national realities, national policies, and national priorities play an important role in the implementation of the SDG Agenda, Ehlers posits that it would likely be possible to determine indicators nationally, where it may be easier to develop a common understanding in the process of balancing site-specific and needs-based interests, including aspects of monitoring capacity.[64]

3.5 Intermittent Conclusion

From the foregoing, the global treaty mechanisms for climate change, biodiversity, and desertification are evolving in a manner that embraces the challenge of land degradation and internalizes implementation of the SDGs. African nations continue to contend with heightened challenges when dealing with biodiversity loss and climate change, with land degradation and low soil resilience as the common factors that negatively impact sustainable development. The development of NBSAPs, and their adoption as formal policies, as required by the CBD, would reinforce mainstreaming of biodiversity concerns across various sectors, including

[61]Boer et al. (2016), p. 62.

[62]Knut Ehlers (2016), p. 81.

[63]Knut Ehlers (2016), p. 81.

[64]Knut Ehlers (2016), p. 81.

land management. However, since the challenge of land degradation is very acute across Africa, it is important that African countries explore how to set specific legal approaches and methodologies that primarily focus on addressing land degradation, such as through LDN, or the implementation of SDG 15.3, from which these can be mutually reinforcing to biodiversity conservation through mainstreaming.

On its part, the UNCCD, in accordance with articles 9 and 10, requires affected Parties to establish National Action Plans (NAPs) to identify the factors contributing to desertification and practical measures necessary and to specify the respective roles of the government, local communities, and land users and the resources available and needed. Boer et al. argue that, apart from these NAPs, the UNCCD has virtually no obligations relating to specific measures.[65] This implies that even in the context of the expanded UNCCD role to support the global implementation of LDN, mechanisms will be necessary, at regional and national levels, to support the identification and setting of targets and indicators that African countries can develop regionally, or at national level, and implement.

4 Assessment of African Treaties and Policy Actions on Soil Health and Land Degradation

In addition to the global treaties, applicable to African states, discussed in Sect. 3, there are a variety of legal instruments within the African continent that govern the management of land and bear implications on addressing land degradation. In this section, the chapter undertakes an appraisal of how Africa-specific treaties and other continental legal and policy approaches address the question of land degradation, including the specific question of soil health as a pathway toward sustainable land management.

4.1 1968 African Convention on the Conservation of Nature and Natural Resources

The 1968 African Convention on Nature and Natural Resources (African Convention) explicitly addresses the challenges of soil and land management in Africa. It sets out a fundamental principle (article 2) for all contracting Parties to undertake to adopt measures necessary to ensure the conservation, utilization, and development of soil, water, flora, and faunal resources in accordance with scientific principles and with due regard to the best interests of the people. Specifically, the African Convention, in article 4, contains dedicated provisions regarding soil, binding contracting Parties to take effective measures for the conservation and

[65] Boer et al. (2016), p. 62.

improvement of the soil, particularly to combat erosion and misuse of soil. To achieve this, Parties are required to establish land-use plans based on scientific investigations, including the classification of land-use capability, and, further, when implementing agricultural practices, to improve soil conservation and introduce improved farming methods, which ensure long-term productivity of the land, and control erosion caused by various forms of land use that may lead to loss of vegetation cover.

In a 2004 report regarding the African Convention,[66] IUCN noted that when it was concluded in Algiers, the 1968 Convention was a milestone as it marked a departure from conservation of natural resources purely for utilitarian purposes but also introduced innovative approaches for the conservation of nature, such as creation of protected areas, making provision for specific natural resources such soil and water, and the integration of environmental concerns in development plans.[67] That IUCN report, however, drew the conclusion that the implementation of the 1968 African Convention had been subpar, especially since it did not make provision for institutional mechanisms to facilitate its implementation, and it did not establish mechanisms to encourage compliance and enforcement.[68] These shortcomings resulted in negotiation and the adoption of the Revised (2003) African Convention on the Conservation of Nature and Natural Resources.

4.2 2003 Revised African Convention on the Conservation of Nature and Natural Resources

The Revised African Convention on the Conservation of Nature and Natural Resources (Revised African Convention) was adopted by the second Heads of States Summit of the African Union, meeting in Maputo Mozambique, on July 11, 2003. In its preamble, the Revised African Convention notes that it is an amendment of the 1968 Algiers Convention, with the intention of expanding on the elements relating to sustainable development. According to IUCN, the 2003 Treaty was designed as a regional agreement, as the principal vehicle through which issues of concern to Africa may be considered and concerted action taken.[69] The objectives of the Convention include enhancing environmental protection and fostering the conservation and sustainable use of natural resources.[70, 71] The Convention notes that States Parties will be guided by its guiding principles, as follows:

[66] IUCN (2004).

[67] IUCN (2004), p. 4.

[68] IUCN (2004), p. 4.

[69] IUCN (2004), p. 6.

[70] Revised African Convention (2003), article 2.

[71] Revised African Convention (2003) article 3.

(a) the right of all peoples to a satisfactory environment favorable to their development;
(b) the duty of states, individually and collectively, to ensure the enjoyment of the right to development;
(c) the duty of states to ensure that developmental and environmental needs are met in a sustainable, fair, and equitable manner.

These principles represent a direct correlation with the concept of sustainable development and an explicit set of duties on states to ensure not only the enjoyment of the right to development but also a balancing of interests between environmental and developmental needs. This approach is valuable where utilization of land and natural resources is concerned, in order to stop, remedy, or rehabilitate land degradation and soil health. The Convention makes specific provision with respect to land and soil management,[72] requiring Parties to take effective measures to prevent land degradation and develop long-term integrated strategies for the conservation and sustainable management of land resources, including soil, vegetation, and related hydrological processes.

In addition, Parties are required to adopt measures for the conservation and improvement of the soil, among other actions, combat its erosion and misuse, as well as the deterioration of its physical, chemical, and biological or economic properties. The Convention further requires Parties to take all necessary measures for the protection, conservation, sustainable use, and rehabilitation of vegetation cover,[73] including through the adoption of scientifically based and sound traditional conservation, utilization, and management plans, taking into account the social and economic needs of the peoples concerned, the importance of the vegetation cover for the maintenance of the water balance of an area, the productivity of soils, and the habitat requirements of species. One significant departure from the Algiers Convention is the establishment of an organizational structure to implement the 2003 Convention, through the establishment of a Secretariat (article 27) and a Conference of Parties (article 26). However, the designation of an organization that will undertake Secretariat functions has to await ratification of the 2003 Convention and its entry into force in order for the first meeting of the Conference of Parties to make that designation. The 2003 Convention has now been ratified, since September 2016, after Burkina Faso deposited the instruments of ratification in June that year. An additional ratification, bringing the total to 16, was provided by Liberia in March 2017. Taking into account the challenges of land degradation and soil governance, the Convention likely requires more of the 54 African nations to ratify. Nonetheless, this is good progress, noting that during the 2016 World Conservation Congress, the IUCN adopted a Resolution[74] "Supporting implementation of the African Convention on the Conservation of Nature and Natural Resources and the African Agenda 2063," calling on Parties to the Revised

[72]Revised African Convention (2003) article 6.

[73]Revised African Convention (2003) article 8.

[74]IUCN (2016). Online: http://www2.ecolex.org/server2neu.php/libcat/docs/LI/WCC_2016_RES_078_EN.pdf.

Convention to make it operational by (1) encouraging further signatories to enable its ratification and (2) establishing a Secretariat and providing resources.

Implementation of the Revised African Convention, which aims for sustainable development, would support the realization of the African Union Agenda 2063, with its objective of a prosperous Africa based on inclusive growth and sustainable development.[75]

4.3 *Maputo Declaration (2003) and the Malabo Declaration*

The Revised African Convention was adopted in 2003, during the same summit that adopted the Maputo Declaration on Agriculture and Food Security,[76] through which African states resolved to take a number of steps aimed at enhancing food security and agriculture and which also targeted soil fertility and land degradation:

(a) revitalizing the agricultural sector, including through special policies and strategies targeted at small-scale and traditional farmers in rural areas and the removal of constraints to agricultural production, including soil fertility, and poor water management;
(b) implementing, as a matter of urgency, the Comprehensive Africa Agriculture Development Programme (CAADP) for agricultural development, at the national, regional, and continental levels, and adopting sound policies for agricultural and rural development, including allocation of at least 10% of national budgetary resources for their implementation;
(c) development of policies and strategies under the African Union and the Regional Economic Communities (RECs), to fight hunger and poverty in Africa.

The CAADP, endorsed by the Maputo African Union Summit in 2003, comprises four pillars:

1. land and water management—extending the area under sustainable land management;
2. market access—improving rural infrastructure and market access;
3. food supply and hunger—increasing food supply and reducing hunger;
4. agricultural research—research and technology dissemination.

With respect to sustainable land management, the CAADP, in 2003, estimated that area under managed water and land development totals some 12.6 million hectares, equivalent to only some 8% of the total arable land, and further that only about 874 million hectares of this land was considered suitable for agricultural production. Of this 874 million hectares, about 83% is restricted by serious soil fertility or other limitations and will need costly improvements and amendments to achieve high and sustained productivity.[77]

[75] Agenda 2063 (2015), p. 2.

[76] Maputo Declaration (2003).

[77] CAADP (2003), p. 24.

While it recognized the challenge of soil fertility and land degradation, the CAADP proposed investments mainly focused on expanding irrigation. It does, however, recognize that the African continent continues to contend with climate variability, and uneven distribution of both surface and subsurface sources of water,[78] a fact that is further acknowledged by the 2014 Malabo Declaration on Accelerated Agricultural Growth and Transformation,[79] adopted by the 23rd Ordinary Session of the AU Assembly in Malabo, Equatorial Guinea. The Declaration stressed the significance of enhancing conservation and sustainable use of all natural resources, including land, water, plant, livestock, fisheries and aquaculture, and forestry, through coherent policies as well as governance and institutional arrangements at national and regional levels.[80] The Malabo Declaration, like the Maputo Declaration, and the CAADP, focus extensively on agriculture and food security but do not address the challenges of soil fertility or land degradation or provide clear options on how to resolve them. This brings into focus the purpose of the Revised African Convention and its helpful correlation of development and environmental management as a duty on states and, further, how the Revised Convention clearly addresses land and soil conservation, as well as other elements of sustainability, including water, conservation areas, procedural rights, among other sustainability elements.

5 Evaluating the Options: Is a Specific African Legal Instrument on Soil Management and Land Degradation Relevant?

Taking into account the types of soils found within the African continent, drivers of degradation, challenges of fertility, climate change, and biodiversity loss, it is apparent that concerted or collective African action is required to focus directly on land degradation, soil health, and sustainable land management. Conceptually, the UNCCD, which places a priority on Africa,[81] has evolved focus from just addressing desertification to the more holistic approach of addressing land degradation. The complexity of land degradation was highlighted during the Rio+20 conference in 2012, and the Rio Outcome document[82] recognized that desertification, land degradation, and drought are challenges of a global dimension and continue to pose serious challenges to the sustainable development of all countries and pose challenges for Africa.[83] More significantly, the Rio Outcome document

[78] CAADP (2003), p. 24.

[79] Malabo Declaration (2014).

[80] Malabo Declaration (2014), preamble.

[81] UNCCD, article 7.

[82] Rio (2012).

[83] Rio (2012), para 205.

recognized the need for urgent action to reverse land degradation and endorsed the need to achieve a land degradation neutral world, in the context of sustainable development.[84]

In practical terms, the key challenges of land degradation in Africa have several commonalities: climate change, biodiversity loss, land-use planning, and desertification. In terms of international law instruments, as reviewed in the foregoing sections, African countries have subscribed to the 1968 African Convention; the UNFCCC; the Paris Agreement; the CBD; the UNCCD; and the 2003 Revised African Convention, which is more progressive with a focus on sustainable development.

The question posed earlier in the chapter must now be answered: is an Africa-specific legal instrument required to address soil health in the context of sustainable land management and in response to land degradation? In a collective sense, African countries have subscribed to the Sustainable Development Goals (SDGs), which set 17 goals and 169 targets that are "integrated and indivisible and balance the three dimensions of sustainable development: the economic, social and environmental," and which all countries, acting in collaborative partnership, will implement. The scope of the SDGs, relevant to the current discussion, includes the following:

- SDG 6.6—By 2020, protect and restore water-related ecosystems, including mountains, forests, wetlands, rivers, aquifers and lakes
- SDG 13.2—Integrate climate change measures into national policies, strategies and planning
- SDG 15.1—By 2020, ensure the conservation, restoration and sustainable use of terrestrial and inland freshwater ecosystems and their services, in particular forests, wetlands, mountains and drylands, in line with obligations under international agreements
- SDG 15.3—By 2030, combat desertification, restore degraded land and soil, including land affected by desertification, drought and floods, and strive to achieve a land degradation-neutral world.

The CBD, through Strategic Plan for Biodiversity 2011–2020 and the Aichi Biodiversity Targets, has set the goal that by 2020, the rate of loss of all natural habitats, including forests, is at least halved and, where feasible, brought close to zero and that degradation and fragmentation are significantly reduced.[85] The Maputo and Malabo Declarations, adopted by the African Union, are effectively focused on food security, and the Malabo Declaration fails to reconcile with the challenge of land degradation or global efforts aimed at mainstreaming land degradation neutrality into key land-use sectors, such as agriculture. At a continental level, a more intensive cooperation, and collaboration, between African states are needed, for instance to identify and determine common indicators for SDG 15.3,

[84]Rio (2012), para 206.

[85]CBD CoP Decision 10/2.

which applies in a regional context, and to support the adoption of LDN strategies at regional or continental level or through binding legal instruments. For this reason, more African specific guidelines on how to approach land degradation strategies are necessary, as well as internalization of African local knowledge, tradition, culture, and heritage into research on options to address land degradation. There is therefore a need for a consolidated action that is focused specifically on African priorities and challenges in order to not only address the challenges but also utilize the global legal targets and pathways set out in the various treaties. Three preliminary options are explored with respect to an Africa-specific legal instrument.

5.1 An Africa-Specific Protocol Under the UNCCD

The UNCCD enjoys a high level of subscription from African countries, and through the development of National Action Plans, many African countries have internalized systems on how to implement the UNCCD. Africa, under the UNCCCD, can therefore pursue implementation of the LDN approach in addressing land degradation, together with related drivers, including climate change and biodiversity loss. Options for taking such actions have been identified, including by the CAADP, which points to the need to implement sustainable land management options to address soil fertility challenges. The CAADP is unique because it is implemented through individual compacts signed with countries for national implementation and monitored by Regional Economic Communities (RECs).[86]

A preliminary proposal is the development and adoption of an African Protocol on Land Degradation, whose scope should include the pertinent objective of resolving land degradation through action to achieve land degradation neutrality. It should integrate, under one instrument, mechanisms for Africa to address degradation, water resource management approaches, sustainable land management, biodiversity loss, and climate change. The integrated LDN targets set under this Protocol would draw from priorities and strategies set under the various international instruments, but the Protocol becomes a "clearing house" through which actions are aligned, and through RECs, the various regions and countries can prioritize actions in local context and circumstances but in a binding manner under international law. This is a constructive approach since according to IUCN,[87] land degradation neutrality is essentially an equation between three processes: degradation, restoration, and sustainable land management. IUCN therefore argues that sustainable land management, as an approach to restore and sustain land resources, has been a central pillar of the UNCCD and the CBD since their inception.[88]

[86]Rampa and van Seters (2013).
[87]IUCN (2015).
[88]IUCN (2015), p. 4.

An African protocol under the UNCCD should therefore focus on the attainment of specific LDN targets, taking into account that achievement of Africa Agenda 2063, or the SDGs, will be difficult if the challenges of land degradation are not resolved, especially relating to agriculture and food security. The continent's exposure to negative impacts of climate change further highlights the challenge, and from the analysis earlier, all of the NDCs reviewed prioritized sustainable land management. Regional Economic Communities can have a direct role of monitoring the implementation and setting of national LDN targets and report on compliance and enforcement to the Conference of Parties of the Protocol. One disadvantage of this approach is that while the protocol is proposed to address Africa-specific problems, the legal mechanism would still be managed by an international body, through the UNCCD Secretariat. Another disadvantage is that the proposal made here includes an enhanced role for RECs in implementation, which may face difficulty finding space within existing UNCCD institutional mechanisms. As active stakeholders, RECs would likely become voting members of the Conference of Parties for the proposed protocol.

5.2 Implementation of the Revised African Convention

The 1968 African Convention was effectively amended through the adoption of the 2003 Revised African Convention. The latter treaty has been adjudged as more progressive and focusing on the elements of sustainable development, including the right to development, procedural rights, and a specific focus on land resources, including land and soil, as well as water. In addition, the 2003 African Convention includes a Secretariat mechanism, as well as a Conference of Parties, both of which provide an institutional or organizational structure for implementation. The 2003 Revised African Convention is now in force. If the Secretariat were promptly established and the Conference of Parties convened, this could be used to frame the question of LDN as a primary issue of convergence for the various biodiversity, degradation, climate change, and other common challenges facing the continent. However, more African countries would need to ratify, and support the institutional mechanisms to enable prompt implementation of the Convention, in order to address the existing challenge of land degradation and soil governance that traverses across Africa. The question of more ratifications is relevant, because enhanced African political commitment is required, considering that there was a 13 year time lapse between when the Convention opened for signature in 2003, and entry into force in 2016. If this is not achieved, it would mean slow or even suboptimal implementation or realization of LDN targets by 16 out of 54 African countries.

The evolving discussion on the appropriate legal mechanisms through which to implement LDN should be addressed in this context, for Africa. First is the role of regional interests within the continent, where contiguous nations face similar challenges of degradation. For instance, the current Intergovernmental Authority on Development (IGAD), established in 1996, succeeded an earlier body, the Intergovernmental Authority on Drought and Development (IGADD), set up to address challenges of degradation and desertification in the horn of Africa region. Therefore, the role of RECs in supporting and coordinating the implementation of LDN by member states, and a system of cooperation among nations and RECs, will remain important. Second, the Conference of Parties should adopt a procedure for the development of specific African (and regional) guidelines on the implementation of interventions to land degradation, including mechanisms to ensure tapping of local, indigenous, and traditional knowledge. This would complement scientific research and inform policy choices.

An assessment of the text of the Revised African Convention discloses that it contains provisions that directly address the challenge of land degradation and development in Africa and, if ratified and implemented optimally, can support endogenous actions and interventions to address land degradation.

6 Conclusion

This chapter was inspired by the conundrum whether, based on the challenges of land degradation facing Africa, it was necessary to develop an Africa- specific legal instrument to address soil health in the context of sustainable land management and in response to land degradation. It is notable that although Africa has a diverse range of soils and land-use systems, very large areas, particularly in West Africa, experience unsustainable systems of land use and erosion, together with widespread low fertility that is caused by nutrient depletion. Further, large parts of the continent are deserts or semi-arid and/or facing ecological damage, and in many instances, unequal distribution of land has relegated a growing population of small-scale farmers onto marginal areas, leading to increased physiological pressure and land resource degradation, including deforestation. The continent also contends with the adverse impacts of climate change, which exacerbate the problem of degradation and increase vulnerability of the socioeconomic and environmental context. Various international instruments have been put in place, including at the global and African levels, but the UNCCD stands out because of the evolution of its mandate from just focusing on desertification to addressing the question of land degradation. Since land is a common factor of production, land degradation unifies various challenges and actions, including biodiversity loss and climate change actions. The chapter therefore urges for consideration of development of an African protocol on sustainable land management under the UNCCD. In the alternative, a more endogenous approach is proposed—for the 2003 Revised African Convention to be implemented fully and with enhanced political commitment by African governments, which could allow achievement of the goals, including setting of LDN targets through the Conference of Parties, without the need for an additional protocol. However, the Revised African Convention faces two challenges. One, it would need to attain a higher number of ratifications, beyond the current 16, and pursue equitable ratifications across the various African regions. Second, the Convention would need to modify its mandate in order to confer monitoring roles to Regional Economic Communities, in order to allow for more regional focused interventions. A third option is available, which can be implemented together with any of the proposals in Sects. 5.1 and 5.2 above, or undertaken separately. This is the mainstreaming of the soil governance agenda into the work plans, agenda, and discussions of relation treaty mechanisms. For instance, in the foregoing discussion, a random review of NDC's showed that land degradation is a concern for many African countries, and they are proposing to take action under commitments given through the Paris Agreement, in the period to the year 2030. This is also the period within which implementation of the SDGs, including SDG 15.3 is to be undertaken. This approach however requires concerted regional collaboration in order for existing African concerns on soil governance to be presented to the relevant COPs through regional groupings. If such success were to be achieved, for instance, the African land degradation challenge, already highlighted through NDCs, could become a key feature for action under the Paris Agreement, and through NDCs, becomes internationally binding legal commitments for African Countries that have subscribed to, and submitted NDCs. A different example is through the CBD, where mainstreaming of biodiversity across various sectors is a priority, as evident through CBD CoP13, held in Cancun, in December 2016.

References

Algeria (September 2015) Intended Nationally Determined Contribution (INDC). http://www4.unfccc.int/ndcregistry/Pages/All.aspx

African Union (2010) Framework and Guidelines on Land Policy. https://www.uneca.org/sites/default/files/PublicationFiles/fg_on_land_policy_eng.pdf

Boer BW, Ginzky H, Heuser IL (2016) International soil protection law: history, concepts and latest developments. In: Ginzky H, Heuser IL, Qin T, Ruppel OC, Wegerdt P (eds) International yearbook of soil law and policy 2016 (Springer)

Botswana (2015) Botswana intended nationally determined contribution

CAADP (2003) Comprehensive Africa Agriculture Development Programme

CBD CoP Decision X/2 The Strategic Plan for Biodiversity 2011–2020 and the Aichi Biodiversity Targets (2010)

CBD CoP Decision XIII/3 Strategic actions to enhance the implementation of the Strategic Plan for Biodiversity 2011–2020 and the achievement of the Aichi Biodiversity Targets, including with respect to mainstreaming and the integration of biodiversity within and across sectors (2016)

ELD Initiative & UNEP (2015) The Economics of Land Degradation in Africa: Benefits of Action Outweigh the Costs. Available from www.eld-initiative.org

Ehlers K (2016) Chances and challenges in using the sustainable development goals as a new instrument for global action against soil degradation. In: Ginzky H, Heuser IL, Qin T, Ruppel OC, Wegerdt P (eds) International yearbook of soil law and policy (Springer)

Ethiopia (2016) Ethiopia's Forest Reference Level Submission to the UNFCCC. Online. http://redd.unfccc.int/files/2016_submission_frel_ethiopia.pdf

FAO (1982) World Soil Charter. Online: http://www.sepa.gov.rs/download/zemljiste/World_Soil_Charter.pdf

FAO (2012) Voluntary Guidelines on Responsible Governance of Land Tenure

FAO (2015) Revised World Soil Charter. Online: http://www.fao.org/fileadmin/user_upload/GSP/docs/ITPS_Pillars/annexVII_WSC.pdf
FAO and ITPS (2015) Status of the World's Soil Resources (SWSR) – Main Report. Food and Agriculture Organization of the United Nations and Intergovernmental Technical Panel on Soils, Rome, Italy
Ghana (September 2015) Ghana's intended nationally determined contribution (INDC) and accompanying explanatory note. http://www4.unfccc.int/ndcregistry/Pages/All.aspx
Ghana (2017) Ghana's National Forest Reference Level. Online. http://redd.unfccc.int/files/ghana_national_reference__level_01.01_2017_for_unfccc-yaw_kwakye.pdf
IPCC (2006) Guidelines for National Greenhouse Gas Inventories. Online. http://www.ipcc-nggip.iges.or.jp/public/2006gl/pdf/4_Volume4/V4_04_Ch4_Forest_Land.pdf
IPCC (2014) Africa (Niang, I., O.C. Ruppel, M.A. Abdrabo, A. Essel, C. Lennard, J. Padgham, and P. Urquhart). In: Barros VR, Field CB, Dokken DJ, Mastrandrea MD, Mach KJ, Bilir TE, Chatterjee M, Ebi KL, Estrada YO, Genova RC, Girma B, Kissel ES, Levy AN, MacCracken S, Mastrandrea PR, White LL (eds) Climate Change 2014: impacts, adaptation, and vulnerability. Part B: regional aspects. Contribution of Working Group II to the Fifth Assessment Report of the Intergovernmental Panel on Climate Change. Cambridge University Press, Cambridge, pp 1199–1265
IUCN (2004) An Introduction to the African Union on the Conservation of Nature and Natural Resource
IUCN (2015) Land degradation neutrality: implications and opportunities for conservation, Technical Brief Second Edition 27/08/2015. IUCN, Nairobi, p 1
Kenya (2013) Analysis of drivers and underlying causes of forest cover change in the various forest types of Kenya. Ministry of Forestry and Wildlife
Kenya (July 2015) Intentionally Nationally Determined Contributions (Ministry of Environment)
Maputo Declaration (2003) on Agriculture and Food Security, approved by the African Union
Rampa F, van Seters J (2013) Towards the development and implementation of CAADP regional compacts and investment plans: the state of play. Online. http://ecdpm.org/wp-content/uploads/2013/10/BN-49-CAADP-Regional-Compacts-Investment-Plans-Development-Implementation.pdf
Soil Atlas of Africa (2013) Online. http://www.fao.org/3/contents/9d9365e0-a8f0-48dc-8d6f-c33245827719/av020e00.htm
Shikongo (2016) Welcome Note. In: Ginzky H, Heuser IL, Qin T, Ruppel OC, Wegerdt P (eds) International yearbook of soil law and policy. Springer
UNCCD (2016) Science Policy Brief 2/2016 - Land in Balance
UNEP (1982) World Soil Policy. Online: http://library.wur.nl/isric/fulltext/isricu_i34280_001.pdf
Zambia (2016) Zambia's Forest Reference Emissions Level Submission to the UNFCCC. Online. http://redd.unfccc.int/files/2016_submission_frel_zambia.pdf

International Forest Regulation: Model for International Soil Governance

Anja Eikermann

1 Introduction

While the first volume of the International Yearbook of Soil Law and Policy[1] was concerned with the concepts of soil governance at the international, regional, and national levels and the concept of sustainability, the volume at hand focuses on soil and sustainable agriculture. The issue of sustainable agriculture is well suited to describe the area of conflict for natural resources: basically, they are necessary for human subsistence. This mere fact alone legitimizes efforts for their conservation. However, it is their utilization that is to some extent necessary for human subsistence, putting conservation and utilization apparently in a constant opposition. A mediating concept that seeks to strike a balance between utilization and conservation is the concept of sustainable use. The concept of sustainability has definitely found its way onto the international political agenda and is firmly anchored within the international environmental discourse. Nevertheless, despite alarming developments, it largely seems to fail to win full recognition. Sustainability needs a broker or negotiator, a tool to support its implementation and enforcement. International law, by its very nature, entails the chance to be such a broker and negotiator required to promote sustainability. However, its abilities in this regard may be

All views expressed in this chapter are solely the opinion of the author. Parts of this chapter are taken from Eikermann (2015), which should be referred to for a more detailed analysis of the arguments.

[1]Ginzky et al. (2016).

A. Eikermann (✉)
German Federal Hydrographic and Maritime Agency, Hamburg, Germany
e-mail: anja.eikermann@gmx.de

H. Ginzky et al. (eds.), *International Yearbook of Soil Law and Policy 2017*,
International Yearbook of Soil Law and Policy,
https://doi.org/10.1007/978-3-319-68885-5_22

and have been challenged.[2] The effectiveness of international environmental law is hard to estimate.[3] Nevertheless, international law has been an important tool used in the protection of the environment in general and in the conservation of natural resources already for various decades. Specific instruments include multilateral environmental agreements (MEA),[4] such as CITES,[5] the Ramsar Convention,[6] or the UNFCCC,[7] covering various natural resources such as specific species, habitats, or natural processes in a variety of ways, establishing trade restrictions, nature reserves, or financial incentives.

Creating a stand-alone international soil convention, however, failed.[8] In this regard, soils share the fate of forests on the international legal and political agenda. Just like climate change, biodiversity, and desertification, forests were bound to be the sole subject of an international convention in the run-up to the United Nations Conference on Environment and Development (UNCED) in Rio in 1992. However, this endeavor failed and has failed since.[9] The reasons for failure have been attributed particularly to the opposing—*i.e.*, economic and conservationist—interests concerning forests; to the structural differences of forests compared to, for example, biodiversity or climate change as a legal subject of an international convention[10]; and, finally, to what is perceived as the fragmentation of forest-related processes at the international level.[11] Nevertheless, the "quest"[12] for an international forest convention remains on the international agenda, promoting governance options, as well as legal approaches,[13] thus suggesting the general

[2]See for example with a particular focus on the work of the UN in this regard Birnie et al. (2009) or refer to Bodansky (2010).

[3]For the most influential studies in this regard see Miles et al. (2001), Brown Weiss and Jacobson (1998) and Young (1999).

[4]See with further references Birnie et al. (2009), pp. 84 *et seq.*

[5]Convention on international trade in endangered species of wild fauna and flora, Geneva, 1 July 1975, UNTS, Vol. 993, p. 243.

[6]Convention on Wetlands of International Importance especially as Waterfowl Habitat. Ramsar, 2 February 1971, UNTS, Vol. 996, p. 245.

[7]United Nations Framework Convention on Climate Change, New York, 9 May 1992. UNTS, Vol. 1771, p. 107.

[8]See in this regard particularly Boer et al. (2016), pp. 49 *et seq.*

[9]For analyses of the reasons for the failure of a forest convention in 1992 see *inter alia* Hönerbach (1996). See also Davenport (2005) and Lipschutz (2000).

[10]The structural differences are particularly related to the perception of biodiversity and climate change as common concerns of humankind which is not the case for forests and soils. See Hönerbach (1996), pp. 83 *et seqq.*

[11]See Tarasofsky (1996), pp. 687 *et seq.*; Hönerbach (1996); Brunnée and Nollkaemper (1996). Within the context of international forest law and policy fragmentation has largely been used to describe the divergence of a multitude of international institutions and instruments of different legal nature governing international forests. See in more detail below Sect. 5.

[12]In style of Humphreys' "quest for a global forest convention", Humphreys (2005).

[13]See for example van Asselt (2012), exploring "autonomous interplay management" in accordance with Oberthür (2009); Krohn (2002); Schulte zu Sodingen (2002); Mackenzie (2012); for an overview see Giessen (2013).

will and need to strike a balance between conservation and utilization by international regulation.

The need for soil protection law at the international level is widely acknowledged.[14] At the same time, it has been submitted that a concerted international legal mechanism is required, and options such as the International Soil Convention or a protocol on sustainable use of soils linked either to the CBD[15] or the UNCCD[16] have been proposed.[17] However, these options considered focus on a "single instrument-approach," *i.e.* a stand-alone legal agreement on soils. It is argued that these approaches are likely to fail just as a stand-alone international forest convention would if the current fragmented "regime" of international soil governance is not thoroughly included in the calculations. Accordingly, this chapter will explore the comparability of forests and soils as potential subjects of international legal regulation and explain the challenges that have to be met given, the particular characteristics of these subjects. Against this background, it will subsequently present a possible option for international soil regulation.

2 Forests and Soils as Subjects to International Legal Regulation: A Conceptional Challenge

Complexity is used as a major argument against an international soil convention.[18] Soils are deemed to be too complex with regard to their legal status—private *vs.* public; furthermore, there are very different types of soil resources, and the economic, cultural, and political circumstances in the various countries differ too severely.[19] This very complexity of soils indeed poses a challenge to the international regulation of natural resources of any kind, but it does not oppose international regulation. It is rather required to identify and use those factors that contribute to the complexity of shaping the concept of soils as a subject of international regulation. Taking into account the example of forests, specific categories of criteria can be identified that contribute to tailoring the concepts needed for international legal regulation: (1) ecosystem services and functions, (2) preferences of specific services and functions by different stakeholders, and (3) threats.

[14]See the rationales put forward by Boer et al. (2016), pp. 67 *et seq.*

[15]Convention on Biological Diversity, Rio de Janeiro, 5 June 1992. UNTS, Vol. 1760, p. 79.

[16]United Nations Convention to Combat Desertification in those Countries Experiencing Serious Drought and/or Desertification, Particularly in Africa, Paris, 14 October 1994, United Nations, Treaty Series, Vol. 1954, p. 3.

[17]Boer et al. (2016), pp. 56 *et seq.*; Wyatt (2008), pp. 200 *et seqq.*; Hannam and Boer (2002), pp. 74 *et seqq.*

[18]See also Markus (2015), pp. 217 *et seqq.*

[19]*Cf.* Boer et al. (2016), p. 69.

Forest ecosystem functions and services[20] may be categorized as follows: providing resources such as fuelwood, industrial wood, and nonwood products; providing ecological services like water protection, soil protection, and health protection; providing biospheric regulation services like biodiversity conservation and climate regulation; providing social services such as ecotourism and recreation; and providing spiritual, cultural, and historical amenities.[21] Taking into account the ecosystem services and functions provided for by soils, their essential importance for human subsistence and well-being becomes even more perceived. In comparison to forests, soil is a "genuine 'nexus' resource."[22] Soils built up the foundation for all other terrestrial ecosystems, such as forests. Thus, the ecosystem services and functions connected to forests rely on soils and, thus, may be perceived as being an integral part of the ecosystem services and functions of soils. Soils have a role to play in various ecosystem processes—the carbon cycle, the nutrient cycle, the water cycle—and they provide habitat for species and a considerable gene pool.[23]

This diversity of services and functions legitimizes protection on the one hand.[24] On the other hand, the scope of these services and functions in turn simultaneously constitutes a challenge with a view to the scope of a possible international instrument. Not all of the services and functions provided for by forests or soils are valued and utilized equally by all stakeholders, and the relevance attributed to a given function is largely a matter of local proximity and interest. Thus, a very different significance might be attributed to one and the same function depending on whether it is perceived from an economic, sociopolitical, or ecological point of view or from a local or global perspective. As forests and soils vary across the globe and in time, not all forest or soil ecosystems provide equally for the same quantity and quality of services and functions, and a single forest or soil does not provide its services and functions steadily. Moreover, not all of the services provided by a forest or a soil can be provided by the same forest or soil at the same time. The vast amount of services and functions and their diversity in, as well as knowledge gaps with regard to, monetary value[25] brings about opposing interests by stakeholders attaching diverse preferences to the subject "soil."

[20]Note that ecosystem services in general and forest ecosystem services in particular are categorized slightly different in available surveys. The approach taken here largely follows the Millennium Ecosystem Assessment, Hassan et al. (2009). For an overview see Eikermann (2015), pp. 15 *et seq.*

[21]Hassan et al. (2009), p. 600.

[22]Wolff and Kaphengst (2016), p. 129, taking the term from Weigelt et al. (2014).

[23]In detail see FAO (2015); Hannam and Boer (2002), pp. 9 *et seqq.*

[24]On the interlinkages between ecosystem services and human well-being see Hassan et al. (2009).

[25]See most prominently the so-called TEEB-Studies, available http://www.teebweb.org/our-publications/all-publications/. Accessed 14 April 2017. For example: European Communities (2008) The Economics of Ecosystems and Biodiversity—An Interim Report; The Economics of Ecosystems and Biodiversity: Ecological and Economic Foundations, Earthscan, London and Washington, 2010; With regard to soils see Etter et al. (2016).

Consequentially, an international regulation becomes a matter of hierarchy and allocation, *i.e.* priorities. International regulation is likely to be initiated and driven by those beneficiaries that support an allegedly high-ranking concern. Society tends to prefer those ecosystem functions that provide for more developmental gain.[26] This may lead to the neglect of connected ecosystem functions. Thus, with various stakeholders involved, it is rather difficult to agree on a common set of priorities that complicates international regulation. In this regard, the dichotomy of "public vs. private" adds yet another layer of complexity. Basically, just as forests as such, *i.e.* in their shape of a conglomerate of trees, are a territorially bound, immovable natural resource, soils are as well. Therefore, the services and functions they provide are predominantly at the disposal and for the benefit of those stakeholders in close proximity or, even more, only for those who legally own the forest or the soil and have the authority to exclude other stakeholders from utilization. However, some forest and soil ecosystem services are clearly providing a public good to a local, transboundary, or even global audience.[27] It thus appears to be fully legitimate to also perceive forests and soils as public goods, just as it seems to be compelling to categorize forests and soils as private goods. This leads to the perception of so-called hybrid goods.[28] However, this categorization does not exclude forests or soils as a subject of an international regulation, but an international regulation of any kind needs to take this specific character into account, particularly when it comes to implementation.

An additional layer of complexity is added when looking into the threats to forests or soils. Attempting to tailor the concepts for international regulation requires to focus not only on what constitutes the subject but also on the threats that have to be included in the scope of a regulation. In simple terms, two general sets of threats to forests can be distinguished: deforestation and forest degradation. While deforestation concerns the *quantitative* dimension of forests, forest degradation relates to the *qualitative* dimension of forests. The European Commission has identified eight "soil threats": erosion, organic matter decline, salinization, compaction, landslides, contamination, sealing, and the decline in soil biodiversity.[29] As with forests, these threats have a qualitative and a quantitative dimension. Even though, estimating the extent or impact of threats in numbers is difficult,

[26]With regard to forests see in particular Humphreys (2006), pp. 216 *et seqq.*

[27]See Hooker (1994), pp. 836 *et seq.* Note particularly, that climate change and biodiversity are recognized as embodying a "common concern". See on the concept in general for example Brunnée (2007); with regard to soils refer to Boer et al. (2016), pp. 65 *et seqq.*

[28]On the issue of forests being a public or a private good, see in particular Humphreys (2006). On the issue of forests being a hybrid good Hönerbach (1996), pp. 83 *et seqq.* and Hooker (1994), pp. 825 *et seqq.*

[29]See for a good overview European Commission, Communication from the Commission to the Council, the European Parliament, the European Economic and Social Committee and the Committee of Regions, Thematic Strategy for Soil Protection, Brussels 22.09.2006, COM(2006) 231 final.

specific drivers can be idenitfied. Certain drivers can be identified.[30] They combine demographic, social, ecological, economic, technological, political, cultural, religious, climatic, and biophysical factors and may be natural or human induced. They have different effects on a spatial and temporal scale. Drivers have to be distinguished with regard to direct or indirect effects. Furthermore, drivers interact with each other and are themselves subject to change.

This multifunctionality of soils and forests and its complex consequences have been used as an argument against international legal regulation. However, fundamentally different conclusions may be drawn. Soils and forests merit international regulation just because of their multifunctionality and because the interlinkages have not yet been fully discovered and understood. An international regulation entails the opportunity to assure that the diverse interests in soils and forests resulting from their multifunctionality do not lead to a prioritization of one soil or forest function, which would likely be the economically more profitable one, to the detriment of other soil or forest functions and, hence, to the detriment of human well-being in general and particularly of stakeholders dependent upon the most likely economically undervalued functions. Therefore, international regulation is needed to ensure that the full range of functions remains available to all stakeholders, both locally and globally.[31] An international regulation of any kind needs to be able to adapt to the changing parameters inherent to natural systems and to the threats they encounter. Furthermore, an international (legally-binding) regulation on soils or forests will have a decisive role to play in establishing a prerogative of interpretation. The way soils will be used in the future will be shaped by the way soils are perceived—which in turn requires scientific research. Shaping the concept of soils and its sustainable use, taking into account the criteria named above, by way of an international regulation will provide a valuable source for perception and awareness.

3 International Institutions and Instruments Covering Forests and Soils: A Multitude of Competences

The diversity in priorities accorded to the specific ecosystem services and functions by the various stakeholders involved is furthermore reflected in the multiple institutions that govern forests and soils at the international level and in the—nonlegally binding—instruments that have been developed to regulate forests and soils.[32]

[30]A driver is any natural or human-induced factor that directly or indirectly causes a change in an ecosystem. A direct driver unequivocally influences ecosystem processes. An indirect driver operates more diffusely, by altering one or more direct drivers. See Hassan et al. (2009), p. 74.

[31]See the exact observation made for forests as well, Eikermann (2015), p. 29.

[32]Note the following section does not present the development of international political forest and soil processes in its entirety, but focuses on the overlapping developments. In more detail on the evolution of international forest processes see Eikermann (2015), pp. 31 *et seqq.*, on the evolution

"The international forest regime is disconnected and multi-centric; it has developed at different speeds and in different directions, rather than strategically and holistically along a common front."[33] This often cited quote describes the essence of the institutional challenge for international legal forest regulation. Since the nexus between forest conservation for utilization has been recognized and has gained relevance on the international political agenda, multiple political processes have been involved in attending to this issue. In the era up until 1990, with a particular focus on the United Nations Conference on the Human Environment (UNCHE) held in Stockholm in 1972 and the resulting Declaration of the United Nations Conference on the Human Environment, the Stockholm Declaration,[34] the different processes taking up forests as a concern reflect the lingering tension between conservation and development. The Stockholm Conference lay some of the foundations for the subsequent UN environmental agenda and set up the environmental principles that were to keep international politics and international law occupied until today.[35] The Stockholm Declaration, however, did not pay specific attention to the issues of forests[36] or soils.[37] Nevertheless, despite the lack of a formal acknowledgment of these issues, the Stockholm Declaration merits consideration with regard to the fundamental recognition of environmental liability, given the elaborations on the relation between the sovereign right of States to exploit their own resources pursuant to their own environmental policies and the responsibility thereby not to cause harm to States or areas beyond their national jurisdiction in its Principle 21—a principle that is reiterated in the discussions on the international regulation of forests and soils.

With the pretalks leading up to the UNCED in Rio in 1992, the issue of forests and soils was newly introduced and perceived on the international agenda. While a legally binding consensus on forests failed[38] and the issue of soils was only indirectly covered by the UNCCD,[39] the UNCED brought about Agenda 21, with its Chapter 11 on "Combating Deforestation,"[40] Chapter 10 on "Integrated

of international soil processes see Boer et al. (2016), pp. 51 *et seqq.*, both with further references and more detailed explanations.

[33]Humphreys (2006), p. 213.

[34]Declaration of the United Nation Conference on the Human Environment, Stockholm, 16 June 1972, UN Doc. A/Conf.48/14/Rev.1; 11 ILM 1416 (1972).

[35]Cf. for example Birnie et al. (2009), pp. 48 *et seq.*

[36]Kasimbazi (1995), p. 75.

[37]See Principles 2 and 5 of the Stockholm Declaration; *Cf.* Boer et al. (2016), p. 51.

[38]For analyses of the reasons for the failure of a forest convention in 1992 see inter alia Hönerbach (1996). See also Davenport (2005) and Lipschutz (2000).

[39]See also Boer et al. (2016), pp. 52–53 and for more detail below Sect. 4.

[40]Report of the UN Conference on Environment and Development, Rio de Janeiro, 3–14 June 1992. Annex II: Agenda 21, UN Doc. A/CONF.151/26 (Vol. II), 13 August 1992.

Approach to the Planning and Management of Land Resources,"[41] Chapter 14 on "Promoting Sustainable Agriculture and Rural Development,"[42] and the "Non-Legally Binding Authoritative Statement of Principles For A Global Consensus on the Management, Conservation and Sustainable Development of All Types of Forests"—the Forest Principles.[43] In the aftermath of the failure of the international forest convention, there was a general confusion about how to proceed with forest issues and the missing attribution of these issues to a single institution. At the international level, the attention for forests shifted to the newly established Intergovernmental Panel on Forest (IPF) and its successors,[44] taking considerable attention away from the FAO, which—until then—perceived itself to be the responsible forest institution and which also has a major role to play with regard to soils.[45] Additionally, regional attempts to create a legally binding agreement tried to fill the void for forest and soil regulation. With the World Summit on Sustainable Development in Johannesburg in 2002, the 2012 United Nations Conference on Sustainable Development (Rio+20 Conference) and their related declarations, as well as the "2030 Agenda for Sustainable Development" with the Sustainable Development Goals that entail goals and targets for forests and soils,[46] a third layer of processes is added to the international processes covering soils and forests.

Disregarding the differences in the evolution of international forest and soil processes and the different institutions involved, international soil and forest processes share the effect these international political processes exercised: the adoption of these instruments marked a significant paradigm shift[47] with regard to international environmental governance and, thus, for forest and soil governance and collaterally equipped both with a new impetus. The instruments referred to briefly above established and further supported the notion that these issues cannot be dealt with in isolation from developmental issues. They initiated and supported research with regard to the multiple roles inherent to forests and soils, nowadays known as ecosystem services and functions, but also with regard to the actual causes of deforestation, as well as forest and soil degradation. These instruments

---

[41]Report of the UN Conference on Environment and Development, Rio de Janeiro, 3–14 June 1992. Annex II: Agenda 21, UN Doc. A/CONF.151/26 (Vol. II), 13 August 1992.

[42]Report of the UN Conference on Environment and Development, Rio de Janeiro, 3–14 June 1992. Annex II: Agenda 21, UN Doc. A/CONF.151/26 (Vol. II), 13 August 1992.

[43]Report of the UN Conference on Environment and Development, Rio de Janeiro, 3–14 June 1992. Annex III: Non-Legally Binding Authoritative Statement of Principles for a Global Consensus on the Management, Conservation and Sustainable Development of all Types of Forests, UN Doc. A/CONF.151/26 (Vol. III), 14 August 1992.

[44]Intergovernmental Forum on Forests (IFF) and United Nations Forum on Forests (UNFF).

[45]See Boer et al. (2016), p. 56, with particular reference to the Global Soil Partnership and the Revised World Soil Charter.

[46]See goal 15, United Nations General Assembly, Seventieth Session, No. 11688, Agenda items 15 and 116, Resolution adopted by the General Assembly on 25 September 2015, 'Transforming our world: the 2030 Agenda for sustainable development', A/RES/70/1.

[47]See Desai (2011), p. 18.

thus have a strong knowledge- and awareness-building function. Despite the reiteration of the principle of state sovereignty over natural resources in all of the instruments, the achievement of these instruments lies in the introduction of the common interest element into the discussion, even though it has not been formally, *i.e.* legally, accepted. Thus, even though they lack legally binding consensus,[48] they at least point out a consensus of concern and a shared responsibility.

The multifunctional character of forests and soils is generally acknowledged; however, there seems to be no solution as to how to effectively transfer this awareness into effective international regulation. Forest and soils share the lack of an explicit competence for a single institution for forest or soil matters and consequentially the lack of a coherent debate and interpretation of these concerns. These processes are substantially fragmented, *i.e.* there is no one institution holding an exclusive competence but several institutions working independently from one another on the same subject matter with overlapping and conflicting competences and with different objectives, causing a multiplication of efforts and, eventually, wasted resources. Additionally, the fundamental objectives, rules, and principles are scattered among several instruments working on these issues independently, under their own agenda with different overall aims and purposes, representing very different priorities.

4 International Law Covering Forests and Soils: Diverging Objectives

While the previous section has been focused on the international institutions, processes, and mechanisms that cover forests and soils but have not been transformed into law, the following examinations turn to the existing international law that covers forests and soils at least collaterally. The repeated and continuing failure to transform the extensive amount of political processes and negotiations into international law is a clear indication of the existence of a cluster[49] of fragmented international law that solely relates indirectly to forests and soils. There is an apparent lack of an explicit, overall recognition of a common interest with regard to forests and soils. Therefore, international law is unable to access the regulation of these issues directly. Nevertheless, both matters are already collaterally regulated—and thus transformed into law—by a variety of international treaties.[50]

[48]On the issue of "soft law" in this regard see Boer et al. (2016) and Eikermann (2015), pp. 137–138 with further references.

[49]Note that the term "cluster" is not used as the technical term as manifested by Konrad von Moltke in Moltke (2001), but simply to describe the uncoordinated and fragmented collectivity of instruments relating—directly or indirectly to forests. It is precisely not employed to describe a concerted, homogenous system.

[50]For a more detailed elaboration on the effect of these treaties on forests see Eikermann (2015), pp. 61 *et seqq.* with further references. For a detailed list of international treaties relating to soils

With regard to forests, three—respectively six—thematic contexts of international treaties have a bearing on the regulation of forests. These contexts are trade, traditional nature conservation, and the Rio-Conventions-context which may be subdivided into the thematic contexts of biodiversity, climate change, and desertification. The trade complex relates to CITES, the International Tropical Timber Agreements (ITTA 1983, 1994, and 2006),[51] and the international law of the WTO (especially the GATT 1994).[52] The traditional nature conservation complex covers the WHC[53] and the Ramsar Convention. Finally, the Rio-complex entails the CBD and its accompanying protocols, the UNFCCC and the Kyoto Protocol,[54] and the UNCCD.[55]

With regard to soils, it has been noted that "[t]he majority of multilateral environmental agreements (MEAs) relate either directly or indirectly to land and soil issues."[56] Soils and forests overlap in terms of the relevant international treaties with regard to the Rio-complex.[57] Additionally, the nature conservation conventions in their capacity as habitat conservation conventions are of major relevance to soils. While soil as such is not primarily viewed as a tradable resource with regard to the conventions referred to above, the resources produced by soils are in turn tradable, and as has been established before, soils have a monetary value. Therefore, it may be argued that the treaties related to forests have a bearing on soil as well and thus have to be added to the list of treaties relevant to soils.

Thus, on the one hand, there is elaborate international environmental law that regulates forests and soils. While it has been submitted that the international institutions and their related documents capture a variety of fundamental principles and objectives with regard to soils and forests, each of the treaties relating to forests and soils captures one or more forest or soil functions. Thus, even if there is no specific international forest or soil law as such, these treaties form an indispensable part of an overall *indirect* international forest and soil law. However, the

see Hannam and Boer (2004), pp. 95 *et seqq*. The following enumeration and itemization is not exclusive and solely exemplary.

[51]International Tropical Timber Agreement 1983, Geneva, 18 November 1983. UNTS, Vol. 1393, p. 67; International Tropical Timber Agreement, 1994 (adopted Geneva, 26 January 1994, entered into force provisionally on 1 January 1997, in accordance with Article 41(3)), 1955 UNTS 81; International Tropical Timber Agreement, 2006 (adopted Geneva, 27 January 2006, entered into force 7 December 2011), UN Doc. TD/TIMBER.3/12.

[52]General Agreement on Tariffs and Trade 1994, UNTS, Vol. 1867, p. 187.

[53]Convention for the protection of the world cultural and natural heritage, Paris, 16 November 1972, UNTS, Vol. 1037, p. 151.

[54]Kyoto Protocol to the United Nations Framework Convention on Climate Change, Kyoto, 11 December 1997. UNTS, Vol. 2303, p. 148.

[55]United Nations Convention to Combat Desertification in those Countries Experiencing Serious Drought and/or Desertification, Particularly in Africa, Paris, 14 October 1994, United Nations, Treaty Series, Vol. 1954, p. 3.

[56]See Wyatt (2008), p. 180 with further reference to Montgomery (2007). See for a more detailed elaboration Hannam and Boer (2002), pp. 59 *et seqq*.

[57]See for example Wyatt (2008).

preeminent feature of the international treaties pertaining to forests and soils is that they lack specific norms on forests and soils. Their broad and general ambits are neither created for nor oriented toward the regulation of forests and soils in general but only capture the regulation of forests and soils according to their specific prerequisites, *i.e.* within the framework of their specific purposes and objectives. Hence, the treaties of the indirect international law, on the one hand, facilitate the prioritization of specific functions within the framework of the respective treaty. On the other hand, the lack of forest- respectively soil-specific regulation within the treaties leads to the lack of forest- and soil-specific implementation of these treaties. "While a number of [MEAs] contain elements that can assist in achieving sustainable use of soil, it is contended that none are sufficient in their own right to meet the requirements of international environmental law in relation to soil."[58]

The conclusion to be drawn is that even if each of the treaties individually contained the potential to contribute to the regulation of one or more forest or soil functions and thus indirectly have a positive effect on one or more ecosystem functions, these treaties are not in a position to address the multifunctionality of forests or soils, the related interests, and the threats imposed particularly by human behaviour on the functioning of forests and soils for the benefits of human well-being.

5 Options for Regulation and the Impact of Fragmentation

Considering the need for regulation on the one hand but also considering the shortcomings of the international processes dealing with forests and soils, as well as the shortcomings of the indirect international forest and soil law on the other, the question remains of *how* to regulate the concerns of forests and soils. Turning to options of international regulation against this background, an obvious answer is a stand-alone international convention as it has been pursued over the decades. However, the *need* and the *feasibility* of a stand-alone international convention as the means to achieve the required regulation have to be considered.

5.1 *The Impact of the* Status Quo*: The Current International Regime on Forests and Soils*

All the instruments[59] on forests and soils, when taken together, provide for a complex, multilayered set of values, objectives, principles, obligations, guidelines,

[58]Hannam and Boer (2002), p. 59.

[59]The term "instruments" is to be understood in a nontechnical way, referring simply to the single elements considered and analyzed within the framework of Sects. 3 and 4. Following this

recommendations, rules of procedure, decisions, resolutions from international and nongovernmental organizations, treaty organs, standard-setting and certification businesses. This creates a "multi-instrument approach"[60] to international regulation that entails considerable substance, particularly fundamental principles and objectives for international regulation.[61] The nonlegally binding instruments foresee the need to equitably support and put into effect the social, economic, ecological, cultural, and spiritual interests involved and thus acknowledge the multifunctional character of the subject matter.[62] Furthermore, they fill the concept of sustainability with substance and thus provide for a performable instrument required for implementation.[63] Additionally, it has to be considered that the nonbinding processes and instruments address the threats to forests and soil and thus provide for the basis to develop specific measures to counter these threats.[64]

The legally binding instruments—though they apply to forests respectively soils only indirectly—cover the multifunctional character of the two subject matters as each treaty serves a different function or even various functions while simultaneously serving the differing stakeholders' interests. Even more important, these agreements come with a permanent and reliable treaty infrastructure. The relevant treaty organs safeguard the "living character" of the treaties, thereby allowing them to develop in accordance with technical and scientific developments, or political changes, and provide and create opportunities for cooperation.[65] The treaty infrastructure furthermore provides for the crucial financial infrastructures. In addition, the international treaties provide for regulation with regard to monitoring, assessment, and reporting, as well as mechanisms for the settlement of disputes, compliance, and enforcement mechanisms. Hence, regarding content, it may be asserted that—considered as a whole—the indirect international law on forests respectively soils already in existence provides the substantial elements for international regulation of these subjects.[66]

---

approach, the term "multi-instrument-approach" is used to refer to the composite body of all components. For further information see Eikermann (2015), p. 146, fn. 59.

[60]See Eikermann (2015), pp. 145 *et seqq.*

[61]The concept of sustainable forest management is just one, but an important example in this regard.

[62]See above Sect. 3 with regard to the fundamental principles provided for by the UN driven environmental processes.

[63]For example the criteria and indicators for sustainable forest management, *cf.* Humphreys (2006), pp. 116 *et seqq.*

[64]*Cf.* with regard to forests Eikermann (2015), pp. 136 *et seqq.* for a more detailed comparison of the "multi-instrument-approach" with elements for an ideal substance for international forest regulation. With regard to soils see Hannam and Boer (2002, 2004).

[65]See in more detail Eikermann (2015), pp. 164 *et seqq.*; with regard to international cooperation see Wolfrum (2011).

[66]For the required minimum elements of international forest regulation see Eikermann (2015), pp. 136 *et seqq.*; with regard to soils see Hannam and Boer (2002, 2004).

However, the interrelation of treaties has to be considered. While international treaties are in general established independently from one another, they do not and cannot operate in isolation.[67] This is particularly true within the field of international environmental law. The separation of international environmental treaties is already precluded to a large degree by the factual interaction of biological and ecological components. Apart from the factual interdependence of the regulated subject matter of international (environmental) treaties, these treaties furthermore interrelate—whether in a conflictive, synergistic, or duplicative way—in legal and political aspects.[68] While this interrelation as such does not necessarily imply either a conflictive or a synergistic interrelation, the indirect international forest and soil law does not form a comprehensive, homogenous whole. Each treaty is construed and predetermined to pursue its own objective. No indication is given with regard to achieving a balance between treaties pursuing opposing objectives. Furthermore, these treaties are neither forest- nor soil-specific. They lack detailed substance on the fundamental principles and objectives required for an ideal international soil or forest regulation as they are put forward by the specific political processes. Thus, forest- and soil-specific implementation of these treaties is significantly impeded and subordinated to the implementation for the achievement of the actual treaty objective.

Therefore, the multiinstrument approach has considerable shortcomings. The essential contents—*i.e.* particularly the fundamental principles, objectives, measurable parameters—of the regulation are, firstly, scattered among instruments of different legal nature, *i.e.* nonlegally binding instruments and legally binding international treaties. Secondly, the contents are scattered disproportionately. While the fundamental principles and objectives for a comprehensive regulation are stipulated in nonlegally binding instruments, the legally binding instruments lack forest- respectively soil-specific detail and thus create a regulatory and particularly an implementation gap. Additionally, there is no indication as to how these objectives might be achieved simultaneously, as a common vision or goal. In turn, the soft law character of an instrument is not necessarily an impediment to its effectiveness. Nevertheless, the characterization of an instrument as hard and soft law will necessarily result in a difference in terms of actors' compliance. Consequentially, the fundamental principles and objectives for a comprehensive regulation, stipulated in nonlegally binding instruments, remain formally inadequate. The multiinstrument approach is thus a sufficient approach with regard to its contents—

[67] A feature of international law addressed by the Vienna Convention on the Law of Treaties (VCLT), 23 May 1969. UNTS, Vol. 1155, p. 331; and particularly by the United Nations General Assembly, Fragmentation of international law: difficulties arising from the diversification and expansion of international law, Report of the Study Group of the International Law Commission Finalized by Martti Koskenniemi, 58th session, Geneva, 1 May–9 June and 3 July–11 August 2006, UN Doc. A/CN.4/L.682.

[68] See in general Matz (2006). The line between legal and political interrelations or interdependencies is often hard to draw, see Wolfrum and Matz (2003), p. 12. On the interrelation of treaties in forest matters see Eikermann (2015), pp. 150 *et seqq.*

the principles, objectives, and procedures already in existence. However, the content is unpractical, given its fragmented structure. In conclusion, the existing legally and nonlegally binding instruments act independently from one another within their own remit. They do follow a common, superior goal and are generally not intended to be mutually supportive, *i.e.* to further the coherent and consistent implementation of goals outside their remit.

5.2 The Impact of Fragmentation

5.2.1 The Concept of Fragmentation in International Law

While the fragmentation of international law was traditionally understood as referring to the proliferation of treaties and thus, in a specification, as well as resulting in the proliferation of legal fora, particularly for the settlement of disputes, it has developed to be understood as referring to treaty congestion and, finally, in rather broad terms, as referring to the divergence of values, objectives, or rationales of treaties that generally may be merged under a common topic.[69] The traditional narrow definition of the fragmentation of international law as causing normative conflicts does not embrace the nature and development of international law today, particularly in the field of international environmental law. International treaties allow for protocols and amendments to be able to react swiftly to changing conditions.[70] Thus, even though the same topic may be addressed, new rules and decisions pertaining to it do not emanate from a single, uniform process but are rather scattered among a variety of processes. A comprehensive framework for regulation is subjected to significant insecurity caused by the living character of the treaties that prevents a permanent status.[71]

The classical tools of international law to manage the interrelations of treaties are conflict tools.[72] They are aimed at the resolution of normative conflicts and the prioritization of one norm over another. Just as the traditional concept of the fragmentation of international law no longer does justice to this case, the classical tools of conflict resolution are not apt to resolve the treaty interrelations at hand. However, international law is able to use means of institutional cooperation and

[69]See particularly Matz (2006). For further detail with regard to forests see Eikermann (2015), pp. 158 *et seqq.*

[70]See Pauwelyn (2008), para. 17.

[71]The case of the unknown outcomes of rule development has been termed "a blind spot in the fragmentation debate" by van Asselt (2012), p. 1253.

[72]United Nations General Assembly, Fragmentation of international law: difficulties arising from the diversification and expansion of international law, Report of the Study Group of the International Law Commission Finalized by Martti Koskenniemi, 58th session, Geneva, 1 May–9 June and 3 July–11 August 2006, UN Doc. A/CN.4/L.682; Matz (2006).

coordination to address the particular cases defined by the lack of a clear normative conflict and divergence in underlying rationales.[73]

5.2.2 Fragmentation in International Soil and Forest Governance

This gives rise to the fundamental awareness that the concept of an international environmental convention in the traditional sense,[74] or similar approaches, such as an amendment of or a protocol to an existing indirect international treaty,[75] do not seem to lend themselves to an international forest or soil regulation. An international convention, respectively a comparable option as referred to above, might just add another layer to the already uncoordinated cluster of instruments, causing an increase of fragmentation, legal uncertainty, and ineffectiveness. It has to be considered that there is no tool in international law that would equip an international forest or soil convention with the authority to be the supreme treaty in international forest or soil regulation matters. A resolution of ambiguities between existing treaties as they exist today seems hard to achieve under these circumstances.[76]

The existing international instruments already available for the regulation of forests and soils will persist. A new, stand-alone international convention would be "born" into this multilayered system. That means that whatever solution is developed, it will have to deal with the complexity of the existing regime.[77] This does of course not imply that a new international stand-alone treaty is bound to fail. There are different possible options to cope with this type of fragmentation. The CBD itself was "born" very prominently into the nature conservation regime that had already been in existence. It also provides for a rather innovative[78] conflict clause in its Article 22.[79] Additionally, some MEAs have concluded Memoranda of Understanding to promote and achieve coordination and cooperation with other MEAs or initiated special cooperative work programs.[80] Another solution could be to rather

[73]Matz (2006).

[74]With regard to forests see for an overview over the advantages and disadvantages provided for by an international treaty approach to forests see inter alia Tarasofsky (1996), p. 682; Brunnée (1996), pp. 49 et seqq.; Humphreys (2005), p. 2; Mackenzie (2012), p. 251; With regard to soils see for example the pros and cons weighed by Wyatt (2008), pp. 200 *et seqq.* or Boer et al. (2016), pp. 56 *et seqq.*

[75]Such an approach would also include a legal format for REDD. For the options with regard to forest amendments and protocols see Tarasofsky (1996), p. 673; Boyd (2010); Levin et al. (2008); Mackenzie (2012); Srivastava (2011); van Asselt (2011). In terms of soils see Wyatt (2008), pp. 203–204 or Boer et al. (2016), pp. 56 et seqq.

[76]See in general Matz (2006).

[77]Alter and Meunier (2009), p. 21.

[78]Matz-Lück (2008), para. 44.

[79]See in detail Matz (2006).

[80]On these approaches see Eikermann (2015), pp. 164 *et seqq.* with further references.

"embrace the complexity."[81] The current initiatives available are full of potential, but "[...] they require a more effective approach to coordination if they are ultimately to improve forest conditions and livelihoods as well as achieve their own goals."[82]

5.3 Coordinative Approaches

The question remains as to how to achieve the goal of a rather concerted, coordinated approach to international regulation. There are already approaches to create a more effective coordination among the instruments of international environmental law.[83] These approaches range from the concept of "clustering,"[84] the "principle of interlinkages"[85] with the idea of a World Environmental Organization,[86] to more specific approaches in the context of forest governance like "policy patching"[87] or "autonomous interplay management."[88] All these approaches share the notion to rather accept the fragmented character of international environmental law and policy and strive for options to coordinate this fragmentation.

Considering in this regard an international legally binding approach, it could furthermore take the form of an "international coordination convention."[89] This means a convention that does not aim at the overall substantive regulation of forests or soils as such but at the coordination and cooperation of the all the processes involved and the existing international law.[90] Such an approach deviates from the concept of a traditional treaty and treaty cooperation and coordination in that it does not promote the cooperation advanced from within the single treaties but provides an external legal "framework" for cooperation. An international coordination convention could provide, on the one hand, for the fundamental principles and general objectives for regulation. On the other hand, an international coordination convention needs to provide explicitly for the objective to provide a framework for coordination. An international coordination convention provides for an independent treaty structure and the organs, such as a Conference of the Parties (COP), a secretariat, scientific or technical advisory bodies, that come with it. These organs

[81]A description lent from Rayner et al. (2010).

[82]Rayner et al. (2010), p. 16; See with regard to soils also Wyatt (2008), pp. 205–206.

[83]For an overview in more detail see Eikermann (2015), pp. 170 *et seqq.*

[84]von Moltke (2001), p. 5.

[85]Chambers (2008), p. 247.

[86]Chambers (2008), p. 249.

[87]Rayner et al. (2010), pp. 93 *et seqq.*

[88]van Asselt (2012).

[89]For further details regarding this concept see Eikermann (2015), pp. 170 *et seqq.*

[90]On the advantages and disadvantages arising from the merging of treaties see von Moltke (2001), p. 4.

play a major role in the promotion of the overall objective on the one hand. On the other hand, they provide for a permanent dialogue structure, participation, capacity building, monitoring, assessment, and reporting.[91] Ultimately, however, the focus of a coordination convention is on coordinated implementation of the existing treaties. In this respect, a coordination convention could provide for the specific features required to implement the indirect international forest or soil law in a "forest friendly" respectively "soil friendly" manner. The concept of this type of convention is not to establish a hierarchy or to grant itself priority over other relevant international treaties of the same issue area—which is unfeasible anyways—but to address the mutual supportive realization of all relevant agreements in the light of a common theme. Taking into account the example of the VCLT, an international coordination convention has to be understood as to tie its parties on the one hand to its own provisions. On the other hand, it intervenes in the relation of other treaties as and when required with regard to its own objective. Given the current lack of tools to address the fragmentation in international environmental law and policy today, an international coordination convention might serve as such a tool of conflict avoidance and resolution.[92] An international coordination convention furthermore provides for the necessary framework for concerted discussion. It allows for the involvement of the multiple actors already concerned, and it may ensure that forest and soil issues do not fall behind for the benefit of the actual treaty objectives of international treaties only relating indirectly to forests and soils.[93] Additionally, as indicated before, an international coordination convention is able to fill the role of establishing a prerogative of interpretation, particularly leading to forest- and soil-specific implementation of rules and procedures.

6 Conclusions

Forests and soils are well comparable subjects for international regulation. Both subjects are characterized by their multifunctionality, and the need for their international regulation is compelling with regard to the equilibrium that needs to be achieved between conservation and utilization for human well-being globally. The

[91]See on the role of treaty organs for example Brunnée (2002) and Eikermann (2015), pp. 170 *et seqq.* with further references.

[92]"A hallmark of the regime complex is a shift in the locus of action—away from elemental regimes and toward legal inconsistencies that tend to arise at the joints between regimes, and away from formal negotiations and toward the more complicated processes of implementation and interpretation." Raustiala and Victor (2004), p. 306.

[93]Developments pointing in the direction of more coordinative approaches for forest governance are the Legally Binding Agreement on Forests in Europe, see Eikermann (2015), pp. 37–39, 173–176; Jürging and Giessen (2013); as well as the rather recent developments under the UNFF, ECOSOC, Report of the United Nations Forum on Forests on its 2017 special session, 8 February 2017, UN Doc. /2017/10–E/CN.18/SS/2017/2.

issue of regulating forests and soils internationally has developed over decades and created multilayered regimes consisting of nonlegally binding instruments that provide for the valuable principles and objectives for sustainable forest management and the sustainable use of soils and an indirect international law that relates to forests, as well as soils, but follows its own objectives, affecting forests and soils rather collaterally either positively or negatively. Creating a new stand-alone international convention or a protocol to an international convention has to be considered against this background. The existing international instruments available for the regulation of forests and soils will persist, whether a stand-alone international treaty or not. A new instrument therefore has to be able to deal with this particular kind of fragmentation of international law and policy and take into account all the approaches that already exist. It is therefore suggested to explore means to cooperate and coordinate the valuable *status quo* that has already been created and that formed a certain degree of consensus. The idea of an international coordination convention might provide some additional perspective to further the discussion and achieve the overall objective of sustainability.

References

Alter KJ, Meunier S (2009) The politics of international regime complexity. Perspect Polit 7: 13–24

Birnie PW et al (2009) International law and the environment. Oxford University Press, Oxford

Bodansky D (2010) The art and craft of international environmental law. Harvard University Press, Cambridge, MA

Boer BW, Ginzky H, Heuser IL (2016) International soil protection law: history, concepts and latest developments. In: Ginzky H, Heuser IL, Qin T, Ruppel OC, Wegerdt P (eds) International yearbook of soil law and policy. Springer, Cham

Boyd W (2010) Ways of seeing in environmental law: how deforestation became an object of climate governance. Ecol Law Q 37:843–916

Brown Weiss E, Jacobson HK (eds) (1998) Engaging countries: strengthening compliance with international environmental accords. MIT Press, London

Brunnée J (1996) A conceptual framework for an international forest convention: customary law and emerging principles. In: Canadian Council on International Law, global forests & international environmental law. Kluwer Law International, London, pp 41–78

Brunnée J (2002) COPing with consent: law-making under multilateral environmental agreements. Leiden J Int Law 15:1–52

Brunnée J (2007) Common areas, common heritage, and common concern. In: Bodansky D et al (eds) The Oxford handbook of international environmental law. Oxford University Press, Oxford, pp 550–573

Brunneé J, Nollkaemper A (1996) Between the forests and the trees – an emerging international forest law. Environ Conserv 23:307–314

Chambers WB (2008) Interlinkages and the effectiveness of multilateral environmental agreements. United Nations University Press, Tokyo

Davenport DS (2005) An alternative explanation for the failure of the UNCED forest negotiations. Global Environ Polit 5:105–130

Desai BH (2011) Forests, international protection. In: Wolfrum R (ed) Max Planck encyclopedia of public international law, online edition. www.mpepil.com. Accessed 14 April 2017

Eikermann A (2015) Forests in international law – is there really a need for an international forest convention? Springer, Cham

Etter H, Gerhartsreiter T, Stewart N (2016) Economics of land degradation: achievements and next steps. In: Ginzky H, Heuser IL, Qin T, Ruppel OC, Wegerdt P (eds) International yearbook of soil law and policy. Springer, Cham

FAO (2015) Status of the World's Soil Resources Main report. http://www.fao.org/3/a-i5199e.pdf. Accessed 14 April 2017

Giessen L (2013) Reviewing the main characteristics of the international forest regime complex and partial explanations for its fragmentation. Int For Rev 15:60–70

Ginzky H, Heuser IL, Qin T, Ruppel OC, Wegerdt P (eds) (2016) International yearbook of soil law and policy. Springer, Cham

Hannam I, Boer B (2002) Legal and institutional frameworks for sustainable soils: a preliminary report. IUCN Environmental Policy and Law Paper No. 45. https://portals.iucn.org/library/sites/library/files/documents/EPLP-045.pdf. Accessed 14 April 2017

Hannam I, Boer B (2004) Drafting legislation for sustainable soils: a guide. IUCN Environmental Policy and Law Paper No. 52. https://portals.iucn.org/lbrary/sites/library/files/documents/EPLP-052.pdf. Accessed 14 April 2017

Hassan R et al (eds) (2009) Ecosystems and human well-being: current state and trends: findings of the condition and trends working group, The millennium ecosystem assessment series, vol 1. Island Press, Washington

Hönerbach F (1996) Verhandlung einer Waldkonvention Ihr Ansatz und Scheitern, Discussion paper FS-II 96-404. Wissenschaftszentrum, Berlin. http://bibliothek.wz-berlin.de/pdf/1996/ii96-404.pdf. Accessed 14 April 2017

Hooker A (1994) The international law of forests. Nat Resour J 34:823–877

Humphreys D (2005) The elusive quest for a global forests convention. Rev Eur Community Int Environ Law 14:1–10

Humphreys D (2006) Logjam: deforestation and the crisis of global governance. Earthscan, London

Jürging J, Giessen L (2013) Ein "Rechtsverbindliches Abkommen über die Wälder in Europa": Stand und Perspektiven aus rechts- und umweltpolitikwissenschaftlicher Sicht. Natur und Recht 35:317–323

Kasimbazi ED (1995) An international legal framework for forest management and sustainable development. Annu Surv Int Comp Law 2:67–97

Levin K et al (2008) The climate regime as global forest governance: can reduced emissions from Deforestation and Forest Degradation (REDD) initiatives pass a 'dual effectiveness' test? Int For Rev 10:538–549

Lipschutz RD (2000) Why is there no international forestry law: an examination of international forestry regulation, both public and private. UCLA J Environ Law Policy 19:153–180

Mackenzie CP (2012) Future prospects for international forest law. Int For Rev 14:249–257

Markus T (2015) Verbindlicher internationaler Bodenschutz im Rahmen der Alpenkonvention. ZUR 4:214–221

Matz N (2006) Wege zur Koordinierung völkerrechtlicher Verträge: völkervertragsrechtliche und institutionelle Ansätze. Springer, Berlin

Matz-Lück N (2008) Biological diversity, international protection. In: Wolfrum R (ed) Max Planck encyclopedia of public international law, online edition. www.mpepil.com. Accessed 26 May 2017

Miles EL et al (2001) Environmental regime effectiveness: confronting theory with evidence. MIT Press, Cambridge

Montgomery DR (2007) Dirt: the erosion of civilizations. University of California Press, Berkeley

Oberthür S (2009) Interplay management: enhancing environmental policy integration among international institutions. Int Environ Agreements Polit Law Econ 9:371–391

Pauwelyn J (2008) Fragmentation of international law. In: Wolfrum R (ed) Max Planck encyclopedia of public international law, online edition. www.mpepil.com. Accessed 14 April 2017

Raustiala K, Victor DG (2004) The regime complex for plant genetic resources. Int Organ 58: 277–310

Rayner J et al (eds) (2010) Embracing complexity: meeting the challenges of international forest governance. A global assessment report, prepared by the global forest expert panel on the international forest regime, IUFRO world series, vol 28. Vienna
Schulte zu Sodingen B (2002) Der völkerrechtliche Schutz der Wälder: nationale Souveränität, multilaterale Schutzkonzepte und unilaterale Regelungsansätze. Springer, Berlin
Srivastava N (2011) Changing dynamics of forest regulation: coming full circle? Rev Eur Community Int Environ Law 20:113–123
Tarasofsky R (1996) The global regime for the conservation and sustainable use of forests: an assessment of progress to date. Zeitschrift für ausländisches öffentliches Recht und Völkerrecht 56:668–684
van Asselt H (2011) Integrating biodiversity in the climate regime's forest rules: options and tradeoffs in greening REDD design. Rev Eur Community Int Environ Law 20:139–150
van Asselt H (2012) Managing the fragmentation of international environmental law: forests at the intersection of the climate and biodiversity regimes. J Int Law Polit 44:1205–1279
von Moltke K (2001) On clustering international environmental agreements, IISD. http://www.iisd.org/sites/default/files/publications/trade_clustering_meas.pdf. Accessed 14 April 2017
Weigelt J, Müller A, Beckh C, Töpfer K (eds) (2014) Soils in the nexus – a crucial resource for water, energy and food security. München
Wolff F, Kaphengst T (2016) The UN Convention on biological diversity and soils: status and future options. In: Ginzky H, Heuser IL, Qin T, Ruppel OC, Wegerdt P (eds) International yearbook of soil law and policy. Springer, Cham
Wolfrum R (2011) International law of cooperation. In: Wolfrum R (ed) Max Planck encyclopedia of public international law, online edition. www.mpepil.com. Accessed 14 April 2017
Wolfrum R, Matz N (2003) Conflicts in intersnational environmental law. Springer, Berlin
Wyatt AM (2008) The dirt on international environmental law regarding soils: is the existing regime adequate? Duke Environ Law Policy Forum 19:165–207
Young OR (ed) (1999) The effectiveness of international environmental regimes – causal connections and behavioral mechanisms. MIT Press, Cambridge

The Sustainable Management of Soils as a Common Concern of Humankind: How to Implement It?

Harald Ginzky

1 Introduction

Fertile soils are a precondition for a sustainable development. Without fertile soils, security of food production is at risk. Fertile soils are needed for the production of plants for renewable energy too. In addition, fertile soils are the second-largest biological sequester of carbon. Finally, soils host an almost infinite biodiversity.[1]

Considering these services, current figures of the ongoing or even the acceleration of land degradation are alarming. About one third of world's soils are already affected.[2] An area of the size that Italy has been, and will be, if nothing happens, constantly degraded per year.[3] As soils are the fundamental component of land, all figures mentioned are valid for the degradation of soils, too.[4]

Soils are immobile and thus local.[5] Nevertheless, there is a strong argument that international regulation and cooperation are needed to ensure sustainable management[6] of soils worldwide.[7] International cooperation, including regulation,

[1]Ginzky (2016), p. 2.

[2]Ginzky (2016), p. 3.

[3]Linz and Lobos (2016), p. 197.

[4]Boer et al. (2016), p. 49.

[5]However, erosion by wind might cause that soil elements could find its way to the higher atmosphere or spots in a great distance.

[6]The term "sustainable management" should be understood to include the protection of soils which might require—under certain circumstances—that specific uses of soils are excluded in certain areas.

[7]Boer et al. (2016), p. 67.

H. Ginzky (✉)
German Environment Agency, Dessau, Germany
e-mail: harald.ginzky@uba.de

H. Ginzky et al. (eds.), *International Yearbook of Soil Law and Policy 2017*,
International Yearbook of Soil Law and Policy,
https://doi.org/10.1007/978-3-319-68885-5_23

is in general justified for three reasons. First, the relevant topic causes transboundary effects. Second, the topic could be a potential cause of conflict between states, and third, international cooperation seems to be more effective for solving the emerging problems than pure national approaches.[8] Strong arguments underline that all three reasons are valid with regard to land and soil degradation.[9]

The chapter takes the need for an adequate international regulation for soils as the starting point. It shows that the sustainable management of soils has to be regarded as a common concern of humankind from a scientific point of view and is at least politically agreed. It then asks how to provide for appropriate measures. To this end, it analyzes the existing international regime for soils as fragmented and insufficiently coordinated. In addition, regimes on other topics regarded as common concern of humankind, in particular the Convention on Biological Diversity (CBD), are considered, and whether instruments applied for CBD could be applicable for the implementation of the sustainable management of soils. As a next step, the chapter analyzes whether the approaches applied under the regime on Deep Seabed Mining (DSM) under Part XI of UNCLOS could be used as a model. The deep seabed beyond national jurisdiction, the so-called Area, has been declared as common heritage of mankind.[10] After clarifying why and to which extent this regime could be used as model, some conclusions are drawn with regard to how to design international obligations with regard to the sustainable management of soils. The chapter concludes with an outlook.

2 Sustainable Management of Soils as a Common Concern of Humankind

The following section will analyze why sustainable management of soils has to be regarded as a common concern of humankind. To this end, the concept of a "common concern of humankind" has to be explained. The article shows whether this concept applies to sustainable management of soils both from a scientific, political and a legal perspective. Then the challenges with regard to such an effective implementation are described. This section finally analyzes to what extent the existing international regimes provide for an effective implementation.

---

[8]Ginzky (2016), p. 7.

[9]Boer et al. (2016), p. 68.

[10]UNCLOS uses the term "common heritage of mankind" which could be questioned as it triggers a gender bias.

2.1 Theoretical Contents

The concept of "common concern of humankind" has been developed by international lawmaking accompanied by academic debate. Several multilateral environmental treaties have included formulations that incorporate the basic idea of this concept. Not surprisingly, the formulations differ in the various treaties that have been adopted at different points of time. Formulations used are "the interest of the nations of the world,"[11] "in the interest of mankind,"[12]or "an irreplaceable part of the natural systems of the Earth which must be protected for this and the generations to come."[13] Most prominently also, UNFCCC[14] and CBD[15] have used the wording "common concern of humankind."[16] Despite the different wording, the contents seem to be as more or less equivalent.

Scholars have come to a condensed understanding of the main content of the "common concern of humankind." For example, *Birnie, Boyle and Redgwell* state:

> If "common concern" is neither common property nor common heritage, and if it entails a reaffirmation of the existing sovereignty of states over their own resources, what legal content, if any, does this concept have? Its main impact appears to be that it gives the international community of states both a legitimate interest in resources of global significance and a common responsibility to assist in their sustainable development. Moreover, insofar as states continue to enjoy sovereignty over natural resources and the freedom to determine how they will be used, this sovereignty is not unlimited or absolute, but must now be exercised within the confines of the global responsibilities set out principally in the Climate Change and Biological Diversity Conventions, and also in the Rio Declaration and other relevant instruments.[17]

In other words, from a more theoretical perspective, the concept of "common concern of humankind" primarily refers to global problems or challenges that "inevitably transcend the boundaries of single states and require collective action in response."[18] The relationship to other international principles or concepts such as "global commons" or "common heritage of mankind" is not thoroughly clear. It could, however, be said that whereas the principle of common heritage of mankind or the concept of global commons refers to specific resources, the principle of common concern of humankind deals with global problems or challenges that transcend national boundaries.[19]

---

[11]International Convention for the Regulation of Whaling, 1946.

[12]Antartic Treaty 1959.

[13]Convention on International Trade in Endangered Species of Wild Fauna and Flora 1973.

[14]Framework Convention on Climate Change 1992.

[15]Convention on Biological Diversity 1992.

[16]For a more holistic review of relevant international treaties, see Boer (2014), p. 295.

[17]Birnie et al. (2009), p. 130.

[18]Shelton (2009), p. 83.

[19]Bowling et al. (2016), p. 3 and Sanchez Castillo-Winckels (2017), p. 135.

The principle of common concern of humankind is not limited to environmental challenges. It can be traced to humanitarian and human rights law, which reflects "the notion of common concerns or a global set of values and interests independent of the interests of states."[20] Furthermore, the principle has also been adopted by the "Convention for the Safeguarding of the Intangible Cultural Heritage."[21]

The challenges may affect resources under the jurisdiction of states.[22] The principle of common concern of humankind does generally not question the right of states to exercise their sovereignty over these resources. The principle of common concern of humankind, however, states that this right is not absolute as these resources are of global significance. "Global significance" could be assumed if the resources, *inter alia*, provide ecological services that could be of relevance with regard to specific challenges in the interest of humankind, like the eradication of poverty, migration, or avoidance of conflicts. If so, states could be obliged to undertake actions, including the implementation of conservation and protection strategies, to achieve sustainable development.[23] The sovereign right of states over their domestic resources is thus limited by the responsibility to act to this end.[24]

The conceptual idea of something being declared as a common concern of humankind is that it requires as a minimum the establishment of an effective international cooperation.[25] In practice, as this designation is usually agreed upon by an international instrument, the specific cooperative means are put normally in place by the same international treaty. The level of cooperation is to be decided by the international agreement.

2.2 Sustainable Management of Soils as a Common Concern of Humankind

Soils are undisputedly immobile and thus local. They inevitably refer to a national jurisdiction.[26] The question to be answered is: do they provide any transboundary ecological service in the interest of humankind?

Soils provide the following services, which are indispensable for sustainable development:

---

[20] Shelton (2009), p. 83.

[21] Convention text under: https://ich.unesco.org/en/convention. Further information with Sanchez Castillo-Winckels (2017), p. 142.

[22] This is the case although some of the treaties mentioned above refer to Global commons.

[23] Boer (2014), p. 299.

[24] Boer et al. (2016), p. 66.

[25] Bowling et al. (2016), p. 3 and Cottier et al. (2014), p. 298.

[26] The Antarctic is the one and only exemption.

- basis for food production and production of plants for renewable energy;
- sequester for carbon;
- host for biodiversity;
- vital role in the world's biological cycle by, *inter alia*, storage for nutrients or filter for groundwater from hazardous substances;
- cultural and biological archive.

To a certain extent, all these services could trigger transboundary implications on aspects in the interest of humankind, such as climate change, hunger, poverty, migration, and political or even military conflicts.[27]

Thus, although soils are immobile and local, often privately owned, and always a resource that falls under a national jurisdiction, they provide transboundary ecological services that are in the interest of humankind. *Anja Eikerman* has branded the very illustrative term "hybrid good" for pointing out this duplicity with regard to forests.[28] The term "hybrid good" seems to be also a good "brand" for soils. Thus, from a factual point of view, sustainable management of soils could be regarded as a common concern of humankind.[29]

Legally speaking, the status is not that clear. The UNCCD[30] does not state that desertification or land degradation is a common concern of mankind, neither does the CBD nor the UNFCCC. However, recent developments, in particular three major decisions, must also be taken into account. All these decisions are, however, not legally binding. First, the Outcome Document of Rio+20 states that "desertification, land degradation and drought are challenges of a global dimension and continue to pose serious challenges to the sustainable development of all countries."[31] As this document was jointly adopted by all state representatives, it expresses their common view. Second, Target 15.3 of the Sustainable Development Goals demands that states should strive to achieve a "land degradation neutral world" by 2030.[32] The Sustainable Development Agenda was adopted in September 2015 by the UN General Assembly and could therefore also be regarded as the common opinion of the world's community. The objective of a "land degradation neutral world" implicitly emphasizes the transboundary importance of the ecological services of soils. Third, the decisions adopted by the UNCCD Conference of the Parties in 2015 have underlined the increasing perception of the

---

[27]Ginzky (2016), p. 7; Boer et al. (2016), p. 68; see also Flasbarth (2016), p. 17 mentioning that even the Syrian conflict has been at least partly caused by land degradation processes.

[28]Eikermann (2014), p. 4.

[29]The discussion has been so far focusing whether land degradation is a common concern of mankind. However, as soils are the fundamental component of land and degradation is the contrary to sustainable management, this previous debate is also relevant for soils or the conservation of fertile soils.

[30]UN Convention to combat desertification in Countries Experiencing Serious Droughts and/or Desertification, particularly in Africa 1994.

[31]United Nations (2012) The Future We Want, Para 2005.

[32]United Nations (2015).

world community that land degradation is a global issue. Parties to the UNCCD adopted several decisions with regard to the implementation of the objective of a "land degradation neutral world."[33] By this, the UNCCD has established itself as the international lead organization for "land degradation neutral world" issues and stressed the importance to avoid or at least reduce the ongoing land degradation in the interest of humankind.

All these decisions underline that the world's community perceives the sustainable management of soils as a common challenge. At least from a political point of view then, the fundamental idea or concept of "common concern of humankind" that state sovereignty is limited by a state's international responsibility is now being acknowledged in relation to land degradation by the community of states and, by this also, to sustainable management of soils. As the decisions are all—formally seen—nonlegally binding instruments, it could not be stated that sustainable management of soils has been adopted as a common concern of humankind in a legally binding manner. Nevertheless, as international law usually evolves, an agreement at a political level has already been achieved.

2.3 *Challenges from a Soil Perspective with Regard to Implementation*

Defining the specific measures in order to actually implement sustainable management of soils factually faces many challenges.

First, soils differ significantly with regard to their components, structure, and functions. There are an enormous number of different types of soils. Thus, each soil type carries its own characteristics, which would have to be taken into account to determine appropriate sustainable management measures.

Second, the kind of use for a specific area of soil is important with regard to the appropriate way of management. For example, an area of soil that is used for food production requires different standards of management than an area used for recreation, for example.

Third, soils are vulnerable to different types of threats. The EU Thematic Strategy for Soils of 2006[34] has differentiated eight types, including soil erosion, decline in organic matter, local and diffuse contamination, sealing, compaction, and decline in biodiversity, salinization, floods, and landslides.[35] Each of these threats requires a different management strategy.

[33]Ginzky (2016), p. 19 and Boer et al. (2016), p. 62 and Minelli et al. (2016).

[34]EU Commission (2006). Additionally, eutrophication is nowadays discussed as a further threat category.

[35]EU Commission (2012) with further information on the status of degradation with regard to the various threats.

Fourth, the political, economic, and technical realities are very different in the states worldwide. These different circumstances are most relevant concerning the political priorities that societies might choose.[36] Thus, also these aspects are decisive in order to determine which specific action is appropriate and needed.

2.4 *Status Quo of International Soil Protection Law with Regard to an Effective Implementation*

The following section will analyze the status of international law on soils, in particular whether it foresees effective provisions to trigger the implementation of actions concerning the sustainable management of soils. The analysis will focus on the provisions of international treaties, as well as on the UN Sustainable Development Goals, in particular on the objective to achieve a "land degradation neutral world."

2.4.1 Treaty Law Provisions

First of all, it has to be said that there is no international treaty specifically dealing with soils.[37] The UNCCD, CBD, and UNFCCC contain provisions relevant to the sustainable management of soils but only to a limited extent and in a quite general fashion. All three treaties apply almost universally.[38] The following could be stated with regard to these three regimes.

The UNCCD could be regarded as the most relevant regime for the management of soil as it expressively refers to "land." However, it has to be stated that the UNCCD—formally taken—only deals with the so-called drylands, which occupy about 40% of the terrestrial surface.[39] Moreover, the UNCCD does not contain any soil-specific actions. Affected countries are only obliged to implement National Action Programs.[40] Although the UNCCD is established as the leading regime for the implementation of the objective of "land degradation neutral world" (LDN), by various decisions taken by the Conference of the Parties in 2015,[41] there are no legally binding obligations on specific soil-related measures.

[36]Boer et al. (2016), p. 69.

[37]Ginzky (2016), p. 11.

[38]"Almost" because US are not a Party to CBD.

[39]Boer and Hannam (2003), p. 153. The determination of whether an area falls within the scope of application of UNCCD depends on the "ratio of annual precipitation to potential evapotranspiration" (Article 1 (g) UNCCD).

[40]Ginzky (2016), p. Hannam and Boer (2002), pp. 62–63.

[41]Boer et al. (2016), p. 62. See also Wunder et al. (2017) in this volume.

The scope of application of the CBD could, in general, encompass all soil functions.[42] However, despite some programs that might have a positive benefit for soils, like the program on agricultural biodiversity, the CBD is nonspecific with regard to both soils and the implementation of soil management requirements.[43]

The UNFCCC primarily deals with climate change issues. With regard to soils, the UNFCCC provisions on accounting, maintaining, and restoring sinks and reservoirs are important. The UNFCCC does not, however, establish specific requirements concerning the management of soils.[44] The Paris Agreement signed in December 2015 has set ambitious quantitative targets in an attempt to hold the increase of global average temperature "well below 2 degree."[45] An ad hoc Working Group was established to provide guidelines for the accounting of "nationally determined contributions."[46] It has to be awaited whether the guidelines will provide requirements for soil and land management.

To sum up, soil-related provisions in international treaties are fragmented and uncoordinated.[47] Neither the UNCCD nor the CBD or the UNFCCC foresees soil-specific provisions, in particular with regard to implementation.

2.4.2 Objective of a "Land Degradation Neutral World"

Target 15.3 of the UN Sustainable Developments Goals requires states to strive to achieve a "land degradation neutral world" by 2030. The neutrality concept demands that the balance of ongoing land degradation and land restoration has to be zero.[48] Several aspects have to be settled before the objective could be implemented, such as the geographical and temporal scale of the neutrality concept, as well as indicators for measuring the status quo, land degradation, and restoration processes.[49]

There are no specific provisions on how the objective is to be implemented. The UNCCD has launched a series of pilot projects in several countries and with a reasonable amount of funding being provided. The prime objective of these projects is to assist states in determining their voluntary targets with regard to the LDN implementation. Thus, the projects are limited to the determination of a nation-specific voluntary target but fail to consider the implementation of specific soil-related measures.

[42]Ginzky (2016), p. 18. Hannam and Boer (2002), pp. 63–64.

[43]Wolff and Kaphengst (2016).

[44]Boer et al. (2016), p. 58. See also Streck and Gay (2016).

[45]Article 2.1 (a) Paris Agreement.

[46]Paris Decision, Para 31.3.

[47]Hannam and Boer (2002), p. 62.

[48]Ehlers (2016), p. 75.

[49]Ehlers (2016), p. 80 and Wunder et al. (2017) in this volume.

2.5 Intermediate Conclusions

To sum up, it could be said that the sustainable management of soils is to be regarded as a common concern of humankind from a factual perspective. Legally speaking, it is under development, politically agreed, but the legally binding adoption is still pending.

With regard to determining appropriate actions, several challenges have to be dealt with: a large variety of soil types, the different uses of soils, the different types of soil threats, as well as the variations of political, economic, and societal circumstances.

Finally, the existing international law and non-legally binding instruments do not foresee clear and specific commitments for states to implement a national level of the sustainable management of soils.

3 Analysis of Regime of Other Topics Being Regarded as a Common Concern of Humankind

Next, the existing legal regimes dealing with common concerns of humankind, in particular the CBD, should be analyzed as to whether they provide a basis for an effective implementation of measures for sustainable management of soils as a common concern of humankind.

The CBD has declared "the conservation of biological diversity" as a common concern of humankind. Interestingly, during the preparation work of the CBD text, different concepts were discussed with regard to the conservation of biological diversity. "Common resource" and "common heritage of mankind" were withdrawn as they were considered to interfere too strongly with national sovereign rights over the local biodiversity resources. Thus, it was finally agreed to regard the "conservation of biological diversity" as a common concern of humankind.[50]

Concerning the aspect of effective implementation, the following characteristics of the CBD could be identified[51]:

- strong emphasis of national sovereignty of Parties;
- different responsibilities of states according to their capabilities;
- international financial cooperation[52];
- application of the precautionary approach;
- importance of in situ conservation by states applying, inter alia, Environmental Impact Assessment (Articles 8 and 14 CBD);

[50] See Bowling et al. (2016), p. 7 with further references to the preparatory work.

[51] See Bowling et al. (2016), p. 9.

[52] Morgera and Tsioumani (2010), p. 28.

- awareness of the need for additional scientific information and knowledge;
- sharing of benefits derived from the exploitation of genetic resources.[53]

It could to be asked whether and to which extent these conceptual approaches of the CBD can be applied to soil. In this regard, it has to be emphasized that the "conservation of biological diversity" is similar to "sustainable management of soils" because both concepts refer to resources that are normally under national jurisdiction.[54]

In fact, most of the CBD instruments are applicable to soil management. Soil, like components of biological biodiversity, is locally bound, and the sovereign rights of states seem to be essential too. Different responsibilities also seem to be adequate as the conditions strongly vary between the states, depending on the interpretation of the principle. Here it should be understood to require that the specific circumstances within a state are considered. Following from that, it requires a certain degree of flexibility as to how to implement an international obligation. Thus, it is not appropriate to demand the same level of protection and engagement from all states with regard to sustainable management of soils.

The following three instrumental requirements—the application of a precautionary approach, effective in situ conservation, and the need for additional scientific information and knowledge—are also appropriate for the sustainable management of soils.

An environmental impact assessment seems to be always useful if an activity could potentially cause detrimental effects. The in situ conservation seems to be also an appropriate tool for soils as the key functions of soils should be maintained. The provision and exchange of soil-related knowledge and data would be without doubt useful.

However, sharing of benefits as a general approach[55] seems to be not applicable with regard to soils. There is a crucial difference between the exploitation of genetic resources linked to biological diversity and the use of soils. The genetic information—of which the deriving benefits should be shared—is in fact not locally bound but abstracted from the local resource and carries long-lasting values. The genetic information could be important for various different purposes, for the development of medicine or enhancement of food production. Using soil primarily relates to the local bound resource. An economic value beyond the soil per se does not—or not to that extent—exist, except perhaps the soil function of being an archive for cultural

[53]Text of the respective protocol under: https://www.cbd.int/abs/doc/protocol/nagoya-protocol-en.pdf.

[54]Biological diversity could also be found beyond areas under national jurisdiction (see Article 4 (b) CBD). However, concerning the resources being under national jurisdiction both concerns pose similar questions and challenges.

[55]However, payment for sustainable land management to mitigate climate change could be an option. Although being an economic instrument payment is distinct from the benefit sharing concept.

or biological heritage. Moreover, this archive function usually does not entail relevant economic gains that could in fact be shared.

4 Deep Seabed Mining (DSM) Under Part XI of UNCLOS

The next section will analyze whether the regime of Part XI of the UN Convention of the Law of the Sea (UNCLOS) provides helpful conceptual approaches with regard to the implementation of actions that might be made applicable also for sustainable management of soils.

Part XI has declared the deep seabed as common heritage of mankind. The so-called Area is the part of the seabed beyond the continental shelf that usually extends to 200 nautical miles from the coastline.[56]

The principle of "common heritage of mankind" contains five elements, according to UNCLOS Part XI. First, UNCLOS sets out that no state may claim sovereign rights over parts of the Area or its resources. Thus, any kind of appropriation of its resources is not allowed (Article 137 UNCLOS). Acquisition of mineral resources deriving from mining operations is thus only allowed in line with the procedures foreseen by Part XI. Second, UNCLOS establishes the "International Seabed Authority" (ISA) through which "States Parties shall organize and control activities in the Area, particularly with a view to administering the resources of the Area" (Article 157 UNCLOS). ISA is based in Kingston, Jamaica. Third, the Area may only be used for peaceful purposes (Article 141 UNCLOS). Fourth, UNCLOS requests ISA to adopt appropriate provisions "to ensure effective protection of the marine environment from harmful effects" arising from mining operations (Article 145 UNCLOS).[57] Fifth, ISA has to ensure that financial and other economic benefits derived from activities in the Area are equitably shared among States Parties (Article 140 UNCLOS) (so-called benefit sharing).[58] This was agreed upon to avoid that only the frontrunner from developed countries would profit from exploitation activities. Thereby, the interests of developing countries should be guaranteed.

[56]States could extend their continental shelf up to 350 nautical miles via an application with the Commission on the Limits of the Continental Shelf, Article 76 UNCLOS.

[57]Jaeckel (2015).

[58]The conceptual approach of benefit sharing was taken on by CBD. Parties to CBD have adopted the stand-alone Nagoya-protocol in this context https://www.cbd.int/abs/doc/protocol/nagoya-protocol-en.pdf.

4.1 Comparability of the Principles of Common Heritage of Humankind and Common Concern of Humankind

Law comparison is by nature somewhat limited with regard to providing "compelling conclusions." Conclusions drawn from a law comparison could therefore never be absolute, in particular if not only different legal approaches to one specific problem (for example, protection of whales in different states) are compared but also the comparison refers to different topics and the respective legal approaches (here: sustainable management of soils with deep seabed mining). It has to be discussed whether why and to which extent such a comparison is justified. First of all, it has to be stated that there is a fundamental difference between the DSM regime, which deals with a topic declared as "common heritage of mankind," and the regime on soil management. Whereas neither sovereign rights may be declared over the Area or its mineral resources nor is appropriation legally possible, soils are under the jurisdiction of sovereign states.

However, a comparison seems to be justified mainly for two main reasons. First, both topics require sustainable management of the respective resource. With regard to deep seabed mining, the principle of common heritage of mankind, according to UNCLOS, requires both the promotion of the mining operations and an effective protection of the environment. Both objectives have to be complied with according to the principle of common heritage of mankind. For soils, the sustainable management is the key tool to maintain the transboundary ecological services.

Moreover, both the Area and its resources on the one hand and soils on the other hand provide services in the interest of mankind. Whereas for the Area it is primarily the provision of minerals, soils provide ecological services that are essential for, inter alia, climate mitigation, food production, and the eradication of poverty and hunger.

Finally, it could be questioned why the DSM regime was chosen for this comparison, although there are other regimes dealing with topics declared as common heritage of mankind. One example is the "Agreement Governing the Activities of States on the Moon and Other Celestial Bodies," which declares the moon and its resources as common heritage of mankind (Article 11).[59] However, the DSM regime seems to be more informative as it has already developed several implementation mechanisms due to current economic development of mining operations at the bottow of the oceans.[60]

[59] http://www.unoosa.org/oosa/en/ourwork/spacelaw/treaties/moon-agreement.html.

[60] ISA has adopted a bulk of regulations concerning exploitation and exploration activities. See Ginzky and Damian (2017), p. 328.

4.2 *Implementation by Sponsoring States*

Although in general ISA is in charge of administrating the mineral resources in the Area, UNCLOS directs a specific role to the so-called sponsoring states with regard to the implementation of appropriate measures.

UNCLOS requires that a contractor, before applying for a license to run a specific mining operation, shows that its entity is sponsored by a certain state—called then the sponsoring state (Article 153 UNCLOS).[61] The "contractor" must either be a national of this state or be effectively controlled by it.[62]

UNCLOS establishes two obligations of the sponsoring state, which could be concluded as being two different conceptual approaches on how to ensure implementation: first the due diligence obligation, second the so-called direct obligations.

4.2.1 Due Diligence Obligation

States Parties in general are obliged to ensure that mining operations by a contractor who is a national of a certain state or effectively controlled by it "shall be carried out in conformity with this Part" (Article 139 I UNCLOS). A failure in carrying out this obligation should entail liability (Article 139 II UNCLOS).

However, the liability is not absolute; there is no strict liability.[63] Article 139 II UNCLOS states:

> A State Party shall not however be liable for damage caused by any failure to comply with this Part by a person whom it has sponsored under article 153, paragraph 2(b), if the State Party has taken all necessary and appropriate measures to secure effective compliance under article 153, paragraph 4, and Annex III, article 4, paragraph 4.

According to the Advisory Opinion of the Seabed Disputes Chamber (SDC) of the International Tribunal of the Law of the Sea (ITLOS),[64] the due diligence obligation could be described as follows.

First, a sponsoring state has to take all necessary and appropriate measures to ensure compliance by those contractors sponsored by them.[65] To this end, they have to adopt laws, regulations, and administrative measures that may not be less stringent than the regulations and standards adopted by ISA.[66]

[61] Markus and Singh (2016), p. 356.

[62] ISA could establish its own "Enterprise" to run mining operations. In that case no sponsorship is required (Article 153 UNCLOS). Up to this date this "Enterprise" is not put in place.

[63] Ginzky and Damian (2017), p. 327; Markus and Singh (2016), p. 356.

[64] Advisory Opinion, under: https://www.itlos.org/fileadmin/itlos/documents/cases/case_no_17/17_adv_op_010211_en.pdf at paragraph 189.

[65] ITLOS Advisory Opinion at paragraph 186. See also Tanaka (2015).

[66] ITLOS Advisory Opinion at paragraph 240. As for the time being only exploration activities have been conducted the existing ISA legislation only regulates exploration activities—the

In addition, ITLOS SDC requires that the measures taken by the sponsoring state are risk adequate. ITLOS SDC has made this statement, in particular, with regard to the different challenges posed for an effective protection of the marine environment by exploration and exploitation activities.[67] Moreover, this means that the measures taken must correspond with the specific mining operation.

Finally, ITLOS SDC has emphasized that a sponsoring state has a kind of a proactive responsibility to ensure compliance. The Advisory Opinion of DSC referred to a judgment of the International Court of Justice and a commentary of the International Law Commission requiring "a certain level of vigilance in their enforcement" or "to exert its best possible efforts to minimize the risk."[68]

It has to be emphasized that ITLOS SDC has jurisdiction on all disputes with regard to the due diligence obligation of sponsoring states (Article 190 UNCLOS). Other states, ISA, and even another contractor are allowed to sue a claim with ITLOS SDC on potential failures of a sponsoring state to comply with its obligations.

Conceptually, the due diligence obligation is quite general and vague, only requiring that the ISA minimum standards have been met and the sponsoring state's actions are risk adequate and proactive. However, in combination with the control mechanism by the DSC, this seems to work out as an incentive for sponsoring states "to do their best."

4.2.2 Direct Obligations

In addition, ITLOS SDC has identified some so-called direct obligations of the sponsoring state, as follows:

> the obligation to assist the Authority in the exercise of control over activities in the Area; the obligation to apply a precautionary approach; the obligation to apply best environmental practices; the obligation to take measures to ensure the provision of guarantees in the event of an emergency order by the Authority for protection of the marine environment; the obligation to ensure the availability of recourse for compensation in respect of damage caused by pollution; and the obligation to conduct environmental impact assessments.[69]

These obligations are found in the UNCLOS provisions, in provisions of the mining code adopted by ISA, or in customary international law. These direct obligations contain specific requirements as, for example, to conduct an Environmental Impact Assessment, to apply a precautionary approach, or to have in place an emergency response and compensation scheme.

---

so-called mining code. See ISA has launched a process to develop and adopt regulations with regard to exploitation. See Ginzky and Damian (2017).

[67]ITLOS Advisory Opinion at paragraph 117.

[68]ITLOS Advisory Opinion at paragraphs 114 and 116.

[69]ITLOS Advisory Opinion at paragraph 122.

The conceptual approach for an effective implementation is here just to define precisely the specific international obligations that have to be implemented by the states.

4.3 Conclusion for Implementation Approaches of Common Interest of Humankind for Soils

The following conclusions are based on the assumption that there is a need for international cooperation and regulation with regard to the sustainable management of soils. The section that took into account the conceptual approaches under the CBD has provided arguments that certain CBD instruments would be transportable, such as the following:

- strong emphasis of national sovereignty of Parties;
- different responsibilities of states;
- international financial cooperation;
- application of the precautionary approach;
- importance of in situ conservation by states applying, inter alia, Environmental Impact Assessment (Articles 8 and 14 CBD);
- awareness of the need for additional scientific information and knowledge.[70]

However, it has still to be answered how to do it. In this regard, the comparison with the DSM regime could be informative. As explained, the DSM regime has put in place two different but relevant approaches for the management of soil.

The first approach is to put in place a quite general obligation combined with a control system: in the case of DSM, the due diligence obligation and the option of a juridical check by other states, ISA, or other contractors.

It seems to be that such an approach would encounter strong challenges with regard to sustainable management of soils. Such a general obligation could be the responsibility to ensure the objective of a "land degradation neutral world" or "sustainable management of soils." The basic challenge would be the broad diversity of circumstances, given the vast variety of soil types and of soils threats and real conditions in the different nations. It would therefore be extremely demanding for a juridical control body to determine what "sustainable management of soil" is actually required for a specific situation.

The second approach is the determination of the specific obligations via internationally binding law, as it is the case for DSM for the so-called direct obligations. It seems to be that this approach would be more appropriate also for the sustainable management of soils. This would allow the international determi-

[70]See Sect. 3.

nation of which specific obligations should be complied with, taking into account the various soil threats and permitting thus what soil threats should be internationally regulated.

By this, the conclusion of the analysis of the DSM regime that measures have to be risk adequate should be considered too. Hence, the international community would have to decide which soil threats have to be tackled and by which measure. However, as circumstances differ in the various states, a certain degree of flexibility has to be ensured.

5 Concluding Remarks and Outlook

This chapter concludes that sustainable management of soils is a common concern of humankind from a scientific perspective. The scientific arguments could be summarized by the term "hybrid good," stating the soils, although being locally bound nevertheless, provide crucial transboundary ecological services. From a legal perspective, it has to be said that the sustainable management of soils has not been accepted as a common concern of humankind.

In addition, it has be shown that certain general approaches for conservation of biological diversity would be transportable for the sustainable management of soils, inter alia, the sovereign rights of states or different responsibilities of states.

Finally, taking into account the two conceptual approaches for the implementation of required measures according to the DSM regime, it is concluded that an approach that would try to determine the specific obligations of states internationally would be more appropriate. However, the obligation would need to provide an appropriate level of flexibility in order to deal with the very different challenges in the various states.

Assuming that an international regulation is a reasonable objective for soil, a debate is needed on whether a new stand-alone treaty or additional regulations under either the CBD or the UNCCD would be the best option. Furthermore, it has to be discussed which soil threats deserve international regulation and how to balance the level of obligation with the need of flexibility in order to deal with national particularities. These are all very difficult questions that involve scientific legal and political considerations.

The questions thus in fact require discussion between lawyers and scientists on the one hand. On the other, decisions will also follow political priorities. Thus, the debate must be initiated by civil society and among policy makers.

Acknowledgements The author thanks Ralph Bodle, Beth Dooley, Ian Hannam, Irene Heuser, Till Markus and Pradeep Singh for very constructive comments. Errors and omissions are the sole responsibility of the author.

References

Birnie P, Redgwell C, Boyle A (2009) International law and the environment, 3rd edn

Boer B (2014) Land degradation as a common concern of humankind. In: Lenzerini F, Vrdoljak F (eds) International law for common goods, pp 289–307

Boer B, Hannam I (2003) Legal aspects of the sustainable soils. Rev Eur Community Int Environ Law 12(2):149–163

Boer B, Ginzky H, Heuser I (2016) International soil protection law: history, concepts and latest developments. In: International yearbook of soil law and policy, pp 49–72

Bowling C, Pierson E, Ratté S (2016) The common concern of humankind: a potential framework for a new international legally binding instrument on the conservation and sustainable use of marine biological diversity in the high seas. White Paper, 1–15, under: http://www.un.org/depts/los/biodiversity/prepcom_files/BowlingPiersonandRatte_Common_Concern.pdf

Cottier T, Aerni P, Karapinar B, Matteotti S, de SEphibus J, Shingai A (2014) The principle of common concern and climate change. Archiv des Völkerrechts 52:293–324

Ehlers K (2016) Chances and challenges in using the sustainable development goals as a new instrument for global action against soil degradation. In: International yearbookof soil law and policy, pp 73–84

Eikermann A (2014) Der Wald im internationalen Recht: Defizite, Regelungsoptionen und Mindestanforderungen. Rechtsgutachten im Auftrag des Bundesamtes für Naturschutz

EU Commission (2006) Communication: Thematic Strategy for Soil Protection, COM(2006)231 final

EU Commission (2012) Report: The implementation of the Soil Thematic Strategy and ongoing activities, COM (2012) 46 final

Flasbarth J (2016) Soils need international governance: a European perspective for the first volume of the international yearbook of soil law and policy. In: International yearbook of soil law and policy, pp 15–20

Ginzky H (2016) Bodenschutz weltweit – Konzeptionelle Überlegungen für ein internationales Regime. Handbuch Boden, pp 1–32

Ginzky H, Damian H (2017) Bergbau am Tiefseeboden – Standards und Verfahren für einen effektiven Schutz der Umwelt. Zeitschrift für Umweltrecht, pp 323–331

Hannam I, Boer B (2002) Legal and Institutional Frameworks for sustainable use of soils. IUCN Environmental Policy and Law Paper No. 45.

Jaeckel A (2015) An environmental management strategy for the international seabed authority? The legal basis. Int J Mar Coast Law 30:93–119

Linz F, Lobos I (2016) Boden und Land in der internationalen Nachhaltigkeitspolitik – von der globalen Agenda zur lokalen Umsetzung. Zeitschrift für Umweltrecht, pp 195–199

Markus T, Singh S (2016) Promoting consistency in the deep seabed: addressing regulatory dimensions in designing the international seabed authority's exploitation code. Rev Eur Community Int Environ Law 25:347–362

Minelli S, Erlewein A, Castillo V (2016) Land degradation neutrality and the UNCCD: from political vision to measurable targets. In: International yearbook of soil law and policy, pp 85–104

Morgera E, Tsioumani E (2010) Yesterday, today and tomorrow: looking afresh at the convention on biological diversity. Yearb Int Environ Law 21:3–40

Sanchez Castillo-Winckels N (2017) Why "Common Concern of Humankind" should return to the work of the International Law Commission on the Atmosphere. Georgetown Environ Law Rev 29:131–151

Shelton D (2009) Common concern of humanity. Environ Policy Law 39:83–96

Streck C, Gay A (2016) The role of soils in international climate change policy. In: International yearbook of soil law and policy, pp 105–128

Tanaka Y (2015) Obligations and liability of sponsoring states concerning activities in the area: reflections on the ITLOS advisory opinion of 1 February 2011. Neth Int Law Rev 60:205–230

United Nations (2012) General Assembly, Resolution adopted by the General Assembly on 11 September 2012, 66/288, The Future We Want

United Nations (2015) General Assembly, Seventieth Session, No. 11688, Agenda items 15 and 116, Resolution adopted by the General Assembly on 25 September 2015, 'Transforming our world: the 2030 Agenda for sustainable development', A/RES/70/1, p 1

Wolff F, Kaphengst T (2016) The UN convention on biological diversity and soils: status and future options. In: International yearbook of soil law and policy, pp 129–148

Wunder S, Kaphengst T, Freilih-Larsen A (2017)Implementing land degradation neutrality (SDG 15.3) at national level: general approach, indicator selection and experiences from Germany. International yearbook of soil law and policy

Development of Soil Awareness in Europe and Other Regions: Historical and Ethical Reflections About European (and International) Soil Protection Law

Irene L. Heuser

1 Historical Milestones of Soil Protection Law

The soils of Middle and Southern Europe developed in a process of several thousands of years: they were formed since the late Ice Age and after Ice Age in the course of 14,000–16,000 years.[1] Due to geomorphologic stability that prevailed in Europe under natural forest coverage in the Holozän up to the influence by farming on the vegetation and the land, soils with special profiles developed.[2] During the past centuries, they were regarded particularly with respect to their production function; this was especially due to the danger of famines and the lack of food. Soils have been used for agricultural and also engineering purposes for nearly 10,000 years.[3] When large-scale deforestation first began in Central Europe in the Younger Stone Age and agriculture boasted its first great revolution, soil resources seemed endless; the realization that soil resources on earth are finite only happened recently.[4] However, the aspects of the use of the soils and also their protection were already far more often considered in former times[5] than we would expect due to the subordinated position that the soils have in relation to other environmental media nowadays. For example, there already existed the latent danger of human-caused

[1]Wissenschaftlicher Beirat, Welt im Wandel (1994), p. 49.

[2]Bork et al. (1999), p. 16.

[3]Richter and Markewitz (2001), p.5.

[4]Blume et al. (2016), p. 581.

[5]Oldeman et al. (1992), p. 1.

I.L. Heuser (✉)
Ministry of Justice, Europe and Consumer Protection, Potsdam, Germany

IUCN World Commission on Environmental Law, Specialist Group on Sustainable Soils and Desertification, Kleinmachnow, Germany
e-mail: Irene.Heuser@googlemail.com

H. Ginzky et al. (eds.), *International Yearbook of Soil Law and Policy 2017*,
International Yearbook of Soil Law and Policy,
https://doi.org/10.1007/978-3-319-68885-5_24

soil erosion in the pre-Christian cultures of the Mediterranean area and in the country of Mesopotamia due to the spread of agriculture with increasing deforestation.[6]

In the *Antiquity*, Empeklodes of Akragas (500–430 BC) set up the teaching that the elements fire, air, water, and soil form "the four root forces of all things."[7] Even earlier than the Greek philosophy, the *Indian* teaching also refers to the soil as the solid element and additionally includes ether as the fifth of the basic elements of which the world is formed. According to verse I.I.2 Chandogya Upaniśad, one of the oldest vedic texts that is variously dated to have been composed by the eighth to sixth century BCE in India[8] is "esam bhutanam prthivi rasha" ("The essence of all things (is) the soil (the earth)").[9] An early quotation of the Rig Ved, another ancient Indian collection of vedic hymns, deals with the importance of the law of nature:

> Upon this handful of soil our survival depends. Husband it and it will grow our food, our fuel and our shelter and surround us with beauty. Abuse it and the soil will collapse and die, taking humanity with it.[10]

Considering the fact that this text is about 3500 years old, it reveals a wisdom that is still very modern and unique, especially in the context of environmental law and ethics. So the principles of soil conservation and sustainable management were already entrenched in prehistoric India.[11] The land originally was under tree cover, but as the human settlements expanded, trees were cleared to make way for cultivation.[12] Sustainability was ingrained in the thought processes of early Indians as evident from the teachings of Vedas. For example, the Atharva Ved hymn 12.1.35 reads:

> Whatever I dig out from you, O Earth! May that have quick regeneration again; may we not damage thy vital habitat and heart.

Coming back to *Europe*: after the first settlements of farmers on woodland soils, the *Greek* introduced the plough and spread the farming into steeper terrain. They already knew about the fertilizing properties of manure and compost, but it is not clear how widely such practices were followed.[13] The agricultural activities led to widespread erosion around 2000 BC. Platon described the condition of the soil in Attica with the following remark:

> Compared to what it was once, our country is like the skeleton of a body drained by disease.[14]

[6]Heine (1986), p. 116 et seq. (p. 117).

[7]Concerning Greek mythology: Pavan (1970), p. 4 et seq.

[8]Olivelle (2014), pp. 12–13.

[9]Lad and Frawley, Introduction to the Chapter "The manifestation of consciousness into plants."

[10]Cf. Johnston and Watson (2005), Fixen (2007).

[11]Kumar (2008), p. 299 et seq. (p. 299).

[12]Kumar (2008), p. 299 et seq. (p. 300).

[13]Montgomery (2007), p. 50.

[14]Gesellschaft für Technische Zusammenarbeit (2000), p. 1.

This proves that in these times, soil degradation was so strong that whole civilizations (like the Greeks or later the Romans) were—at least partly—concerned with the decrease of soil productivity; another example is the stone terraces of the Incas in the Andes, which were also built to prevent soil erosion.[15] Regional climate changes cannot explain the boom-and-bust pattern of human occupation in ancient Greece. Extensive Bronze Age soil erosion coincides with changing agricultural practices that allowed a major increase in human population. In the fourth century BC, the Greek historian and philosopher Xenophon wrote in his treatise "Oikonomikos" (better known under the title "Oeconomicus") about the two fallow-field systems.[16] Three centuries later, the famous Roman politician and writer Marcus Tullius Cicero considered Xenophon's Oeconomicus as being so important that he translated it into Latin.[17] Another famous Latin prose writer, Marcus Portius Cato (Cato the Elder), wrote a book about agriculture and stated that the price of land depends upon the quality of soils; he even recognized 21 classes of soils and made notes about their protection necessities.[18]

The symbolically powerful image of farming and the soil[19] has been incorporated in religious celebrations of fertility since Antiquity. The quoted view of Platon will be even more crucial considering the fact that North Africa as the "grainary" of the Roman Empire once was a territory with 600 flourishing cities. In 330 BC, North African grain had relieved the Greek famine. Nowadays, this area to a large extent is a desert. American soil experts from the United States Department of Agriculture found live olive trees thought to be 1500 years old, confirming that a drying climate was not responsible for the collapse in North African agriculture.

As in Mesopotamia, we recognize a similar development in China: the history of Chinese agriculture provides another example of dryland farmers from the uplands moving down onto floodplains as the population exploded.[20] Then about 60 years ago, Mao and a huge group of people made their way to find new land in Western China where they found barren, dry, alkalized, or even saline land.[21] Furthermore, in China, since the 1950s, sand drifts and expanding deserts have taken a toll of nearly 700,000 ha of cultivated land, 2.35 million hectares of rangeland, 6.4 million hectares of forests, woodlands, and shrub lands.[22]

As did the Greeks before, the *Roman* philosophers were aware of the fundamental problems of soil erosion and loss of soil fertility. But unlike Plato, who simply described evidence for past erosion, Roman philosophers exuded confidence that

---

[15]CoE, European Soil Resources, p. 13.

[16]Kutilek and Nielsen (2015), p. 179.

[17]Kutilek and Nielsen (2015), p. 179.

[18]Kutilek and Nielsen (2015), p. 179.

[19]As the classical Greek myths about Demeter who was asked to take care of the soil, see Kutilek and Nielsen (2015), p. 2.

[20]Montgomery (2007), p. 43.

[21]Blume et al. (2016), p. 581.

[22]UNCCD: http://www2.unccd.int/frequently-asked-questions-faq.

human ingenuity would solve any problems. But keeping soil on the Roman land became increasingly problematic, and therefore, as Rome grew, agriculture kept up by expanding into new territory.[23] After the Roman Empire collapsed, many fields north and west of the Alps reverted to forest or grass. But due to the fact that soils do not forget, the utilization of these past centuries, even if it was more than thousand years ago, is sometimes still visible today (for example, there exists an aerial photo of the former Via Claudia Augusta near Königsbrunn in Bavaria).

In the eleventh century, famers worked less than a fifth of England. Less than 10% of Germany, Holland, and Belgium were ploughed annually in the *Middle Ages*. In Europe, over the years, the intervention into the soil balance, in particular soil erosion, increased by means of constant extension of the agricultural land and the increasing deforestation.[24] By about AD 1200, Europe's best soils had almost been cleared of forest. By the close of the thirteenth century, new settlements began ploughing marginal lands with poor soils and steep terrain.[25] These years of human settlement are characterized by further ecosystem degeneration and subsequent soil erosion. Paracelsus, a Swiss scientist and alchemist, made interesting observations about soil formation and protection in the sixteenth century.[26]

At the time of the "Ancien regime," after the death of King Louis XIV of France, particularly strict and rigid soil utilization regulations existed because the consequence of insufficient resource management was an inadequate agricultural production that had negative consequences in the form of famines. Technological innovation played a key role in the spread and adaptation of people to new environments. Agricultural "improvers" came to prominence in the seventeenth century once the landscape was fully cultivated. In other regions of the world, especially in the Global South, this development took place much later.

So far, the hardly existing political consideration of soil protection in Europe is to be attributed (among other things) to the fact that after the abolishment of the "common land" called "Allmende"[27] (which was the property of the village community) in the *eighteenth century*, the soils were not equally regarded as common heritage like the other environmental media water and air. For example, in England, by the end of the eighteenth century, common fields had almost disappeared from the landscape. Even nowadays, the proportion between private and common property still varies very much in all regions and legal regimes of the world.[28)]

In the *nineteenth century*, with the liberal land policy following technical, economic, and political developments, the attitude toward the soil changed.[29] Problems of soil contamination of a larger extent began in the course of the

[23]Montgomery (2007), p. 58 et seq.

[24]Wey (1982), p. 21.

[25]Montgomery (2007), p. 91.

[26]Wallander (2014), p. 22.

[27]Wey (1982), p. 22.

[28]For details concerning property law tools, please read the next article in this IYSLP of Owley.

[29]Kocher (1983), p. 144 (p. 150).

industrial revolution about 200 years ago in England. Since the middle of the nineteenth century, it spread to Central Europe and brought a new quality in the conflict between the expansion of human needs and the stability of nature, in particular the soil balance.[30] Karl Marx, who is known for having had a focus on worker's rights, also saw commercialized agriculture as degrading to the soil:

> All progress in capitalistic agriculture is a progress in the art, not only of robbing the worker, but of robbing the soil. All progress in increasing the fertility of the soil for a given time is a progress towards ruining the more long-lasting sources of that fertility.[31]

Due to the constant increase in population in many European regions (especially after founding the Greater German Reich in the year 1871), the need for agricultural products increased substantially and led to an expansion of the land use on so far uncultivated soils and an intensification of the use of farmland.[32] There are several indications that land decline was greatly accelerated by human settlement. Iceland is a striking example for a sensitive environment that has been damaged by unsustainable land use and natural forces during the last millennium—and it also reflects the successes and failures in soil conservation and "healing the land" activities since the establishment of the world's oldest Soil Conservation Service in 1907.[33]

At the same time, the knowledge about the soils gradually increased, as, for instance, the publication of the first scientific journal exclusively concerned with the soil science (*Pochvovedenie*)[34] in the year 1899 verifies. The potential effects of agriculture on soil and water have been known for centuries and were almost certainly among the first environmental impacts to be recognized, especially in the first half of the last century. In the *twentieth century*, in many European States legislation was used in order to solve specific soil problems, in particular soil erosion.[35] After World War I, under the impression of famines in Europe, the increase of agricultural production was considered the most important task; this shows, for instance, the obligation of the landowner toward the community to cultivate the soil in accordance with Article 155 of the Constitution of the Weimar Republic.[36] However, the forms of use of land in the European countries varied substantially,[37] and legislation was used for many years in order to control land-use activities, to indirectly influence soil management systems,[38] and to regulate specific soil problems.

---

[30]Wey (1982), p. 24, 29, 31 et seq.

[31]Montgomery (2007), p. 109.

[32]Henneke (1986), p. 7.

[33]Arnalds and Runolfsson (2005), p. 67 et seq.

[34]Cf. Yaloon (1999), p. 22 et seq. (p. 22).

[35]Boer and Hannam (2003), p. 149 et seq. (p. 154).

[36]Cf. Holzwarth et al. (2000), p. 7.

[37]Oldeman et al. (1992), p. 1.

[38]Kurucz (1993), p. 467.

In comparison to Europe, the *United States of America*'s first legal acts dealing with soil conservation were already issued in the past century following the enormous dust bowls in the mid-west in the 1930s.[39] The dust bowl provided a vivid example of an agriculturally based environmental catastrophe, and it has been immortalized in literature, films, etc. (for example, in the "Grapes of Wrath" by John Steinbeck or in Rachel Carson's "Silent Spring"). The US Congress declared soil erosion to be a topic of national emergency[40] and took numerous steps in order to repair the soil damage. In this context, for instance, the first US "soil bank" was created,[41] and in the year 1933, the "U.S. Soil Conservation Service" was founded by President Roosevelt,[42] who has formed the saying: A nation that destroys its soils, destroys itself.

US environmental protection legislation started quite early and has made enormous progress in dealing with air and water pollution. American industry has responded to the environmental challenges in an outstanding degree. The Clean Air Acts of the 1960s and 1970 and the Clean Water Act are examples of the law being lined up to curb unsustainable conduct. But in contrast to water and air protection laws, the USA is not getting the law right in the field of soil protection. Soil programs currently only address soil erosion and contamination but not nutrient loss or other fundamentals essential to sustainable soils.[43] The Obama Administration did not initiate a fundamental shift in the understanding of soils as valuable resources that play a major role in sustaining life—although it seems that President Obama tried hard to introduce the means to combat climate change. There are strong doubts whether the Trump Administration will realize the change we need, especially after the recent statements of the current President about the Paris Agreement.

Meanwhile, soil degradation is a serious *global problem*. In many areas of the world, soil is being irreversibly lost and degraded as a result of increasing and often conflicting demands from nearly all economic sectors. Global warming is not the only problem facing agricultural and other types of land. According to Monique Barbut, Executive Secretary of the UNCCD, only 7.8 billion hectares of land is suitable for food production globally, about 2 billion hectares is already degraded, 500 million hectares of these has been totally abandoned.[44] We are losing around 12 million hectares of land each year. These lands could be restored to fertility for future use.

---

[39]Narjes (1984), p. 5 et seq. (p. 13).

[40]OECD, Integrated Policies, p. 16.

[41]Narjes (1984), p. 5 et seq. (p. 13).

[42]Wissenschaftlicher Beirat (1994), p. 158.

[43]Futrell (2009), p. 10077 et seq.

[44]Barbut, Speech at the global observance of the World Day to Combat Desertification at the EXPO Milano on 17 June 2016: "Land degradation is a growing threat to global security", http://www.unccd.int/en/media-center/Press-Releases/Pages/Press-Release-Detail.aspx?PRId=64.

In the *European Union*, more than 150 million hectares of soil is affected by erosion, in Southern Europe predominantly by water erosion and in Central and Western Europe by wind erosion. In the most extreme cases, it leads to desertification. Western Europe is highly urbanized, and competition for the land available results in soil sealing, the covering of the ground by an impermeable material.[45] Moreover, soil contamination remains a problem especially in Western Europe (application of sewage sludge, landfill of waste, etc.). There are probably 3.5 million potentially contaminated sites in the EU that affect water and soil. In Central and Western Europe, soil degradation problems are similar, although there is less soil sealing. Forty-five percent of the European soils have a low content of organic matter. There are great differences with respect to the soils' physical, chemical, and biological properties. More than 320 major soil types have been identified in Europe.[46] Like soil formation, soil degradation rates depend on soil properties, the local climate, organisms, topography, and other factors.[47]

This state of the European soils is a result of increasing and often conflicting demands from nearly all economic sectors. The allocation of land, for centuries being considered as ancient law (*nomos*) of the earth,[48] was completed in the *1970s* by the "soil care" as a further "terrestrial law of nature" (*Urgesetz*). While in earlier centuries the soil was regarded predominantly as a resource for food production, as well as a raw-material supplier,[49] today also the ecological soil functions are more and more often recognized. Despite the realization that there is a need for action to protect the soils, concrete and efficient legal instruments are still lacking on European Union and international levels.[50]

2 Phases of EU Soil Protection Law

The environment was a latecomer to the policy agenda of European integration: the design of the European Economic Community (the treaty dates from 1957) was driven primarily by the quantitative dimensions of building the Common Market, with relatively little attention paid to its qualitative aspects. Since the beginning of the 1970s, the European environmental policy increasingly has committed itself to cleaning air and water. Nevertheless, insights into the necessity of soil protection were delayed substantially by the point of view that soil degradation is essentially about problems of countries of the Global South and only minimally affects

[45] Guidelines on best practice to limit, mitigate or compensate soil sealing were made public by the European Commission on 12th April 2012.

[46] KOM (1992) 23 endg. p. 1 et seq., Vol. III, p. 30.

[47] Montgomery (2007), p. 20.

[48] Kimminich et al. (1996), p. 315 et seq.

[49] Blum (1988), p. 22.

[50] Boer and Hannam (2003), p. 149 et seq. (p. 154).

industrial nations.[51] When the forest damage occurred at the beginning of the 1980s in Europe, the importance of the soils for the forest ecosystem and the effects of soil acidification were obvious,[52] and the existing need for action was slowly realized. On the basis of gathering information about the condition of the soils in EU Member States, some of these countries introduced legal mechanisms to protect the soils in the 1980s and 1990s. Meanwhile, some of the current 28 Member States have specific legislation on soil protection. However, these laws often cover only one specific threat, such as soil contamination, and do not always provide a coherent protection framework.[53]

Only in recent years, the problem of full soil protection has attained more awareness. In a nutshell, we could say that the development of EU soil protection law occurred in three phases[54]: until the adoption of the Single European Act of February 17, 1986, soil protection was not explicitly considered within the context of EC Agricultural Policy or the initial EC Environmental Policy. This changed in a second phase of development with the inclusion of the environment chapter in the Single European Act and the beginning of a common policy on the environment. It led to particular aspects of soil protection being considered with varying degrees of intensity in the Community's policies. Environmental requirements should be integrated into the definition and implementation of other policies (like Common Agricultural Policy, Research Policy, Transport Policy, or Regional Policy). The principle of precaution, the preventative principle, and others formulated in Article 191 (2) of the Treaty of the Functioning of the EU (TFEU, as part of the Lisbon Treaty)[55] have become internationally accepted guiding principles in human interaction and have a major importance in the context of soil protection. Furthermore, any proposed legislation has to take into account the complexity of environmental situations in the various regions of the Community.

The Sixth Community Environment Action Programme (EAP) adopted in 2002 by the European Parliament and the Council marks the beginning of the third phase of development and the most important to date. In its 2002 Communication "Towards a Thematic Strategy on Soil Protection," the Commission identified eight key soil threats: organic matter decline, soil biodiversity loss, soil erosion, and soil contamination as priority aspects, as well as the additional aspects of soil sealing, soil compaction, salinization, and floods and landslides. A specific thematic strategy for soil protection was presented in 2006, together with the Proposal for a Soil Protection Framework Directive. The Soil Thematic Strategy is part of a plan

[51]Hero, GAIA 6 (1997), p. 205 et seq. (p. 208).

[52]Kloepfer and Franzius, UTR Vol. 27 (1994), p. 179 et seq. (p. 236).

[53]COM (2006) 231 final of 22.09.2006, p. 1 et seq.

[54]Heuser (2005), p. 33 et seq.

[55]According to Art. 191 (2) TFEU "Union policy on the environment shall aim at a high level of protection taking into account the diversity of situations in the various regions of the Union. It shall be based on the precautionary principle and on the principles that preventive action should be taken, that environmental damage should as a priority be rectified at source and that the polluter should pay."

to protect and preserve natural resources.[56] Among the elements being assessed and explored are the obligation to identify risk areas for soil erosion, organic matter decline, compaction, salinization and landslides, and the possibility of requiring Member States to establish programs of measures to combat the risks in those areas. The draft Soil Protection Framework Directive could have filled the gap that a classification and provisions for the treatment of contaminated land are completely missing in EU law at this point. It proposed setting a common definition of contaminated sites, a soil status report, and a common list of potentially soil-polluting activities that could be used for Member States to identify the contaminated sites on their territory and to establish a national remediation strategy. Moreover, preventive requirements have also been considered in the draft Directive, e.g., to limit soil sealing.

The EU Proposal for a Soil Framework Directive was in abeyance for many years. Some Member States (like Austria or Germany) even refused to take the floor in the Environment Council in Brussels. Unfortunately, the proposal for a Soil Protection Framework Directive was withdrawn by the Juncker Commission in 2014. At least it is clear that the legal base for soil protection provisions on EU level is Article 192 (1) TFEU; it requires majority voting in the Council. In a crucial session in December 2007, EU Environment Ministers failed to find a compromise due to the fear of administrative burdens for the Member States and the principle of subsidiarity according to Article 5 (3) TEU (Treaty on European Union) as a (false) counterargument; the subsidiarity principle is often used as an excuse in these discussions. With respect to the strong local dimension of soils, action should mainly be taken at the local level. But, nevertheless, there are good reasons in favor of EU wide measures for soil protection, in particular the multifunctionality of soils; the transboundary consequences of, for example, soil erosion and contamination; and the close correlation between soil degradation and other environmental problems (like climate change). Failure to protect the soils will undermine not only sustainability but also the long-term competitiveness of the EU.[57]

At present, there is no extensive, independent legal instrument for soil protection in EU Environmental Law. The current Seventh Environment Action Programme ("General Union Environmental Action Programme to 2020," with the title "Living well, within the limits of our planet"), which entered into force on January 17, 2014, recognizes that soil degradation is a serious challenge. It mentions in one of the action areas that "there are also topics which need further action at EU and national level, such as soil protection and sustainable use of land"; other instruments for resource efficiency are also supported.[58]

[56]It includes a Communication assessing the current situation and outlining the strategy's objectives (COM (2006) 231 of 22.09.2006, p. 1), a detailed Extended Impact Assessment and a Proposal for a Soil Framework Directive addressing the key soil threats (COM (2006) 232 of 22.09.2006, p. 1).

[57]Concerning the subsidiarity principle see Heuser (2006), p. 573 et seq.

[58]European Commission, "Living well, within the limits of our planet", http://ec.europa.eu/environment/pubs/pdf/factsheets/7eap/en.pdf, p. 2.

Several EU policies and fields of environmental law are contributing to soil protection—the environmental provisions preventing pollutant entries via the pathways of air, wastes, dangerous substances, pesticides, and sewage sludge, which have an effect on the soil because they clearly provide limiting values or other regimes to minimize pollutants. But these policies often have other aims and other scopes of action. Especially in relation to its biological and physical protection, soil has been a secondary consideration to broader environmental legislation. All in all, there are many positive approaches in existing EU law, but these do not detract from the impression of a mosaic or "patchwork" at present.

In contrast to climate change where consensus has been reached in the Paris Agreement on December 12, 2015, the worldwide crisis in soil degradation unfortunately receives relatively little public attention. In June 2008, the EU hosted a high-level conference on soil and climate change in Brussels, paying attention to the fact that healthy soils play a major role in carbon fixing, whereas soil degradation leads to the transfer of massive amounts of carbon fixed in soils to the atmosphere, contributing to greenhouse warming. Soil management plays a major role in climate change mitigation and adaption: 70 billion tons of carbon is stored in our soils, so even small losses can have huge effects on our emissions of greenhouse gases. Organic matter plays another fundamental role supporting soil fertility, retaining water, sustaining biodiversity, and regulating the global carbon cycle. But organic matter is in decline. The former Environment Commissioner Stavros Dimas called on the Council "to reconsider the need to protect this most precious resource through European legislation. A soil Framework Directive would increase soil protection and safeguard crucial functions like carbon sequestration."

At the moment, the EU pursues the objectives of soil protection with a set of mostly indirect instruments that form the "acquis communautaire" of soil protection. There are still many hurdles to take before a specific EU soil protection regulation will be adopted.

3 Soil Protection in International Law

At a global level, environmental policy is also still far from achieving comprehensive, legally relevant protection of the soils. Over the last 40 years, 30% of the world's arable land has become unproductive. More than a tenth of the earth's land area is desertifying. The economic losses are about $ 400 billion a year worldwide.[59] Soil degradation and desertification are happening not just in Africa but all over the world.

These problems in their entirety are not conceptionally tackled by the law of international organizations in an obligatory form but only in the form of so-called *pré-droit* or soft law. These *non-binding instruments* can be completed within a

---

[59]Futrell (2009), p. 10077 et seq.

shorter time frame because they are not mandatory and do not require ratification.[60] There is an increasing recognition of the role of international environmental law to overcome the global problem of soil degradation. But none of these instruments are sufficient to meet the requirements of international environmental law in relation to soil. The following soft law instruments on international level make reference to soils[61]: in 1981, the Food and Agriculture Organization (FAO) adopted the World Soil Charter. In 1982, the United Nations Environment Programme (UNEP) published the World Soils Policy. Both documents were prepared as conjunctive instruments to encourage international cooperation in the rational use of soil resources.[62] Furthermore, in 1982, the World Charter for Nature stated that the productivity of the soils shall be maintained or enhanced through measures that safeguard their long-term fertility and the process of organic decomposition and prevent all forms of degradation. They were put in concrete terms by the UNEP Environmental Guidelines for the Formulation of National Soil Policies of 1983. Agenda 21, the Action Plan of the 1992 UN Conference on Environment and Development in Rio, identified concrete steps to integrate environment and development. In 1996, the Nairobi Declaration pointed out that deforestation, soil degradation, and desertification had reached alarming proportions, seriously endangering living conditions. Furthermore, several special conferences deal with problems of soil protection, e.g. at the IUCN World Conservation Congress 2008, a specific legal workshop focused on the protection and sustainable use of soils.

At *regional level*, the Council of Europe achieved international attention to the problems of soil protection by the adoption of the first "European Soil Charter"[63] already in the year 1972, which went far beyond other regimes of international law. The Charter described the soil as one of the most precious goods of mankind and formulated first standards relevant for European soil protection policies. Increased awareness of the importance of soil protection is reflected by the "Revised European Charter for the protection and sustainable management of soil" of May 28, 2003. However, these achievements are subject to the reservation of their binding nature and the requirement of putting it in concrete terms. As the COE's provisions are instrumentally limited, they are able to develop suggestions for EU or international law only to the extent that they contain particularly concise or progressive instruments that could systematically and conceptionally influence the view de lege ferenda.

Investigation of the World Commission on Environmental Law (WCEL) of the International Union of the Conservation of Nature (IUCN) into provisions within existing international and regional legislative instruments indicates that existing

[60] Hannam and Boer (2002), p. 36 et seq.

[61] Please see Boer et al. (2016), p. 49 et seq.

[62] Hannam and Boer (2002), p. 61.

[63] European Soil Charter 197 (Recommendation for the Member States), Council of Europe, Strasbourg, Ref.: B (72) 63; published in: Council of Europe, Feasibility Study, p. 40 et seq.

binding instruments (also the three Rio conventions)[64] are insufficient as a framework to meet the objective of sustainable use of soil. In this respect, the UN Convention to Combat Desertification (UNCCD) of June 17, 1994 (which entered into force on December 26, 1996), forms the framework for international law activities within the range of the fight against soil erosion and desertification and sets up rules for the establishment of strategies; national, regional, and subregional action programs; international cooperation; research; financial and institutional mechanisms; etc. Although the geographic focus is limited due to the definition of "desertification," the Sixth Conference of the Parties adopted a regional annex for Central and Eastern European countries that enables the CCD to be applied more widely. At the moment, the UNCCD, as it currently stands, cannot be considered an effective instrument for the protection and sustainable use of soils. Some Mediterranean and most new Member States are affected parties and are therefore in the process of adopting regional and national action programs to combat desertification.

Other binding instruments might influence the protection and use of the soils, especially the Convention on Biological Diversity (CBD) and the UN Framework Convention on Climate Change (UNFCCC). The Kyoto Protocol,[65] adopted in 1997 (coming into force in 2005), determined mandatory emission reduction targets for developed countries for the first time[66] and highlights that soil is a major carbon store that must be protected and increased where possible. The Paris Agreement, which was adopted in December 2015, is a further step forward with regard to climate change policy.[67] Carbon sequestration in agricultural soils by certain land management practices can contribute to mitigating climate change. In addition, with the growing importance of agrobiodiversity within the context of the CBD, the sustainability of soils will become an increasingly important subject for debate. In 2002, the COP to the CBD decided "to establish an International Initiative for the Conservation and Sustainable Use of Soil Biodiversity" as a cross-cutting initiative, regarding soil biodiversity as a provider of essential ecosystem goods and services.

It is likely that soil will become a much more important part of the debate within the UNFCCC meetings, particularly in relation to the function of soil as a carbon sink. There is an urgency to improve the synergies between the existing conventions, especially UNFCCC, CBD and UNCCD, to provide benefits for the soils. In general, a treaty can have a lawmaking effect because it may be a means by which states lay down new rules of international law. With population growth until 2050 reaching more than 9.6 billion people, the harvested yields must be increased by about 40% in order to maintain a food supply comparable to today's situation[68]).

---

[64]Concerning CBD and UNFCCC see Heuser (2005), p. 350 f., and Hannam and Boer (2002), p. 62 et seq.

[65]Text is available at http://unfccc.int/resource/docs/convkp/kpeng.pdf.

[66]The first commitment period ended in 2011.

[67]Text available at http://unfccc.int/resource/docs/2015/cop21/eng/l09r01.pdf.

[68]Blume et al. (2016), p. 5.

Therefore, the idea of a future international soil protection convention or at least a protocol to the CBD or the UNCCD, which would contribute to a more intensified dealing with soil protection problems in all its facets, points into the right direction. A "Protocol for the Protection and Sustainable Use of Soil" as prepared by the Specialist Group on Sustainable Soils and Desertification of the IUCN World Commission on Environmental Law (WCEL) is currently being discussed.

At the international level, some hard law documents, like the CBD, take an ecosystem approach, and all three Rio conventions are binding instruments that might influence the use of the soils, but there is no specific or comprehensive soil or land protection convention yet. No legal document does include anywhere near a sufficient range of legal elements needed to protect and manage soil in a sustainable way yet.

At *regional level*, the Convention concerning the Protection of the Alps entered into force in 1995; the Alpine Convention Soil Protection Protocol (ACSPP)[69] was adopted in 1998. In order to achieve the general objective of protection of the Alps, the parties have a specific obligation for soil protection. The Protocol is based on an ecosystem perspective, and it aims to reduce the quantitative and qualitative damages to the soils. It seeks to preserve the ecological functions of soil, prevent soil degradation, and ensure a rational use of soil in that region. Due to its legal and administrative measures for soil protection, the coordination mechanisms, and the specific instruments to special soil uses and threats, the Alpine Convention with its ACSPP recognizes the important role and ecological values of the Alps and makes a binding legal instrument for the European soils available. In its proposal for a Council Decision, the European Commission promised that several elements of the Protocol may be included in a Community policy on soil protection, e.g. monitoring requirements, identification of risk zones for erosion, flooding and landslides, an inventory on contaminated sites, and the establishment of harmonized databases.[70] The Protocol could also help to implement appropriate measures in other countries at national level as any approach to soil protection has to take into account the diversity of regional and local conditions that exist in the Alpine region.

Meanwhile, the crucial role of soils in promoting sustainable development has received increased attention at a political level in the years before and after the Rio +20 Conference. Land resources are degrading at an alarming pace, affecting sustainable development. There were a number of relevant activities conducted by UNEP, UNCCD, FAO, and other actors on international level prior to this conference. The Rio+20 Outcome Document "The Future We Want" deals (in paragraphs 205 to 209) with "DLDD" (desertification, land degradation, and drought). According to paragraph 206:

[69]Proposal for a Council Decision (presented by the Commission) on the conclusion, on behalf of the European Community, of the Protocol on Soil Protection, the Protocol on Energy and the Protocol on Tourism to the Alpine Convention, COM (2006) 80 final of 02.03.2006, p. 1 et seq.

[70]Proposal for a Council Decision, COM (2006) 80 final of 02.03.2006, p. 1 et seq.

> we recognize the need for urgent action to reverse land degradation. In view of this, we will strive to achieve a land-degradation-neutral world in the context of sustainable development.

The Rio+20 Outcome statement is now included in the *2030 Agenda for Sustainable Development* "Transforming Our world" and its Sustainable Development Goal (SDG) 15. Agenda 2030 came into effect in January 2016 and will guide the decisions over the next 13 years, taking into account different national realities, capacities, and levels of development and respecting national policies and priorities. According to the Resolution adopted by the General Assembly on September 25, 2015: "We resolve to take further effective measures and actions, in conformity with international law, to remove obstacles and constraints..." The text of SDG 15 "to halt and reverse land degradation" appears ambitious, maybe even demanding, although it does not specify when this goal should be achieved. SDG target 15.3 states:

> By 2030, combat desertification, and restore degraded land and soil including land affected by desertification, drought and floods, and strive to achieve a land-degradation neutral world.

The UNCCD defines land degradation neutrality as "a state whereby the amount and quality of land resources necessary to support ecosystem functions and services and enhance food security remain stable or increase within specified temporal and spatial scales and ecosystems."[71] Target 15.3 uses the restrictive words "to strive to achieve a land-degradation neutral world" (LDNW). It has received wide political support, and the UNCCD invited parties to formulate national voluntary targets to achieve LDN and to integrate LDN targets into their UNCCD National Action Programmes. Through this LDN Target Setting Programme, which became operational in spring 2016, the UNCCD's operational arm—the Global Mechanism—is supporting countries in their national voluntary LDN target-setting processes. The main objective is to enable country parties to define national baselines and to identify voluntary targets and measures to achieve LDN by 2030. In March 2016, the UN Statistical Commission approved a draft global indicator framework intended for the follow-up and review of progress toward the SDGs at the global level. SDG indicator 15.3.1, and its subindicators, which are recognized as suitable metrics for monitoring and reporting on restoration, combatting desertification, and achieving land degradation neutrality, reads as follows: "the proportion of land that is degraded over total land area." The three subindicators are land cover and land cover change, land productivity, and carbon stocks above and below ground.

This process of target setting at the international and national levels demands a high level of awareness of the fundamental role of soils. In the process of defining land degradation neutrality, it is anticipated that there will be more focus on the development of both national and international legal regimes of sustainable use of soil over the coming years.

---

[71] http://www2.unccd.int/sites/default/files/documents/LDN%20Scientific%20Conceptual%20Framework_FINAL.pdf, p. 1.

Several countries, including the USA, Japan, Canada, Australia, Brazil, and countries of the Global South, have already established soil protection policies that include legislation, guidance documents, monitoring systems, identification of risk areas, inventories, remediation programs, and funding mechanisms for contaminated sites for which no responsible party can be found. Such policies ensure a comparable level of soil protection to the approach endorsed by the EU Soil Protection Strategy. But most countries and international organizations still do not have an extensive, independent legal instrument for soil protection. Therefore, to sum it up, international law currently fails to take into consideration all of the implications of soil and tends to disregard it as such. What we need is an ethically relected concept to protect the soils.

4 Toward a New Soil Ethics

Starting point for ethical considerations is the philosophical discipline "environmental ethics," which considers the moral and ethical relationship of human beings to the environment.[72] It results in increased awareness of how the rapidly growing world population is impacting the environment, as well as the environmental consequences.[73] Ecological integrity is defined as the "the capacity of an ecosystem to support and maintain a balanced, integrated, adaptive community of organisms having a species composition, diversity in functional organization comparable to that of similar, undisturbed ecosystems in the region."[74] Soil ethical considerations motivate action in three distinct contexts; acts of single individuals such as farmers, collective management of soils as common pool resources, and finally the justification of public policy.[75] Caring for the ecological integrity of the soils is an adequate response to the current soil threats.

Failures in soil protection mainly stem from the fact that soil is undervalued by society as a resource. Policy makers on all levels mostly do not think of soil as a living resource, a complex blend of organic material and minerals necessary for our life. Many think of it as property—instead of regarding it as an ecological resource (which is no less vital and no less vulnerable than air and water[76]) and of taking appropriate measures to protect it. What are the reasons why soil protection has not been realized as consequently as other fields of EU environmental law? Which

[72]Definition in: DSST Environmental Science: Study Guide & Test Prep Chapter 18 (Ethical and Political Processes of the Environment), http://study.com/academy/lesson/environmental-ethics-human-values-definition-impact-on-environmental-problems.html.

[73]Concerning Agriculture and Environmental Ethics: see Thompson (2017).

[74]Cited from Mackey (2008), p. 61 et seq. (p.76; see also his general explanation of the Earth Cahrter and its values and statements, p. 64 et seq.).

[75]Thompson (2011), p. 31–42 (p. 31).

[76]Futrell, William J., Passing the torch, The Environmental Forum, 2009, p. 40 et seq. (p. 43).

ethical considerations, e.g. whether property rights should be defined to include ecological limitations of the soils, should be the basis for the law?[77]

In the EU, most of the Member States have special *regimes on property* and their guarantee. Due to the process of reception of Roman law, soil is not a free good, but mostly in private property. This also explains the different scope of environmental media in political initiatives since an access to the soils protected by national (civil) law was made more difficult (due to the regularly constitutionally founded property situation in the EU Member States). Although Article 345 TFEU[78] states that "The Treaties shall in no way prejudice the rules in Member States governing the system of property ownership" (meaning that the EU leaves the national property regime untouched), it has become clear from case law developed by the European Court of Justice that this does not mean that property law as such cannot be repealed by European legislation. Property is but a particular type of institutional arrangement. It is an arrangement that defines the rights and the duties of all members. With only rare exceptions, any right must be matched by a correlative duty. Therefore, most constitutions meanwhile define property as a combination of individual entitlement and social responsibility.[79] Roman land law strengthened the assignment of the land to legal persons to order the land seizure and laid down the relevant legal relationships. Since Roman land law, there exists a differentiation between the environmental media to the extent that air and water were not regarded as personal property in the legal sense and therefore were not accessible to influences according to private law, whereas the soils were not equally regarded as common heritage (*res communes omnium*). Therefore, the individual owner or—to a smaller extent—the beneficiary was incorrectly assumed to protect the soils.[80]

One of the most significant features of European rural communities from the Middle Ages to the nineteenth century was the ubiquity of common lands, especially waste lands, and the diversity of management systems.[81] In the rationalist and individualistic discourse of the Enlightenment, the achievement of the common welfare was incompatible with the survival of the "commons goods." Under this ideological aegis, the nineteenth century witnessed the culmination in Europe and Latin America of the dismantling of communal property and its privatization. Toward the end of that century, different currents of intellectual criticism of economic liberalism, both on the left and right sides of the political spectrum, vindicated the consideration of communal property as a positive thing, understanding it as an example of an assumed primitive communism and proposing its reinstatement with the social reform schemes.[82] Common land was used extensively in England, Wales, and many former British colonies; in Spain; in

---

[77]Bosselmann, in: Westra/Bosselmann/Westra, p. 319 et seq. (p. 328).

[78]Commented by Kingreen, in: Calliess and Ruffert (2016) Art. 345 AEUV, Ref. 2 et seq.

[79]Bosselmann, in: Westra/Bosselmann/Westra, p. 319 et seq. (p. 328).

[80]Von Lersner, NuR 1982, p. 201 et seq.

[81]Berasin and Miguel (2008), p. 162 et seq. (p. 166).

[82]Berasin and Miguel (2008), p. 162 et seq. (p. 163).

Scandinavia (*Allemensratten*); in Germany (*Allmende*); and in the Alpine countries, especially in Switzerland. Similar common property regimes are to be found in southern Asia (e.g., India and Nepal) and Latin America (e.g., Mexico).

The metaphor of the "tragedy of the commons" (published by Garret Hardin in the journal "Science" in 1968) illustrates how free access and unrestricted demand for a finite resource ultimately structurally dooms the resource through overexploitation. This occurs because the benefits of exploitation accrue to individuals or groups, each of whom is motivated to maximize use of the resource to the point in which they become reliant on it, while the costs of the exploitation are distributed among all those to whom the resource is available. The consequence might be to specifically advocate the privatization of commonly owned resources. Of course it would be worse if the land was not owned by anybody because then it would be used without any regard to the disadvantages resulting. But soil protection is not about privatization of communally managed resources but about the opposite, that is, the obligation of private persons to care about a common good or better a common heritage. In effect, private use repeatedly resulted in worse outcomes than compared to the previous commons management—and we see that in the context of soils as well! So the solution lays in soil protection obligations.

However, the property situation does not constitute the only reason for the legal neglect of the soils. Over the time of nearly 10,000 years, ever since soil has been used, humanity has developed an intimate relationship with the soil. Environmental consciousness and natural resource and biodiversity conservation were intrinsic features of many *religious rituals and practices* all over the world. This is well illustrated in our most ancient texts such as the Bible, the Torah, or the Rig Ved. The last one is the oldest of the four ancient Indian scriptures, written between 1500 and 2500 BC, and it speaks poetically about this relationship.[83] Another example is a Persian proverb saying:

> God will not ask thee thy race, nor thy birth.
> Alone he will ask of thee,
> What hast thou done with the land I gave thee?

Taking up these old religious texts and what they mean in the context of soil protection, Professor Ronald Engel presented at the World Forum on Soils, Society and Global Change in Iceland in September 2007 the thesis that "It may well be that a reluctance to appreciate the importance of soil in our lives is due to a reluctance to accept the full spiritual implications of our participation in this reality."[84] This convincing statement reveals the fact that, in modern industrial society, most people do not relate to the soil any more. We have an expression in German called "Bodenständigkeit," meaning "being rooted in the soil," which underlines that relation and also its continuation to decrease. The most ancient texts have acknowledged the importance of soils in our lives, and in many languages, soil has entered

[83]See quote on page 2. In another sutra the Rig Ved stated: "Harness the plows, fit the yokes, now that the womb of the earth is ready to sow...", see Richter and Markewitz (2001), p. 5 et seq.

[84]Ronald Engel on the World Forum on Soils, Society and Global Change (2007).

into the symbolism of culture and religion in the context of "earth" and "ground" and "foundation." In theology, we speak of the ultimate reality as the "ground of being." "Human," "humility," and "humus" come from the same ancient Indo-European root meaning "soil." For instance, the Hebrew name of the first man, Adam, is derived from the word "Adamah," which means "earth" or "soil."[85] The Koran too alludes to humanity's relation to soil:

> Do they not travel through the earth and see what was the end of those before them? (...) They tilled the soil and populated it in greater numbers (...) to their own destruction (Sura 30:9).

Nowadays, only for a few people, soil fertility plays a major role for the living and the economic status. Nevertheless, the lack of sensuous experience does not necessarily constitute one of the substantial *reasons for ignoring the consequences of soil degradation*—because climate change and the loss of endangered species are still less easily to experience.

In addition, there is also the rather banal reason that the soil is often equated with "dirt," and therefore soil protection is not very popular. There is also a lack of information: the fact that the imbalance between soil degradation (which takes decades) and soil formation (which needs centuries until many thousands of years) was not sufficiently considered yet. Building up soil layers of 30 cm requires a period of 1000 to 10,000 years. Therefore, changes in soil ecosystems happen over decades and occur at a pace that is relatively slow in relation to human events.[86] Soil is and should be considered as a valuable, nonrenewable resource.

Another reason is that soil damage is not always clearly and immediately perceptible but often comes to light through damage to the other elements (water, air, flora, fauna). The effects of air and water pollution usually become noticeable after a relatively short period of time and affect human beings more directly than soil. The entry of pollutants into the soils and their ecological consequences are perceived only with considerable temporal delay, which is due to the storage and buffering capacity of soils. The fact that most soil degradation usually becomes noticeable after a relatively long period of time and that it affects human beings more indirectly than air and water pollution is one of the main reasons why the seriousness and urgency of the situation of soils is underestimated.

And finally, soil contamination and soil degradation do not produce any direct concern since it is regarded predominantly as a local or regional problem. Unlike for the protection of rare animal species, it is difficult to emotionally convey the problem of degrading soils because they do not evoke any "panda bear effect."

Therefore, any commitment for the protection of a dirty thing like the soil was regarded as a bit weird in former times (and maybe still today...). Even Charles Darwin was declared to have a decaying mind when he published his last and least-known book one year before he died in 1882. It is called "The formation of

[85] Montgomery (2007), p. 20.

[86] Richter and Markewitz (2001), p. 31.

vegetable mould, through the action of worms, with observations on their habits" and focused on how earthworms transform dirt and rotting leaves into soil. Darwin documented a lifetime of what might appear to be trivial observations: how the ground cycles through the bodies of worms and how worms shaped the English countryside. Had he just become crazy or discovered something fundamental about our world? Something he felt compelled to spend his last days conveying to posterity?

Darwin already found out that soil is a dynamic system that responds to changes in the environment. We now know that soil performs a multitude of environmental, economic, and cultural *soil functions*. According to Article 1 of the former draft EU Soil Framework Directive, they include food and other biomass production, storage, filtration, and transformation of many substances, including water, carbon, nitrogen, etc. Soil also has a role as a habitat and gene pool and a carbon pool. It acts as a provider of raw materials and serves as a platform for human activities, landscape, and heritage. Soils also serve as archives of geological and archeological heritage and give information about how people in prehistoric times lived. Soils support many different forms of life on Earth.[87]

The previous use of the medium soil in Europe led to an endangerment of the various (in particular the natural) soil functions. But this is similar in other countries of the world, especially in Africa. For example, there exists an old Kenyan proverb that says:

> Treat the Earth well. It is not inherited from your parents;
> it is borrowed from your children.

This Kenyan proverb seems to reflect what is considered part of the *Principle of Sustainability*. In the European Union, sustainable development is a fundamental objective since the Treaty of Amsterdam of 1997 (it entered into force on May 1, 1999). It seeks to improve the quality of life for human beings without increasing the use of natural resources beyond the capacity of the environment to supply them indefinitely. These requirements of sustainable development were already introduced by the Brundtland Commission,[88] and sustainable development is now being considered as a principle of international law.[89] Incorporating the principle into the EU Treaty (now Article 3 (3) 2 TEU) has made it a fundamental justification for the existence of the Union ("It shall work for the sustainable development of Europe based on (...) a high level of protection and improvement of the quality of the environment"). In 2001, the European Council adopted the EU Sustainable Development Strategy, which provides a long-term vision that involves combining a dynamic economy with social cohesion and high environmental standards. The renewed EU Sustainable Development Strategy (for an enlarged EU of currently

[87]Blume et al. (2016), p. v.

[88]The "Brundtland Report" with the ofiicial title "Our Common Future" was published in 1987 and is the outcome of the work by the World Commission on Environment and Development. It laid out the concept of sustainability as containing environmental, economic and social aspects.

[89]Bosselmann (2008), p. 319 et seq. (p. 325).

28 Member States, after the "Brexit" (the withdrawal of the United Kingdom from the European Union) in 2019 maybe just 27) sets out a single, coherent strategy on how the EU will more effectively live up to its long-standing commitment to meet the challenges of sustainable development.[90] It reaffirms the need for global solidarity and recognizes the importance of strengthening the work with partners outside the EU. Environmental sustainability mainly means meeting the present needs of humans without endangering the welfare of future generations. In the context of soil protection, the focus is on minimizing soil degradation, halting and reversing the processes. Soil sustainability means the ability of the soils to continue to function properly: it requires that human activity only uses soil resources at a rate at which they can be renewed naturally.

While international hard law is still a long way from developing a coherent soil protection approach, SDG 15 and its target 15.3, "Striving for a LDNW" are promoting soil integrity in the context of sustainable development: avoiding land degradation through land-use planning that fully accounts for the potential and resilience of land resources (reduction of the annual rate of land degradation to a certain amount), adopting sustainable land management policies and practices in order to minimize current land degradation, and rehabilitating and restoring degraded lands (an annual amount of land degraded in order to avoid net loss of land-generated ecosystem services provided to society). So in terms of principles of national and international environmental law, this is about the precautionary and preventative principle, as well as the polluter-pays principle.

Since the permanent blockage of the Soil Framework Directive in the Environment Council by some Member States, it is not sure whether and when there will be a comprehensive legal basis for sustainable use and protection of soils in the European Union that meets the requirements of the Brundtland Commission. According to the Progress Report on the Sustainable Development Strategy 2007,[91] soil quality continues to deteriorate with climate change exacerbating both greenhouse gas emissions from soil and threats such as erosion, landslides, salinization, organic matter decline, etc. Several economic activities are still causing soil pollution in Europe, particularly those related to inadequate waste disposal and losses during industrial operations and agricultural activities. Implementing preventive measures introduced by the legislation already in place is expected to limit the inputs of contaminants into the soil in the coming years, for example of toxics etc.

[90]Council of the European Union, Document No. 10917/06 of 26 June 2006, Note from General Secretariat to Delegations, Review of the EU Sustainable Development Strategy (EU SDS).

[91]Communication from the Commission to the Council and the European Parliament, Progress Report on the Sustainable Development Strategy 2007, COM (2007) 642 final of 22.10.2007, p. 1 et seq. (p. 8). According to this Progress Report key EU initiatives to foster resource conservation and biodiversity include the Thematic Strategy on Soil Protection. The EU target of halting the loss of biodiversity by 2010 and contributing to a significant reduction in the worldwide rate of biodiversity loss will not be met unless substantial additional efforts are made.

The basis of the *Precautionary Principle* according to Article 191 (2) TFEU is that science cannot predict absolutely how, when, or why adverse impacts will occur or what their effect may be on humans or ecosystems. Despite the lack of proof, there is often enough information to identify and avoid a serious risk that can lead to unacceptable high costs. According to Principle 15 of the Rio Declaration, the lack of full scientific certainty is not to be used as a reason for postponing measures to prevent damage. The *Preventative Principle* according to Article 191 (2) TFEU or the principle "that preventive action should be taken" is often linked to the precautionary principle and promotes the prevention of environmental harm as an alternative to remedying harm already caused.

These principles (principle of sustainable development, precautionary and preventative principle) form the basis for all activities and combine the mentioned ethical approaches and seem to show the way to a new form of governance for ecological integrity.[92] Which ethical considerations, e.g. whether property rights should be defined to include ecological limitations of the soils, should be the basis for the law? But concerning the soils, it still seems to lack practical concretion. An integrated understanding of the concepts of ecology, esthetics, and ethics was what Aldo Leopold (American forester and wildlife ecologist, 1887–1948) called the *Land Ethic* in his "Sand County Almanac" (1949). He wrote that the ethical obligation that the members of a natural system have is to "preserve the health of the system by encouraging the greatest possible diversity and structural complexity and minimizing the violence of man-made changes." He considered ecology a moral mandate (instead of "the biotic community," we may extrapolate "the soil community"):

> A thing is right when it tends to preserve the integrity, stability, and beauty of the biotic community. It is wrong when it tends otherwise.

The primary motivation of Leopold's career was what he called "the oldest task in human history (is) to live on a piece of land without spoiling it." [93] He communicated his vision:

> That land is a community is the basic concept of ecology, but that land is to be loved and respected is an extension of ethics.

Ron Engel has argued that a sacred covenant between humans and all life is the source of universal responsibility for the present and future well-being of humans and the living world.[94] Applied to the context of soil protection, our focus should be on the protection of the ecological functions of the soils for their own inherent value. Therefore, we need to preserve ecologically stable and healthy soils and to restore the soils that are already degraded (with due regard to certain criteria) in order to ensure that future generations will have at least the same quality of soil (soil sustainability). We also need equal access to soil resources for all people. We need

---

[92]Bosselmann (2008), p. 319 et seq. (p. 324).

[93]Engel (2008), p. 277 et seq. (p. 286, 288).

[94]Engel (2008), p. 277 et seq. (p. 281).

appropriate ecological soil standards and need to include knowledge and traditional land-use practices of indigenous and local people. And we also need to establish new legal instruments of soil protection like education, participation, monitoring, etc. The challenge facing modern agriculture is how to merge traditional agricultural knowledge with modern understanding of soil ecology to promote and sustain the intensive agriculture needed to feed our citizens. All in all, we need to develop a new soil ethic, valuing especially the ecological functions of the soils, in order to reach a high level of soil integrity.

5 Conclusion

Soil integrity is one of the basic and most neglected components of our human culture and the global ecosystem. Soil is a limited natural resource. It is a very dynamic system that performs a multitude of functions: environmental, economic, social, and cultural, which are central to both human well-being as well as the health of the environment in general. It plays a vital role in the earth's ecosystem, being the fundamental basis for terrestrial biodiversity, supplying the vast majority of the world's food and being more broadly recognized as having an essential function in regulating the global climate. Any use of soil should take into account its multiple ecological functions, with a view to their conservation. Soil variability is very high, and enormous differences exist worldwide in its state both within individual profiles and between soils. These diverse conditions and needs should be taken into account by individuals and by politics as they require different specific solutions. What we need is a shift in our way of thinking, away from the view of the soil only as a productive commodity.

Soil is essentially (on human timescales) nonrenewable in that the degradation rates can be rapid, whereas the formation and regeneration processes are extremely slow. Currently, at least a quarter of the world‘s land area is either highly degraded or undergoing high rates of degradation. The historic development of soil awareness and especially our hesitation to sufficiently protect the soils against further damage from industrial activities, inadequate agricultural and forestry practices, urban and industrial sprawl, or construction works reflect our lacking readiness to assume responsibility for this so-called third media. In the EU (and in many other parts of the world), the fields of biological, chemical, and physical soil protection are still "poor nephews" compared to other environmental policy making and legislation.

Although soil degradation is regarded as a serious problem in Europe, the current situation in the Environment Council where many EU Member States (e.g., Germany and the Netherlands) refuse to even discuss a Draft Soil Framework Directive arises out of the "traditional" view of soil as personal property whose landowner cannot be submitted to any legal regime or legal obligations. Given the serious nature of soil degradation across Europe, the need for regulations on soil protection is obvious: the view that soil is much more than just a piece of property will hopefully and soon lead to a comprehension of the urgent necessity of soil protection measures in the European Union and most other countries in the world. Soil is a common heritage, and its protection is in the public interest.

As a society, we need to acknowledge the importance of soil for sustainability as a whole and especially in the context of the implementation of the 2030 Agenda and its SDG 15 and its relevance also for other SDGs. For instance, soil quality is also important for the climate, having a crucial impact on the carbon absorption capacity. Where there is a risk of serious damage to one of the ecological functions of soil and when there is scientific uncertainty as to the extent of future deterioration, the precautionary principle should be applied.

When we consider the essential role that soils play in the evolution of the biosphere and the maintenance of life, we must conclude that reconciling human existence with ecological integrity would also mean that we need a paradigm shift in our thinking about the soils to ensure that future generations will have at least the same quality of soil. Our societal task would be to tread new paths toward the preservation of ecologically stable and healthy soils and to develop a new, ecology-based soil protection ethic.

References

Arnalds A, Runolfsson S (2005) A century of soil conservation in Iceland. In: SCAPE/Agricultural University of Iceland, Strategies, Science and Law for the Conservation of the World's Soil Resources, AUI Publication No. 4, p 67 et seq. (International Workshop, Selfoss, Iceland, September 14–18, 2005)

Berasin L, Miguel J (2008) From equilibrium to equity, the survival of the commons in the Ebro Basin: Navarra form the 15th to the 20th centuries. Int J Commons 2:162 et seq

Blume H-P, Brümmer GW, Fleige H et al (2016) Scheffer/Schachtschnabel, soil science. Springer, Berlin

Blum WEH (1988) Problems of soil conservation, Steering Committee for the Conservation and Management of the Environment and Natural Habitats (CDPE). Council of Europe, Strasbourg

Boer B, Ginzky H, Heuser I (2016) International soil protection law: history, concepts and latest developments. In: IYSLP 2016, p 49 ff

Boer B, Hannam I (2003) Legal aspects of sustainable soils: International and national. Rev Eur Commun Int Environ Law 12:149 ff

Bork H-R, Dalchow C, Frielinghaus M (1999) Bodenentwicklung, Bodenzerstörung und Schutzbedürftigkeit von Böden in der Vergangenheit. In: Frielinghaus M, Bork H-R (eds) Schutz des Bodens, Bonn

Bosselmann K (2008) In: Westra L, Bosselmann K, Westra R (eds) Reconciling human existence with ecological integrity. Earthscan, London, p 319 et seq

Calliess C, Ruffert M (2016) EUV/AEUV, Das Verfassungsrecht der Europäischen Union mit Europäischer Grundrechtecharta, 5th edn. Beck C.H., München

Commission of the European Communities, Communication from the Commission to the Council, the European Parliament, the European Economic and Social Committee and the Committee of the Regions, Thematic Strategy for Soil Protection, COM (2006) 231 final of 22.09.2006, p. 1 et seq

Engel JR (2007) Our covenant with earth: the contribution of soil ethics to our planetary future, world forum on soils, society and global change Selfoss, Iceland, September, 2007

Engel JR (2008) What Covenant sustains us? In: Westra L, Bosselmann K, Westra R (eds) Reconciling human existence with ecological integrity. Earthscan, London, p 277 et seq
Fixen PE (2007) Potential bio fuel influence on the fertilizer market. International Plant Nutrition Institute, Brookings
Futrell WJ (2009) New action for soil protection. Environ Law Rep 39:10077 et seq
Gesellschaft für Technische Zusammenarbeit, Den Boden nicht verlieren, Bonn 2000
Hannam ID, Boer BW (2002) Legal and institutional frameworks for sustainable soils: a preliminary report. IUCN Environmental Policy and Law Paper No. 45
Heine G (1986) Ökologie und Recht in historischer Hinsicht. In: Lübbe H, Ströker E (eds) Ökologische Probleme im kulturellen Wandel. Paderborn, S. 116 ff
Henneke H-G (1986) Landwirtschaft und Naturschutz
Heuser IL (2005) Europäisches Bodenschutzrecht – Entwicklungslinien und Maßstäbe der Gestaltung. Berlin
Heuser IL (2006) Milestones of soil protection in EU environmental law. J Eur Environ Plan Law (JEEPL) 3:190 ff
Holzwarth F, Radtke H, Hilger B, Bachmann G (2000) Bundes-Bodenschutzgesetz/Bundes-Bodenschutz- und Altlastenverordnung, Handkommentar, 2nd edn. Berlin
Johnston AE, Watson CJ (2005) Phosphorus in Agriculture and in relation to water quality. Agricultural Industry Confederation, Peterborough
Kimminich O, von Lersner H, Storm P-C (1996) Handwörterbuch des Umweltrechts (HdUR), 2nd edn
Kommission der Europäischen Gemeinschaften, Für eine dauerhafte und umweltgerechte Entwicklung, Ein Programm der Europäischen Gemeinschaften für Umweltpolitik und Maßnahmen im Hinblick auf eine dauerhafte und umweltgerechte Entwicklung, KOM (1992) 23 endg., Vol II, April 3rd 1992, p 1 ff. (cited: KOM (1992) 23 endg., Vol II, p 1 ff.)
Kocher G (1983) Das moderne Bodenrecht – Rückschritt in die Geschichte? In: Weimar R (ed) Die Ordnung des Bodens – heute und morgen. Frankfurt a.M. Bern, New York, p 144 ff
Kumar BM (2008) Forestry in ancient India: some literary evidences on productive and protective aspects. Asian Agri-Hist 12(4):299 et seq
Kurucz M (1993) Land protection. Conn J Int Law 8(2):467
Kutilek M, Nielsen D (2015) Soil. Dordrecht
Mackey B (2008) The Earth charter, ethics and global governance. In: Westra L, Bosselmann K, Westra R (eds) Reconciling human existence with ecological integrity. Earthscan, London, p 61 et seq
Montgomery DR (2007) Dirt, the erosion of civilizations
Narjes K-H (1984) Bodenschutz in der Europäischen Gemeinschaft. In: Bureau Européen de l'Environnement/European Environmental Bureau. Seminar on Soil Protection in the European Community, Brussels, p 5 ff
Oldeman LR et al (1992) Manuel sur la conservation des sols en Europe (PE-S-SO (92)5). Strasbourg
Olivelle P (2014) The early Upanishads. Oxford University Press, Oxford
Pavan M (1970) The defence of soil in nature conservation. Rom
Richter DD, Markewitz D (2001) Understanding soil change
Thompson PB (2011) Sustaining soil productivity in response to global climate change: science, policy, and ethics. Wiley, Hoboken, pp 31–42
Thompson PB (2017) The spirit of the soil. London
Wallander H (2014) Soil, Cham
Wissenschaftlicher Beirat der Bundesregierung Globale Umweltveränderungen, Welt im Wandel - Die Gefährdung der Böden, 1994
Wey K-G (1982) Umweltpolitik in Deutschland
Yaloon DH (1999) On the importance of international communication in soil science. Eurasian Soil Sci 32:22 ff

INSPIRATION: Stakeholder Perspectives on Future Research Needs in Soil, Land Use, and Land Management—Towards a Strategic Research Agenda for Europe

Detlef Grimski, Franz Makeschin, Frank Glante, and Stephan Bartke

1 Introduction

The way we steward the management of our soils is central to ensuring a safe and sustainable future.[1] There are at least three significant challenges that need to be considered:

1. soils are multifunctional, but they cannot serve all human demands at the same location at the same time[2];
2. overall, soils are limited; and
3. our knowledge about the formation and functioning of soils is incomplete.

Research and innovations are needed to address these challenges in order to support farmers, regulators, and other stakeholders in implementing more sustainable land-use management and governance. The application of research findings and knowledge can improve our approaches to securing soils and land for the following generations, for competitive economies, and for healthy landscapes.

The European Commission has picked up this challenge within its Horizon 2020[3] (H2020) research program and initiated a project that has been aiming at strategically structuring research on soil, land use, and land management in

[1]UN (2014).

[2]Adhikari and Hartemink (2016), Baveye et al. (2016), Blum (2005).

[3]https://ec.europa.eu/programmes/horizon2020/.

D. Grimski (✉) • F. Glante • S. Bartke
German Environment Agency, Dessau-Roßlau, Germany
e-mail: detlef.grimski@uba.de; frank.glante@uba.de; stephan.bartke@uba.de

F. Makeschin
Dresden International University, Dresden, Germany
e-mail: makeschin@t-online.de

H. Ginzky et al. (eds.), *International Yearbook of Soil Law and Policy 2017*,
International Yearbook of Soil Law and Policy,
https://doi.org/10.1007/978-3-319-68885-5_25

Europe. The project *INtegrated Spatial PlannIng, land use and soil management Research AcTION – INSPIRATION*[4] was funded as a coordination and support action to develop a Strategic Research Agenda (hereinafter SRA) on soil and land-use issues.

At the time of writing this article, development of this SRA has entered its third and final stage (status as of April 2017). The public release of the first version of the INSPIRATION SRA is scheduled to take place on World Soil Day 2017. Throughout the following article, INSPIRATION's methodological approach is briefly outlined, and the research fields identified for soil and land management are presented in summary.

2 The INSPIRATION Project

2.1 Scope and Workflow

The overall objective of the SRA has been to draw a picture detailing how to inform and facilitate sustainable soil and land-use management that will meet current and future societal needs. According to the definition of sustainability,[5] land-use management also needs to be environmentally friendly, socially acceptable, and economically affordable, e.g., minimizing the consumption of natural resources, such as uncontrolled land take for settlement and transportation purposes. Coordinated by the German Federal Environment Agency (UBA), partners from 17 countries (see Fig. 1) are developing the SRA from 2015 to 2018. As regards the societal challenges, the H2020 program of the European Commission addresses seven societal challenges. They are aimed at fostering a greater understanding of Europe by providing solutions and supporting inclusive, reflective European societies with innovative public sectors in the context of unprecedented transformations and growing global interdependencies.[6] The INSPIRATION project is particularly addressing societal challenge no. 5 "Climate action, environment, resource efficiency and raw materials," which therefore was significant for developing the SRA.

The general scope of INSPIRATION was to

1) formulate, consult on, and revise an end-user oriented SRA;
2) scope out models for implementing the SRA; and
3) prepare a network of public and private funding institutions willing to jointly fund the execution of the SRA.

[4]www.inspiration-h2020.eu.

[5]http://www.unece.org/oes/nutshell/2004-2005/focus_sustainable_development.html.

[6]http://ec.europa.eu/programmes/horizon2020/en/h2020-section/europe-changing-world-inclusive-innovative-and-reflective-societies.

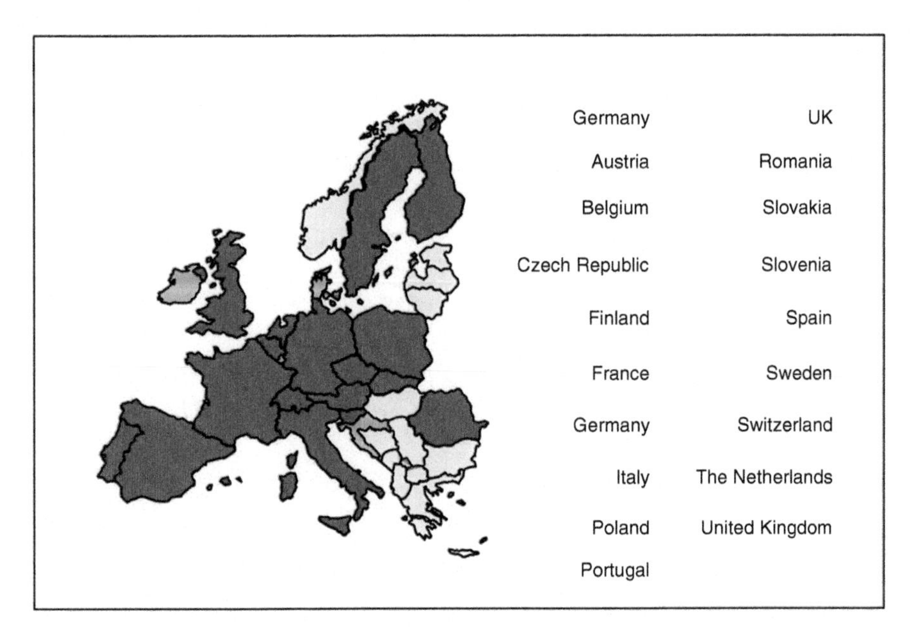

Fig. 1 INSPIRATION partner countries

The INSPIRATION SRA is intended to be attractive for research funding by public and private parties and to ensure that knowledge gained is widely applied so that SMEs and other industries will innovate and the EU Member States will become greener, more socially cohesive, smarter, and more competitive.

To achieve this goal and to address societal challenges, broad stakeholder involvement was essential. The INSIRATION approach, therefore, was to develop the SRA from the bottom up based on research needs and knowledge gaps identified and expressed by land and soil stakeholders working as funders, scientists, policy makers, public administrators, consultants, et cetera. They were involved as national stakeholders from each of the 17 INSPIRATION partner countries. In each country, a National Focal Point (NFP) was responsible for collecting the research needs through interviews with the national stakeholders and funneling them into the agenda of their respective national workshops. Each workshop was conducted with ca. 20 national stakeholders in order to verify, modify, complement, and prioritize the suggested research needs into nationally agreed research demands. This process was the first phase of the project as depicted on the left side of Fig. 2, which presents the overall workflow of the project.

In the second project phase, the identified national research demands were collated, reviewed, and synthesized in order to identify transboundary research demands. The national research issues were structured according to a set of predefined integrated themes within the INSPIRATION Conceptual Model[7] (see Sect. 2.2). Continuous involvement of national stakeholders was also a key priority

[7]Cf. Makeschin et al. (2016).

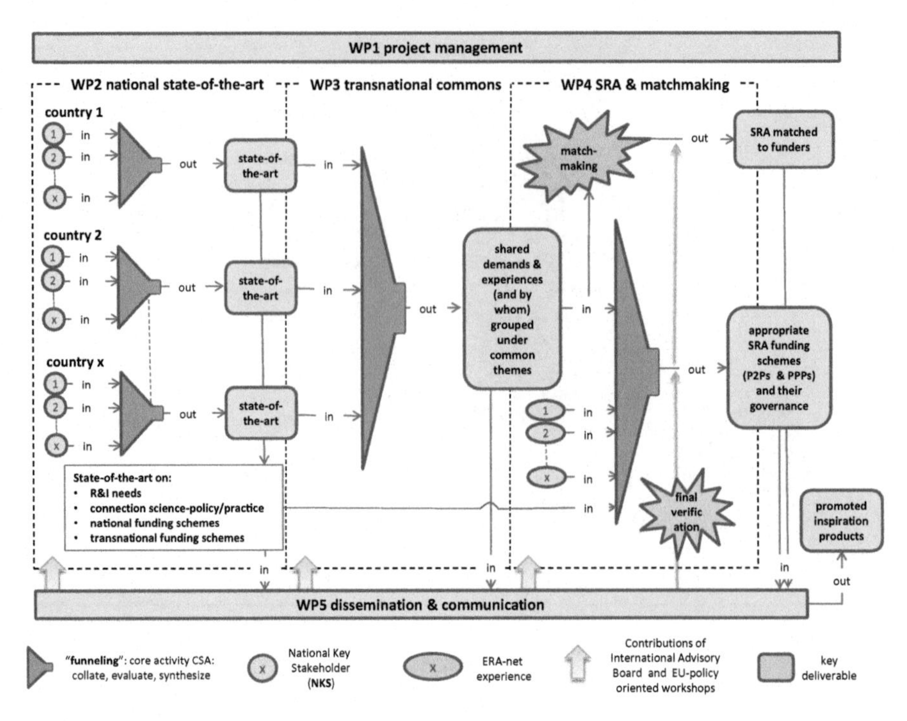

Fig. 2 INSPIRATION workflow

during this phase of the project in order to ensure that all addressed research demands were taken thoroughly into account for further processing.

The third phase of the project has been devoted to scoping out and developing the trans-country and trans-discipline SRA. Verification through dialogue and discussion with relevant funding bodies across Europe followed (which at the time of writing this article is still ongoing). Priorities and preferred models for the implementation of specific components in multilateral European research initiatives were to be identified accordingly.

The entire project was supervised by an International Advisory Board (IAB) composed of experts representing private and public institutions dealing with land and soil management.

2.2 *INSPIRATION's Conceptual Model*

As the baseline for structuring future research needs, INSPIRATION developed a Conceptual Model (see Fig. 3). The basic idea for the Conceptual Model was that land and the soil/sediment/water system (SSW-system) are goods and natural capital stocks that have to be used in a way that maximizes nondepletion of our ecosystems. However, it has to be considered that there are manifold drivers that

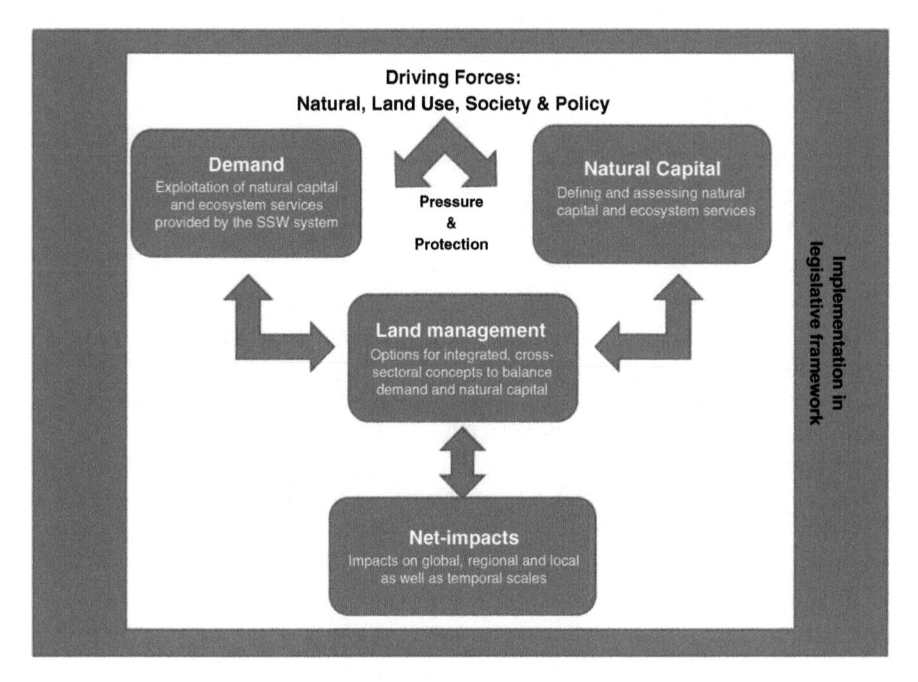

Fig. 3 INSPIRATION's Conceptual Model, adapted from Makeschin et al. (2016)

affect natural resources, their state, as well as their use and may eventually lead to the degradation of whole ecosystems. Intensive and unsustainable land use can have significant impacts on ecosystems and their ecological functions. Additionally, ecosystems are affected by natural drivers like extreme climate events, particularly due to climate change (flooding, droughts, etc.).

Furthermore, there are conflicting interests regarding land use among many relevant stakeholders in society, such as farmers, land planners, developers, industry, residents, and so forth, regarding the productivity of areas and/or protecting natural resources, for instance. Consequentially, each stakeholder has demands regarding the use of natural resources according to his or her specific interest. As a rule, there are mainstream viewpoints within this paradigm of 'Either-Or': maximizing the land's economic benefits on the one hand and protective regulations to conserve environmental functions on the other. Thus, sustainable management of agricultural, forest, and urban land resources, as well as the conservation of biological diversity or natural capital ideally has to follow integrated, interdisciplinary, and cross-sectorial concepts in order to address the different demands of all stakeholders. Namely, the economic, societal, administrative, and political impacts from different land management approaches, which steer and govern land use in the broad sense, including stocks of and goods from natural resources, have to be considered. Thus, the net impacts on a local, regional and global, as well as temporal scale are significant drivers and determinants of crucial importance.

These challenges must be tackled to benefit from the land and the SSW-system and to avoid depletion of our natural capital and resources. Sustainable land management must, therefore, seek to balance the demand and the supply, with the latter being based on the resources provided by our natural capital. As an integral part of such a sustainable soil management model, the net impact, meaning the local to global footprint of human land management decisions, must be minimized.

Therefore, INSPIRATION's Conceptual Model was the basis for identifying and structuring cross-country and cross-sectorial research demands. The four integrated themes of the Conceptual Model were as follows:

Demand: what does society demand from natural capital and ecosystem services, including the SSW-system?

Natural capital: what does nature, including the SSW-system, have to offer, and which determinants sustain the system?

Land management: what are options for integrated, cross-sectoral land management to balance societal demands and natural capital needs?

Net impacts: what are the impacts of different options for managing natural capital, including the SSW-system, on global, regional and local, as well as different temporal scales?

2.3 Collating and Structuring the Identified Research Needs

In the first phase of the project, NFPs collectively contacted more than 450 stakeholders, who identified pressing national research topics and research questions from their points of view. In total—from all 17 countries—more than 200 research topics with circa 1200 research questions were collected.[8]

For the collation, review, and synthesis of these research demands, the following stepwise approach was applied:

- All identified research demands were assessed with regard to the four integrated themes of the Conceptual Model (demand, natural capital, land management, net impact).
- Within each of these integrated themes, areas of specific research themes were identified and clustered as Clustered Thematic Topics (CTTs) (see Table 1).
- Furthermore, Integrated Research Topics (IRTs) addressing cross-thematic topics were identified (see Sect. 3.2).
- Verification and consultation of CTTs and IRTs happened through several consultation steps:

[8]Brils et al. (2016).

Table 1 Clustered Thematic Topics (CTTs) of the four integrated themes

Demand	Natural capital	Land management	Net impact
CTT-D1: The 4Fs: food, feed, fiber, (bio)fuel **CTT-D2**: Regulating ecosystem services **CTT-D3**: Urban/infrastructure land **CTT-D4**: Water **CTT-D5**: Geological (and fossil) subsurface resources **CTT-D6**: Areas where natural hazards are prevented **CTT-D7**: Health and quality of life (living environment)	**CTT-NC1**: Quantity and quality of soils, health of soils, soil carbon, greenhouse gases **CTT-NC2**: Biodiversity, organismic and genetic resources **CTT-NC3**: Water, water cycle **CTT-NC4**: Pollutant degradation, filtering, and immobilization capacity **CTT-NC5**: Prevention of erosion and mud slides, natural hazards **CTT-NC6**: Geological resources **CTT-NC7**: Intrinsic values of soils and landscapes	**CTT-LM 1**: Governance, management mechanisms, instruments, and policy **CTT-LM 2**: Climate change challenges for land management **CTT-LM 3**: Land as a resource in urban areas (sustainable urban land management) **CTT-LM 4**: Land as a resource in rural areas (multifunctionality of rural areas)	**CTT-NI 1**: Developing impact assessment methodology **CTT-NI 2**: Understanding and assessing impacts of drivers **CTT-NI 3**: Trade-off analysis and decision support **CTT-NI 4**: Science–society–policy interface

- a workshop with all 17 NFPs;
- two workshops with the national key stakeholders; four stakeholders from each of the 17 partner countries took part in this consultation process;
- an online survey with national stakeholders and NFPs regarding prioritization of CTTs and IRTs.

3 Transnational Research Topics

The research topics and research questions identified through the process described above cover a broad spectrum of further research—basic research, as well as applied research, and respecting implementation needs. As described above, they were structured along CTTs under the four integrated themes of the Conceptual Model and complemented by overarching IRTs. Altogether, INSPIRATION identified 22 CTTs and 17 IRTs as relevant tasks for future research to address societal challenges in Europe.[9] They are described in more detail below.

[9] Makeschin et al. (2016).

3.1 Clustered Thematic Topics (CCTs)

3.1.1 Demand

Sustainable handling and management of natural resources is indispensable for providing for the needs of a growing and affluent population and coping with other societal challenges, such as climate change. Land use, as a constrained market due to finite supply, cannot be considered only from the angle of productivity. Societies demand a lot of different ecosystem services from land: the production of food, fiber, or wood and also to act as a carbon sink, water buffer, and biodiversity archive. Demand for the goods and services provided by natural resources is driven by the total final consumption of our societies. Research in the context of demand must focus on "demand" for the SSW-system services.

The first aim is to understand the links between consumption and the use of SSW-system services. The need is to quantify and map in time and space the systemic impacts on the nexus of SSW-resources from final consumption of their goods and services.

A second area relates to the abilities of soil and land to adapt to consumption on one hand and to control the demand for SSW-system services on the other hand. The need is to find "more resource-efficient" methods of consumption linked to direct demand for bio-sourced goods and space for the built environment. Soils serve as the production site or primary geo-resource for the production of biomass—a key element in the bio-economy (the so-called four Fs). Biomass may be in the form of agricultural goods for direct consumption or conversion to diverse food products or materials (like oils), feed used in animal husbandry, various fibers, and the growing field of organic renewable energy, e.g., nonfossil fuels. However, soils and land are increasingly under demand for both production of biomass and to provide a platform for infrastructure. Therefore, it must be recognized that the availability of fertile soils is shrinking, while various demands for land are growing.

Under the demand perspective—at least in a society committed to sustainable development—sediment and water ecosystem functions are also strategic research needs. Such functions are important for regulating and maintaining soil and land quality and combating against potential threats due to the demand for soil and land, i.e., coping with the consequences of land uses. Furthermore, there is demand for land for human settlement areas, for surface, as well as for subsurface infrastructure. The demand for landfill sites is also constantly increasing—all affecting the social, economic, and environmental quality of cities and regions and affecting the soil and urban climate. Research must be conducted in recognition of these conflicting goals and interests with regard to the demands for different land uses. It must provide solutions for European cities so that they can rise to the challenge of urban development, such as using empty spaces between buildings, brownfields regeneration, multifunctional and temporal uses, densification, and retrofitting older constructions. Research in the demand field must also pay special attention to the following:

- water resources affected by agricultural land use, e.g., high-density livestock breeding, runoff from arable systems, irrigation, conversion of agricultural land to settlements, and ongoing climate change;
- geological subsurface resources, like peat, gravel, sand, lignite, and other materials that are needed for economic development;
- achieving efficient mitigation against and resilience to natural hazards and disasters (flooding, landslides, etc.); and
- indicators and tools for assessing the role of the living environment as it affects and is itself affected by the demand for natural resources (and a high standard of living).

3.1.2 Natural Capital

Natural capital is the economic metaphor for the limited stocks of physical, chemical, and biological resources found on earth and the capacity of ecosystems to provide goods and services. Natural capital includes all kinds of natural resources, such as the subsurface, landscape, groundwater and surface water, atmosphere, as well as all living organisms. Natural capital provides society with a wide range of goods and services, which are often considered to be free of charge (e.g., crop irrigation by rainfall, pollination by insects, pollutant degradation by microorganisms). These ecosystem services may be grouped into four major categories:

1) supporting services, which form the basis for all other ecosystem services;
2) provisioning services, which include all products generated by ecosystems;
3) regulating services, which include the benefits from the proper regulation of processes; and
4) cultural services, which are the nonmaterial benefits people get through recreation, aesthetic perception, spiritual cognition, etc.[10]

A number of goods and services provided by natural capital can hardly be quantified or expressed in economic values; thus, they are difficult to communicate to society. Consequently, a decrease in the quantity and quality of natural capital is often ignored. Therefore, there are several research questions for natural capital that arise, such as follows:

- Which soil functions drive various ecosystem services?
- What are the soil's potential and limits for sequestration within the carbon cycle?
- How should soil health be assessed and monitored?

In summary, future research needs on natural capital are related to the following areas:

---

[10]http://cices.eu/supporting-functions/.

- Awareness raising: the significance of natural capital and ecosystem services is still widely unknown or underestimated. Research must contribute to awareness raising and highlight the (positive and fundamental) role of natural capital in protecting and/or restoring ecosystem services.
- Implementation: although it is widely accepted that natural capital must be preserved, there is still a lack of implementation of existing knowledge. Research must contribute to the transfer of know-how into societal action.
- Research is necessary to reveal and understand the underlying and system-related mechanisms of natural capital. Phenomenological descriptions are not sufficient. Research on natural capital must therefore be across sectors and disciplines and also consider the interaction of disciplines (physics, chemistry, biology, etc.) and systems (soil, water, atmosphere, etc.). Reliable predictions within the fourth theme, net impact, will partially rely on a holistic understanding of the processes within natural capital.
- Field research often relies on point measurements (e.g., soil carbon and microbial biomass at a given sampling spot at a given time). Research must contribute to extrapolating these data in space and time at the landscape level. Issues such as geostatistics and spatial and temporal variability are important to accomplishing that need.
- Within the context of increasing risk of weather extremes due to climate change (rain, draught, temperature, etc.), research must provide solutions on how to better assess highly dynamic systems under stress instead of "normal" steady state conditions (e.g., considering resistance and resilience or new equilibrium states).

3.1.3 Land Management

Land management is the process of managing the use and development of land resources in urban and rural areas. It covers all activities related to the management of land from an environmental, economic, and social perspective. The future role of land management must be more focused on balancing the demand for and the supply of resources and natural capital in urban and rural areas. Land management includes the institutional capacity of local, regional, and national governments to identify and protect vulnerable areas, ensure long-term productive potential of agricultural land (cropland, rangeland, forests), enhance adaptation to climate change, provide strategies to reduce urban sprawl (which is related to land sealing and land degradation), as well as minimize urban structure fragmentation and foster the reuse and reintegration of degraded, derelict, or abandoned sites into new uses. The fundamental challenge related to land management is to achieve integration between different policy levels and various stakeholders involved in this process and to ensure sustainable land-use management by introducing appropriate instruments for solving land-use conflicts.

Basically, research is justified by a general lack of integration among stakeholders and policy levels. Furthermore, social science is underrepresented in land management processes. Research must therefore contribute to more integration across stakeholder and policy boundaries. Research on governance and management mechanisms can contribute to the enhancement of existing instruments and the implementation of innovative mechanisms to keep land in socially and environmentally beneficial use. Demonstration projects and trans-boundary information and experience exchange can make administrative procedures more efficient, highlight and stimulate innovative solutions, and provide knowledge about the institutional capacities that are required to carry out holistic and systemic approaches. A general requirement for more effective spatial planning is that environmental and societal objectives should be identified at an early stage of the planning process—including urban–rural interaction. Research must contribute to developing proper guidelines and handbooks for land-use planning institutions. Also, climate change presents a serious challenge for urban and rural areas. Research must help to better understand the relation of land management/land use and its impacts on meteorological phenomena. The development of spatial planning instruments is essential in order to cope with land-use impacts on the climate. Finally, it is a remarkable task for research on land management to provide guidance as to how to meet the objectives of the Sustainable Development Goals, particularly Goal 11: "Make cities inclusive, safe, resilient and sustainable."[11] Research needs in this field include impacts of demographic change, economic effects of urban sprawl, nature protection in urban spaces, the role of urban green infrastructure and nature-based solutions, brownfield revitalization, improvement of the quality and efficiency of urban infrastructure, multifunctional uses and flexibility of buildings and infrastructure, and governance of urban planning and design, such as urban agglomeration, polycentric conurbation, and functional urban areas.

3.1.4 Net Impact

Economic prosperity and the well-being of people are underpinned by natural capital, i.e., ecosystems that provide essential goods and services, such as fertile soil, multifunctional forests, productive land and seas, good quality fresh water, clean air, pollination, climate regulation, and protection against natural disasters. The growing demand on productive sites can only be achieved by securing and improving the productivity of the current soil resources, as well as by restoring already degraded sites. Research in this context must mainly provide tools and criteria that can be used in land use and land-use planning to support decision-making tailored to minimize land management systems' net impact on a timely and spatial scale.

Research within the context of the net impact theme is needed, for example, to develop or enhance assessment methods and monitoring indicators, to better

[11] http://www.un.org/sustainabledevelopment/sustainable-development-goals/.

understand impacts and drivers, to understand and clarify synergies and trade-offs between different (societal) goals, and to integrate research-based knowledge into policy making. A fundamental basis to work from is the proper monitoring of changes in the soil-sediment-water system, including the assessment of net impacts on human well-being and economic prosperity. Data for such purposes must be linked across different scales, e.g., from global to local scale, and be based on cheap, efficient, quick, validated, and reliable innovative screening methods for data sampling and analysis. Data collection and data access must be harmonized and standardized. In particular, to deal with different kinds of risks (ecological, technological, and economic), the tools and methods for future assessment should be more integrated, systematic, and comprehensive.

Moreover, the specific roles of the relevant actors in decision-making also must be further explored. How do political and economic interests of stakeholders control the understanding and governance of impacts in the process? Drivers must be better understood and the interdependencies of policy, land-use decisions, ecosystem provisions, and ecosystem changes (e.g., in organic carbon, soil fertility, soil erosion, or water quality) must be clearly elaborated. Sufficient empirical data related to social costs and benefits of urban development, including the relationship between the built environment and human health, is also still missing.

Increasing international trade and the globalization of product supply chains, consumption, and lifestyle patterns play an additional role in shaping land management and politically based land-use decisions. A detailed understanding of the type and impact of such socioeconomic drivers and trade-offs for land use and land-use change and their resulting impacts on natural capital, ecosystem service provision, and human well-being is urgently needed for transparent and evidence-based policy making. More knowledge about the cross-border supply of ecosystem services would also be very helpful in spatial and regional planning, transport planning (especially public transport), and social care services. Furthermore, in many countries, "emerging" contaminants appear on the political agenda (e.g., PFC in Germany). Characterization and assessment regarding their bio-accumulation and bio-dispersion is required—on one hand for decisions on specific policy action and on the other hand to understand the impacts. The impacts of climate change are not yet fully understood (for instance, on soil quality, soil characteristics, soil biodiversity, soil processes, soil subsidence, and ecosystem services), and neither are how land management measures could respond.

Other critical factors for future research are knowledge uptake by policy makers, awareness raising across stakeholder groups, and conflict management between different affected policy sectors. Research must also find ways to communicate across stakeholder groups in a language that is understood by each of them (e.g., how to communicate to farmers in a language compatible with their operation, why and how to produce in a more ecologically sound way without damaging the soil). A consistent land management policy must cope with conflicts arising from the different policy sectors affected, e.g., housing, transport, agriculture and forestry, climate mitigation and adaptation, water management, and nature conservation.

3.2 *Integrated Research Themes (IRTs)*

Whereas the CTTs systematically represent clusters of research topics under the Conceptual Model's four integrated themes, the IRTs integrate specific research topics that bridge several CTTs. Thus, they complement the CTTs in terms of considering cross-thematic topics that are of transnational research need and interest. By their nature, IRTs are particularly designed to stimulate the partner countries to create multinational thematic funding progams. Altogether, 17 IRTs were identified and clustered into five groups (see Fig. 4).

3.2.1 IRT Group "From Information to Implementation"

The three IRTs in this group are designed to merge overarching research needs from integrated monitoring, assessing, and valuing of the results to transferring complex knowledge into manageable tools.

IRT 1: Integrated Environmental Assessment and Soil Monitoring for Europe

Soil monitoring networks have focused on assessing the trends of hazardous compounds in soil, soil biology, erosion, and in some intensive-monitoring sites also fluxes of compounds between soil and groundwater. However, a number of questions still cannot be answered by "classical" sectoral monitoring. Such

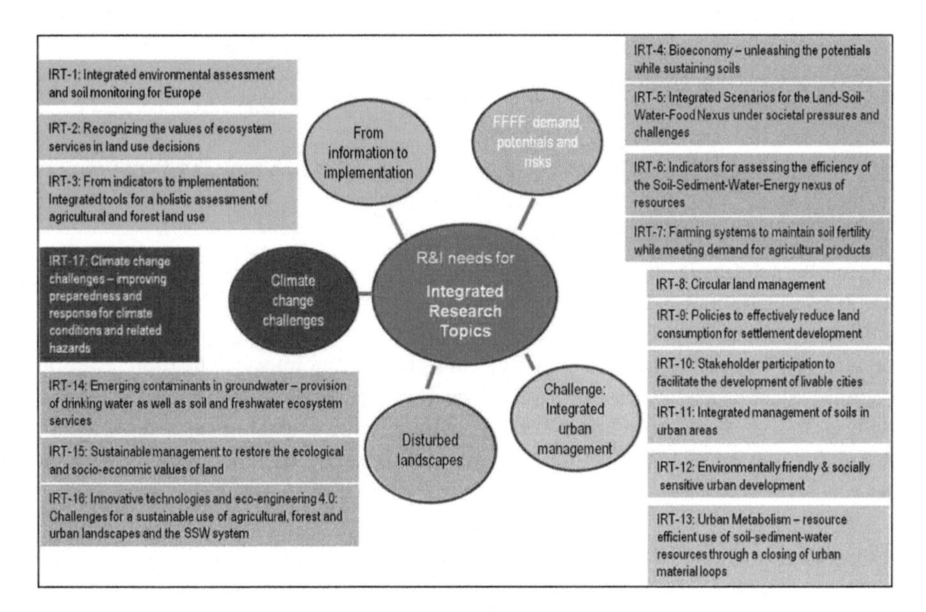

Fig. 4 Integrated Research Topics (IRTs)

questions are related to topics like climate change, food security, SDG implementation and its accomplishment, as well as challenges concerning land-use changes. The challenge for integrated research is to connect established networks and integrate new networks in a way that ensures broad data availability (open data). The goal of this IRT, therefore, is to examine options for a European-wide soil monitoring network that will provide an information and data tool for scientists and decision-makers in order to meet the land degradation neutrality target[12] of the SDGs, define the status and the trends of soils/land-use changes in Europe, assess impacts on soils (chemical, biological, and physical changes of soil functions) and define soil quality (use-related soil quality targets).

IRT-2: Recognizing the Values of Ecosystem Services in Land Use Decisions

Ecosystem services underpin human well-being and economic prosperity. Land use, such as agricultural production or forestry, and land-use change, such as urban development, agricultural intensification, or afforestation, influence the bundle of ecosystem services provided. Few ecosystem services have explicit prices or are traded in markets (e.g., provision services such as crops or timber). However, there are also nonmarketable ecosystem services (e.g., freshwater provision, mitigation of hazardous events, or landscape beauty) that are important for human well-being and hold substantial value for society. There is thus a huge challenge to identify and assess the benefits of such nonmarketable ecosystem services affected by changes in land use and land-use intensity. The goal of this IRT, therefore, is to explore options as to how the importance of the whole range of ecosystem services linked to changes in land use, and land-use intensity can be assessed, integrated, and better recognized in decision making and developing land-use policies. Concepts should be based on recent studies, such as TEEB[13] and CICES[14] systems.

IRT-3: From Indicators to Implementation: Integrated Tools for a Holistic Assessment of Agricultural and Forest Land Use

Currently, administrations, decision makers, and different scientific disciplines work on assessment methodologies in parallel. As a rule, assessment approaches are based on segregated procedures (which focus on ecological, economic, social, or planning aspects) and lack a spatial and cross-disciplinary indication. Natural capital and land-use systems are subjected by diverse disturbances and stressors. Although several scientific indicators are available for an ecological and socioeconomic evaluation of land-use impacts, most of these are still discipline and sector

[12]http://www.un.org/sustainabledevelopment/sustainable-development-goals/.

[13]http://www.teebweb.org/.

[14]https://cices.eu/.

oriented (e.g., having a soil chemical parameter exclusively related to the productivity of sites). Due to the complexity of factors involved in an integrated assessment of land use, appropriate tools for evaluation, planning, commercial, and political decisions are still lacking. Thus, science-based methodologies and assessment approaches are necessary for end users like farmers or forest managers, planners, and decision makers. The goal of this IRT, therefore, is to elaborate end-user friendly tools for an integrated assessment of agricultural and forestry land uses.

3.2.2 IRT Group "FFFF: Demand, Potentials and Risks"

The four IRTs of this group embrace the dimensions, potential, and management options for the 4F-context (food-feed-fiber-[bio-]fuel) under specific consideration of local, regional, and global trans-disciplinary and strategic approaches.

IRT-4: Bio-Economy—Unleashing the Potential While Sustaining Soils

The bio-economy covers those parts of the economy that use renewable biological resources from land and sea—such as crops, forests, fish, animals, and microorganisms—to produce food, materials, and energy. The bio-economy promises a step change where fossil fuels are replaced by sustainable natural alternatives as part of the shift to a postpetroleum society. However, the bio-economy relies crucially on biomass and energy production to be provided by the soil-sediment-water system and the rivers and seas. Soils can provide bio-based resources, but their provisioning needs to be sustained, and an overuse of soils must be prevented as this could deeply impact soil system functions. In particular, the capacities of soils and their sustainable potential to enable a bio-economy with adequate agriculture and forestry are critical. There is a need for research regarding the complex and interrelated factors involved throughout the biomass production and consumption chain in order to better understand risks and environmental impacts (e.g., understanding and minimizing negative externalities), to better cope with varying conditions, and to identify new production options. The goal of this IRT, therefore, is to unleash the potential of soils in order to sustain a bio-economy in Europe by better understanding soil and economic systems.

IRT-5: Integrated Scenarios for the Land-Soil-Water-Food Nexus Under Societal Pressures and Challenges

Societal challenges determine how humans exploit natural resources. Only few integrated scenarios are available that include changes in soil properties, water availability, food, timber, fiber, or bio-energy production. Mutual synergies and trade-offs, what is often referred to as the nexus, remain unknown or unconsidered.

A few integrated scenarios do exist, which include food production, bio-energy and wood biomass production, climate change, and biodiversity, exploring pathways for achieving corresponding global targets. However, the currently available integrated scenarios do not appropriately consider the impacts on soil, water availability, floods and droughts; timber and fiber production; or effects on food, bio-energy, climate, and biodiversity. The goal of this IRT, therefore, is to research exploratory and target-oriented scenarios considering integrated, spatially explicit models that take into account major trade-offs and synergies between ecosystem functions, land use, and societal challenges. External effects of our economy (import of goods, environmental footprint on developing countries) should be taken into account as well.

IRT-6: Indicators for Assessing the Efficiency of the Soil-Sediment-Water-Energy Nexus of Resources

Currently, resource efficiency indicators that are available in the Eurostat scoreboard represent the evolution of the relationship between gross domestic product (GDP) and different inputs, such as energy, water, land, or material resources (including biomass and minerals). Biomass production (food, feed, fiber, fuels—4Fs) is the result of the use of the interconnected resources of soils/sediments, water, and energy. This nexus of resources is not accounted for as such in the indicators. Furthermore, there is a "conceptual gap" in the accounting method for biomass in the "resource efficiency" as most of it is produced by humans. The relationship between the production of biomass and the use of the soil-water-energy nexus needs further investigation.

The goal of this IRT, therefore, is to understand the links between the consumption of our societies and the use of natural resources, like the SSW-system services. There is a need to quantify and map in time and space the use of the nexus of SSW and energy resources related to the final consumption of products and services.

IRT-7: Farming Systems to Maintain Soil Fertility While Meeting Demand for Agricultural Products

A growing world population and increasing demand for food and nonfood agricultural products puts high pressure on farming systems to intensify production. At the same time, it becomes more and more obvious that intensifying conventional farming may be accompanied by severe negative environmental consequences, such as reduced bio- and agrobiodiversity, nutrient leaching to groundwater and rivers, eutrophication of lakes and the sea, and in particular loss of fertile soils due to erosion, nutrient loss, and soil compaction. Although several agricultural production techniques for maintaining soil fertility and reducing environmental impacts have been developed, it is yet unclear if these farming techniques could be scaled up to attain the goal of food security and meet the demand for nonfood products. Moreover, it needs to be better understood what role technology development (e.g., precision farming) might play in reducing environmental externalities of conventional farming systems and in increasing returns from soil-friendly agricultural practices. The goal of this IRT, therefore, is to better understand how sustainable soil management by appropriate agricultural production systems can contribute to sustainable food security, as well as if and how these solutions can be scaled up and widely implemented on farm level.

3.2.3 IRT Group "Challenge: Integrated Urban Management"

The six IRTs of this group encompass the questions about how research can contribute to solving land-use problems involving urban and suburban regions, including interrelations between them.

IRT-8: Circular Land Management

There is ongoing urbanization and, in particular, urban sprawl due to insufficient urban regeneration leading to persistence and new generation of brownfields. Furthermore, there are growing and shrinking cities with different land dynamics and development objectives for all types of land, especially brownfields and other un- or underused parcels of land. Demographic change may lead to new requirements for urban infrastructures. Research is needed to understand the patterns of behavior and interdependencies of the relevant actors in land-related policy areas—on both levels, theoretically and practically. The goal of this IRT, therefore, is to minimize the demand for and the consumption of land by continuously renovating existing settlement structures and overcoming the legacy of the past by reusing and redeveloping abandoned, derelict, and underused land. The modernization of existing settlement structures by circular land management is a continuous task so that knowledge exchange regarding local, regional, and national initiatives and tools must be included in future research and implementation activities.

IRT-9: Policies to Effectively Reduce Land Consumption for Settlement Development

Land take for settlements and transport infrastructure is a main reason for the loss of fertile soils and agricultural land. Land take itself is driven by a range of different factors: changing lifestyle patterns, demographic change, economic developments (e.g., e-commerce, logistics), infrastructure development, trends in property and financial markets, housing policy, regional planning, building codes, as well as agricultural and nature conservation policies. In turn, efforts to prevent land take, such as by fostering inner city development through brownfield redevelopment, often fail because policies and regulations do not effectively address the real drivers for land consumption. The goal of this IRT, therefore, is to better understand the drivers for land take and develop/identify incentives or obstacles for the enforcement of planning and policies so that interventions in property markets and settlement development can become more effective.

IRT-10: Stakeholder Participation to Facilitate the Development of Livable Cities

Urban development and the creation of livable cities involve many stakeholders, such as private households, businesses, planning authorities, land developers, etc. Basically, a transparent and legitimate balance between the different interests of all relevant stakeholders is necessary. Therefore, stakeholder participation is a promising approach in order to identify both mutual benefits and conflicts among different interests. Participation processes and tools may also provide a platform for exchange and communication. However, a wide range of open questions have to be answered to exploit the full potential of participatory processes and to enhance decision-making in terms of legitimacy, acceptance, and local ownership. The goal of this IRT, therefore, is to better understand how stakeholder participation may facilitate urban development and the creation of livable urban spaces, the pros and cons of different participatory approaches in a given context, and how it may best be embedded in the course of planning and project development. Understanding the potential of stakeholder participation will help to ensure the success of urban development and enhance the transparency and legitimacy of decision-making.

IRT-11: Integrated Management of Soils in Urban Areas

Urban soils are created and characterized by the process of urbanization and are an immanent part of urbanized areas. Urban activities may create different types of manmade soils, but all soils situated within cities or urbanized areas should be included in the category of urban soils. Due to the soil's multifunctional role in urban areas, the sound management of this resource is of key importance in urban land management. A typology of urban soils is important to perceive these soils through a wide perspective, including the diversity of their soil functions. It is also important to define the suitability of soils for different urban land uses. Soil characteristics and quality should be taken into consideration in spatial (urban) planning. From the perspective of ecosystem services and the SSW-system, urban soils are an important part of green infrastructure. Especially, soil of high quality should be protected to maintain its biodiversity, habitat provision, and ecosystem services. Also, agricultural use of soil in urban areas must be considered, especially in the context of urban farming and gardening and from the perspective of the global food production market. The goal of this IRT, therefore, is to better understand the role of urban soils and their contribution to the quality of urban spaces and to the health and quality of life for the people living there.

IRT-12: Environmentally Friendly and Socially Sensitive Urban Development

Urban development is confronted with heterogeneous and often conflicting demands. Concerns about urban environmental protection are strongly interconnected with urban development and must be considered in planning and decision-making processes in several ways. Such challenges must be balanced with other challenges in urban development, in particular with social concerns like affordable housing and security of energy and water supply. On one hand, social and environmental needs also have the potential to create synergies with environmental justice. On the other hand, conflicting goals in environmentally friendly and at the same time socially sensitive urban development can be detected; for example, in the field of energy poverty—greener but more expensive renewable energy puts some poverty-ridden households at risk of energy insecurity. Moreover, complexity is added as different societal groups in different cities will not have the same interests, and hence social contexts may differ between regions and across Europe. The goal of this IRT, therefore, is to better understand potential synergies and trade-offs in relation to environmental and social concerns in urban development. Conflicting goals should be identified and more clearly described, sufficient indicators must be developed, and solutions to reduce and dissolve conflicts should be derived.

IRT-13: Urban Metabolism—Enhance Efficient Use of Soil-Sediment-Water Resources Through a Closing of Urban Material Loops

Provision, use, and consumption of resources should be considered with regard to specific products or services. However, a systemic understanding is needed for sustainable development—not least in the case of resources within the soil-sediment-water system. The concept of urban metabolism tries to integrate all urban material flows, stocks, loops, and their internal and external interdependencies in a comprehensive way.

Urban metabolism is the study of material and energy flows arising from urban socioeconomic activities and regional and global biogeochemical processes.[15] The characterization of these flows and the relationships between anthropogenic urban activities and natural processes and cycles define the behavior of urban production and consumption. Urban metabolism is therefore a deeply multidisciplinary

[15] urbanmetabolism.org.

research domain focused on providing important insights into the behavior of cities for the purpose of advancing effective proposals for a more humane and ecologically responsible future. The goal of this IRT, therefore, is to enhance urban resource efficiency, consistency, and sufficiency and to minimize direct and indirect negative environmental impacts that are generated by urban land use. Strategies, tools, and instruments should be developed reflecting the mechanisms of urban material flows, stocks, and loops and their environmental impacts.

3.2.4 IRT "Disturbed Landscapes"

The three IRTs of this group contain elaborated research issues on how to secure, manage, and rejuvenate degraded sites, landscapes, and regions in Europe.

IRT-14: "Emerging Contaminants" in Soil and Groundwater—Ensuring Long-Term Provision of Drinking Water as Well as Soil and Freshwater Ecosystem Services

Deteriorating groundwater quality and reduced soil-related ecosystem services are serious issues in various European countries. Emission of "emerging contaminants," e.g. pesticides used in agriculture, chemical substances used in industrial production or from waste and sewage, may worsen the problem. However, at this moment, it is often unclear what the impacts are of these substances on different temporal and spatial scales, how impacts may be altered if those contaminants are mixed, and what cost-effective strategies may be employed to minimize their discharge or to remediate contamination. The goal of this IRT, therefore, is to better understand the impacts of "emerging contaminants" and to develop cost-effective management opportunities for safeguarding water and soil-related ecosystem services.

IRT-15: Sustainable Management to Restore the Ecological and Socioeconomic Values of Degraded Land

As a consequence of degraded land, soil and landscape functions are harmfully reduced or destroyed; soil and groundwater might even be contaminated. Knowledge about the dimension and grade of degradation is still low, hindering an ecologically sound and economically viable reclamation of these sites and water bodies. In a broader sense, this includes the regeneration of ecological functions even for new types of land uses. The direct impacts of degradation are a major cause for concern; however, the indirect consequences and the loss of services potentially have greater implications for society. The goal of this IRT, therefore, is to develop suitable restoration and rehabilitation approaches along the SSW-system approach to enhance the ecological and socioeconomic values of degraded land and which are appropriate to the site conditions and type and intensity of degradation.

IRT-16: Innovative Technologies and Eco-Engineering 4.0: Challenges for Sustainable Use of Agricultural, Forestry and Urban Landscapes and the SSW-System

Increasing societal demand on land resources and biomass causes land-use pressure and endangers ecosystem functions and the sustainability of land, water, and bio-resources. Classical technologies focus primarily on conventional sectors like agricultural mechanization or landscape engineering. Innovative Key Enabling Technologies (KET)[16] and eco-engineering as the bases for integrated solutions may facilitate a greener economy on a larger scale for farmers, forest managers, and rehabilitation-related SMEs to support future development of sustainable land management. However, societal acceptance for KET is restricted. Thus, understanding and raising awareness for modern sustainable technologies is also a key challenge. The goal of this IRT, therefore, is to develop land use and region-specific manageable, economically viable, and sociologically sound technologies and eco-engineering for agricultural, forestry, and urban areas contributing to a productive and safe environment

3.2.5 IRT Group "Climate Change Challenges"

There is only one overarching IRT in this group covering the preparedness for and response to climate conditions and related hazards.

IRT-17: Climate Change Challenges—Improving Preparedness for and Responses to Climate Conditions and Related Hazards

Climate change is a very complex and challenging issue, which affects urban and rural areas' management at all scales, from global to local. It has been mentioned in almost all national reports of INSPIRATION. This theme is also coherent with the EU Strategy on adaptation to climate change,[17] which sets out a framework and mechanisms for increasing the EU's preparedness for current and future climate impacts. Also, the results and guidelines of COP21[18] should be taken into consideration, especially those related to carbon sequestration in soils, because fertile soils are able to cope with the effects of climate changes. It is important for planning systems and land management practices to take into account the IPCC scenarios with regard to various levels of legal and administrative responsibilities (national,

[16]https://ec.europa.eu/growth/industry/key-enabling-technologies_de.

[17]European Commission (2014).

[18]http://www.cop21paris.org/.

regional, and local). The time frame for all activities related to climate change mitigation and adaptation should be precise (including short-, medium-, and long-term actions). The goal of this IRT, therefore, is to introduce or strengthen climate change aspects within spatial planning and land management practices and to reinforce administrative, technical, and societal preparedness for climate extremes and related hazards.

4 Outlook

As mentioned before, the process of writing the Strategic Research Agenda was ongoing when this article was drafted. The potential impact of this SRA is, however, tremendous. A broad variety of stakeholders identified their research needs as input for the SRA. Therefore, the scope of research topics and the questions that were collected will shape a truly multistakeholder-based research agenda. It will merge individual requirements of EU Member States and bottom-up collected research demands of stakeholders into a consistent SRA. The level of integration of soil and land-use-related topics is remarkable. The SRA will blend research on soil quality, land use, and land management issues, both in urban and in rural areas. This is unique, particularly because of its ambition: structuring research areas toward balancing the demand for and supply of resources and natural capital and reducing the ecological footprint by proper land management methods and tools. With the final public release of the SRA forthcoming, matchmaking with national funding institutions and elaborating implementation models for the SRA are the most challenging remaining tasks for the project. However, the final SRA is expected to be the first milestone in a paradigm shifting process of land and soil-based research policy toward multinational and stakeholder-oriented research funding.

References

Adhikari K, Hartemink AE (2016) Linking soils to ecosystem services: a global review. Geoderma 262:101–111. https://doi.org/10.1016/j.geoderma.2015.08.009

Baveye PC, Baveye J, Gowdy J (2016) Soil "ecosystem" services and natural capital: critical appraisal of research on uncertain ground. Front Environ Sci 4. https://doi.org/10.3389/fenvs.2016.00041

Blum WEH (2005) Functions of soil for society and the environment. Rev Environ Sci Biotechnol 4(3):75–79. https://doi.org/10.1007/s11157-005-2236-x

Brils J et al (2016) National reports with a review and synthesis of the collated information. Final version as of 01.03.2016 of deliverable 2.5 of the HORIZON 2020 project INSPIRATION. EC Grant agreement no: 642372, UBA, Dessau-Roßlau, Germany

European Commission (2014) Adaptation to climate change. Available at: https://ec.europa.eu/clima/sites/clima/files/docs/factsheet_adaptation_2014_en.pdf. Accessed 5 May 2017

Makeschin F, Schröter-Schlaack C, Glante F, Zeyer J, Gorgon J, Ferber U, Villeneuve J, Grimski D, Bartke S (2016) INSPIRATION report concluding 2nd project phase: enriched, updated and prioritised overview of the transnational shared state-of-the-art as input to develop a Strategic Research Agenda and for a matchmaking process. Public version of the final version as of 30.10.2016 of deliverable D3.4 of the HORIZON 2020 project INSPIRATION. EC Grant agreement no: 642372, UBA, Dessau-Roßlau, Germany

UN (2014) Resolution adopted by the General Assembly on 20 December 2013 [on the report of the Second Committee (A/68/444)] - 68/232. World Soil Day and International Year of Soils

Printed by Printforce, the Netherlands